D1628393

Up and Running with Autodesk® Inventor® Simulation 2011

A step-by-step guide to engineering design solutions

Leeds Metropolitan University
17 0549051 1

DEDICATION

I would like to dedicate this book to my late father, whom I dearly miss to this day.

Up and Running with Autodesk® Inventor® Simulation 2011

A step-by-step guide to engineering design solutions

WASIM YOUNIS

AMSTERDAM • BOSTON • HEIDELBERG • LONDON NEW YORK • OXFORD
• PARIS • SAN DIEGO SAN FRANCISCO • SINGAPORE • SYDNEY • TOKYO
Butterworth-Heinemann is an imprint of Elsevier

Butterworth-Heinemann is an imprint of Elsevier
The Boulevard, Langford Lane, Kidlington, Oxford, OX5 1GB
30 Corporate Drive, Suite 400, Burlington, MA 01803, USA

First published 2009
Second edition 2010

Copyright © 2010 Elsevier Inc. All rights reserved

No part of this publication may be reproduced or transmitted in any form or by any means, electronic or mechanical, including photocopying, recording, or any information storage and retrieval system, without permission in writing from the publisher. Details on how to seek permission and further information about the publisher's permissions policies and our arrangement with organizations such as the Copyright Clearance Center and the Copyright Licensing Agency can be found at our website: www.elsevier.com/permissions

This book and the individual contributions contained in it are protected under copyright by the publisher (other than as may be noted herein).

Notices

Knowledge and best practice in this field are constantly changing. As new research and experience broaden our understanding, changes in research methods, professional practices, or medical treatment may become necessary.

Practitioners and researchers must always rely on their own experience and knowledge in evaluating and using any information, methods, compounds, or experiments described herein. In using such information or methods, they should be mindful of their own safety and the safety of others, including parties for whom they have a professional responsibility.

To the fullest extent of the law, neither the Publisher nor the authors, contributors, or editors assume any liability for any injury and/or damage to persons or property as a matter of products liability, negligence or otherwise, or from any use or operation of any methods, products, instructions, or ideas contained in the material herein.

British Library Cataloguing in Publication Data
Younis, Wasim:
Up and running with Autodesk Inventor Simulation 2011:
a step-by-step guide to engineering design solutions.
1. Autodesk Inventor (Electronic resource) 2. Engineering models--Data processing.
I. Title
620' .00420285536-dc22

Library of Congress Control Number: 2010922545

ISBN: 978-0-12-382102-7

LEEDS METROPOLITAN UNIVERSITY LIBRARY
1705490511
1E-B
BP-117473
10-12-10
620.00420285536 YOU

For information on all Butterworth-Heinemann publications visit our website at elsevierdirect.com

Printed and bound in the United States

10 11 12 11 10 9 8 7 6 5 4 3 2

Working together to grow libraries in developing countries
www.elsevier.com | www.bookaid.org | www.sabre.org
ELSEVIER BOOK AID International Sabre Foundation

CONTENTS

FOREWORD

The ability to use digital prototyping as a core design practice has rapidly become a critical initiative for manufacturers of all sizes. To stay competitive in today's global market, manufacturers have to move from a perspective of using 3D design methods for automating the creation of 2D drawings to a perspective of using a 3D model as a complete digital prototype for evaluating form, fit, and function. Critical to making this leap is an understanding of the role and application of simulation in the digital prototyping process. Unfortunately, too many designers and engineers are either unable or unwilling to integrate simulation into their design process. The result is that they are falling behind the best-in-class standards of a rapidly changing manufacturing climate.

The book you are about to read offers a clear path for designers and engineers to begin to perfect their skills using simulation inside Autodesk Inventor®. By using real-world examples to illustrate both the need for and application of simulation, this book is not only a useful learning tool but also a source of inspiration for applying simulation to bringing better products to market faster. Every designer and engineer needs to understand how to use a digital prototype to simulate their product designs before they are real, and every designer and engineer can benefit from reading this book.

The journey to becoming a best-in-class user of digital prototyping requires an understanding of simulation and its application to design problems. This book is an important part of that journey.

Dr. Andrew Anagnost
Vice President, Engineering Design and Simulation Products
Manufacturing Industry Group
Autodesk, Inc.

PREFACE

Welcome to the second edition of *Up and Running with Autodesk® Inventor® Simulation 2011 – A step-by-step guide to engineering design solutions.*

I hope you found the first edition of the book very useful and interesting. I thank you very much for your feedback/suggestions, which have helped me to make the second edition of the book even better. In this edition I have included modal analysis, stress singularities, H–P convergence, and suggestions on how to model and improve results. Simulation theory is now enhanced to include open, closed, and redundant mechanisms – including friction and restitution contact properties. The NEW Frame Analysis functionality is also included in this edition.

This book has been written using actual design problems, all of which have greatly benefited from the use of Simulation technology. For each design problem, I have attempted to explain the process of applying Inventor Simulation using a straightforward, step-by-step approach, and have supported this approach with explanations and tips. At all times, I have tried to anticipate what questions a designer or development engineer would want to ask while he or she were performing the task and using Inventor Simulation.

The design problems have been carefully chosen to cover the core aspects and capabilities of Dynamic Simulation, stress and frame analysis and their solutions are universal, so you should be able to apply the knowledge quickly to your own design problems with confidence.

APPROACH OF THE BOOK

The book basically comprises three sections: Dynamic Simulation (Chapters 1–8), Stress Analysis (Chapters 9–16), and Frame Analysis (Chapters 17–19). Chapters 1, 9, and 17 provide an overview of Dynamic Simulation, stress analysis, frame analysis, and the Inventor Simulation interface and features to give you a good grounding in the core concepts and the software's strengths, weaknesses, and workarounds. Each design problem illustrates a different approach and demonstrates key aspects of the software, making it easier for you to pick and choose which design problem you want to cover first; therefore, having read Chapter 1, it is not necessary to follow the rest of the book sequentially.

The joint creation process, including redundant joints, within Dynamic Simulation is possibly the most powerful but hard to master feature of the software, and, in my experience, one of the areas that most users struggle with. Therefore, this book has a particular emphasis on the joint creation process, and shows all the possible methods of creating joints efficiently. Each of Chapters 2–8 starts by showing which joint is being used to make it easier for you to concentrate on the joints required for your own design problems.

Stress and modal analysis within Inventor has been around for many years but its usage has been limited until now by the single-part stress analysis capability. With the release of 2010, this limitation has been overcome by the inclusion of assembly stress and modal analysis and the unique and powerful parametric optimization function, which is discussed in detail in Chapters 9–16.

Frame analysis is new to this second edition, which now allows the analysis of very large structures. This is made possible by automatically generating beam elements and associated mechanical properties from frames created either from Content Center or Frame Generator. Frame Analysis is also discussed in detail in Chapters 17–19.

This book is primarily designed for self-paced learning by individuals, but can also be used in an instructor-led classroom environment.

I hope you will find this book enjoyable and at the same time beneficial to you and your business. I will be very pleased to receive your feedback, to help me improve future editions. Feel free to e-mail me at younis_wasim@hotmail.com.

ADDITIONAL HELP AND SERVICES

There may be situations in which extra help and advice on the contents of this book or on your own models and designs would be valuable. Please go to my site and follow the instructions on how to become a valued member of the Simulation Community, as well as details on how to access additional help and support services. Membership is free and the site has a wealth of simulation-specific information including an image gallery, tips and tricks, additional tutorials, completed design problem exercises, and much more.

www.vdssolutions.co.uk
(Dedicated Support for the Inventor Simulation Community)

In addition to my web site I have also launched a new Virtual Reality blog which is dedicated to the Autodesk Simulation Community. Here I regularly upload tips, tricks and articles all about Autodesk Inventor Simulation.

vrblog.info
(Blog dedicated to Autodesk Simulation Users)

Wasim Younis

ACKNOWLEDGMENTS

Personal thanks to everyone who provided me with feedback and suggestions on the first edition of this book, with special thanks to Dr J.D. Mather of Pennsylvania College of Technology. Sincere thanks to the brilliant Simulation Quality Assurance, Product Design, and Development Teams at Autodesk for their invaluable support, with huge thanks to Ales Ricar and Alex Plaks, Software Development Engineers at Autodesk.

Most of all, I would like to thank all the companies mentioned below for allowing me to use their innovative product designs and models, without which none of this would have been possible.

Philip Wright – Wright Resolutions Ltd
Adrian Curtis – In-CAD Services Ltd
Jonathan Stancliffe – British Waterways
Kevin Berry – Triple Eight Race Engineering Ltd
Roy Hadfield and Lee Chapman – Unipart Rail
Mark Askew – Sheppee International Ltd
Ian Parker – Halifax Fan Ltd
Matt Cowan – Hallin Marine UK Ltd
Adrian Hartley – Simba International Ltd
Carl Geldard – Planet Platforms Ltd
Adrian Oaten – Aerospace Design Facilities Ltd
Andrew Turner – KONE plc (Escalators, Keighley)

Thanks to the team at Elsevier for invaluable support in getting this book out to you. Finally, I would like to thank my wife Samina, daughter Malyah, and sons Sami and Fasee for their unconditional love, support, and source of inspiration.

The front cover image shows an illustration of an escalator used courtesy of KONE plc (Escalators, Keighley) (*www.kone.com*).

ABOUT THE AUTHOR

Wasim Younis (United Kingdom) is an Inventor Simulation consultant and trainer with more than 15 years of experience in the manufacturing field. He works very closely with Autodesk and Autodesk value-added resellers and users, and was involved with Simulation software when it was first introduced and is well-known throughout the Inventor Simulation community.

He has previously been involved in enhancing and updating the Simulation Autodesk Official Training Courseware, including producing simulation marketing material for Autodesk. He also presents various simulation topics at UK Autodesk reseller events and other key events including AUGI International.

Wasim contributes articles, whitepapers, tips and tricks, and tutorials to various forums most notably to the Autodesk User Group International site *(http://www.augi.com/home/default.asp)* and the Experience Manufacturing site *(http://www.experiencemanufacturing.com/)*. He regularly authors simulation Tips and Tricks articles on his own Virtual Reality blog (*http://vrblog.info*) – a blog dedicated to the Autodesk Inventor Simulation Community.

Wasim has a bachelor's degree in mechanical engineering from the University of Bradford and a master's degree in computer-aided engineering from Staffordshire University.

Currently, he is Director of VDS Solutions *(http://www.vdssolutions.co.uk)*, which provides Inventor Simulation training, support, and consultancy.

HOW TO ACCESS THE BOOK EXERCISE FILES

HOW TO ACCESS THE BOOK EXERCISE FILES

All tutorial files and datasets necessary to complete the book's exercises, plus completed files, can be accessed from various locations including:

1. http://www.vdssolutions.co.uk

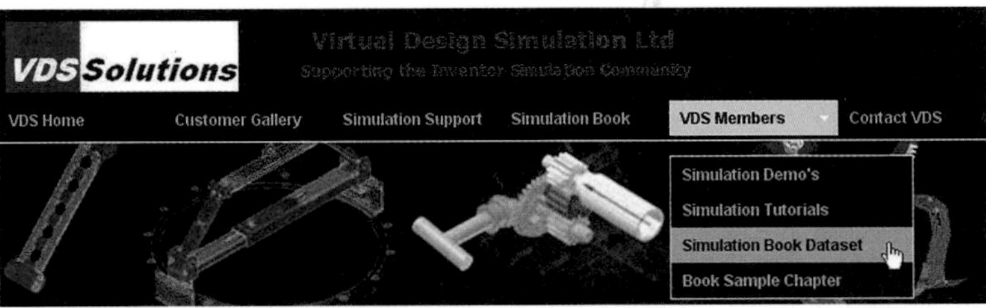

You will need to create a membership account, unless you already have one. Membership is FREE.

2. http://vrblog.info

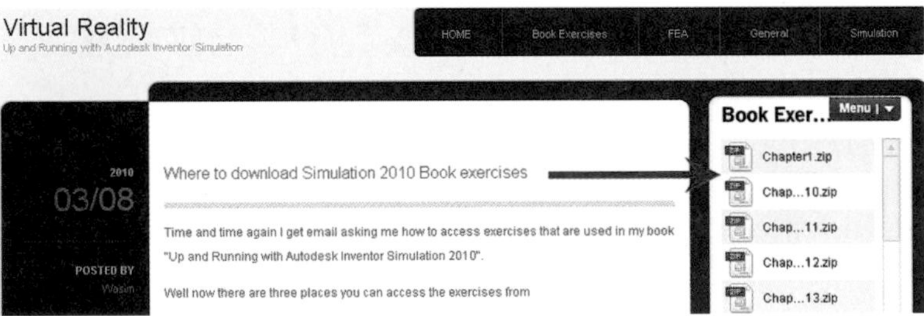

The book exercises are available on the right hand side of the blog (available on all pages/posts). You may need to scroll-down a little to see the exercise files available via Box.net

3. http://www.elsevierdirect.com/companions/9780123821027

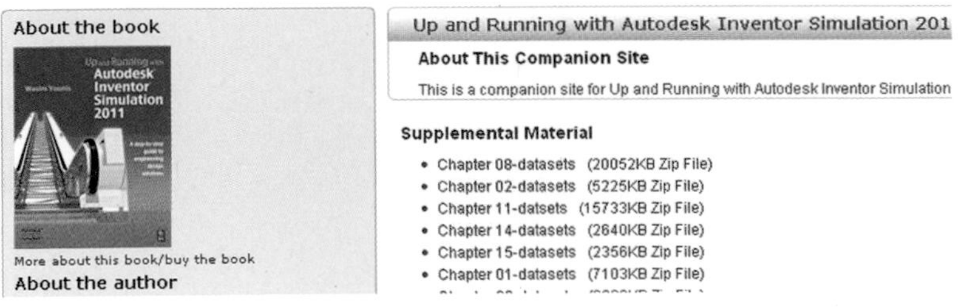

This is the book companion site from the publisher.

NB: There are 19 zip files and they are named by their corresponding chapters.

CHAPTER 1

The Dynamic Simulation Environment

SIMULATION OVERVIEW

During a typical design process, designers go through a series of typical questions, such as: do the parts fit together? Do the parts move well together? Is there interference? Do the parts follow the right path? Even though most of these questions can be catered for by 3D CAD and rendering software, there may be other questions that cannot. For example, designers may want to know the machinery time cycle. Is the actuator powerful enough? Is the link robust enough? Can we reduce weight? All these questions can only be answered by building a working prototype or a series of prototypes. The major issues with this method are that it is timely and costly. An alternative cost-effective method is to create a working virtual prototype by using the Inventor simulation suite. The Inventor simulation suite allows the designer to convert assembly constraints automatically to mechanical joints, provides the capability to apply external forces including gravity, and allows the effects of contact friction, damping, and inertia to be taken into account. As a result of this, the simulation suite provides reaction forces, velocities, acceleration, and much more. With this information, the designer can reuse reaction forces automatically to perform finite element analysis, hence reducing risks and assumptions. Ultimately all this information helps the designers to build an optimum product, as illustrated by the following example.

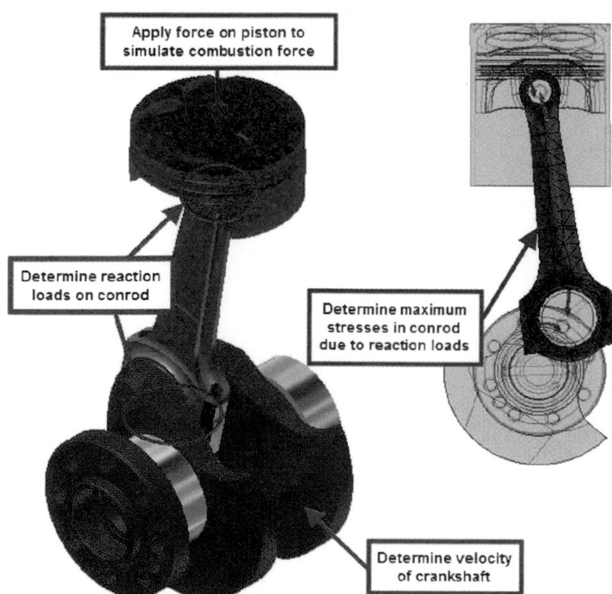

Up and Running with Autodesk Inventor Simulation 2011. ISBN: 978-0-12-382102-7
Copyright © 2010 Elsevier Inc. All rights of reproduction, in any form reserved.

SIMULATION – BASIC THEORY

Simulation enables understanding of the kinematic and dynamic behavior of mechanisms. 'Kinematics' simply refers to the motion of the mechanism, including determining position, velocity, and acceleration, whereas 'dynamics' is the study of masses and inertial forces acting on the mechanism.

$$F = M \times a$$

where
F = external force
M = mass
a = acceleration
This is Newton's Law of Motion, which can also be expressed as

$$F = M \times \frac{dv}{dt}$$

From both equations we can determine acceleration as a function of velocity

$$a = \frac{dv}{dt} = \frac{F}{M}$$

By integrating acceleration we can determine velocity

$$v = \frac{dx}{dt} = \frac{F}{M}t^2$$

By integrating velocity we can determine position

$$x = \frac{1}{2} \times \frac{F}{M}t^2$$

Inventor Simulation 2011 calculates acceleration, velocity, and the position of the component/assemblies at each time step, referred to as image frames within the user interface.

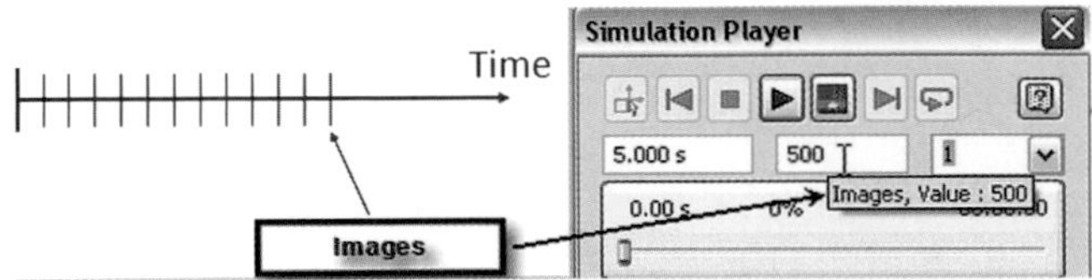

OPEN-AND CLOSED-LOOP MECHANISMS

A mechanism can, furthermore, be conceptually viewed as a set of rigid bodies interconnected to each other by joints that constrain, but not restrict, relative motion between any two bodies. Common joints used in mechanisms include revolution, cylindrical, prismatic, and spherical; a complete list of joints is given on page 17. The slider mechanism below comprises three revolutions, one prismatic, and one fixed (grounded) joint.

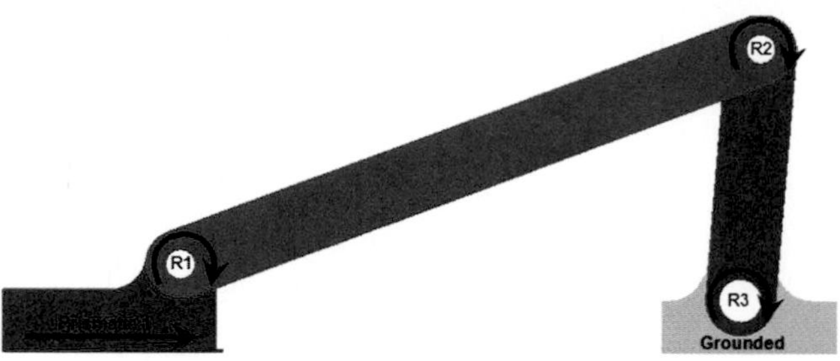

In addition, mechanisms can be generally categorized into open-loop mechanisms and closed-loop mechanisms. The difference is that the joint degrees of freedom (DOF) in open-loop mechanisms are independent of one another whereas in closed-loop mechanisms they are not independent. Extensive information on open- and closed-loop mechanisms is available from standard engineering books; theoretical technical information is beyond the scope of this book. The above slider mechanism is an example of closed loop; another example would be a Whitworth Return Mechanism. On the other hand, the following robot manipulator is an example of an open-loop mechanism, with one spherical, two revolution, and one fixed joint.

REDUNDANT MECHANISMS

Key pieces of information required from a mechanism analysis are the reaction forces and moments, due to acceleration, and inertial and external forces. These reaction forces are unique for a non-redundant model, whereas for a redundant model the reaction forces will not be unique, as explained by the simple shaft bearing example below.

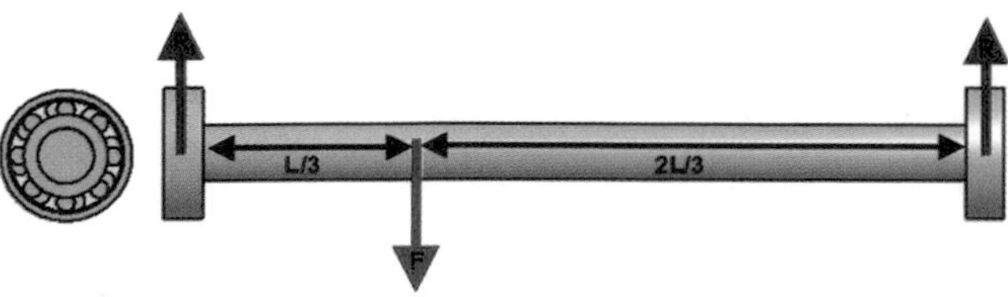

For equilibrium, applied force (F) should be equal to the sum of all reactions at the bearings (RL)

$$F = R_1 + R_2$$

Also, for equilibrium, the sum of all moments should be equal to zero

$$\frac{R_1 L}{3} = \frac{2R_2 L}{3}$$

For a 1 m shaft this becomes

$$\frac{R_1}{3} = \frac{2R_2}{3} \quad \text{or} \quad R_1 = 2R_2$$

Substituting the value of R_1 into the force equation gives us

$$F = 2R_2 + R_2 = 3R_2 \quad \text{or} \quad R_2 = \frac{F}{3}$$

Now substituting R_2 into the force equation gives us

$$F = R_1 + \frac{F}{3} \quad \text{or} \quad R_1 = \frac{2F}{3}$$

For a shaft with two bearings we have two unknowns and two equations, giving us one unique result, as

$$R_1 = \frac{2F}{3} \quad \text{and} \quad R_2 = \frac{F}{3}$$

Now let us consider the same shaft with another bearing in the middle.

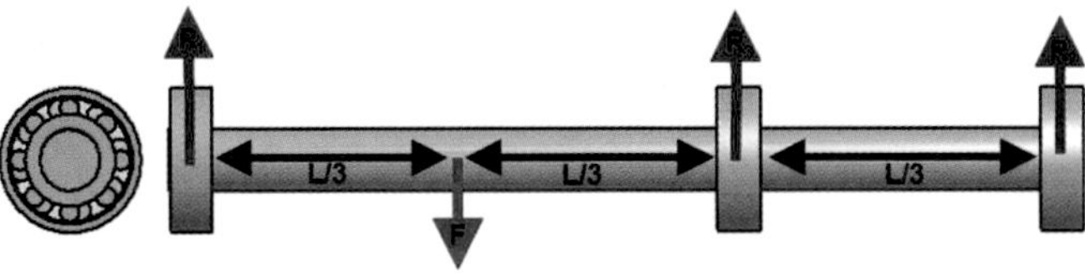

Again, for equilibrium, $\sum F = 0$ and $\sum M = 0$.

$$F = R_1 + R_2 + R_3$$

$$\frac{R_1 L}{3} = \frac{R_3 L}{3} + \frac{2R_2 L}{3}$$

This creates three unknowns and two equations. To determine the reactions, we need to make some assumptions.
Solution 1 – Let's assume R_2 equals 0; then we get

$$R_1 = \frac{F}{2} \quad \text{and} \quad R_3 = \frac{F}{2}$$

Solution 2 – Let's assume R_3 equals 0; then we get

$$R_1 = \frac{2F}{3} \text{ and } R_2 = \frac{F}{3}$$

We can continue to carry on making more assumptions but here it is important to note that adding a third bearing has resulted in there being more than one possible solution.

The reason for this is that, to maintain equilibrium of the shaft, two bearings are sufficient. Adding a third bearing has resulted in over-constraining the shaft mechanism. In reality this may not necessarily happen as the shaft can bend, whereas in the simulation we treat the shaft as rigid, resulting in a redundant model.

The redundancy process is further explained using a door and hinge example later in this chapter.

CONTACT PROPERTIES

To simulate reality as closely as possible, contact properties, including friction and restitution, need to be defined as accurately as possible. In Simulation there are two types of contact, which can be specified as two-dimensional (2D) and three-dimensional (3D). If the contact between two components remains planar before and after contact, and is not normal to the component, then 2D contact should be used. On the other hand, a 3D contact should only be used if the contact between two components does not remain in one 2D plane before and after contact. A contact is defined by two key material properties: restitution and friction.

It is easier to control restitution properties of 2D contact when compared to 3D contact.

In 3D contact, you cannot define restitution by specifying a value between 0 and 1, as it uses elastic stiffness and damping properties to simulate impact and bounce. This means that when two components come into contact there will always be vibration, and thus bounce, meaning that a restitution value of 0 is difficult, if not impossible, to simulate with a 3D contact.

Friction properties can be easily specified in both contacts.

Two-dimensional projected geometry can be used to define a 2D contact.

Restitution

Restitution indicates how the normal velocity between the two components changes during a shock.

$$\text{Restitution} = \frac{\text{Normal velocity after contact}}{\text{Normal velocity before contact}}$$

For example, if we drop a ball, with a restitution value set to 1, the ball will bounce back to its original position and will keep bouncing; in other words, the ball is completely elastic – perhaps made from plastic. On the other hand, if we drop the ball with the restitution value set to 0, the ball will drop without any bounce; in this case, the ball is inelastic – perhaps made out of a lead-like material. The default is set to 0.8 when a contact is created for the first time.

Friction

The coefficient of friction (μ) is the ratio defining the force that resists the motion of one body in relation to another body in contact with it. This ratio is dependent on material properties and most materials have a value between 0 and 1. In Simulation a value between 0 and 2 can be specified. A value higher than 1 allows one to take account of heat, moisture, age, etc. in addition to the coefficient of friction.

For example, if we roll a ball on a table with zero friction then the ball will drop off the table at the other end. If we specify a high enough friction value between the ball and the table, for example any value between 1 and 2, then the ball will eventually stop rolling on the table and hence will not drop off.

SIMULATION WORKFLOW

The process of creating a Dynamic Simulation study involves four core steps.

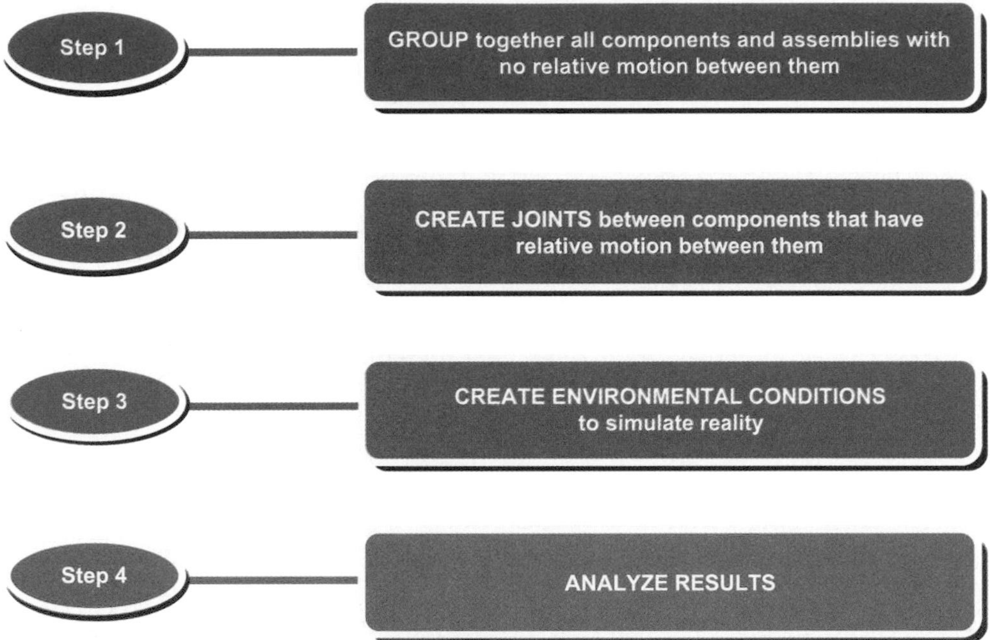

STEP 1: There are **two options** to group components together, and both have their advantages and disadvantages.

Option 1 – Create subassemblies within the Assembly environment.
Disadvantage – Restructuring your subassembly will affect your bill of materials (BOM) database; hence, you may need to create a duplicate for simulation purposes.

Option 2 – Weld components together within the Simulation environment.
Advantage – This method will not alter your BOM database.

STEP 2: The process of creating joints can be broken down into two stages.

Stage 1 – Create standard joints.
Stage 2 – Create nonstandard joints.

Stage 1 – There are **three options** to create standard joints, and, again, each has its own advantages and disadvantages.

Option 1 – Use Automatically Convert Constraints to Standard Joints.
Advantage – This is by far the quickest way to create joints.

Disadvantages

- Can be tedious to go through all joints converted for a large assembly.
- Cannot repair redundancies within the Simulation environment.
- Cannot create standard joints within the Simulation environment, with the exception of spatial joints.

Option 2 – Manually convert assembly constraints.

Advantages

- Can manipulate the type of joint created from constraints.
- Can create standard joints within the Simulation environment.
- Can repair redundancies for all standard joints not created from constraints.

Disadvantage – This method is slower than Option 1.

Option 3 – Create standard joints from scratch.

Advantages

- Complete control over how standard joints are created.
- Can repair redundancies for all standard joints created.

Disadvantages

- This method is the slowest.
- Does not make use of the assembly constraints.

Stage 2 – Comprises creating nonstandard joints that do not make use of assembly constraints and includes the following types of joint:

- Rolling
- Sliding
- 2D contact
- Force

Note: Rolling joints for spur gears, designed using Design Accelerator, can be created automatically.

STEP 3: Once the appropriate joints have been created, the next step is to simulate reality. This can be achieved by applying any of the following:

- Joints – define starting position.
- Joints – apply friction to joints.
- Forces/torque – apply external loads.
- Imposed motion on predefined joints.
  - Position, velocity acceleration (constant values).
  - Input Grapher – create specific motions (nonconstant values).

STEP 4: This is the final step, in which you use the Output Grapher to analyze the results in the joints, including:

- Positions/velocity/acceleration
- Reaction forces
- Reaction torque
- Reaction moments
- Contact forces

The most time-consuming process when creating a Dynamic Simulation study is Step 2 – creating joints – and this can be greatly affected by Step 1 – grouping components. With this in mind, the following approach is suggested for creating joints:

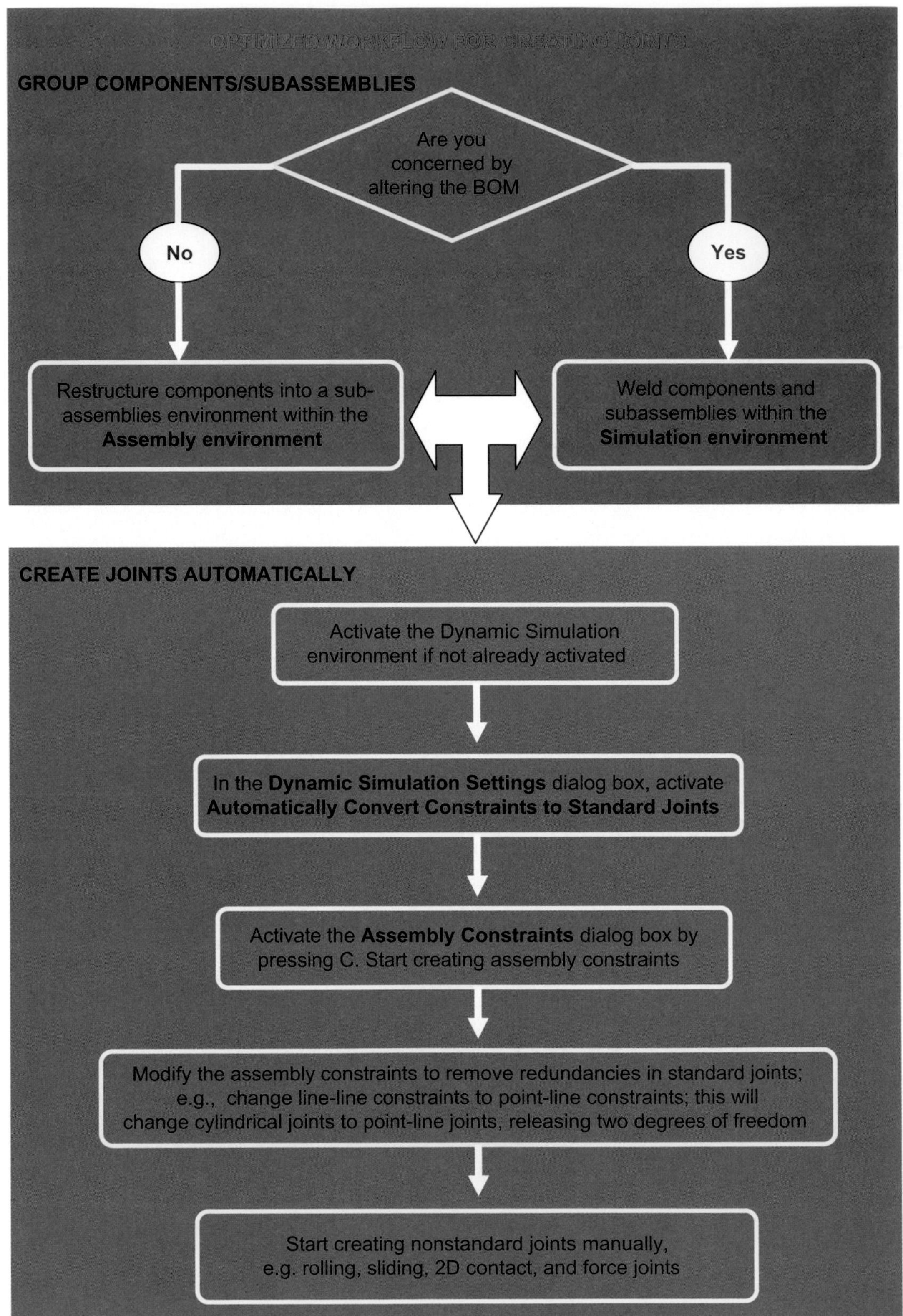
OPTIMIZED WORKFLOW FOR CREATING JOINTS
GROUP COMPONENTS/SUBASSEMBLIES
Are you concerned by altering the BOM
No
Yes
Restructure components into a sub-assemblies environment within the **Assembly environment**
Weld components and subassemblies within the **Simulation environment**
CREATE JOINTS AUTOMATICALLY
Activate the Dynamic Simulation environment if not already activated
In the **Dynamic Simulation Settings** dialog box, activate **Automatically Convert Constraints to Standard Joints**
Activate the **Assembly Constraints** dialog box by pressing C. Start creating assembly constraints
Modify the assembly constraints to remove redundancies in standard joints; e.g., change line-line constraints to point-line constraints; this will change cylindrical joints to point-line joints, releasing two degrees of freedom
Start creating nonstandard joints manually, e.g. rolling, sliding, 2D contact, and force joints

SIMULATION USER INTERFACE

Dynamic Simulation can be accessed from within the Assembly environment via the Analysis tab.

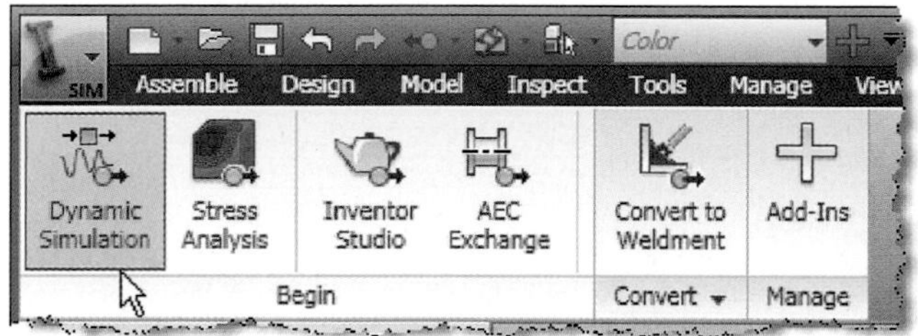

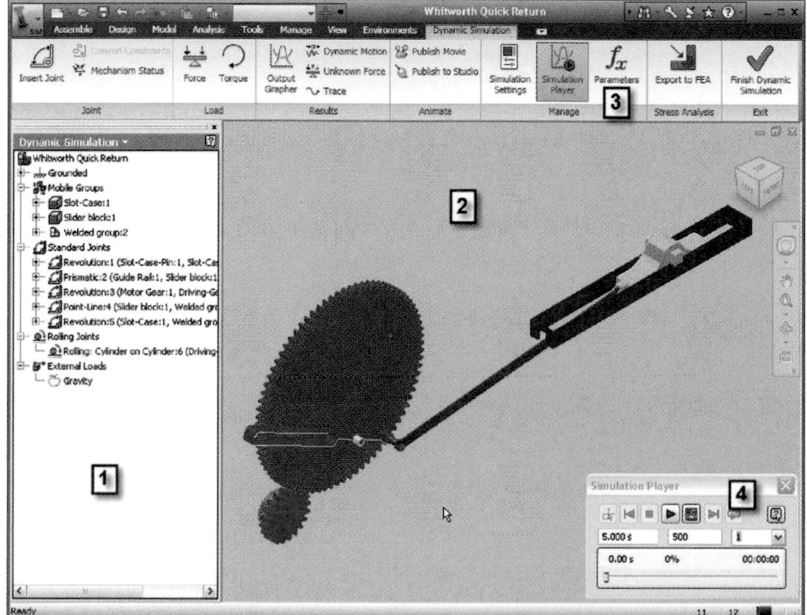

1. **Dynamic Simulation** browser
2. **Dynamic Simulation** graphic window
3. **Dynamic Simulation** panel
4. **Dynamic Simulation Player**

Dynamic simulation browser

Displays the simulations with part or assembly and simulation parameters in a hierarchical view with nested levels of feature and attribute information. You can:

- Copy whole simulations or simulation objects between simulations
- Right click on a node for context menu options
- Expand the folders, select the nodes, and see the selection cross-highlighted in the graphic region

Dynamic simulation graphic window

Display's the model geometry and simulation results. Updates to show current status of the simulation including applying boundary conditions and loads with the help of view manipulation tools.

Dynamic simulation panel

Dynamic Simulation tab	Workflow stage	Description
Insert Joint; Convert Constraints; Mechanism Status (Joint)	Step 2	**Insert Joint** – To create standard and nonstandard Joints. **Convert Constraints** – To create standard joints from selecting assembly constraints between two components. **Mechanism Status** – Used to determine the mobility and redundancy status of the assembly, including repairing redundancies in joints.
Force; Torque (Load)	Step 3	**Force** – Apply external forces on components. **Torque** – Apply external torques on components.
Output Grapher; Dynamic Motion; Unknown Force; Trace (Results)	Step 4	**Output Grapher** – Used to analyze joint results, including positions, velocity, and accelerations. **Dynamic Motion** – Allows the user to check the model before running the full simulation. **Unknown Force** – Used to determine the force, torque, and jack forces for known simulation conditions. **Trace** – Used to calculate the trace and output positions of components and joints in the Output Grapher including position, velocity, and acceleration
Export to FEA (Stress Analysis)	Step 5	**Export to FEA** – Allows the user to transfer reaction loads to the Stress Analysis environment.(Note: FEA - finite element analysis)
Publish Movie; Publish to Studio (Animate)	Step 6	**Publish Movie** – Allows the user to output the motion as a video file. **Publish to Studio** – Allows the user to output the motion to Inventor Studio for producing highly rendered animation/videos.
Simulation Settings; Simulation Player; Parameters (Manage)		**Simulation Settings** – Provides several user options. **Simulation Player** – Provides tools to play the simulation. **Parameter** – The parameters table.
Finish Dynamic Simulation (Exit)		

Simulation player

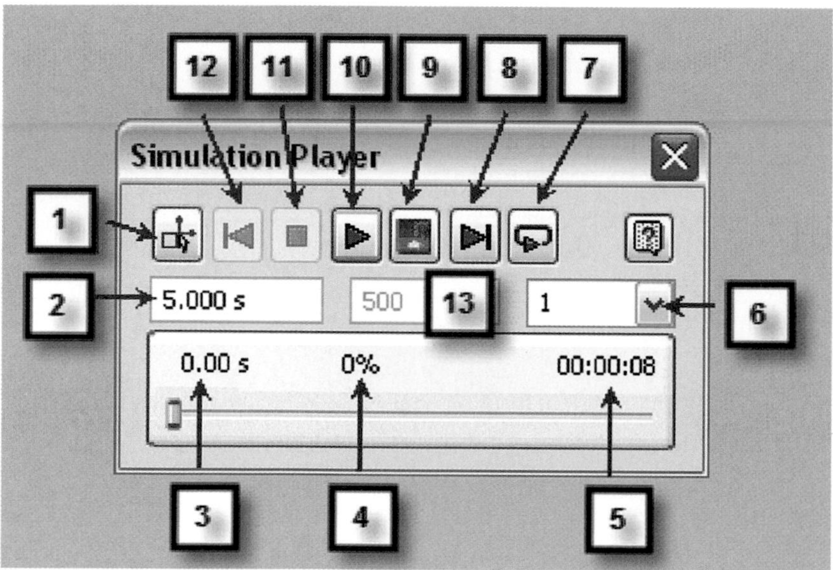

1. **Construction Mode** – After the simulation has finished, the construction mode needs to be selected to continue editing the simulation.
2. **Final time** – Specify the final time of the simulation.
3. **Simulation time** – Read-only value showing the time step during the simulation.
4. **Percentage of realized simulation** – Read-only value displaying the percentage of the simulation completed.
5. **Real time** – Read-only value displaying the actual time elapsed during simulation.
6. **Filter** – Normally set to 1. Can be changed to value other than 1; if set to 10, the simulation will ignore all images between 1 and 10 during simulation playback.
7. **Continuous Playback of simulation.**
8. **Advance to end of simulation.**
9. **Deactivate screen refresh at each time step** – Prevents the screen from refreshing at each time step, which can help speed up simulation.
10. **Play simulation.**
11. **Stop simulation.**
12. **Rewind simulation to beginning.**
13. **Images** – Normally, the higher the number, the more accurate the simulation; however, the simulation will take longer to run.

LEEDS METROPOLITAN UNIVERSITY LIBRARY

Simulation settings

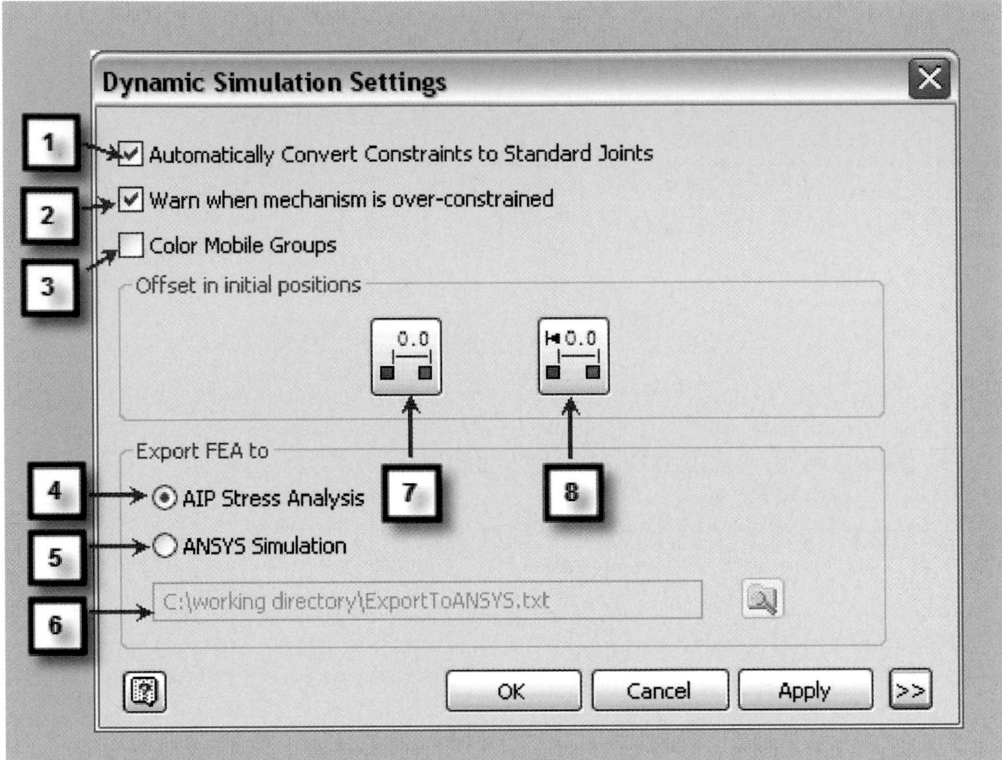

1. **Automatically Convert Constraints to Standard Joint** – If ticked, converts all assembly constraints to standard and rolling joints for spur gears only, if designed using Design Accelerator.
2. **Warn when mechanism is over-constrained** – When an assembly is overconstrained by joints, a warning is displayed.
3. **Color Mobile Groups** – Assigns a predefined color for each mobile component and/or subassembly.
4. **AIP Stress Analysis** – Transfer reaction loads to the Inventor Stress Analysis environment. (Note: AIP - Autodesk Inventor Professional)
5. **ANSYS Simulation** – Prepares a file with all load results for ANSYS DesignSpace.
6. **Location of FEA file** – Where the file containing the load data is saved.
7. **Set initial positions** – Sets all joint positions to 0.
8. **Reset joint positions** – Resets all joint positions to their original positions.

More simulation settings

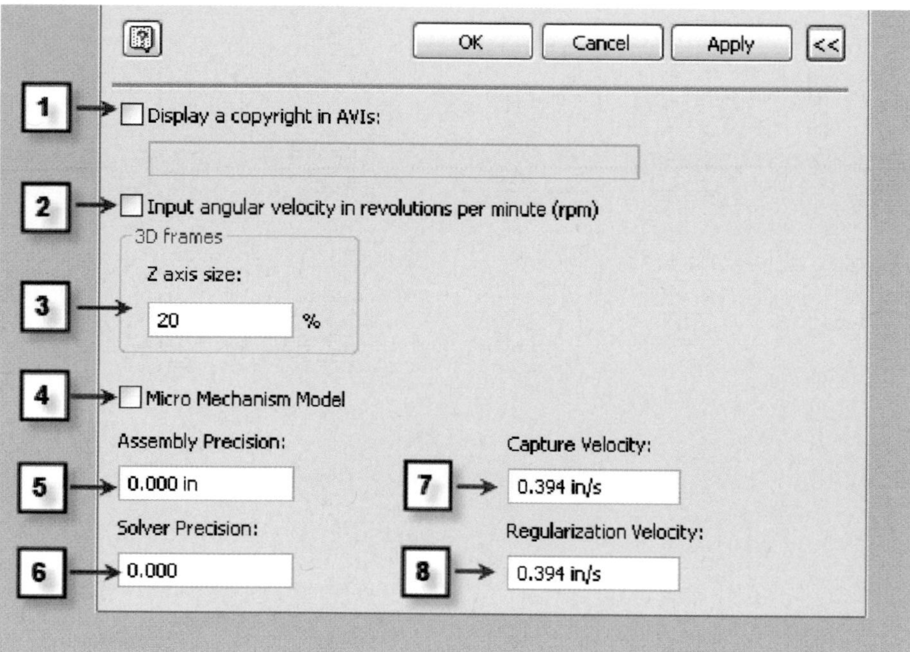

1. **Display a copyright in AVIs** – Displays the information you specify in the generated AVI.
2. **Input angular velocity in revolutions per minute (rpm)** – When selected, allows you to specify input velocity in revolutions per minute.
3. **3D frames** – Sets the length of the assembly Z axis in the graphics window.
4. **Micro Mechanism Model** – Select this when the mass or inertia is greater than 1e–20 kg and 1e–32 kg.m^2; this allows you to work with micro-mechanism models.
5. **Assembly Precision** – Allows the setting of a maximum distance between two contact points. This is only applicable for 2D contact and closed loops.
6. **Solver Precision** – Dynamic equations are integrated using a five-order Runge–Kutta integration scheme.
7. **Capture Velocity** – This is applicable to collision shock and allows the user to limit the number of small bounces before constant contact results. The value can be specified between 0 and 1, with 0 being maximum energy dissipation.
8. **Regularization Velocity** – For 2D contacts a real nonlinear Coulomb friction law is used, and for 3D contacts a regularized Coulomb law is used, to avoid hyperstatic conditions.

JOINTS

Joints are probably the most important aspect of creating a Dynamic Simulation Setting. This section will discuss the following:

- Types of joint
- Joints matrix – a Snapshot of joints used throughout the book
- The process of creating joints
- Redundant joints

Types of joint

In Dynamic Simulation, there are five main categories of joint: standard, rolling, sliding, 2D contact, and force joints. They are discussed in detail in the following sections.

STANDARD JOINTS

Dynamic simulation joints	Equivalent assembly constraints	DOF of joints
Revolution – No translation – Rotation around Z axis	Insert or Axis-axis + point-point	1
Prismatic – Translation along Z axis – No rotation	Face-face + axis-axis	1
Cylindrical – Translation along Z axis – Rotation around Z axis	Axis-axis or Edge-edge	2
Spherical – No translation – Rotation around all axes	Point-point	3
Planar – Translation along X and Z axes – Rotation about Y axis	Face-face or Flush-flush	2
Point-Line – Translation along Z axis – Rotation around all axes	Point-edge (or axis)	4
Line-Plane – Translation along X and Z axes – Rotation about Y axis	Face-edge (or axis)	3
Point-Plane – Translation along X and Z axes – Rotation about all axes	Face-point (also tangent constraint)	4
Spatial – Translation along all axes – Rotation about all axes	Unconstrained	6
Welding – No translation – No rotation	Fully constrained; that is, no DOF between components	0

Standard joints can be automatically converted from assembly constraints by using the **Automatically Convert Constraints to Standard Joints** tool.

With the **Automatically Convert Constraints to Standard Joints** tool activated, the user can continue creating further standard joints by creating more assembly constraints within the Simulation environment.

The contact remains permanent throughout the simulation.

The list of equivalent assembly constraints is not exhaustive.

ROLLING JOINTS

Dynamic simulation joints – There are NO equivalent assembly constraints	
	RI Cylinder on Plane This allows motion between a cylinder and a plane; for example, a gear and a rack.
	RI Cylinder on Cylinder This allows motion between two primitive cylindrical components in opposite directions; for example, spur gears.
	RI Cylinder in Cylinder This allows motion between a rotating cylinder inside a nonrotating cylinder.
	RI Cylinder Curve This allows motion between a rotating cylinder and a rotating cam.
	Belt This creates motion of two cylinders with the same speed. An option allows rotation in the same direction or as a crossed belt.
	RI Cone on Plane This allows motion between a conical face and a planar face.
	RI Cone on Cone This allows motion between two external conical faces; for example, bevel gears
	RI Cone in Cone This allows motion of a rotating conical component within a stationary conical component.
	Screw This is the same as a cylindrical component but also allows the user to specify pitch.
	Worm Gear This allows motion between a worm gear component and a helical gear component.

Rolling Joints can be automatically created for spur gears designed using Design Accelerator.

The primitive surfaces are created by Design Accelerator and need to be made visible to allow them to be selected in order to create rolling joints.

There is no sliding between components and the motion is 2D only.

The contact remains permanent throughout the simulation.

SLIDING JOINTS

Dynamic simulation joints – There are NO equivalent assembly constraints	
	SI Cylinder on Plane This allows sliding between a nonrotating cylinder and a plane.
	SI Cylinder on Cylinder This allows sliding between two primitive cylindrical components, of which one cylinder is nonrotating.
	SI Cylinder in Cylinder This allows sliding between a nonrotating cylinder inside another nonrotating cylinder.
	SI Cylinder Curve This allows motion between a nonrotating cylinder and a rotating cam.
	SI Point Curve This creates motion of a point on one component to stay on a curve, which can be defined by a face(s), edge(s), or sketch(es).

You can select sketches, faces, and edges to create joints.

The primitive surfaces are created by Design Accelerator and need to be made visible to allow them to be selected in order to create rolling joints.

There is no rotation between components and the motion is 2D only.

The contact remains permanent throughout the simulation.

2D CONTACT JOINTS

Dynamic simulation joints – There are NO equivalent assembly constraints	
	2D Contact This allows motion between the curve of the component and the curve of another component.

You can select sketches, faces, and edges to create joints.

The motion is 2D only.

The contact can be nonpermanent throughout the simulation.

FORCE JOINTS

Dynamic simulation joints – There are NO equivalent assembly constraints	
	Spring/Damper/Jack This allows you to create springs, jacks, or dampers.
	3D Contact This allows you to create contacts between two components. It is based on spring–damper forces.

The 3D contact settings are very sensitive to change. Only change, if necessary, when the model is not working.

The 3D contact only takes single components into consideration, even though the subassembly is selected. So, create contacts between all components that have contacts with the subassembly.

Joints matrix – a snapshot of joints used throughout the book

Dynamic simulation joints			Examples using these joints	Design problems using these joints
1		Revolution	All	All
2		Prismatic	2 & 3	4,5
3		Cylindrical	4	1,3,4,5,6,7
4		Spherical	1	3,4,6
5		Planar		7
6		Point-line	2 & 3	3,6,7
7		Line-plane	Not used	
8		Point-plane	4	
9		Spatial		3
10		Welding	?	
11		RI cylinder on plane		1
12		RI cylinder on cylinder	2	2
13		RI cylinder in cylinder	Not used	
14		RI cylinder curve		7
15		Belt	Not used	
16		RI cone on plane		
17		RI cone on cone		2
18		RI cone in cone	Not used	
19		Screw	Not used	
20		Worm gear	Not used	
21		SI cylinder on plane	2	
22		SI cylinder on cylinder	Not used	
23		SI cylinder in cylinder	Not used	
24		SI cylinder curve		7
25		SI point curve		5,7
26		2D contact	1	1
27		Spring/jack		5
28		3D contact	9	3,6

Process of creating joints

The process of creating joints is probably the most time-consuming, especially when you have a large assembly. This process can be drastically enhanced by being able to group components that have no relative motion between them. There are two options to do this.

Option 1 – Restructure components into subassemblies.
Option 2 – Weld components together.

Once the grouping of components has been achieved, there are three options to create standard and some rolling joints.

Option 3 – Automatically Convert Constraints to Standard Joints.
Option 4 – Manually convert constraints to standard joints.
Option 5 – Manually create standard joints.

This section will attempt to explain the above options of grouping components and creating joints by using a series of examples.

- Example 1 – Newton's cradle – Options 1 and 2
- Example 2 – Whitworth Quick Return Mechanism – Option 3
- Example 3 – Slider mechanism – Option 4
- Example 4 – Cam follower mechanism – Option 5

EXAMPLE 1

Newton's cradle – Grouping components

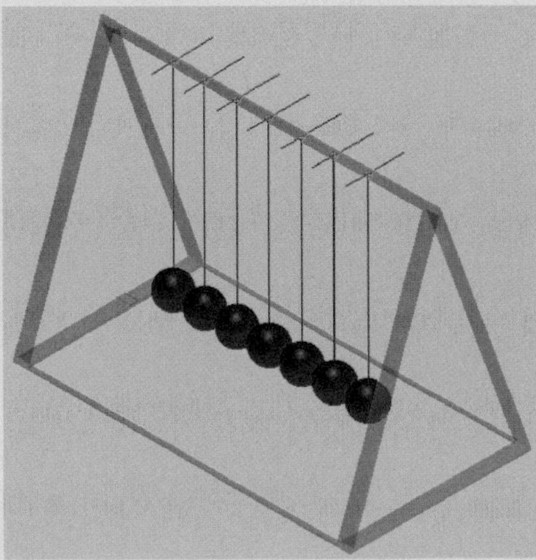

Workflow of Example 1
- Automatically convert constraints to standard joints
- Restructure parts into subassemblies
- Weld parts together
- Lock the joints' degrees of freedom

Joints used in Example 1
- Revolution
- Spherical
- 2D contact

Automatically convert constraints to standard joints

1. Open *Newtons-Cradle1.iam*

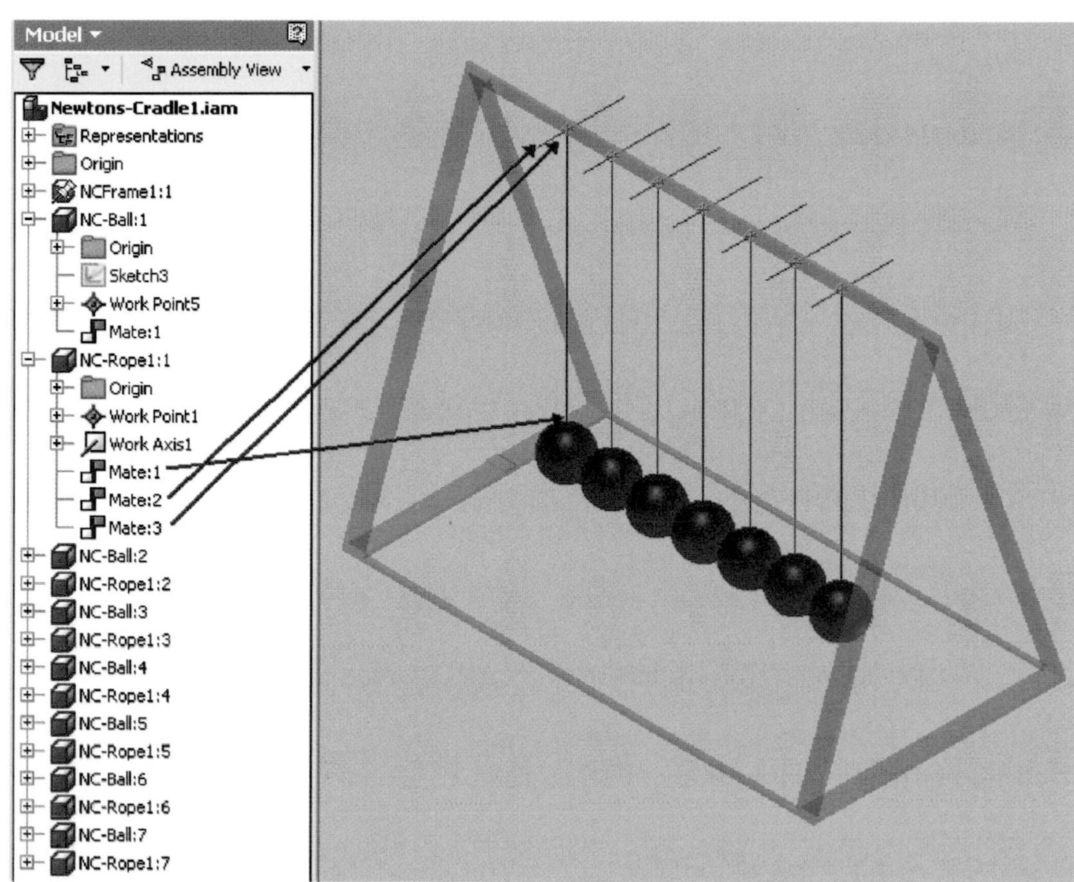

You will see that there are seven balls and seven ropes. There is one point constraint between the ball and rope, and one point and axis constraint between the rope and frame. In total there are three constraints for each pair of ball and associated rope.

2. Select **Environments** tab > **Dynamic Simulation**

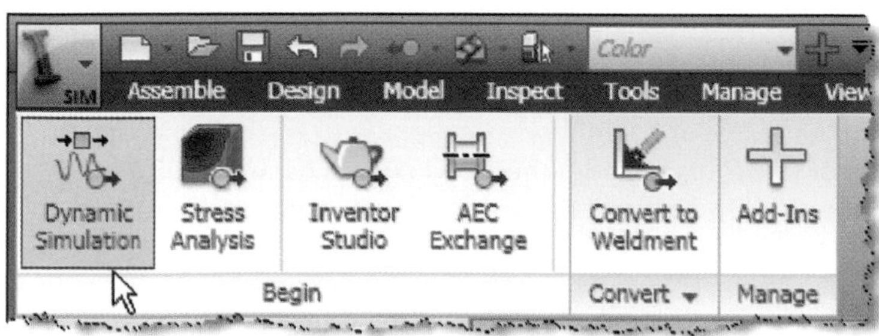

This will activate the Dynamic Simulation environment.

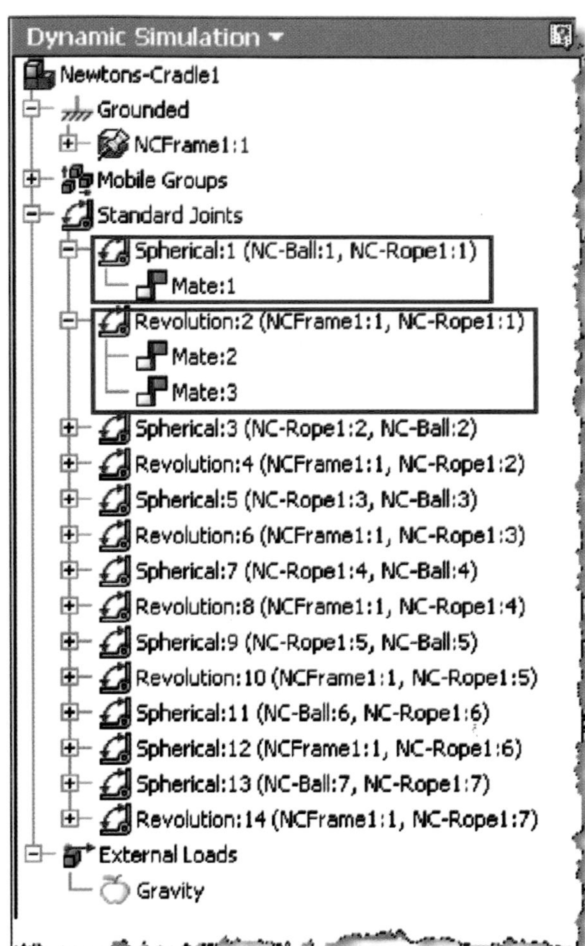

In the **Dynamic Simulation** browser, notice that a spherical joint is created between the ball and rope (Mate:1 – point constraint), and a revolution joint is created between the rope and frame (Mate:2 – point constraint and Mate:3 – axis constraint). This is done seven times so 14 joints are created in total.

To see how Simulation converts constraints to joints, refer to page 14 (standard joints) and see how constraints are converted to joints (this table is not exhaustive). Also note that the number of joints created is not related to the number of constraints.

Another point is that you can create rigid groups automatically by adding more constraints. For example, in this case, by adding a mate constraint between a plane of the ball and one of the ropes, we will lock all degrees of freedoms between these two parts. Dynamic Simulation will then create a rigid group containing the ball and the rope (as if you manually created a weldment).

For the Newton's cradle to work properly, the rope and ball will need to move together such that there is no relative motion between the ropes and balls. With this in mind, we can restructure the ball and rope components into one subassembly; this will hopefully simplify the joints process by reducing the number of joints created.

3. Close the *Newtons-Cradle1.iam* file

Restructure parts into subassemblies

4. Now open the *Newtons-Cradle2.iam* file

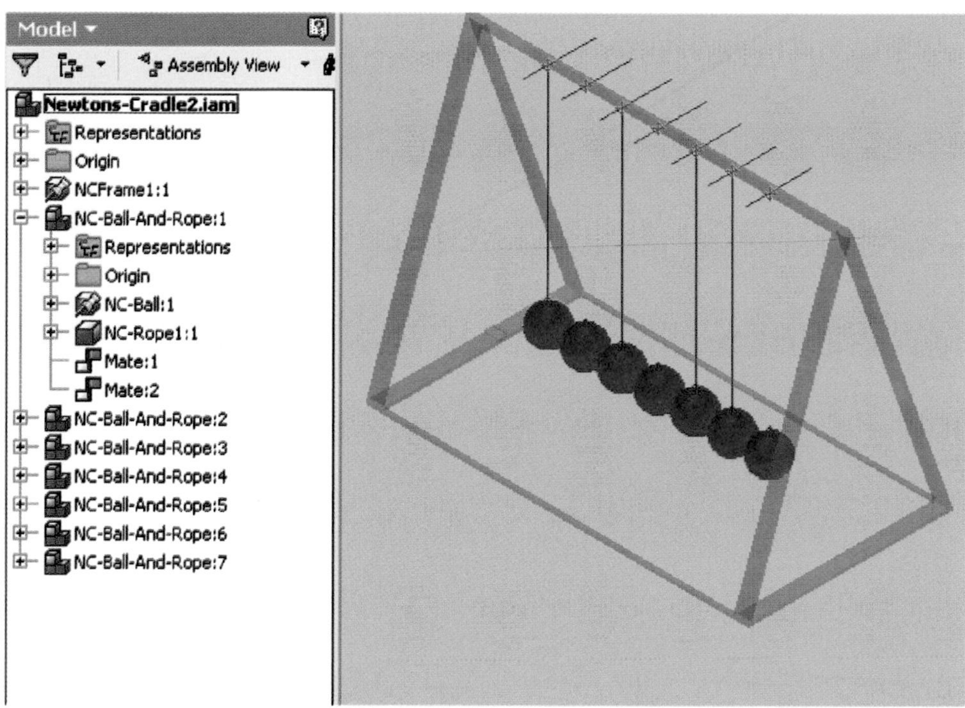

You will now see seven subassemblies, each containing one ball and associated rope. Additionally, there is now one point and axis constraint between the subassembly and frame. In total, there are now two constraints for each subassembly.

5. Select **Environments** tab > **Dynamic Simulation**

In the **Dynamic Simulation** browser notice that there are now seven joints instead of the original 14.

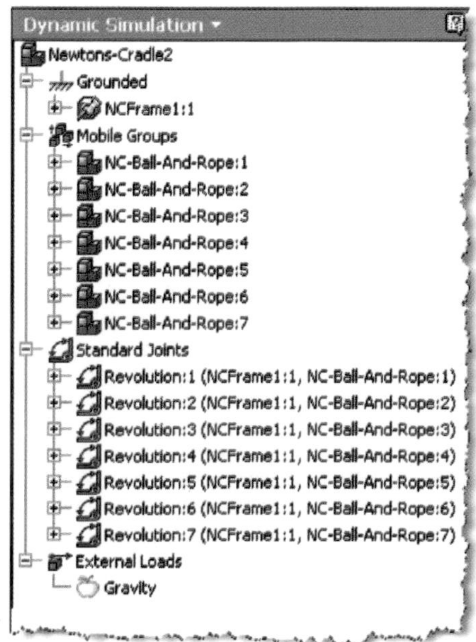

Restructuring your components into subassemblies, like this, will affect your BOM database as now, instead of having seven balls and seven ropes, it will have seven subassemblies. If you do not want to affect your BOM or want to create another assembly for simulation purposes, the only alternative is to weld components together within the Simulation environment before you create the joints, automatically or manually.

6. Close the *Newtons-Cradle2.iam* file

Weld parts together

7. Now open the *Newtons-Cradle3.iam* file
8. Select **Environments** tab > **Dynamic Simulation**

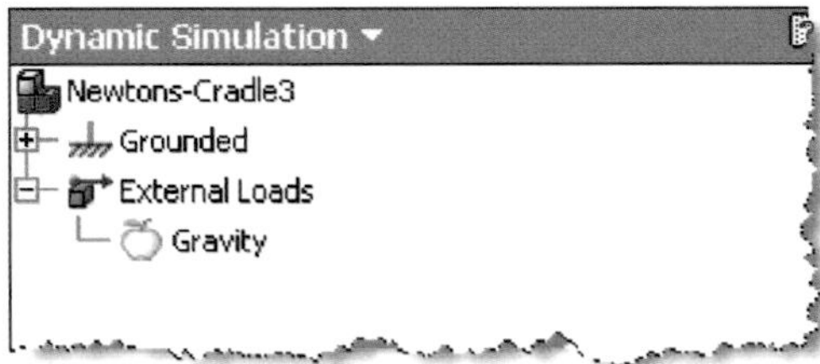

All components are now grounded as no joints have been defined between them, and the **Automatically Convert Constraints to Standard Joints** button is deactivated. This is important because you cannot weld together components that have joints already defined between them.

9. Select NC-Ball:1 and NC-Rope1:1 > Right click and select **Weld parts**

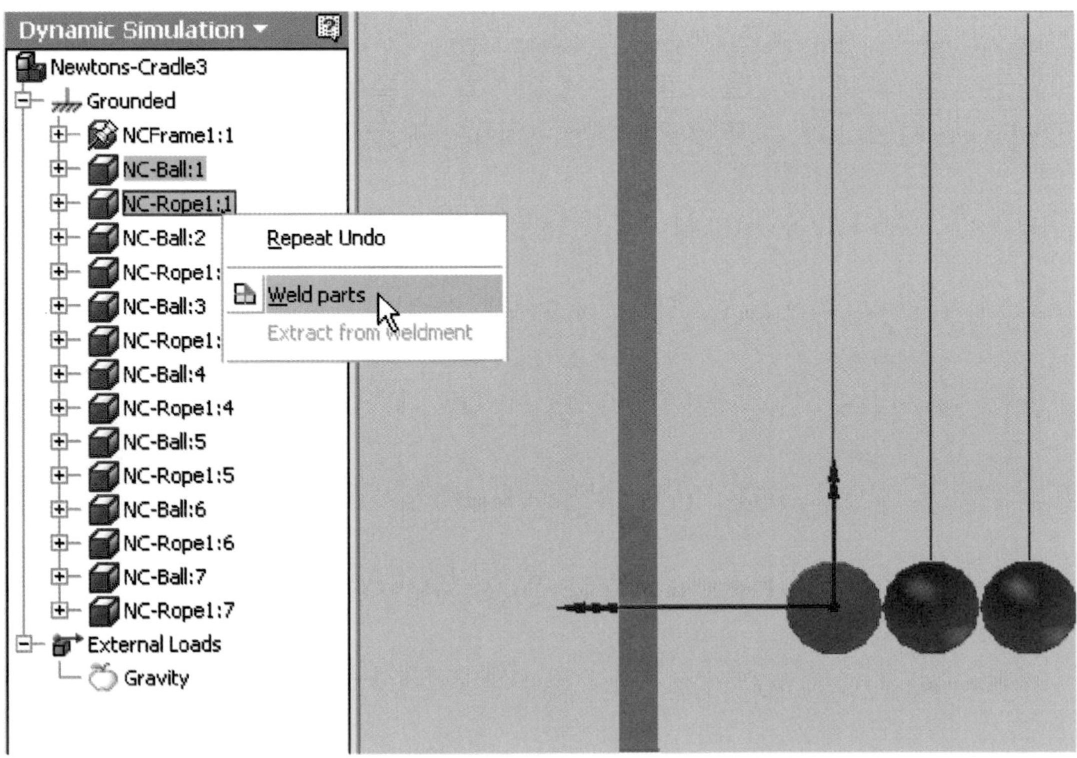

Now notice that both components are welded together as one. This is the same as restructuring components together as subassemblies but does not affect the BOM database.

You cannot use **Automatically Convert Constraints to Standard Joints** as this will not take into account that you have manually welded components together. So, now you will need to convert the joints manually, either by converting constraints to joints or by using the **Insert Joint** tool. We will use the former method, and this is further used in Example 3 – Slider mechanism.

10. Select **Convert Constraints**

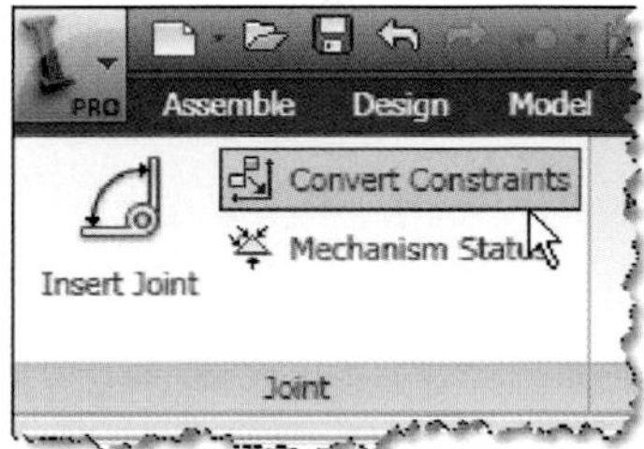

This will enable you to create joints from existing assembly constraints manually.

11. Select the first rope and frame as the two components that need their constraints converted to joints

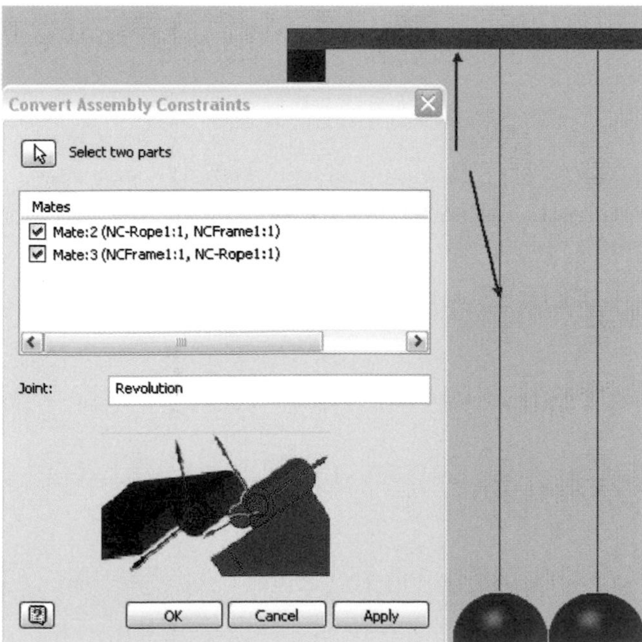

In the **Convert Assembly Constraints** dialog, the two constraints, between the new welded component and the frame, are converted to a revolution joint, as expected.

You will need to repeat steps 9–11 six times to complete the standard joints process for all components.

We had to create two assembly constraints between each subassembly (or welded group) and the frame, to achieve the desired revolution joint. We can achieve this same effect by only using one point constraint between the subassembly and frame, thus eliminating the time needed to create extra constraints. This method is outlined below.

12. Close *Newtons-Cradle3.iam*

Lock the joints' degrees of freedom

13. Open *Newtons-Cradle4.iam*

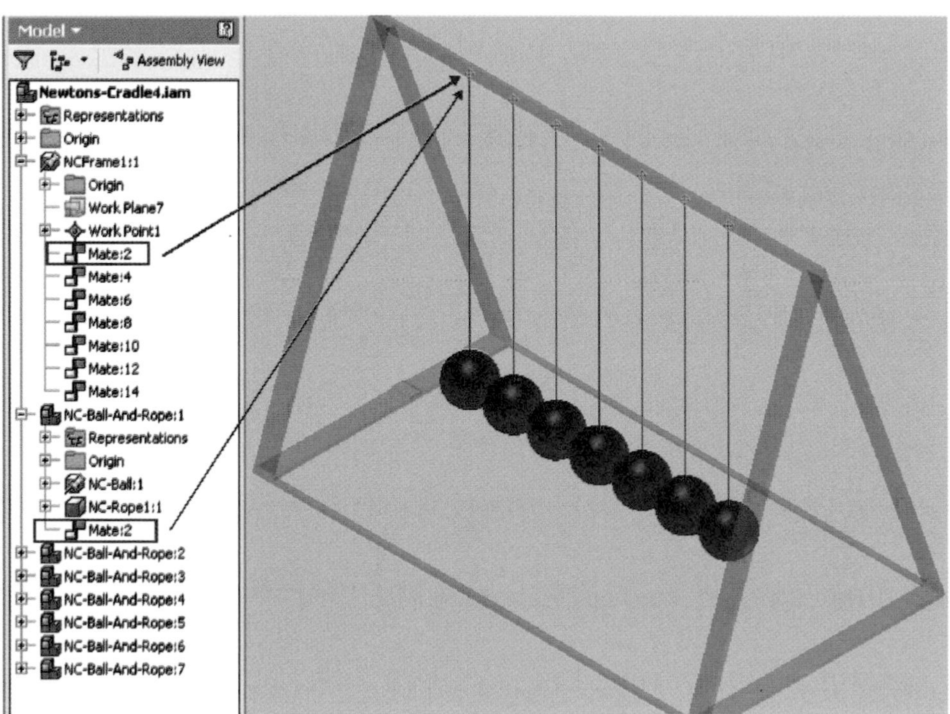

You will see that there are still seven ball and rope subassemblies. The difference now, however, is that there is only one point constraint between each subassembly and frame. In total, there are now seven constraints rather than 14, as in the previous example.

14. Select **Environments** tab > **Dynamic Simulation**

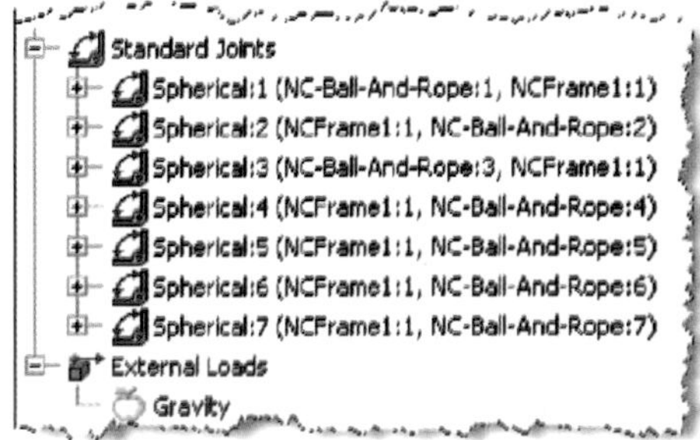

You will now see that seven spherical joints are automatically created, instead of a revolution joint. This is because we only used the point constraint between the two components.

We will now alter these spherical joints to behave like revolution joints without needing to create more assembly constraints.

15. Select all spherical joints > Right click and select **Properties**

We are going to lock two of the three rotational degrees of freedom of the spherical joint, so that it behaves like a revolution joint.

16. Select the **dof 2 (R)** tab in the **Joints** dialog box > Lock the position as shown

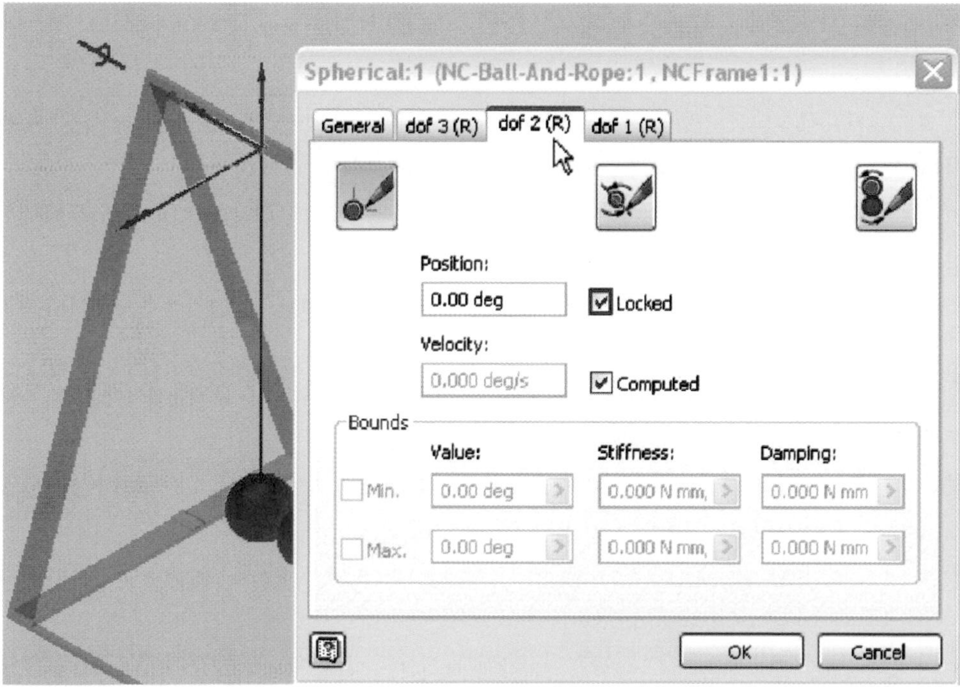

We have just locked the rotation about this degree of freedom as we do not expect the balls to rotate about this axis.

We will further discuss joints properties later within the section on environmental constraints.

17. Select the **dof 1 (R)** tab > Lock the position as shown

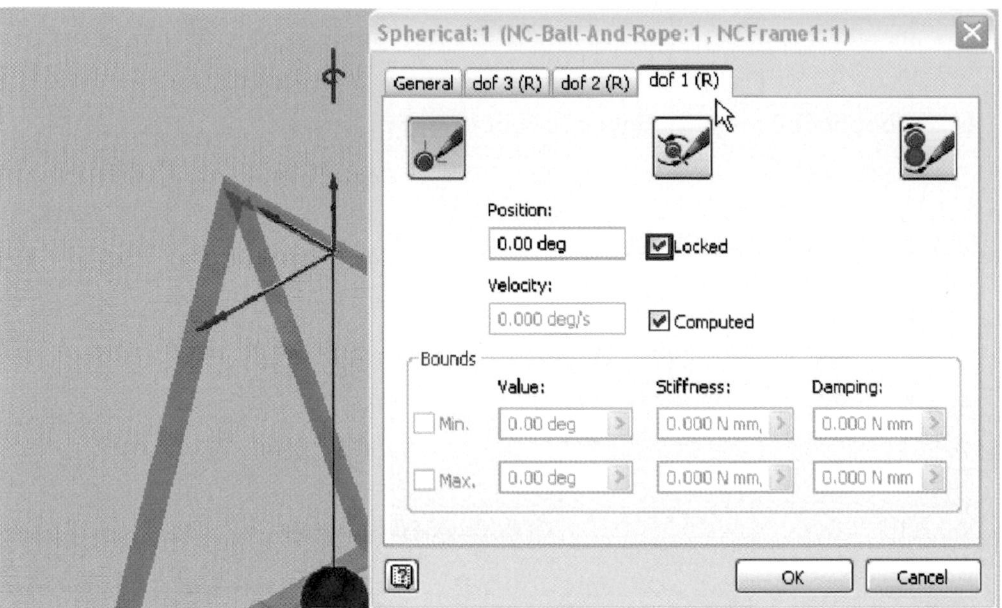

We have just locked another rotation of the spherical joint. This will now make the spherical joint behave like a revolution joint.

18. Click **OK**

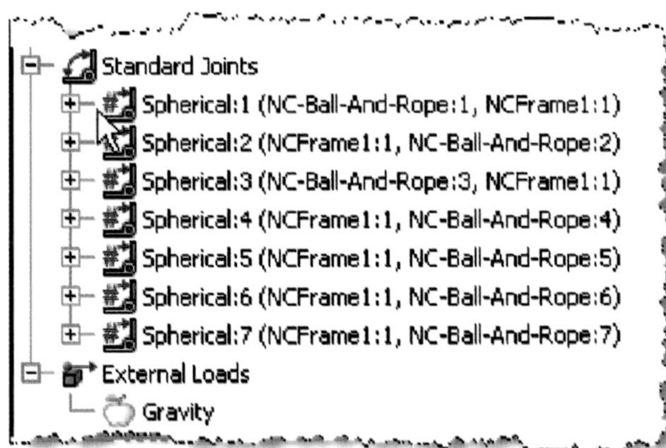

All joints now have a # symbol assigned to them, meaning that they have locked degrees of freedom.

At this point you should be able to appreciate that simply restructuring the components into subassemblies and welding components together can have a significant impact on the number of joints created. We will now discuss how to create joints in more detail.

19. Close *Newtons-Cradle4.iam*

EXAMPLE 2

Whitworth Quick Return Mechanism – Automatic joints

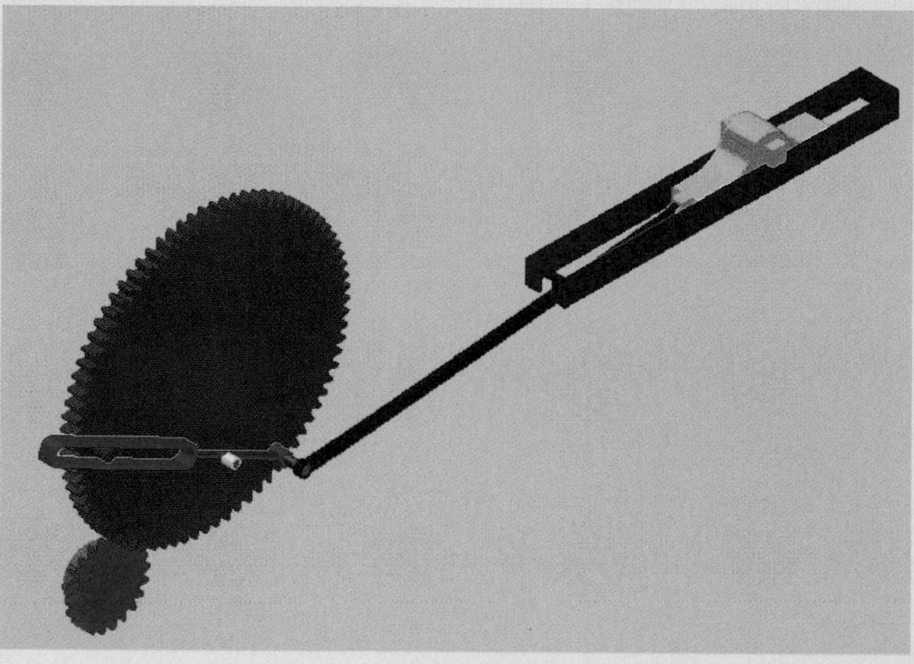

Workflow of Example 2

- Automatically convert standard joints and rolling joints
- Create other nonstandard joints

Note: The only rolling joints that can be converted automatically are between spur gears that have been created using Design Accelerator.

Joints used in Example 2

- Revolution
- Prismatic
- Point-line
- Rolling – Cylinder on cylinder
- Sliding – Cylinder on plane

Automatically convert standard joints and rolling joints

Here, we will create automatic joints from assembly constraints and attempt to analyze how Dynamic Simulation creates joints.

1. Open *Whitworth Quick Return.iam*

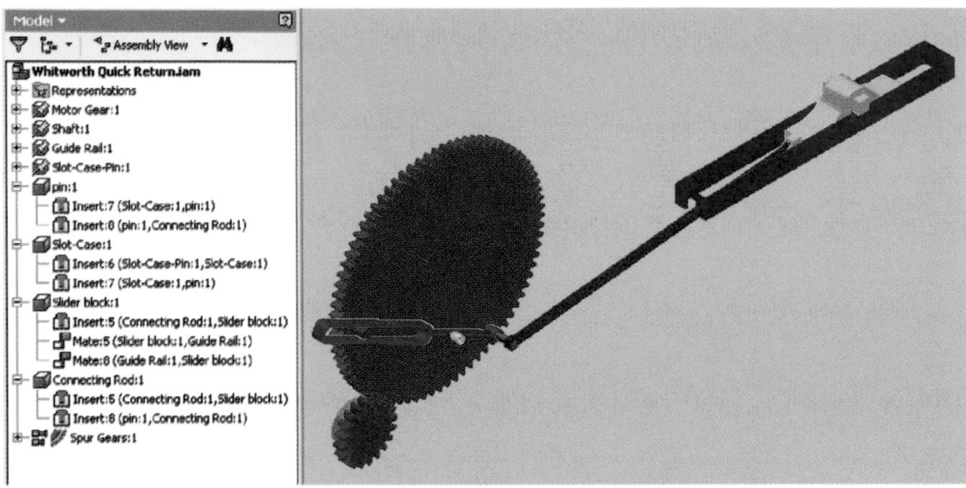

You may need to Rebuild All if the pin is interfering with the slot

In the **Assembly** browser, notice that four parts are grounded and the remaining four components are constrained predominantly using the insert constraint (the equivalent joint is revolution). You may need to expand the components to see the constraints.

2. Select **Environments** tab > **Dynamic Simulation**

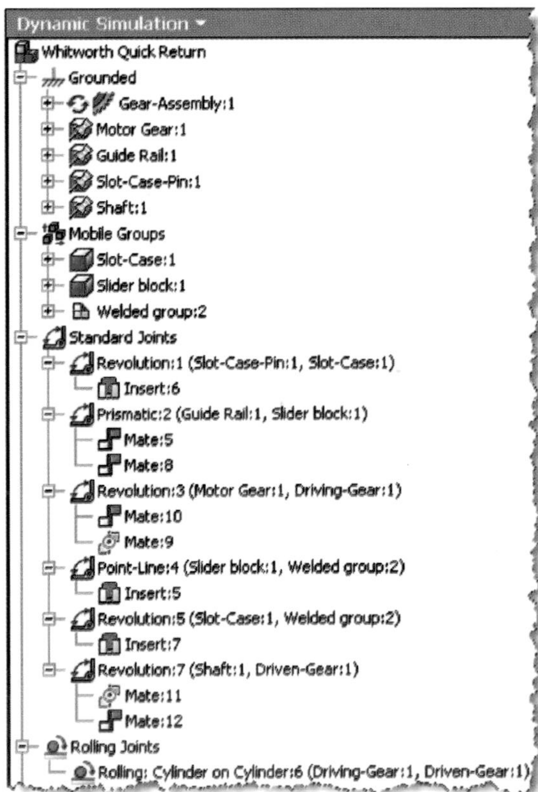

In the **Dynamic Simulation** browser, notice that the components grounded in the Assembly environment remain grounded within Dynamic Simulation. The rest of the components, including the spur gear subassembly, become part of the mobile group; this is because the assembly constraints between these components have been converted automatically to standard and rolling joints.

For Dynamic Simulation to convert spur gear (created by Design Accelerator) constraints to rolling joints, do not alter the hierarchy of the subfolders created by Design Accelerator.

The reason that constraints are created automatically on joints is that **Automatically Convert Constraints to Standard Joints** is activated within the **Dynamic Simulation** settings. We will check this in the following steps.

3. Select **Simulation Settings**

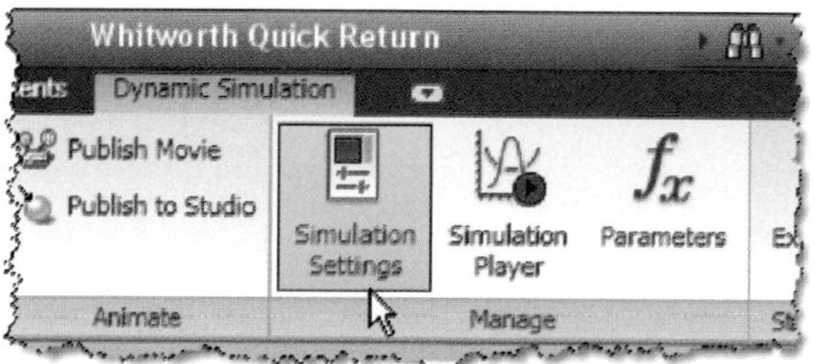

Within the **Dynamic Simulation Settings** dialog box, we can see that **Automatically Convert Constraints to Standard Joints** is activated.

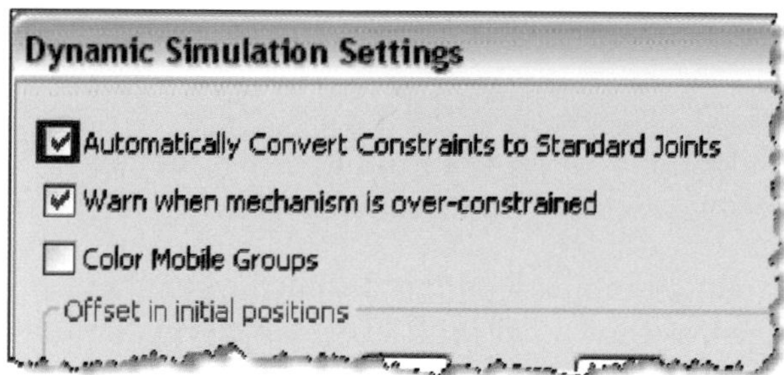

4. Click **Cancel**

Now, we will attempt to analyze the joints converted from assembly constraints.

- Revolution:3 joint – Result is as predicted.
- Welded group:2 – Not as predicted. We were expecting a revolution joint; instead, a welded joint between both components was created, as if they were welded together in reality.
- Point-line:4 joint – Not as predicted. We were expecting a revolution joint; instead a point-line joint was created, producing an extra three degrees of freedom between the two components.

The only solution for this is to avoid creating redundant joints (overconstrained joints). Dynamic Simulation has created the joints and welded bodies as illustrated below. Redundant joints are discussed in detail later in the chapter.

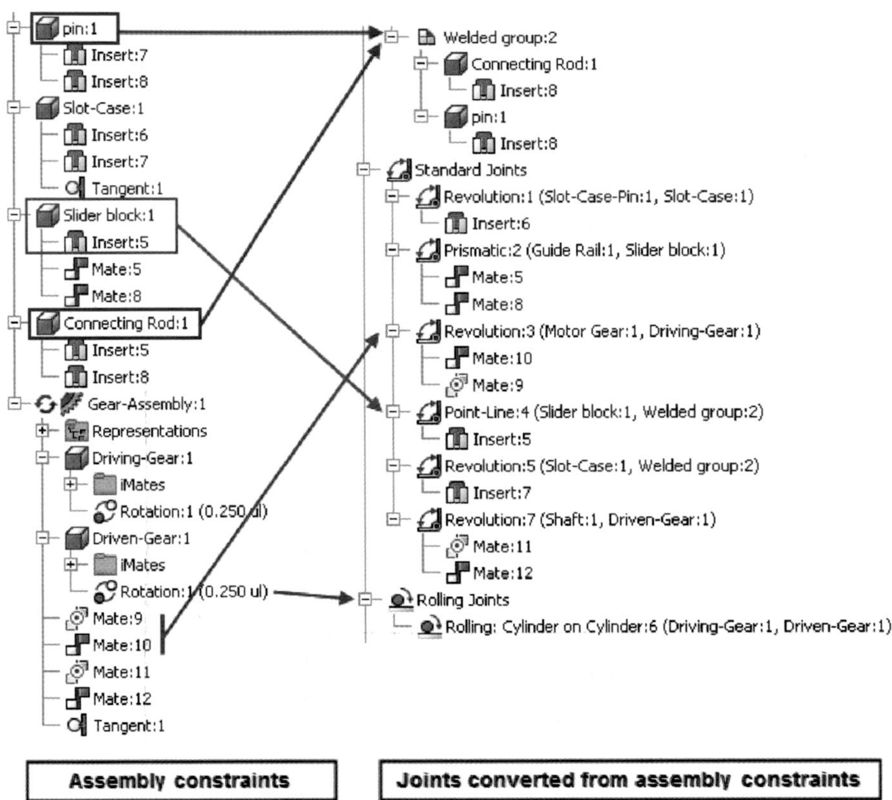

The only way to alter the converted joints is to modify the assembly constraints, and this process can become tedious, especially for a large assembly with redundant joints.

Another option is to manually convert constraints, but this option is disabled when the automatic conversion of constraints is activated, as shown below. This option gives you more control of how joints are created and this process will be illustrated in Example 3.

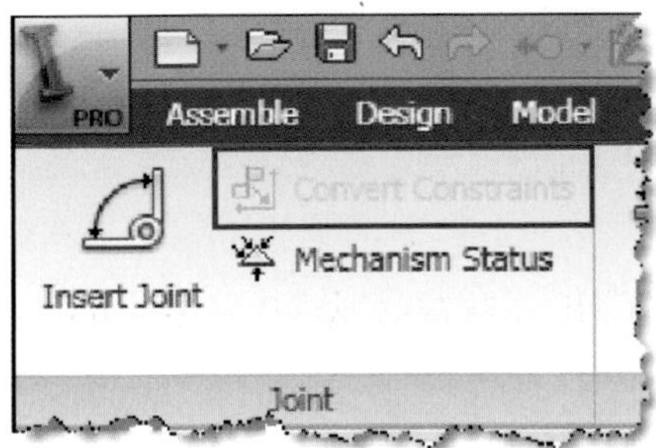

For now, we will continue with Example 2 and create more nonstandard joints.

Create other nonstandard joints

Here, we will create one more sliding joint to complete the mechanism.

5. Select **Insert Joint**

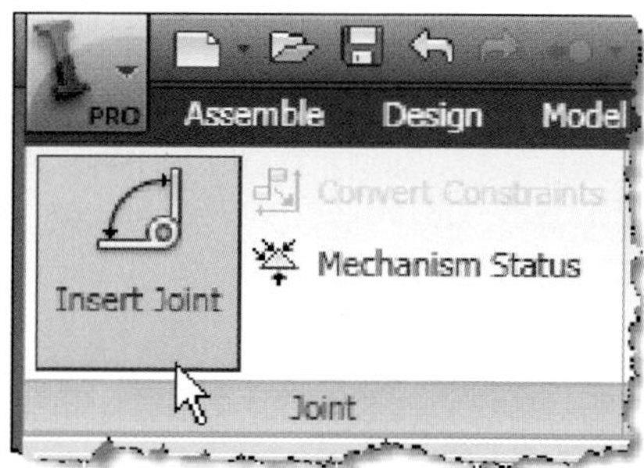

6. Select the **Sliding: Cylinder on Plane** joint from the list

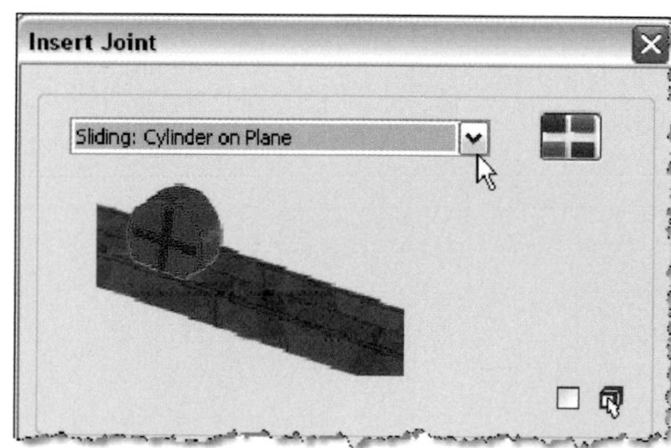

7. To define the plane, select the edge of Slot-Case:1

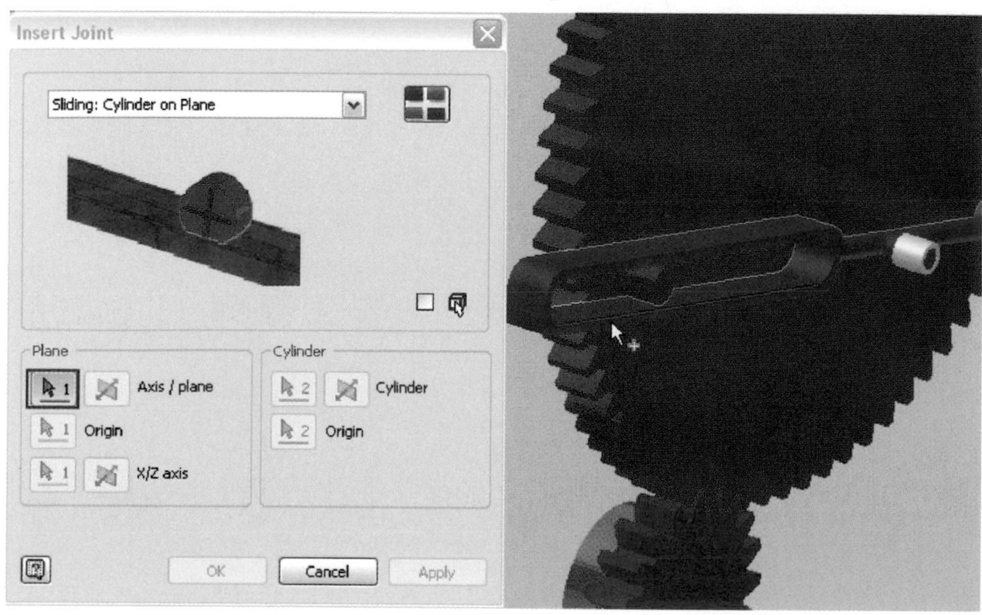

8. Select the **Cylinder** button of Component 2 > Select the edge of spur gear 1 to complete the joint

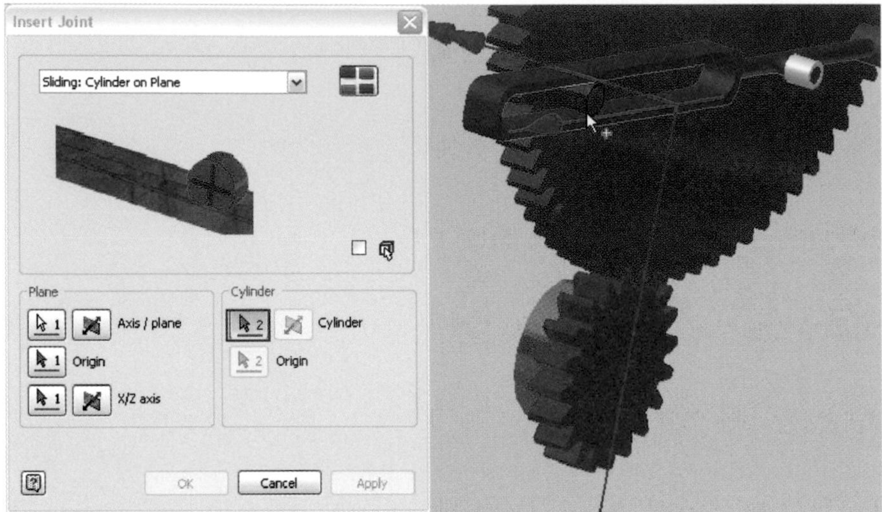

9. Click **OK.** The following sliding joint will be created:

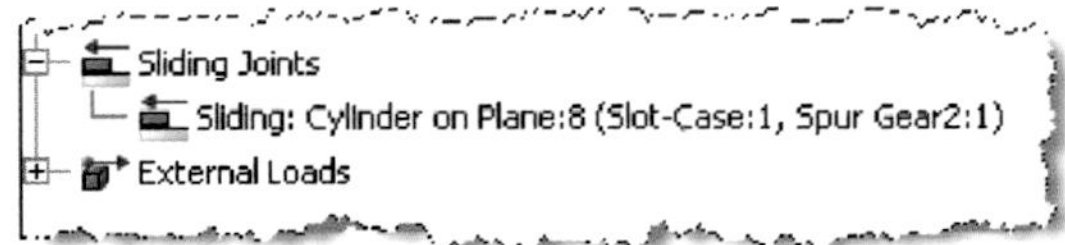

10. Close *Whitworth Quick Return.iam*

EXAMPLE 3

Slider mechanism – Manually convert constraints

Workflow of Example 3

- Manually convert standard joints from assembly constraints
- Repair redundant joints

Joints used in Example 3

- Revolution
- Point-line
- Prismatic

Manually convert standard joints from assembly constraints

Here, we will create joints manually by converting existing assembly constraints.

1. Open *Slider-mechanism.iam*

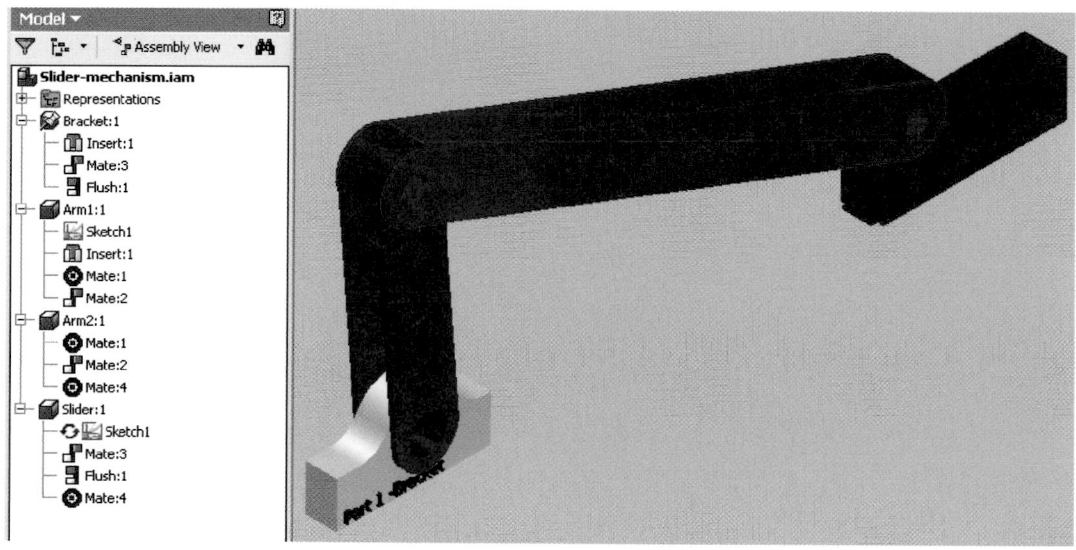

2. Select **Environments** tab > **Dynamic Simulation**

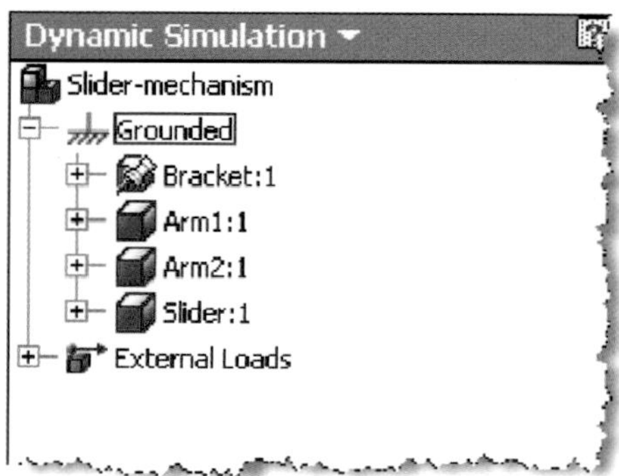

In the **Dynamic Simulation** browser, notice that all components are grounded because automatic conversion of constraints is switched off within Dynamic Simulation.

3. Select **Convert Constraints**

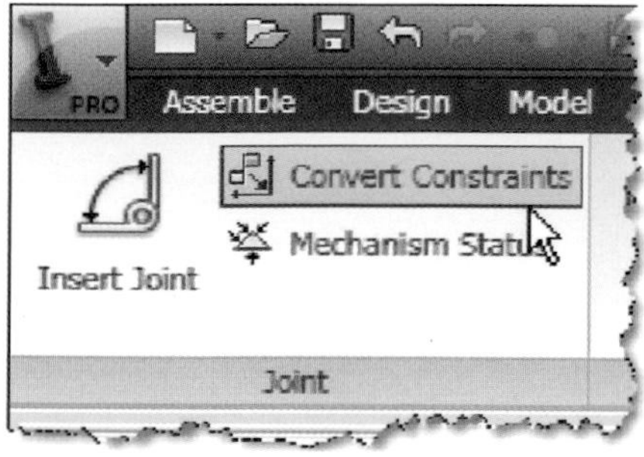

4. In the **Convert Assembly Constraints** dialog box, select Bracket:1 and Arm 1:1

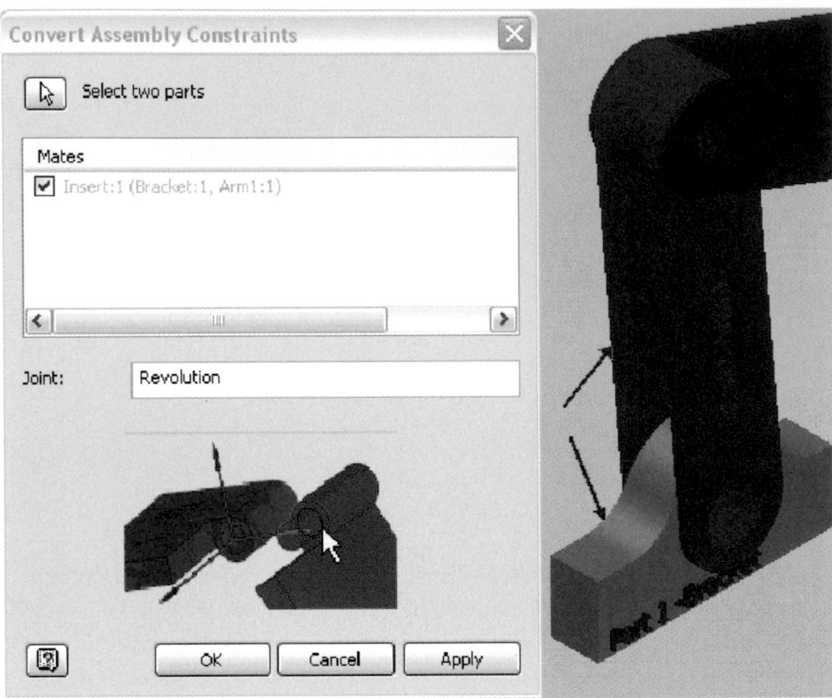

In the **Convert Assembly Constraints** dialog box, the Insert assembly constraint that was used to constrain the two parts in the Assembly environment appears.

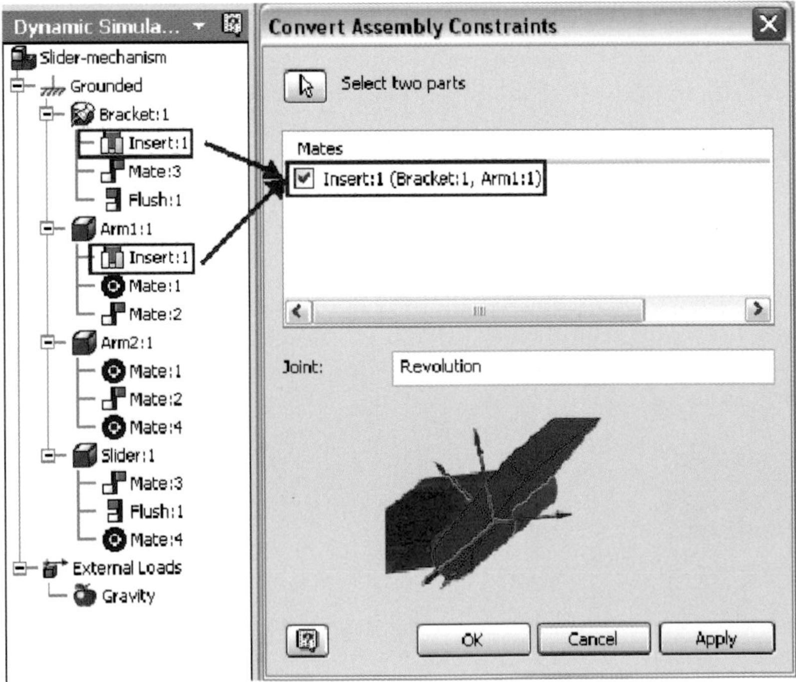

Dynamic Simulation has suggested an equivalent revolution joint.

5. Click **Apply**

Clicking **Apply** instead of **OK** leaves the dialog box open so you can continue converting constraints.

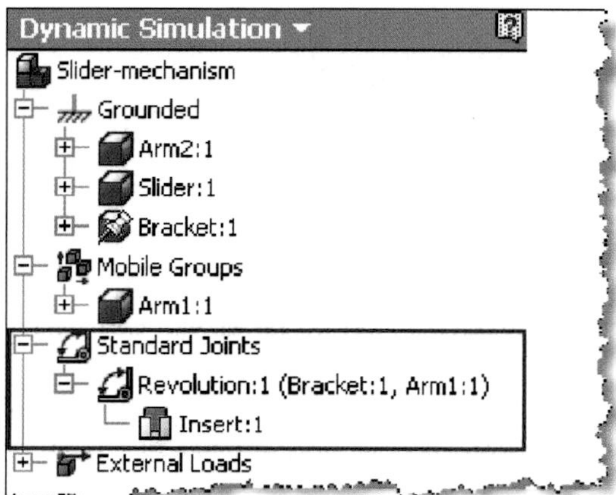

The Revolution:1 (Bracket:1, Arm1:1) joint is created under Standard Joints, as illustrated below. This joint creation process has also made Arm1:1 mobile. The reason that the Arm1:1 component is made mobile is that the Bracket:1 component is grounded from the Assembly environment and hence remains grounded within the Simulation environment.

6. Select the Arm1:1 and Arm2:1 components to create a second joint between the selected components

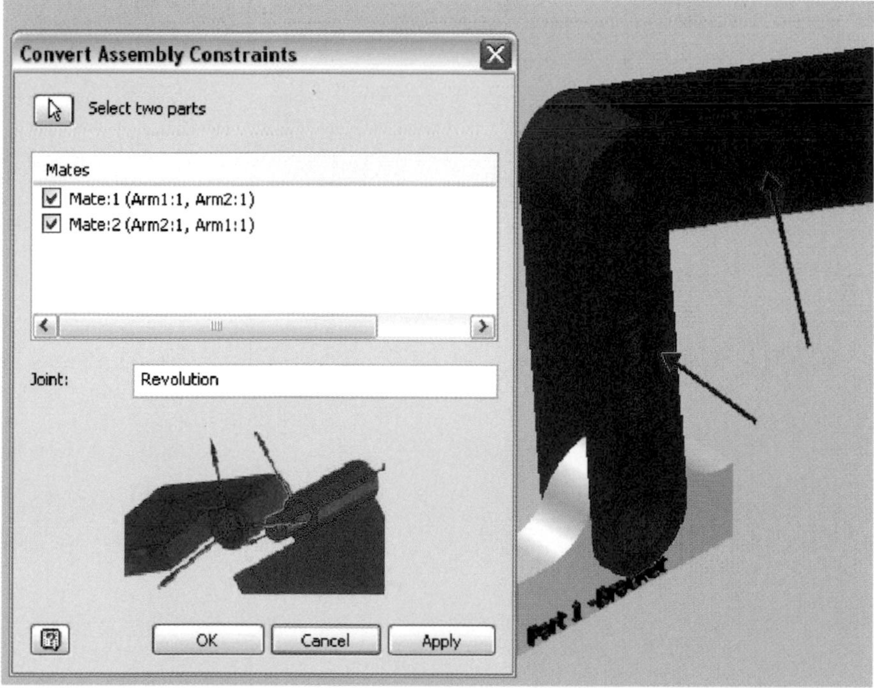

Dynamic Simulation has automatically detected two constraints between these components and as a result has suggested a revolution joint.

7. Now deselect the Mate:2 constraint and see how the joint created now changes to a cylindrical joint

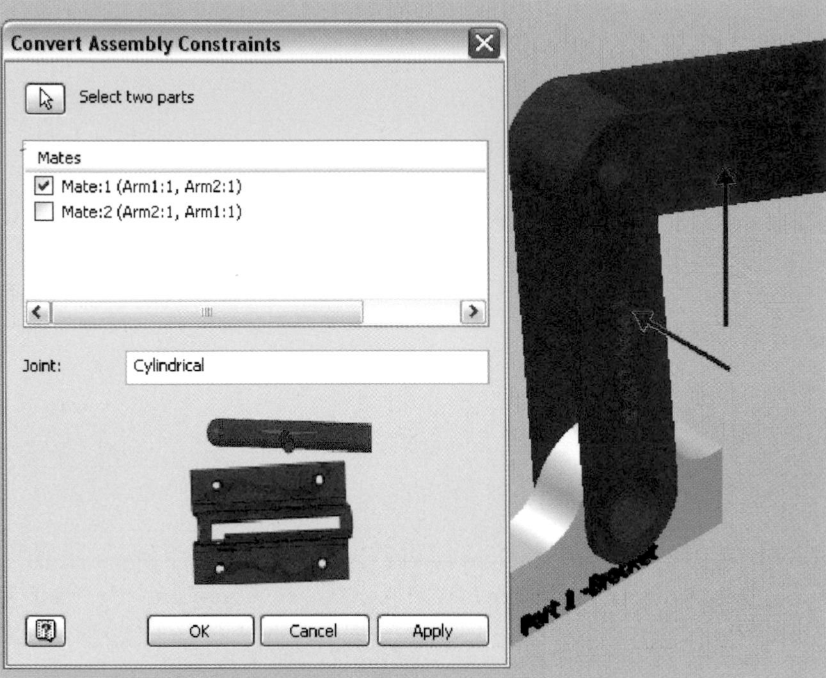

The reason for this is that Mate:1 (Arm1:1, Arm2:1) is an axis-axis constraint that allows rotation about, and translation along, the edge/axis.

8. Now reselect Mate:2 and deselect Mate:1 and see how the joint has now created changes to a planar joint

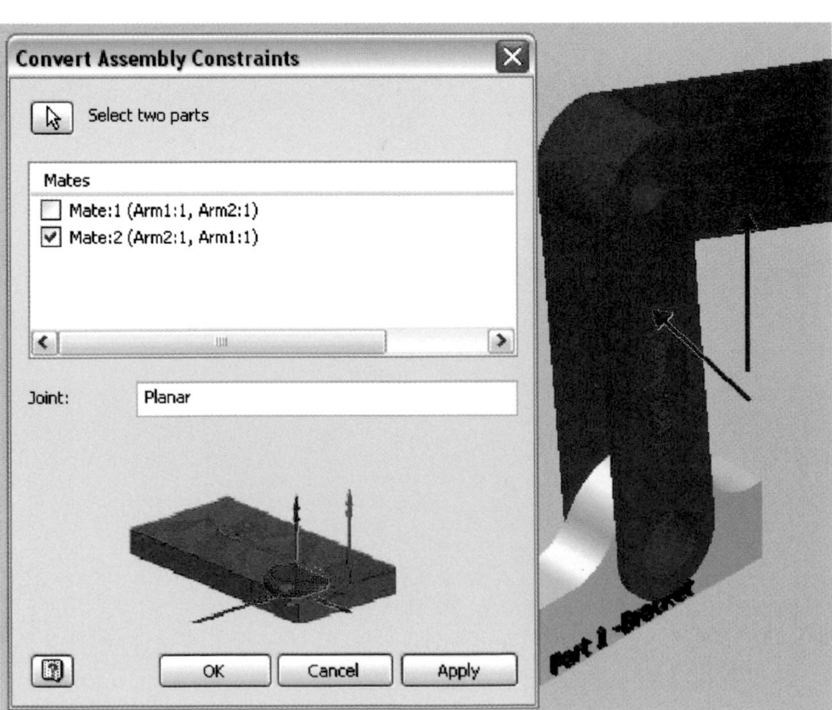

The reason for this is that the Second Mate:2 (Arm2:1, Arm1:1) is a face-face constraint, which allows movement along a 2D plane only.

By selecting both constraints, we are further restricting the motion of the components, for example reducing the degrees of freedom of the selected components.

9. Reselect both constraints > Click **Apply**
10. Select Arm 2:1 and Slider:1 > Click **Apply**

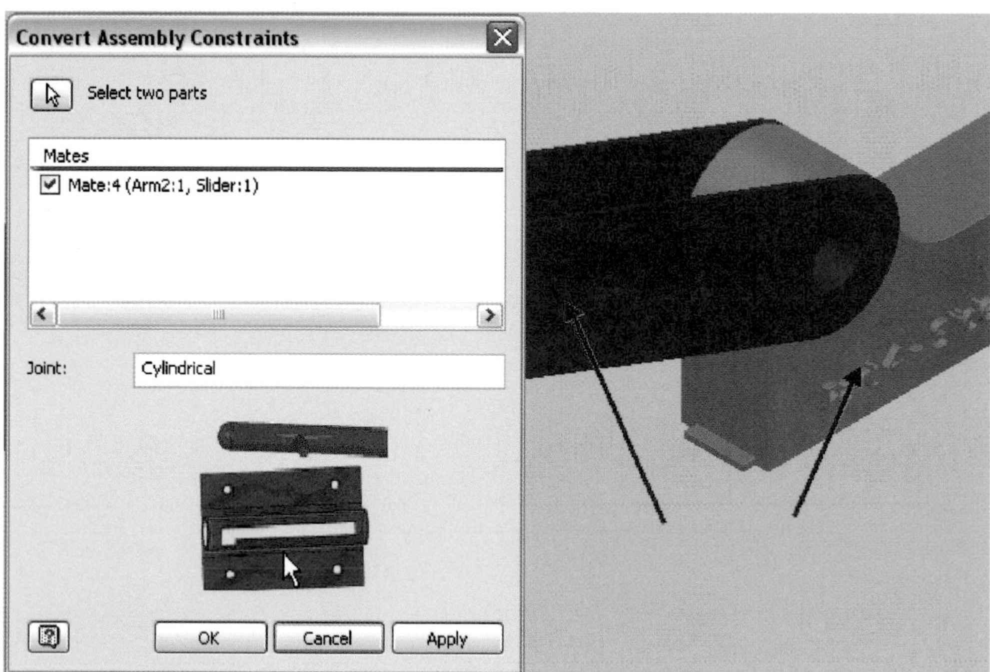

A cylindrical joint is suggested, because Mate:4 is an axis-axis constraint.

11. Finally, select the Bracket:1 and Slider:1 components > Click **OK**

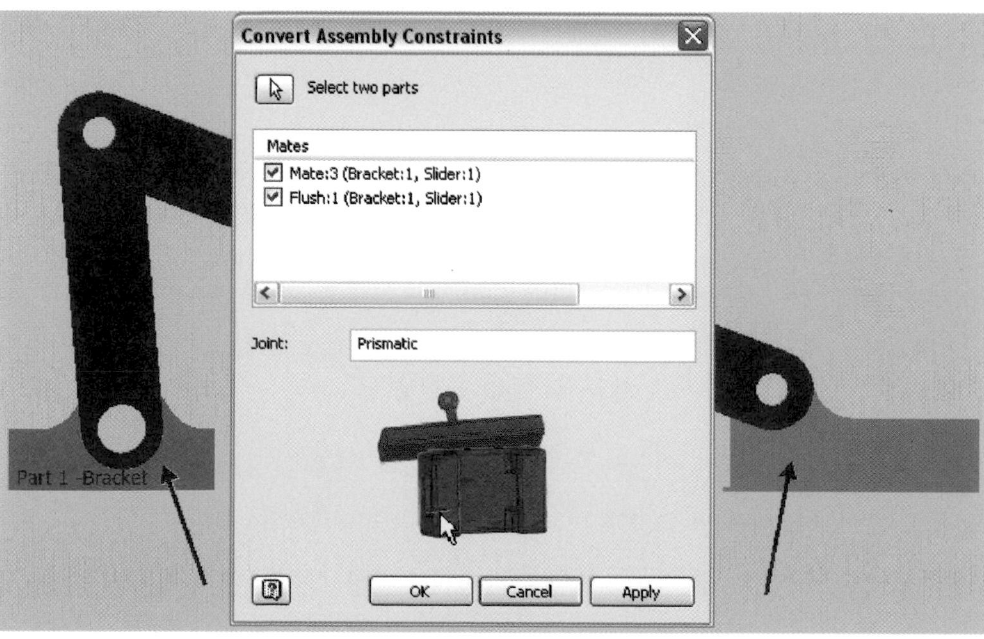

 Mate:3 is an edge-edge constraint and Flush:1 is a face-face constraint; hence, Dynamic Simulation suggests a prismatic joint.

12. Click **OK** to accept the Dynamic Simulation warning

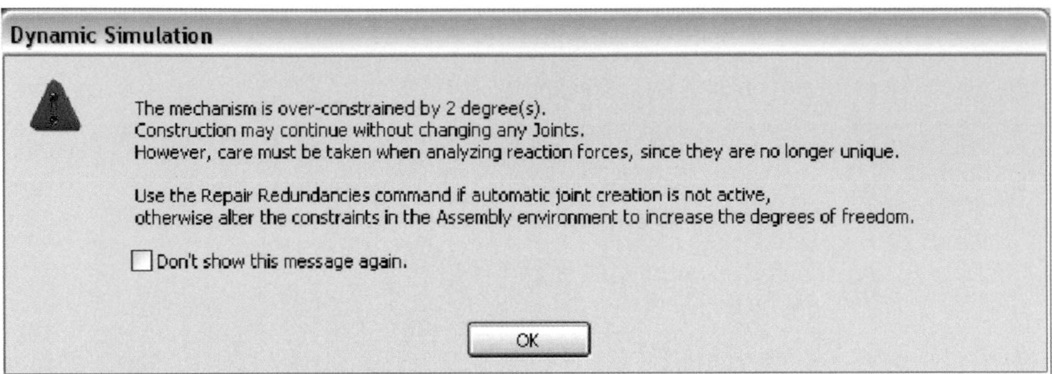

The Prismatic:4 joint is created under Standard Joints, as illustrated below. In total, four joints have been created.

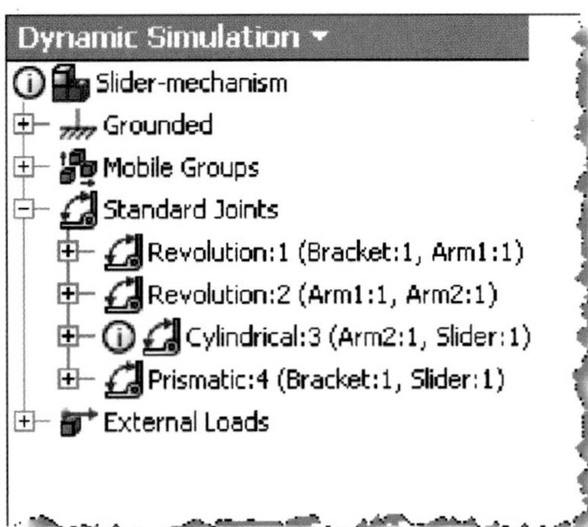

 By accepting the above warning, the Cylindrical:3 joint has now become redundant. This means that the joint is overconstrained by two degrees of freedom, as mentioned in the warning.

The redundancy phenomenon will be discussed later, but, in the meantime, we will determine how we can remove redundancy from this joint.

Repair redundant joints

To resolve the Cylindrical:3 (Arm2:1, Slider:1) joint you can either

a. Alter the joint/constraint to allow for more degrees of freedom;
b. Alter any of the other joints; or
c. Use the **Repair Redundancies** button to automatically resolve the issue.

We will use option a, as option c is not available as we have created joints from assembly constraints.

13. Select **Mechanism Status**

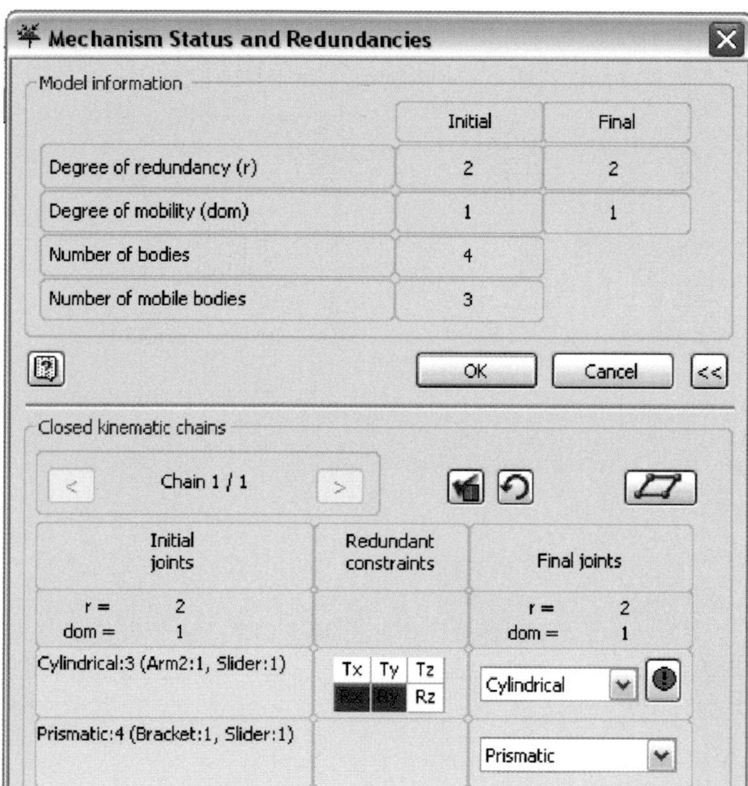

14. Click on the exclamation mark to determine possible solutions

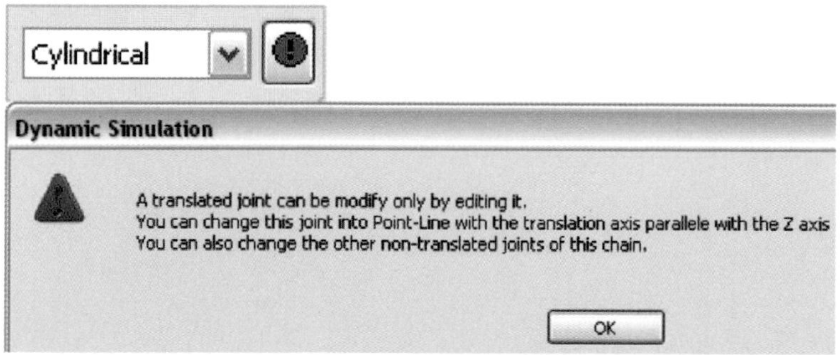

15. Click **OK** twice

As soon as you click on the exclamation mark, a warning will appear suggesting that the joint cannot be automatically repaired as it is a translated joint and can only be repaired by editing the constraints. The warning also suggests that by using a point-line joint the model will have no redundancies. A point-line constraint is basically a spherical joint with one translational degree of freedom.

16. Select Mate:4 > Right click and select **Edit**

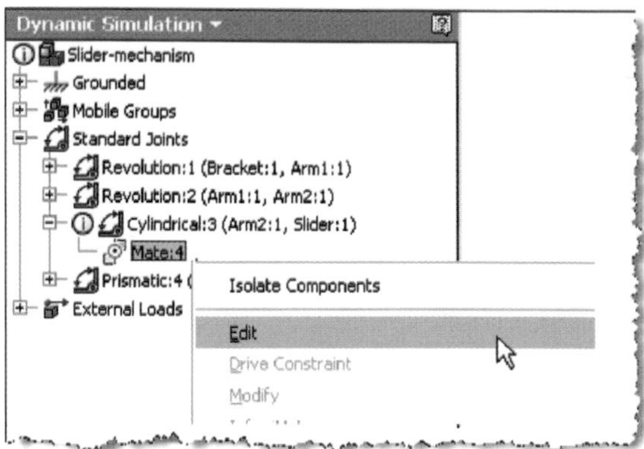

You may need to expand the Cylindrical:3 joint to see the Mate:4 constraint.

17. In the **Edit Constraint** dialog box, click on the second selection > Select the edge of the slider (red) and click **OK**

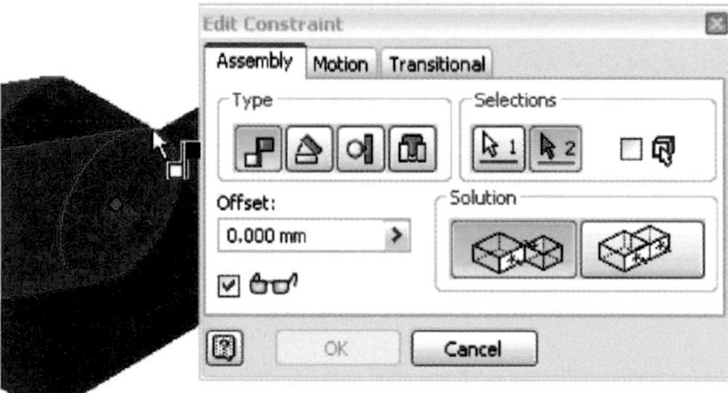

The Cylindrical:3 (Arm2:1, Slider:1) redundant joint has changed to a Point-Line:3 (Arm2:1, Slider:1) nonredundant joint, as illustrated below and suggested by the warning.

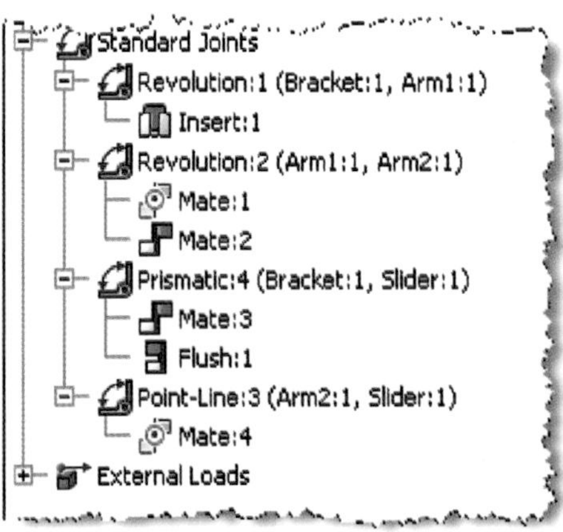

18. Close *Slider-mechanism.iam*

EXAMPLE 4

Cam follower mechanism – Manually create joints

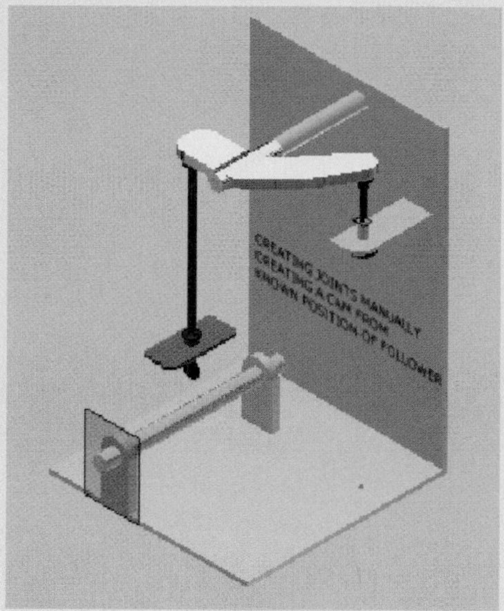

Workflow of Example 4

- Manually create standard joints
- Weld components together
- Continue manually creating joints

Joints used in Example 4

- Revolution
- Cylindrical
- Point-plane

Manually create standard joints

Here, we will create joints manually by using the **Insert Joint** tool. This will hopefully help you to better understand joints.

1. Open *CAM.iam*

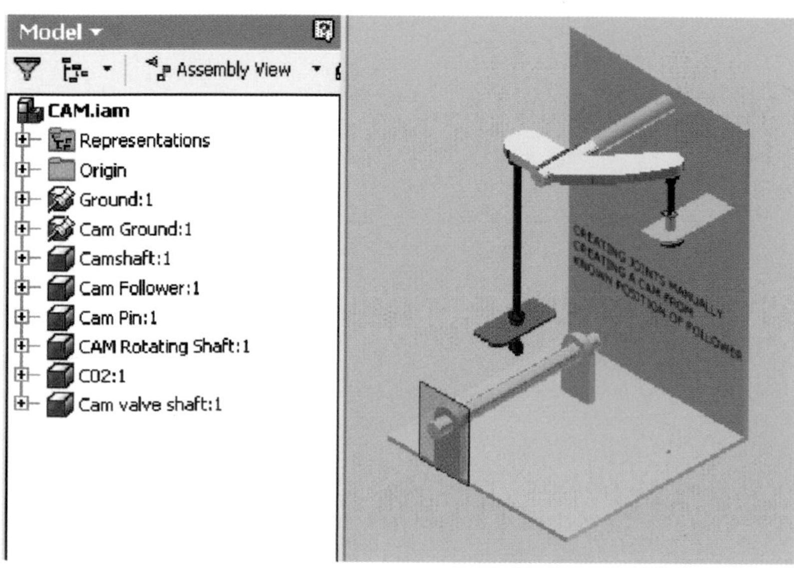

2. Select **Environments** tab > **Dynamic Simulation**

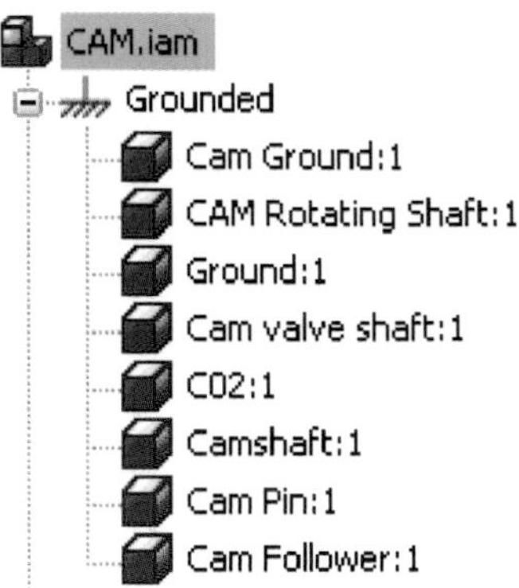

All of the parts are shown under the Grounded node because no joints have been applied or created.

3. Select **Insert Joint** from the **Dynamic Simulation** panel

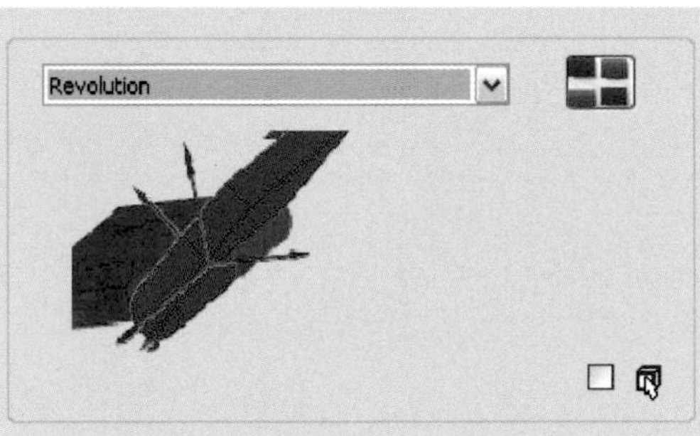

In the **Insert Joint** dialog box, the revolution joint will be the default joint.

4. Select the cam rotating shaft, as shown for the Component 1 selection

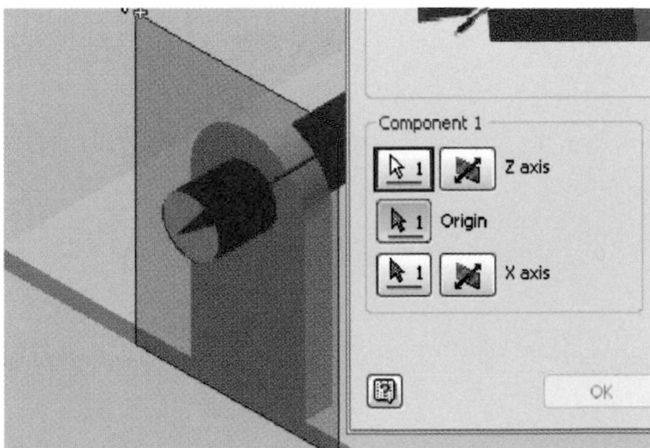

By selecting the component as shown, the origin of the joint is in the middle of the shaft. If we want to maintain the position of the shaft, we also need to specify the origin of the joint.

5. Click on the workplane as shown to define the origin of the joint axis for Component 1

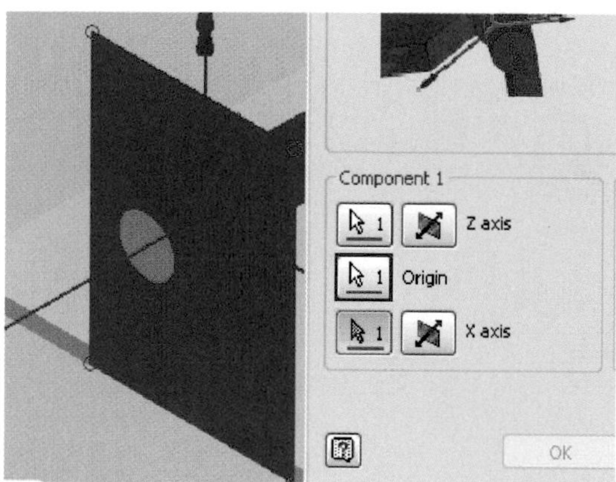

6. Select the Component 2 button and select the top face of the bracket as shown

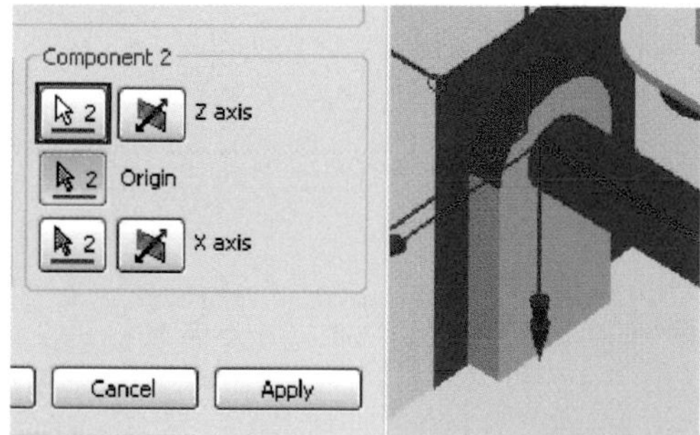

7. Select the edge of the bracket to define the origin of the axes for Component 2

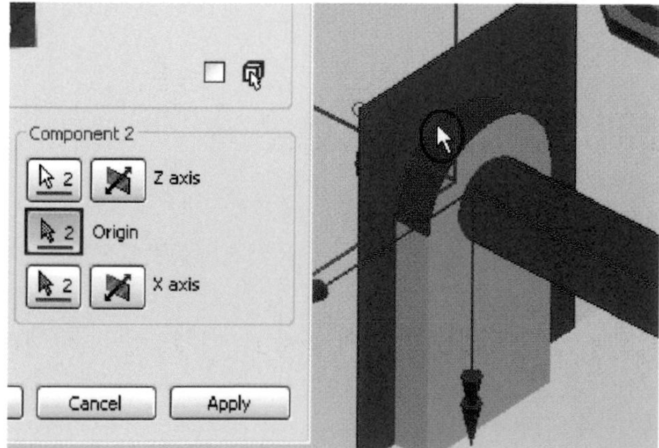

Both axes have the same origin but the Z axes are in opposite directions. We need to flip the direction of the Z axis for Component 2, otherwise the components will not maintain their original positions.

8. Select the **Flip Z axis** button, as shown, to flip the direction of the Z axis. Click **OK**

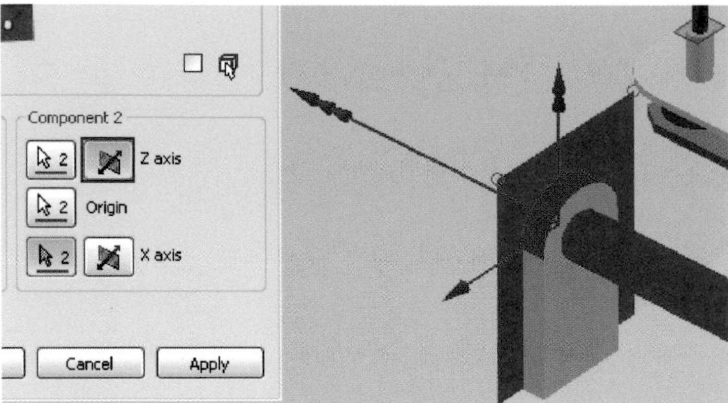

The Revolution:1 joint is created under Standard Joints, as illustrated below.

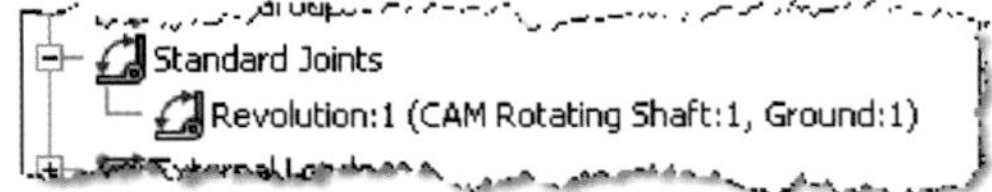

9. Click on the **Insert Joint** tool again to create a revolution joint
10. For Z axis definition of Component 1, click on the edge of Component 1, as shown

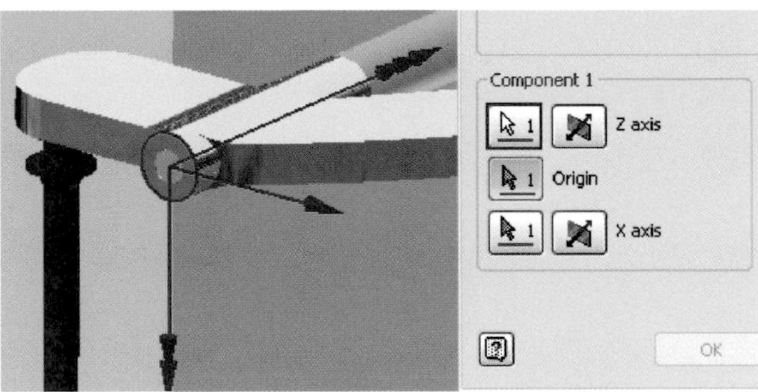

11. Select the Component 2 button and select the edge of Component 2, as shown

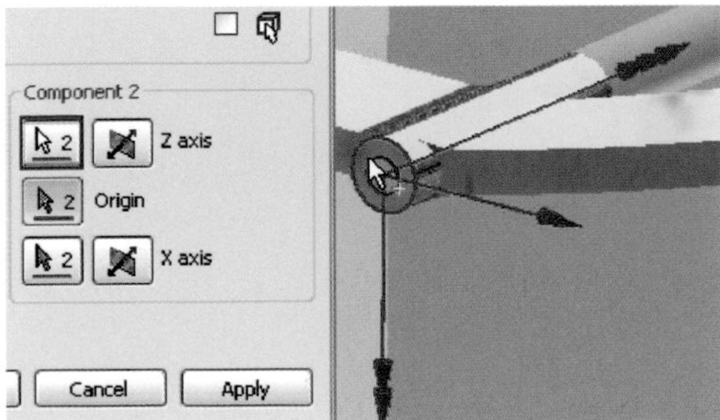

The axis origin and the Z axes align correctly. Selecting the edge of a component also defines the origin.

12. Click **Apply**

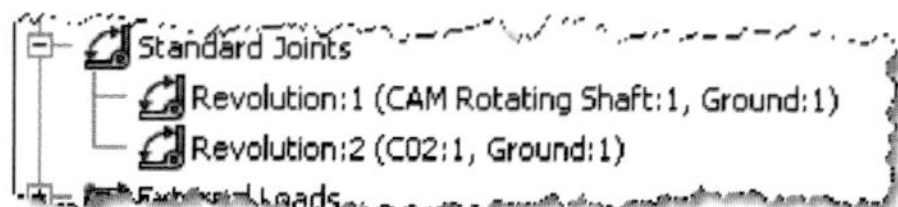

The Revolution:2 joint is created under Standard Joints, as illustrated above.

13. In the **Insert Joint** dialog box, select the cylindrical joint from the pulldown menu, as shown below

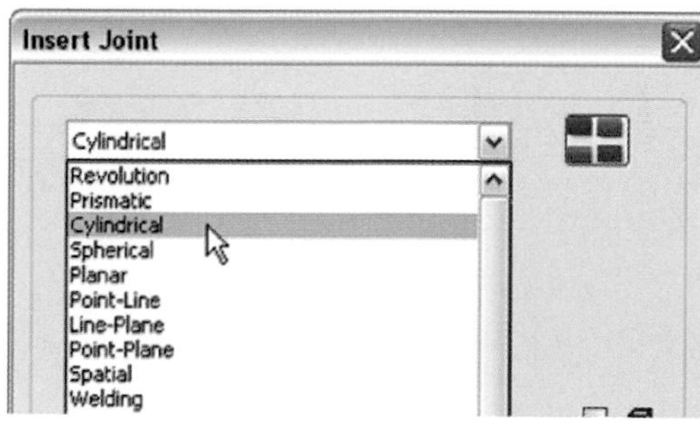

14. Select camvalve face, as shown, to define the Z axis of Component 1

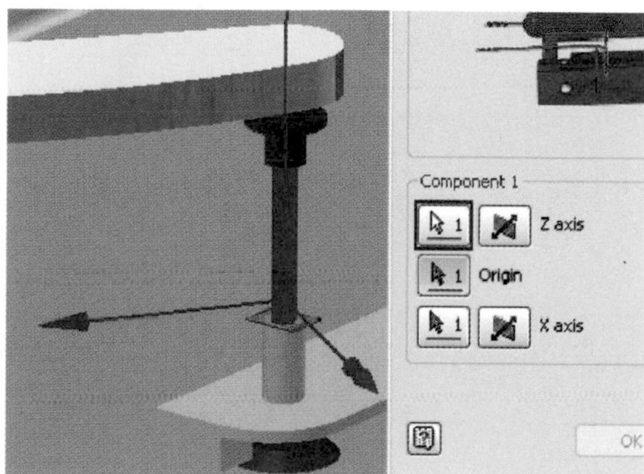

As the camvalve is in the closed position, we also need to define the origin so Component 1 does not move.

15. Select the work plane, as shown, to define the origin of the joint

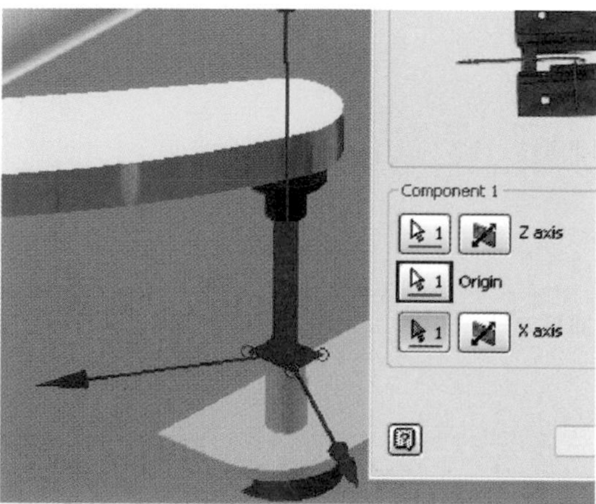

16. Select the Component 2 button > Select the edge of Component 2

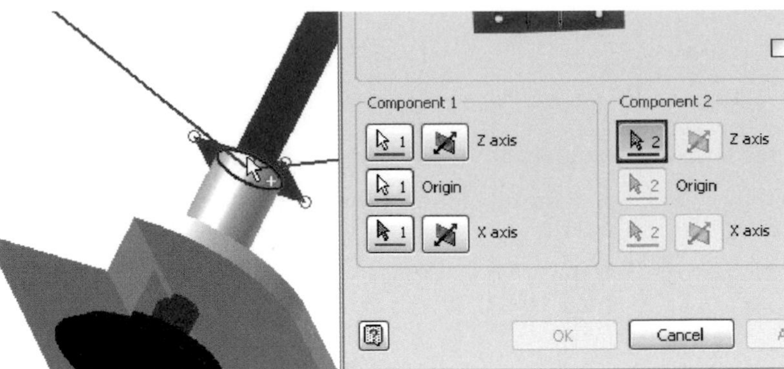

Both axes have the same origin but the Z axes are in opposite directions. We need to flip the direction of the Z axis for Component 2.

17. Select the **Switch Z axis** button, as shown, to flip the direction of the Z axis of Component 2

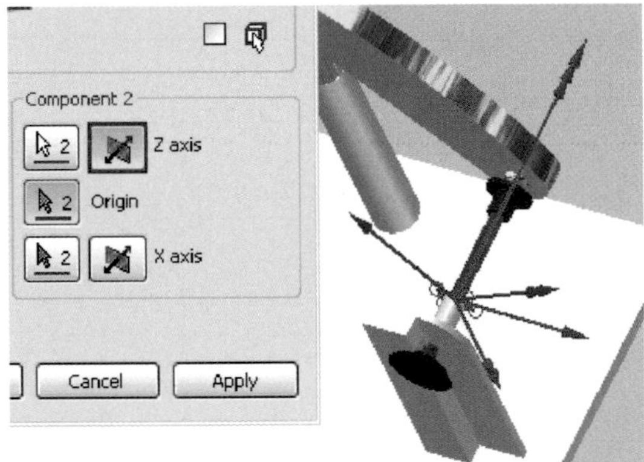

You do not need to manually align the X axes as this will not affect the original position. The software will automatically rotate the cylindrical component to align the X axes to create the joint.

18. Click **OK**

The Cylindrical:3 joint is created under Standard Joints, as illustrated below.

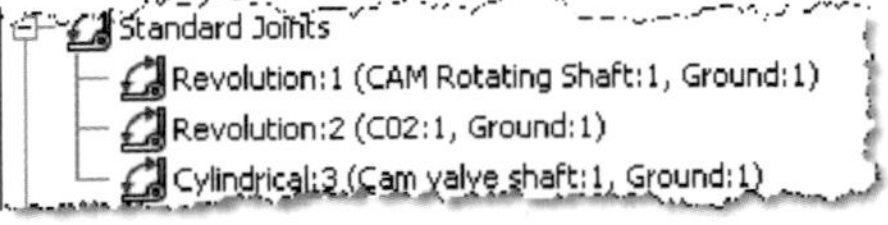

Weld components together

19. Select Camshaft:1, Cam Pin:1, and Cam Follower:1 > Right click, and select **Weld parts**, as shown

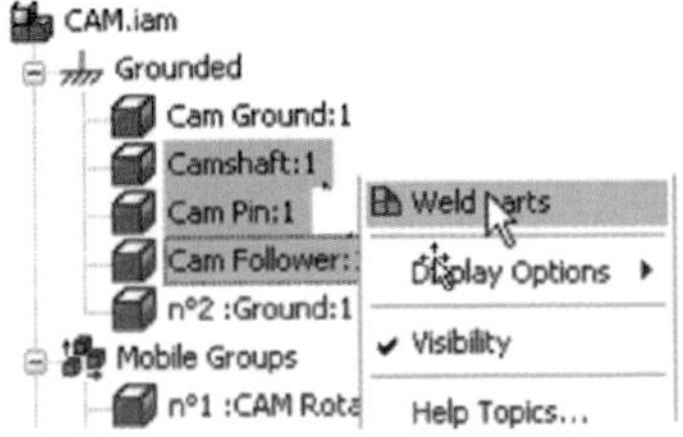

As the selected parts have the same relative motion and we are not interested in the reaction between these components, we can treat them as one part (or as part of a welded group). This is the same as creating subassemblies within the Assembly environment, as Dynamic Simulation will treat subassemblies as one component.

20. Click **OK**. The following welded group is created:

Continue manually creating joints

21. Select **Insert Joint**

The cylindrical joint will now be the default joint, as it was the last created joint.

22. Select the camshaft edge, as shown, to define the Z axis of Component 1

23. Select the Component 2 button > Select the edge of Component 2, as shown

24. Click **Apply**. The following joint will be created:

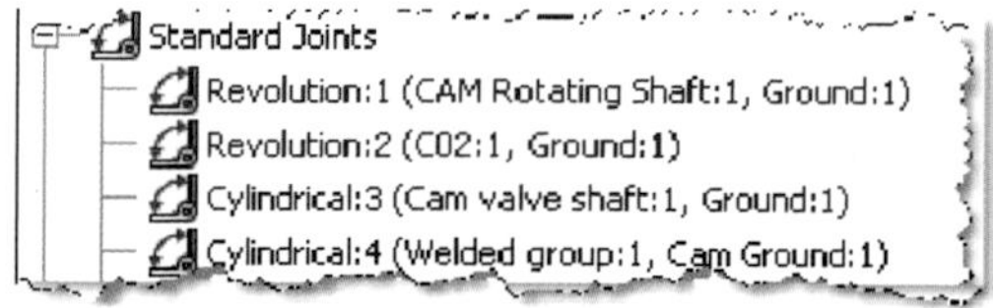

Make sure that the Z axes are pointing in the same direction.

In the following steps, a point–plane joint will be used to simply create contact between the ball and planar faces of the camshaft and cam valve. In reality, a the contact is continuously variable.

25. In the **Insert Joint** dialog box, select the point-plane joint from the pulldown menu

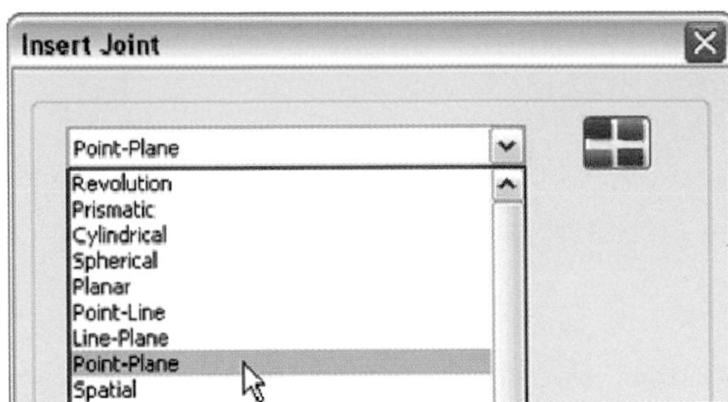

26. Select the cam valve shaft edge, as shown, to define the Z axis of Component 1

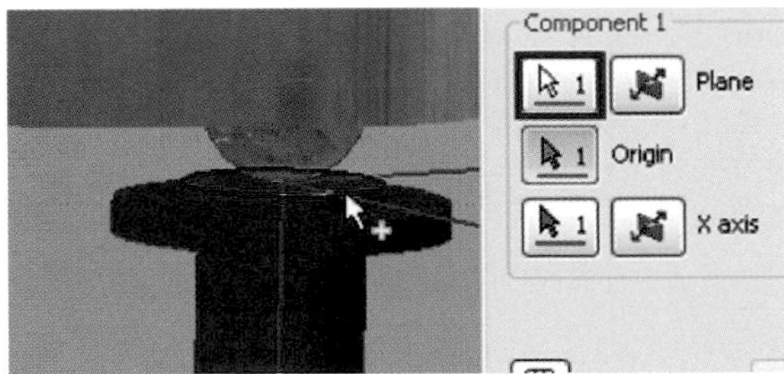

27. Select the Component 2 button > Select the point of Component 2, as shown

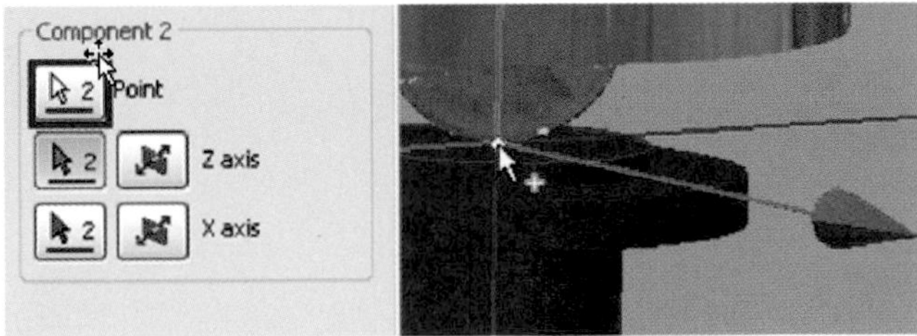

You may need to make the user workpoints visible.

The Z axes of each component are pointing in opposite directions. We need to align them so they point in the same direction.

28. Select the **Switch Z** axis button of Component 2

29. Click **Apply**. The following joint should be created

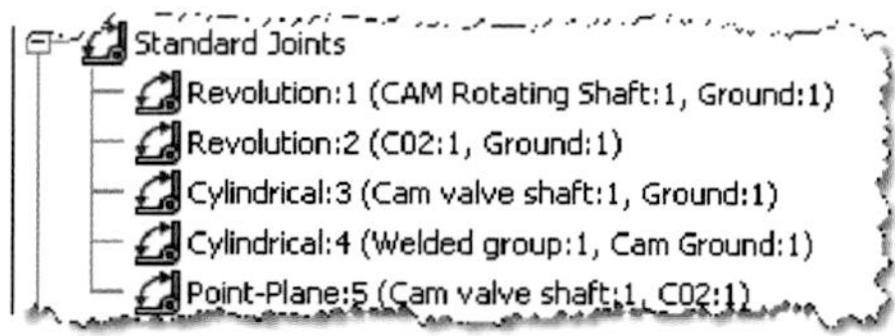

We now need to create a joint on the other side of component C02.

30. Now select the point–plane joint from the pulldown menu, as shown below

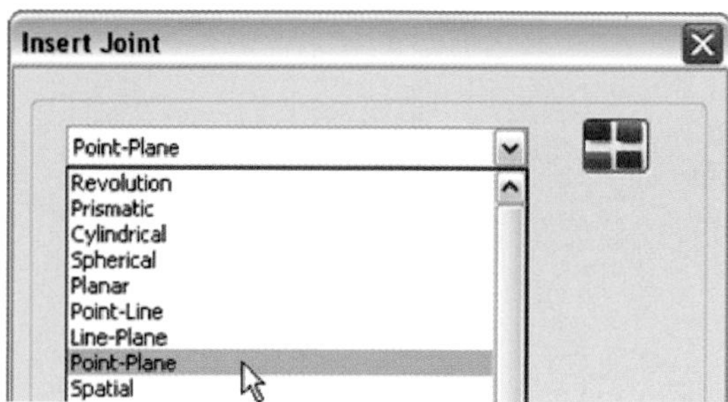

31. Select the camshaft edge, as shown, to define the Z axis of Component 1

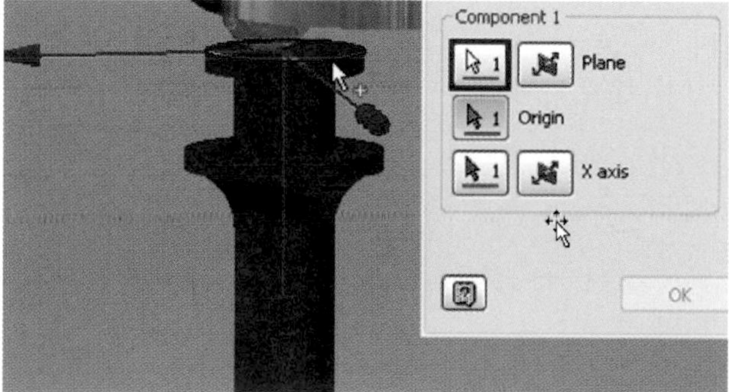

32. Select the Component 2 button and select the point of Component 2, as shown

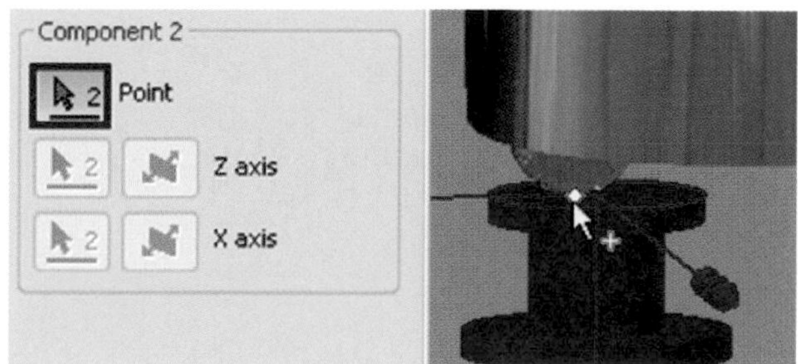

The X axes of each component are pointing in opposite directions. We need to align them so they point in the same direction.

33. Select the **Switch X axis** button of Component 2

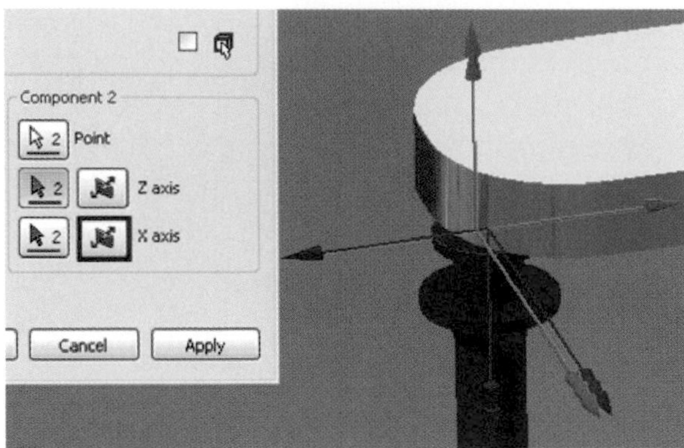

34. Click **OK**. You should now have six joints in total, as follows

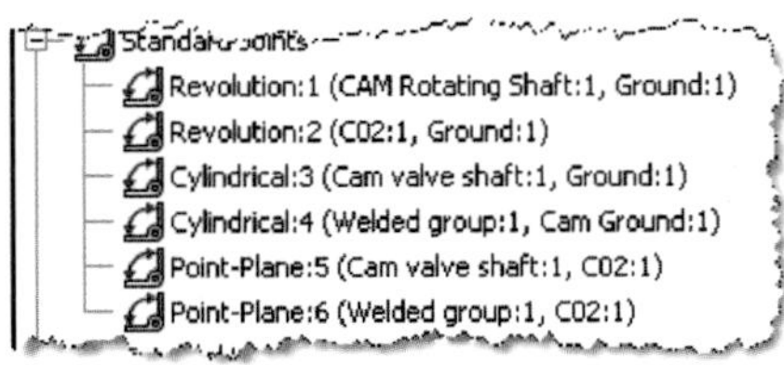

35. Close *CAM.iam*

It is now probably a good idea to explain the concept of redundant joints, why they occur, how we can avoid them, and how to cope with redundant joints if we are unsuccessful in removing them.

Redundant joints

Redundancy, or redundant joints, occurs when an assembly or mechanism is overconstrained as a result of applying joints. This makes the assembly statically indeterminate, which means the assembly has infinite solutions for joint types and thus reaction forces. In order to explain this concept of redundancy, we will use the following door assembly, comprising a door and a frame. Most residential doors have at least two hinges, and some high-security doors will have three hinges, if not more.

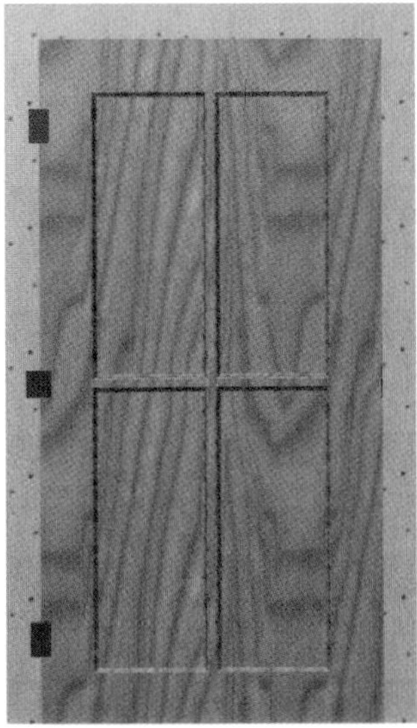

The door example has two components, a frame and a door. Each unconstrained component has six degrees of freedom, and by applying joints we restrict motion by restricting these degrees of freedom. In most door frame examples, the frame is, normally, completely fixed to walls, therefore restricting all degrees of freedom. The equivalent to this in Inventor and Simulation is to ground the component. The door, however, is fixed to the frame by a series of hinges, which restrict the motion of the door to one rotational degree of freedom to enable the door to open and close. The equivalent joint to this hinge action is the revolution joint, as illustrated below.

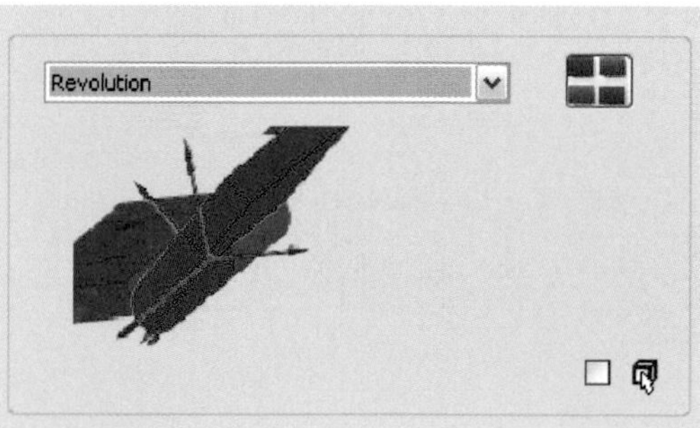

With this in mind, we will attempt to ground the frame and apply a revolution joint at each hinge position between the door and frame. Initially, we will apply two revolution joints in the following example.

1. Open *doorframe.iam*

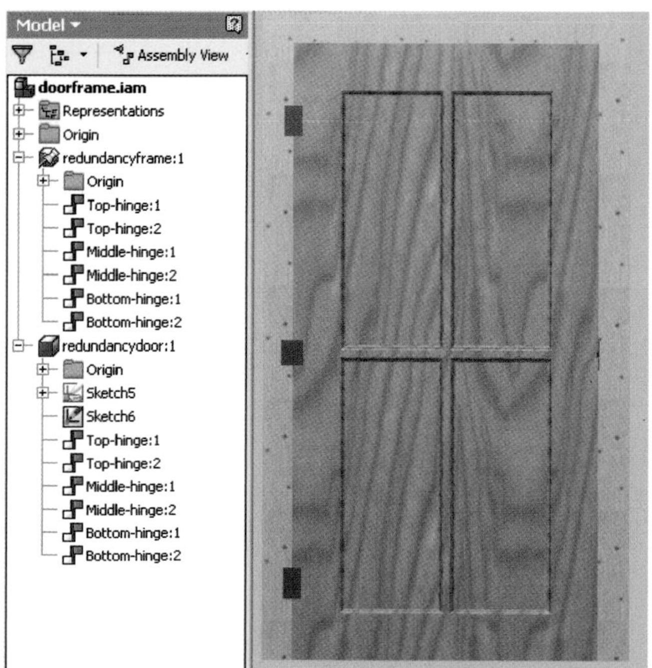

Six constraints have been created, resulting in overconstraining. This is so that you will not need to create any more constraints for this exercise.

The frame is set to ground as a result of it being the first part to be placed in the assembly; hence, there is no need to ground it again. Further, each hinge location has the following two assembly mate constraints:

- Edge-edge constraint
- Point-point constraint

2. Select **Environments** tab > **Dynamic Simulation**

In the **Dynamic Simulation** browser, all components are grounded. At this stage, we can either create joints automatically or convert assembly constraints; obviously, the former is quicker, so we will try that first.

3. Select **Simulation Settings**

4. Within the **Dynamic Simulation Settings** dialog box, select **Automatically Convert Constraints to Standard Joints** > Click **OK**

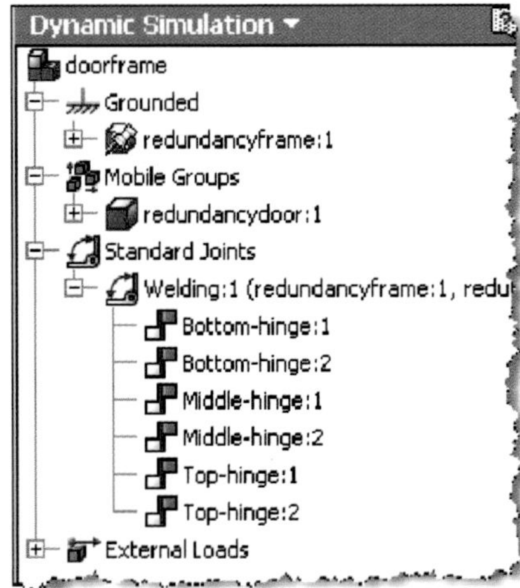

A welding joint has been created as a result of overconstrained assembly constraints (six in total; a maximum of three is normal). As this joint has welded the components together such that there is no relative motion between the components, we need to switch off the automatic conversion of joints and create joints manually from **Convert Assembly Constraints**.

5. Select **Simulation Settings** and clear **Automatically Convert Constraints to Standard Joints** > Click **OK**

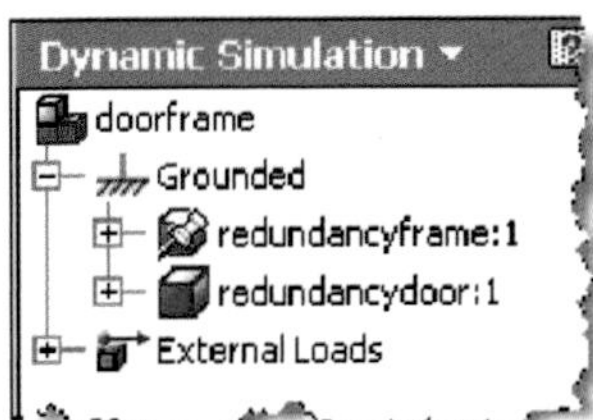

No joints will now be created, as shown here.

6. Select **Convert Constraints**

7. Select the door and frame when the **Convert Assembly Constraints** dialog box appears

In the **Convert Assembly Constraints** dialog box, all constraints between the two components will be selected and no joints will be translated. In addition, the following warning will appear, as six constraints have been applied between the two components.

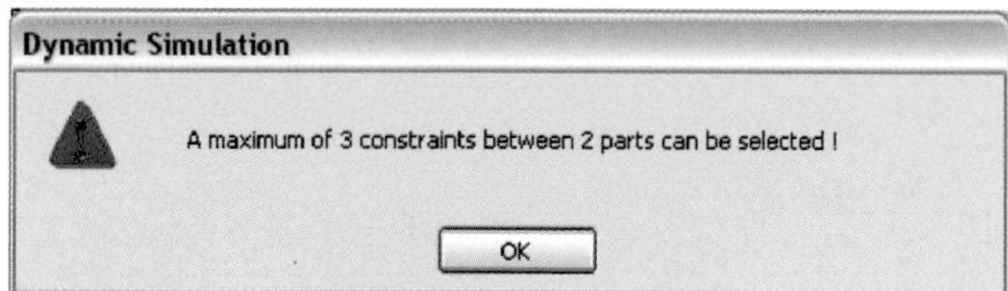

9. Click **OK** to accept the warning

10. Deselect all constraints apart from those of Top-hinge:1 and :2, as shown > Click **OK** to any warning that appears when deselecting constraints

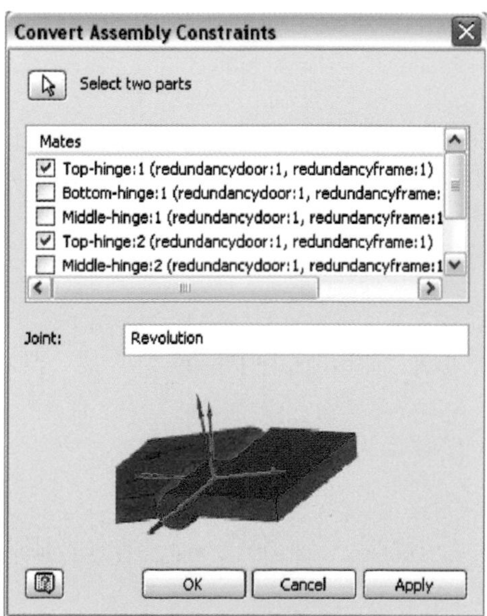

11. Click **OK**

As a result of selecting the edge-edge and point-point constraint, Dynamic Simulation has created a revolution joint, as expected.

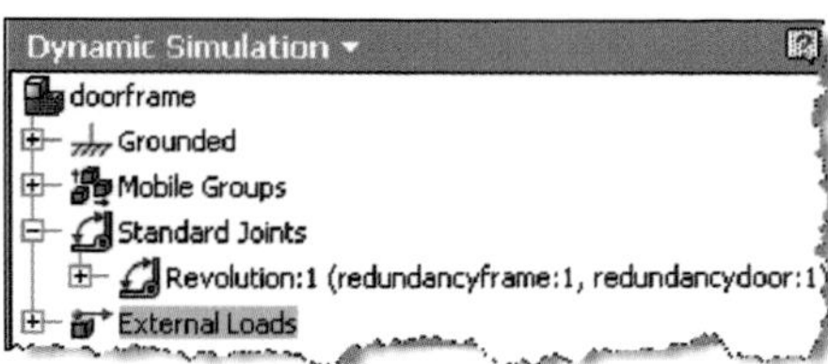

Now, we need to create a revolution joint at the bottom hinge.

12. Repeat Steps 6–9 and select the following constraints:

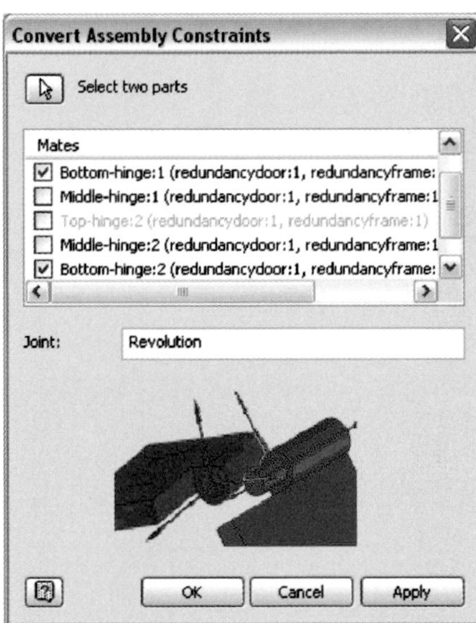

13. Click **OK** > Accept the following warning:

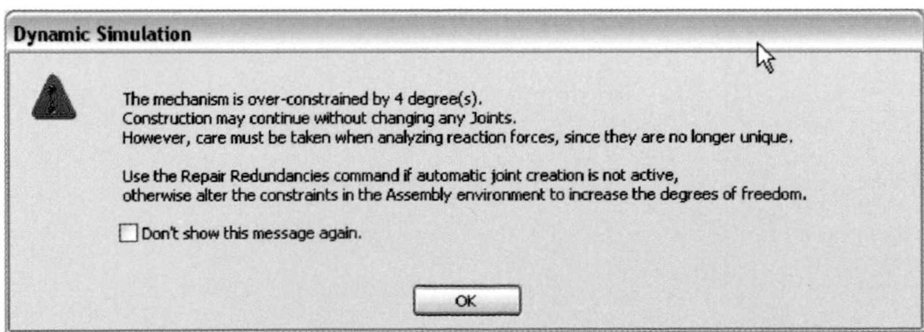

By accepting the warning, we have created a redundant joint, as illustrated below:

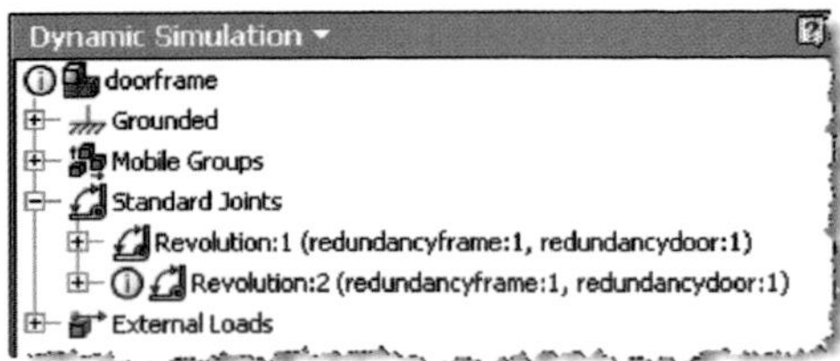

So, the question is, why have we got a redundant joint as most doors have two hinges (i.e. two revolution joints)? Let's examine the table below.

[illegible]	DOF	Joint type	DOF removed	Resultant DOF
Frame	3 rotational; 3 translational	Grounded	−6	0
Door	3 rotational; 3 translational	Revolution:1	−5	1
		Revolution:2	−5	−4
By applying a second joint on the door, we are overconstrained by				−4

Based on the above table, the door only seems to need one revolution joint to work properly as the resultant degree of freedom is one, which allows the door to open and close.

Dynamic Simulation cannot simulate deformity in components when the force causing the deformity is large enough; further, the joints created are perfect in that they do not allow for manufacturing tolerances/imperfections and clearances. For these reasons, the second joint created becomes redundant, and the reaction forces produced will not be unique and, more importantly, can be unpredictable if redundancy in the model is not removed. Apart from the forces, all results are unique and are totally predictable.

At this stage, we have three options.

Option 1 – Simulate the door and frame using one revolution joint.

- Advantage – Will not create a redundant model and the results will be unique.
- Disadvantage – This method will create unnecessary moments that would not occur with two revolution joints.

Option 2 – Simulate the door and frame using two revolution joints.

- Advantages
  - Will produce equal reactions between both joints.
  - Will not induce unnecessary moments as a result of having one joint.
- Disadvantage – This method will create a redundant model.

Option 3 – Simulate the door and frame using two joints that do not result in a redundant model.

- Advantage – Will not create a redundant model and the results will be unique.
- Disadvantage – The success of this model relies on the direction of the loading. This will be explained later.

To go through each of the above options, we will determine the reactions at each joint as a result of gravity and the mass of the door.

As we have already created the two revolution joints (Option 2), we will determine the reactions and moments of these joints first.

14. **Play Simulation** > Select **Output Grapher**

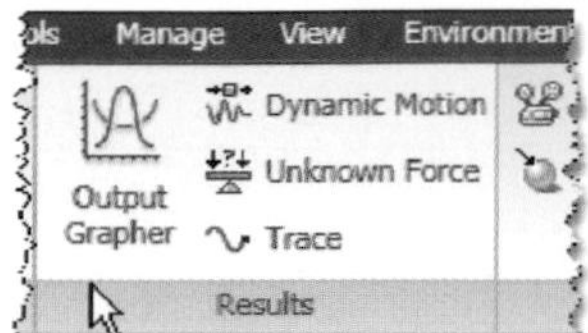

15. Select **Force** for both joints to display the reaction forces at the hinge due to the mass of the door and gravity

Despite the fact that the model is redundant, the reaction forces are correct, as we would expect 50% of the weight to be distributed through each hinge.

The mass of the door is 69 kg.
The gravity is 9.81 m/s^2.
Therefore, weight of door = 69 × 9.81 = 676.89 N.
Half of this is 338 N.

16. Minimize the **Output Grapher**

17. Select **Construction Mode**

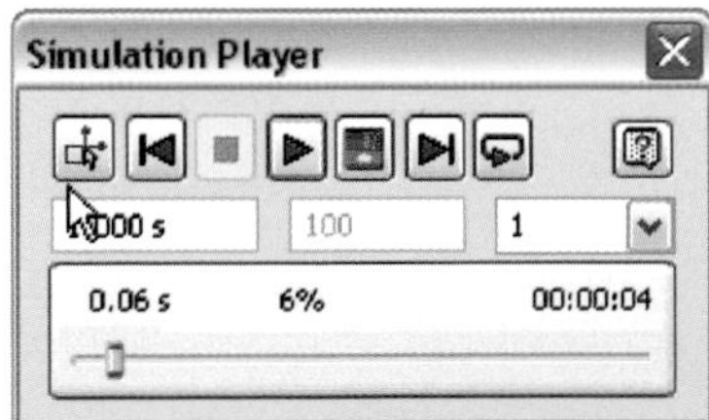

18. Now delete both revolution joints (Option 1)

Now, we will create a nonredundant model by creating one revolution joint.

	DOF	Joint type	DOF removed	Resultant DOF
Frame	3 rotational; 3 translational	Grounded	−6	0
Door	3 rotational; 3 translational	Revolution	−5	1
Total DOF				1

19. Select **Convert Constraints** to create a single revolution joint in the middle hinge

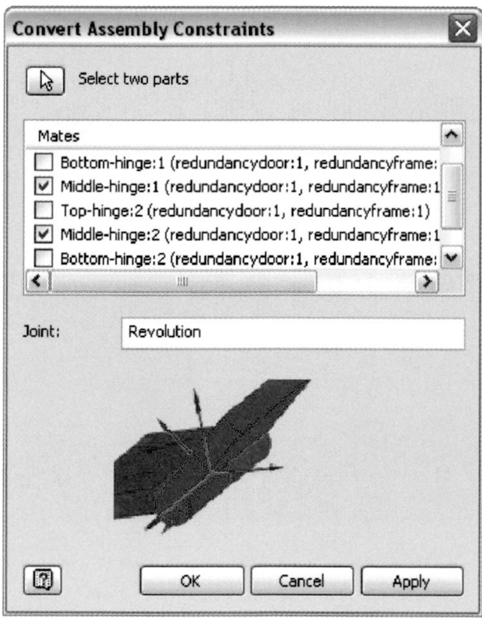

You may need to Click **OK** to the warning messages several times.

20. Click **OK**. The following joint will be created:

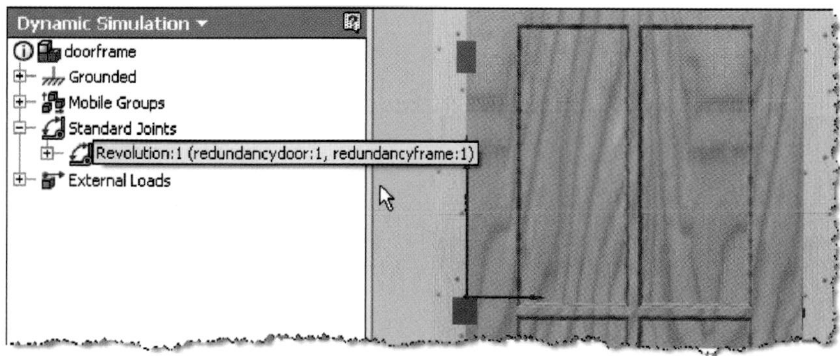

21. Play Simulation > Maximize the **Output Grapher**

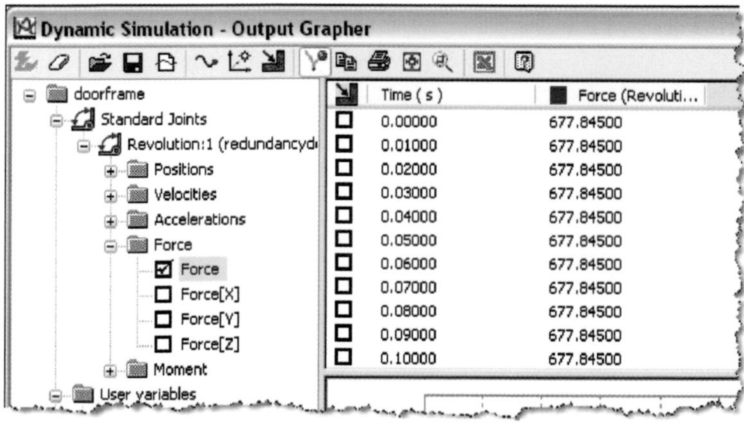

Despite only using one revolution joint, the maximum reaction value is correct at 677 N. Even though the solution does not create any redundancies, it does not represent reality accurately as all doors have at least two hinges or joints. It may also create additional moments that may not otherwise be present.

22. Minimize **Output Grapher** > Select **Construction Mode**

23. Now delete the single revolution joint

Now, we will create a third option by creating one spherical and one point-line joint.

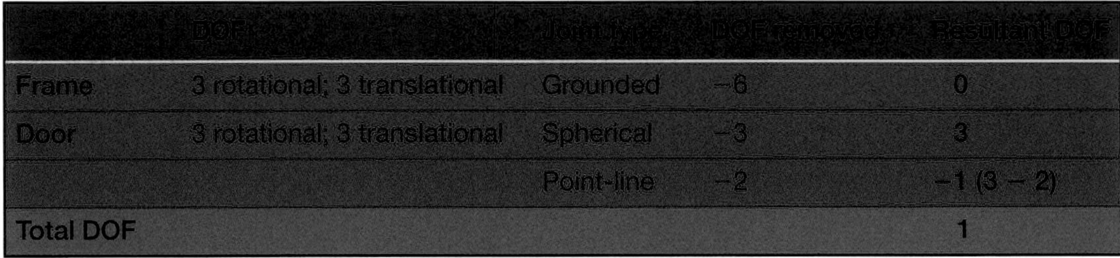

	DOF	Joint type	DOF removed	Resultant DOF
Frame	3 rotational; 3 translational	Grounded	−6	0
Door	3 rotational; 3 translational	Spherical	−3	3
		Point-line	−2	−1 (3 − 2)
Total DOF				1

24. Select **Convert Constraints**

25. Select the door and the frame and for the top hinge select the constraint shown below, then click **Apply**

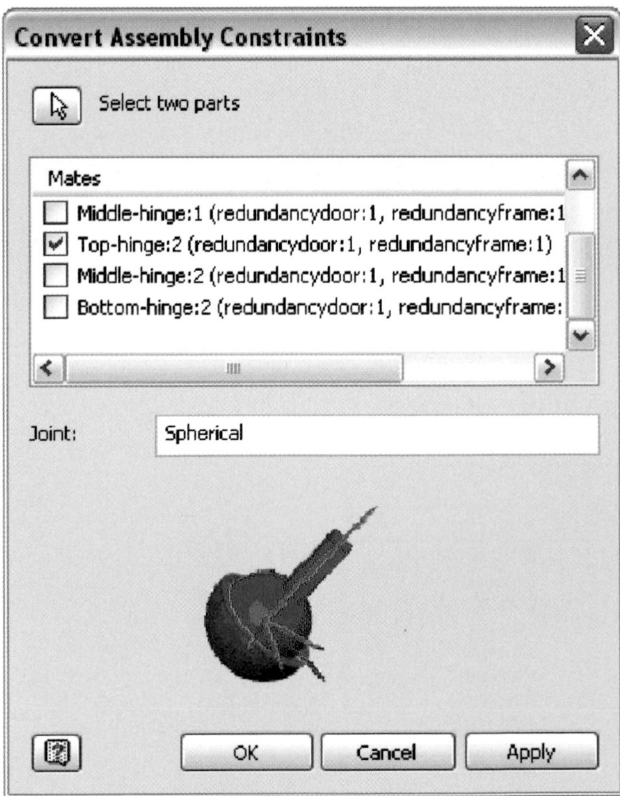

You may need to click **OK** to the warning messages several times.

26. Select the door and the frame and for the bottom hinge select the constraint shown below > Click **OK**

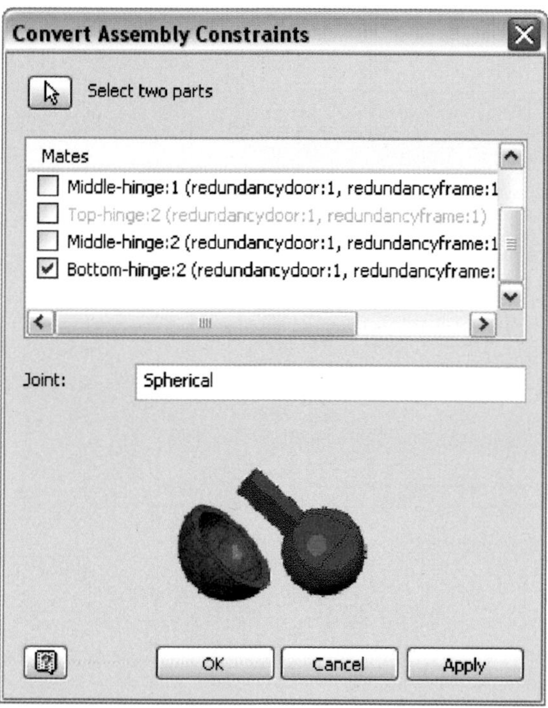

You may need to click **OK** to the warning messages several times.

27. Accept the warning

As a result of accepting the warning, we have a redundant model overconstrained by one degree of freedom.

28. Right click Bottom-hinge:2 > Select **Edit**

29. Reselect Component 2 of the constraint and pick the edge as shown > Click **OK**

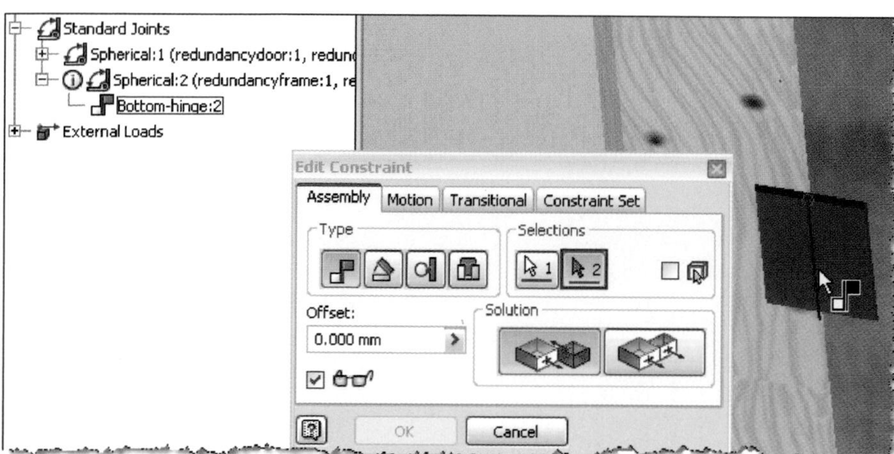

30. Accept the warning > Click **OK**

The joint changes to point-line and the model has no redundancies.

31. Play the simulation

32. Maximize the **Output Grapher** and select the forces to display the results

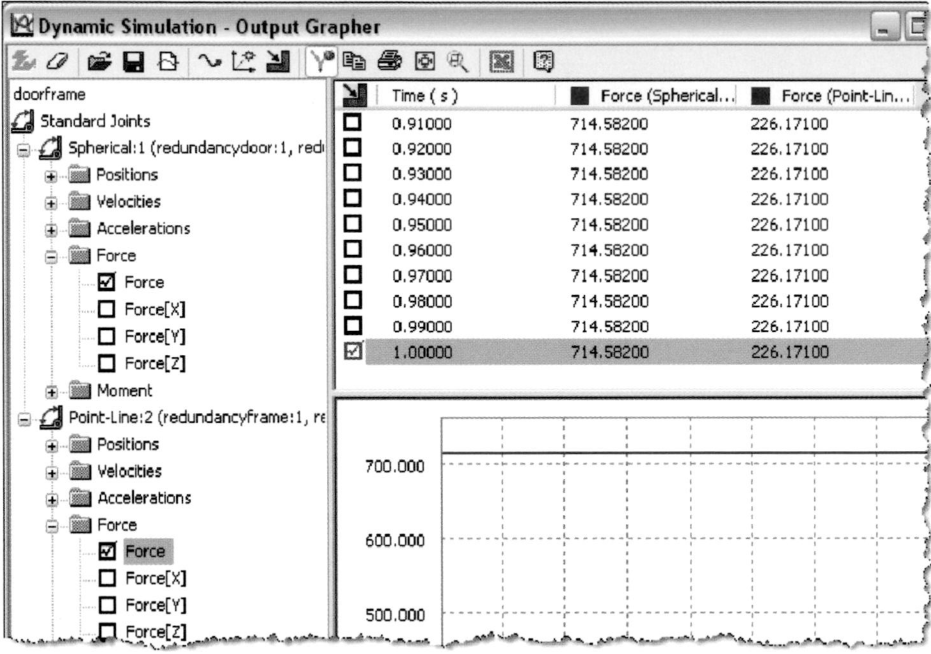

The results are not what we expected as they are not symmetrically distributed through each joint. The reason for this is that the top spherical joint has taken most of the reaction and the lower joint has taken much less. This is because the point-line constraint does not restrict motion in line with the gravity, as it is free to move in that direction. So, even though the model is nonredundant, the values are not equal. Let's try changing the position of the gravity.

33. Minimize **Output Grapher** > Select **Construction Mode**

34. Right click **Gravity** > Select **Define gravity**

35. Change the position of the gravity as shown:

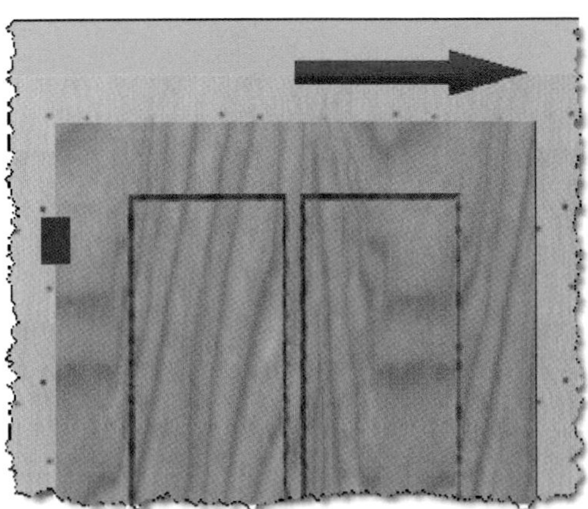

36. Click **OK** > Play the simulation

37. Maximize the **Output Grapher**

38. Close *door frame.iam*

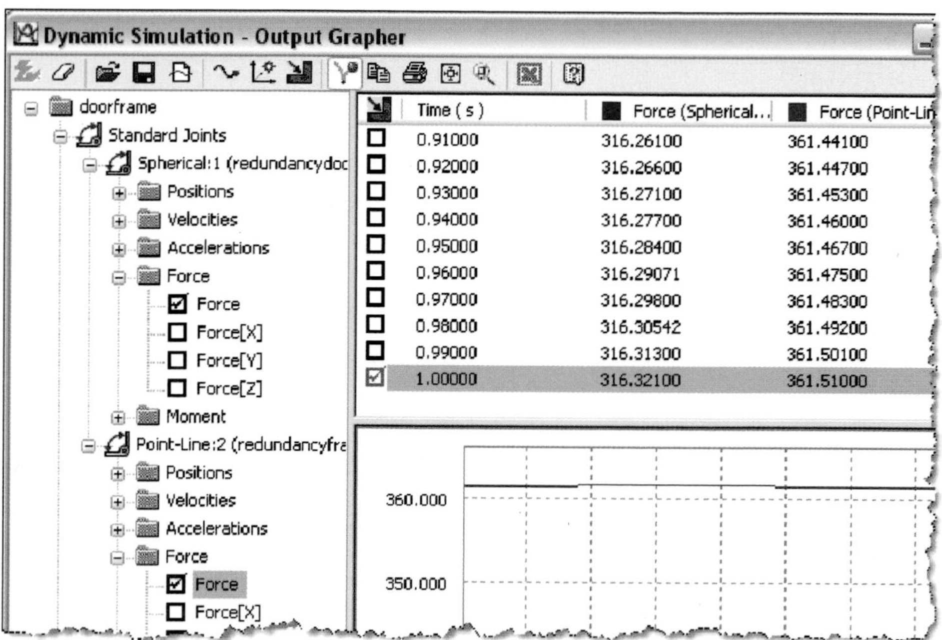

The results are now more reasonable. These results are better than the previous ones because gravity is not in line with the direction of the point-line axial movement. So, keep this in mind when creating nonredundant models/joints.

IN SUMMARY

Option 1 – If the ultimate goal of the simulation is to determine the effects of inertia and dynamics, Solution 1 will suffice.

Option 2 – If the ultimate goal of the simulation is to perform a stress analysis on the door and you want to make use of the load transfer facility, Solution 2 is the most appropriate as it provides even reactions at the joints. In reality, we always use equal strength hinges as we always assume even load distribution.

Option 3 – Is also acceptable, provided the loading is not in the direction of the point-line axial movement.

Option 3 will not work for a three-hinge door.

It is not advisable to use redundant constraints; however, if they are employed, use them with caution as the forces produced will not be unique or even worst case.

SUGGESTED WORKFLOW TO AVOID REDUNDANT JOINTS

Using **Automatically Convert Constraints to Standard Joints** after creating all the constraints is not advisable, especially for a large assembly, as it can become tedious to remove redundancies by altering assembly constraints, as this is the only way to modify joints. With this in view, the following two approaches are suggested.

Option 1 – Automatic joints

As you can create assembly constraints within the Simulation environment, it is advisable to start (or continue) the constraining process within the Simulation environment. The main benefit of this is that, with **Automatically Convert Constraints to Standard Joints** activated, you will see the type of joint created as soon as you place the constraints. This way, it is easier to manage and fix redundant joints as they appear during the constraining process.

Option 2 – Manual joints

Creating standard joints manually within the Simulation environment will provide more control in removing redundancies from within the model. However, this option can be slow and tedious.

The process of Option 2 can be enhanced by additionally using the **Manually Convert Constraints** feature for some joints.

ENVIRONMENTAL CONSTRAINTS

Once the most tedious aspect of creating joints has been completed, the next stage is to simulate external conditions, using a variety of tools available with simulation. This includes

- Applying to external forces/torques/gravity to components/subassemblies
- Redefining joint properties
  - Initializing position
  - Joint friction
  - Contact
  - Imposed motion
    - Constant values
    - Input Grapher

Step 3

The Input Grapher is the most useful and powerful tool in terms of helping to simulate realistic environmental conditions.

Input grapher

The Input Grapher uses a graphical interface to define the laws of dynamic actions, including

- Movement law
- Joint forces
- External forces

Using the graphical interface, various laws can be defined, including

- Linear ramp
- Cubic ramp
- Cycloid
- Sine
- Polynomial
- Harmonic
- Modified sine and trapezoid
- Spline – can import predefined data using a .txt file

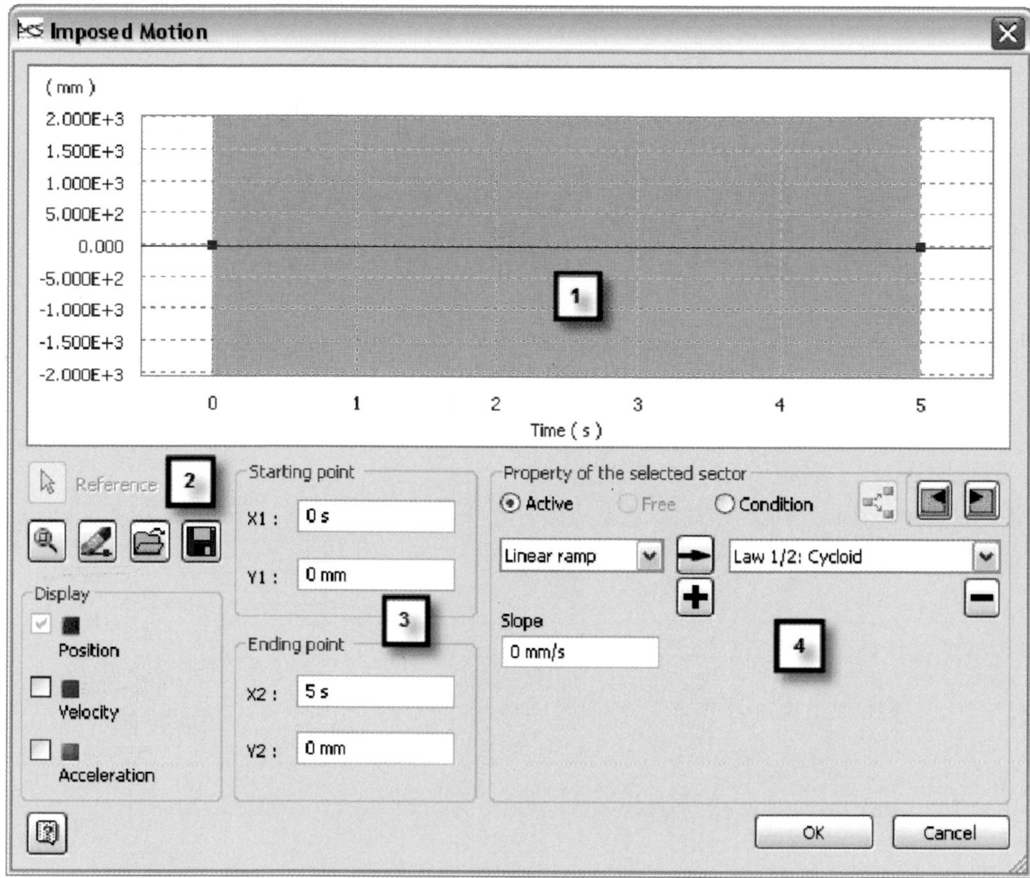

1. The graph area visually displays the laws once they have been defined
   - Scroll the mouse wheel to zoom in and out of the graph
   - Keep the mouse wheel pressed to pan around the graph
   - Double click the mouse wheel to reset the graph to its original state
2. The **Reference** button allows you to select different references/values from the Output Grapher, instead of time, to define laws of motion
3. In this area, you define the start and end parameters of the highlighted sector in the graph
4. In this area, you define the laws of motion using any of the predefined laws or from imported data

The various laws can be combined to create complex laws of motion/dynamics.

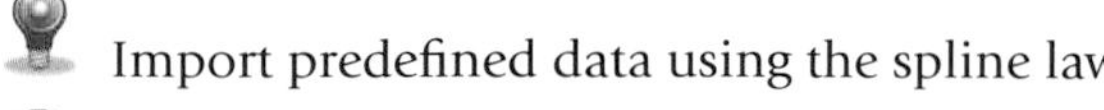

Import predefined data using the spline law.

You can define logic statements using conditions to create more complex laws of motion.

More points can be created in the graph by double clicking the mouse.

The Input Grapher can be accessed from any external force, torque, or imposed motion from within joints.

Joint friction

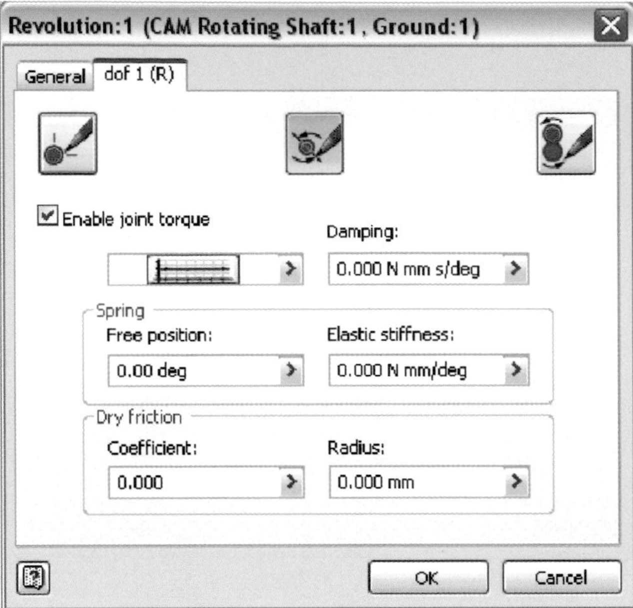

The following shows how Dynamic Simulation calculates the joint torque effort (U)

If a constant value C is specified in the first field, then $U = C$.

If a damping value of $D = 10$ is specified in the damping field, then

$$\mathrm{U} = -D \times v$$

with v = velocity of the DOF ($-$) sign indicating that damping will always oppose motion.

If a free position value of $p0 = 40$ and elastic stiffness $K = 1000$ is specified, then

$$\mathrm{U} = -K \times (p - p0)$$

with p = position of the DOF.

If a coefficient cf = 0.15 and radius = 5 is specified, then

$$\mathrm{U} = -sign(v) \times sqrt(sqr(Fr[X]) + sqr(Fr[Y])) \times of \times R$$

with Fr = force in the dof.

So, in summary, the joint is perfect when **Enable joint torque** is inactive. Once activated, the joint is not perfect and U[i] gives the exact effort (torque if the selected DOF is a revolution, force if it is a translation) inside the DOF.

Environmental constraints (EC) matrix – A snapshot of environmental constraints used throughout the book

	Environmental constraints	Examples using these environmental constraints	Design problems using these environmental constraints
1	Initializing joint positions	5,6	3
2	External gravity		All
3	External force		3,4,6
4	External torque		2
5	Frictional joints	7	6
6	Lock joint degree of freedom	5	3
7	Imposed motion – constant	7	1,2,5,7
8	Imposed motion – Input Grapher	6	3,4,6

The process of creating environmental constraints

This section will explain the different tools available to create environmental conditions, using the same examples that we used in the joints creation process.

- Example 5 – Newton's cradle
  - Initializing position
  - Contact
  - External gravity
- Example 6 – Cam follower mechanism
  - Imposed motion – Input Grapher
- Example 7 – Whitworth Quick Return Mechanism
  - Frictional joints
  - Imposed motion – constant values

EXAMPLE 5

Newton's cradle – Initializing position, contact, external forces

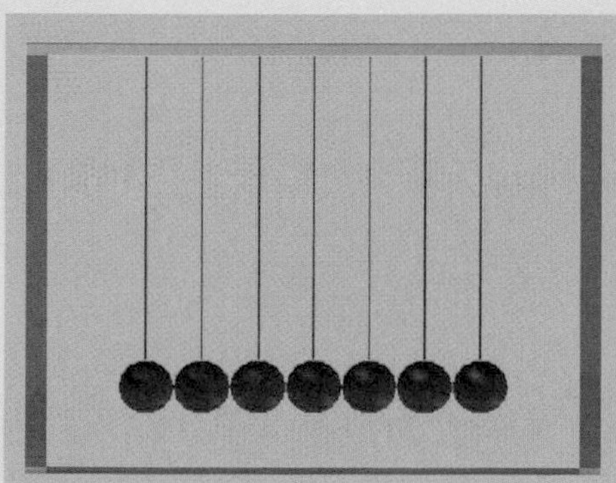

Workflow of Example 5

- Set initial position of first ball
- Set gravity – external forces
- Create 2D contact between balls
- Set contact properties of balls
- Play – simulate Newton's cradle

Set initial position of first ball

1. Open *Newtons-Cradle5.iam*

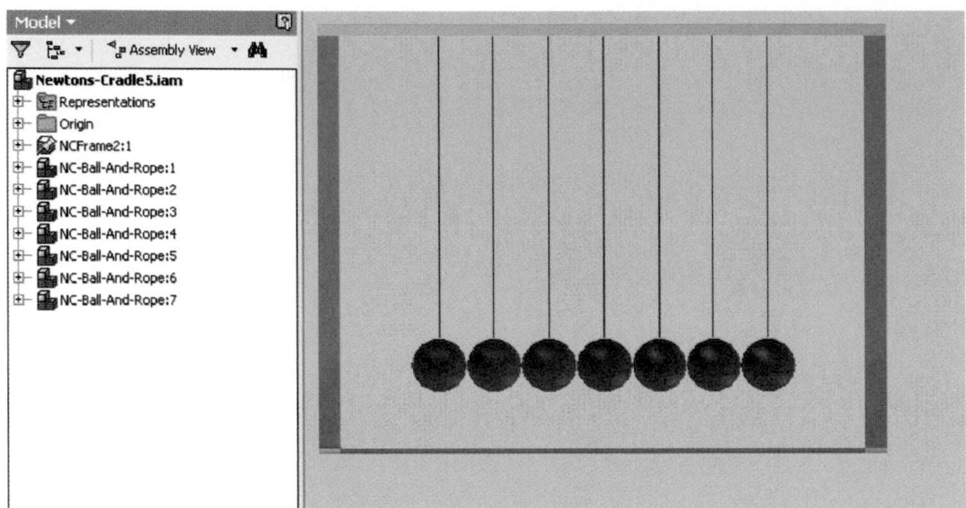

2. Select **Environments** tab > **Dynamic Simulation**

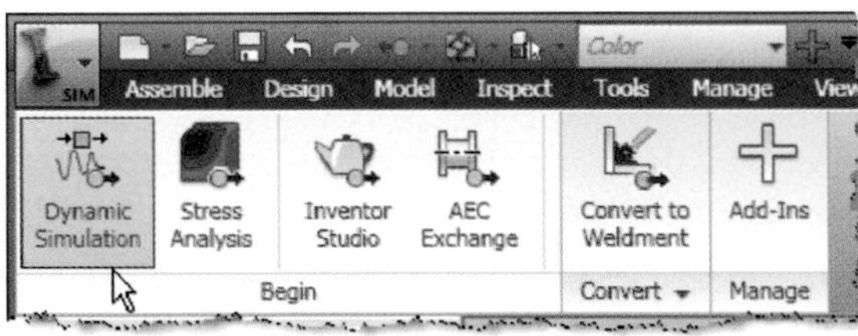

This will activate the Dynamic Simulation environment.

There are seven spherical joints with locked rotational degrees of freedom around the X and Y axes.

3. Double click the first spherical joint
4. Select **dof 3(R)** > Change the starting position of the first ball to 40 deg

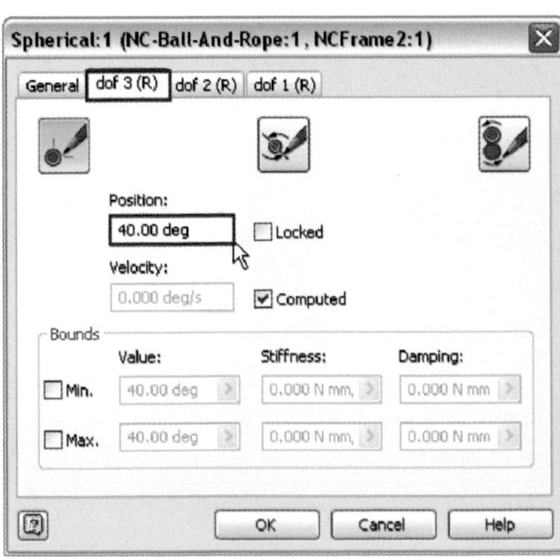

5. Click **OK** > **Play Simulation**

You will notice that the ball will not move; the reason for this is that the gravity is not activated. We will now, therefore, activate gravity.

Set gravity – external forces

6. Select **Construction Mode** in the simulation player to enable you to apply gravity
7. Under **External Loads** in the **Simulation** panel, double click **Gravity**
8. Select **Vector Components** > Specify −9810 in the Y vector value, for gravity to act downward

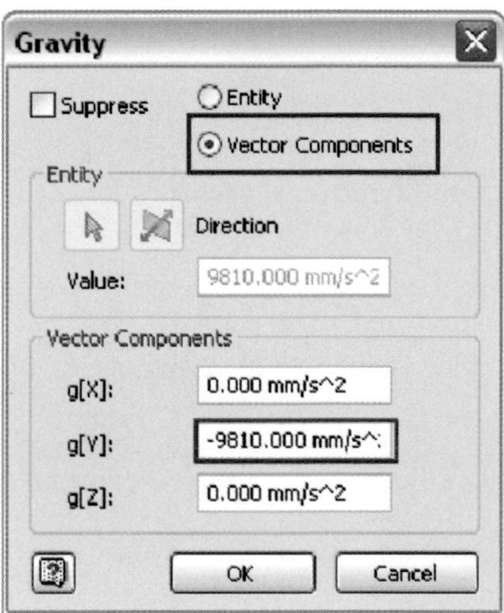

9. Click **OK** > Play the simulation

Even though the ball moves under gravity, the ball does not have any contact with other balls.

Create 2D contact between balls

As the balls do not collide with one another, we need to define contact between them. In Dynamic Simulation, there are two types of contacts: 2D and 3D. 2D contact is used to define contact that occurs only in one plane. 3D contact, on the other hand, is normally used to simulate contact that is not restricted to one plane. It is worth noting that it is much easier to control and simulate 2D contact.

In this design problem, we know that the balls, if operating correctly, will only collide in one plane; hence, will use 2D contact.

To use 2D contact, we need to be able to select closed loops or edges on the ball. So, before we can use the contact, we need to either define projected edges or sketches on the ball or just make the sketches visible, if already defined, where the contact is going to occur.

10. Double click NC-Ball-And-Rope:1. This will activate the Part environment

11. Expand NC-Ball:1 and make Sketch3 visible

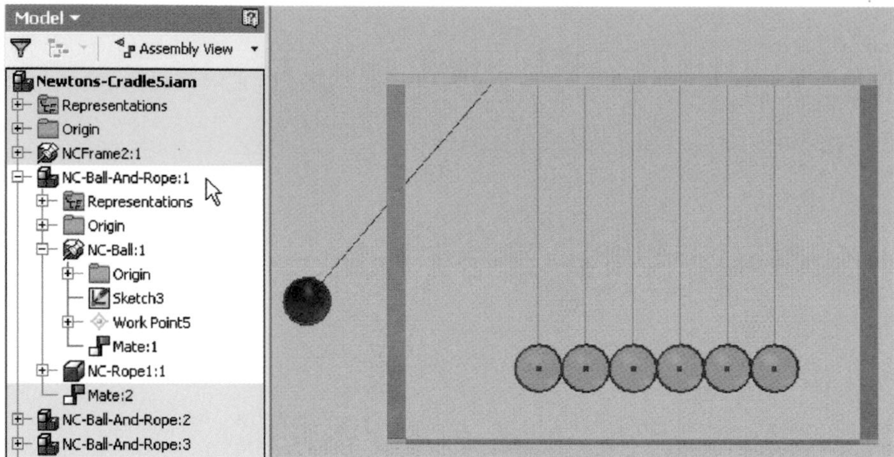

As the ball is an instance, the projected sketch will appear on all balls, as shown.

12. Select **Return** to go back to the Simulation environment

13. Select **Construction Mode** > **Insert Joint**

You may need to select the **Dynamic Simulation** tab to activate the **Simulation** panel.

14. Select **2D Contact** from the pulldown list

15. Select the projected edge of the first ball > Select the projected edge of the second ball

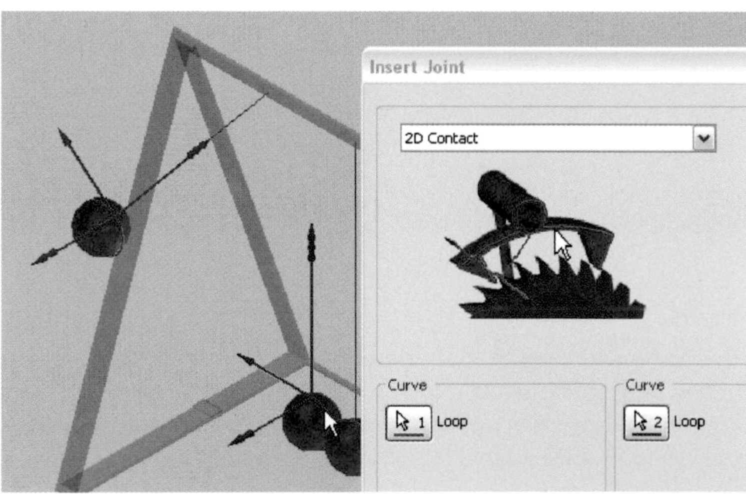

16. Click **Apply**

This creates a contact between NC-Ball-And-Rope:1 and NC-Ball-And-Rope:2.

17. Now, we need to repeat steps 15–16 for the following balls/assemblies:

NC-Ball-And-Rope:2 and NC-Ball-And-Rope:3
NC-Ball-And-Rope:3 and NC-Ball-And-Rope:4
NC-Ball-And-Rope:4 and NC-Ball-And-Rope:5
NC-Ball-And-Rope:5 and NC-Ball-And-Rope:6
NC-Ball-And-Rope:6 and NC-Ball-And-Rope:7

Once complete, you should have six 2D contacts.

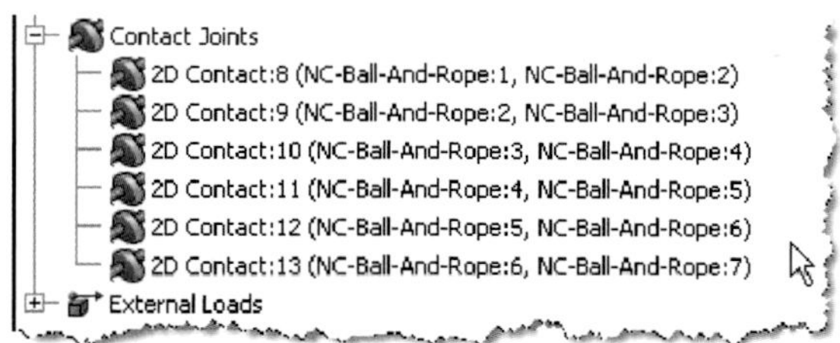

18. Play the simulation

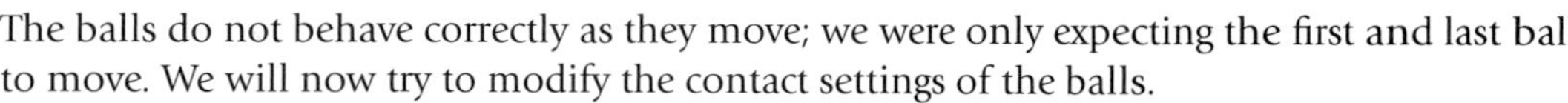

The balls do not behave correctly as they move; we were only expecting the first and last ball to move. We will now try to modify the contact settings of the balls.

Set contact properties of balls

19. Select **Construction Mode** > Double click on the first 2D contact

20. Change the **Restitution** value to 1 and **Friction** to 0. Click **OK**

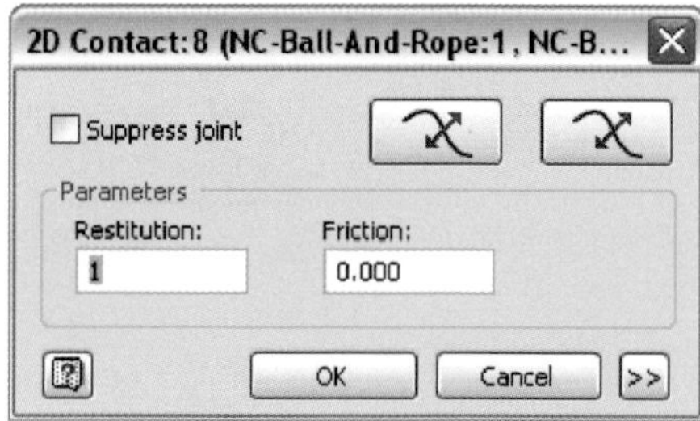

The **Restitution** value is the elastic property of the contact ranging between 0 and 1. A value of 0 will not provide any bounce as there is no elasticity, whereas a value of 1 will provide 100% elasticity and bounce.

A **Friction** value of 0 will provide no friction force on contact, whereas any value above 0 will provide friction force on contact.

With values specified as above, we will assume the balls are made out of chrome, with zero friction and 100% bounce. This means the balls will always bounce back to their original offset position, and bounce forever.

21. To quickly alter the rest of the contact joints, select the second contact joint and, with the Shift key pressed, select the last contact joint. Right click and select **Properties**

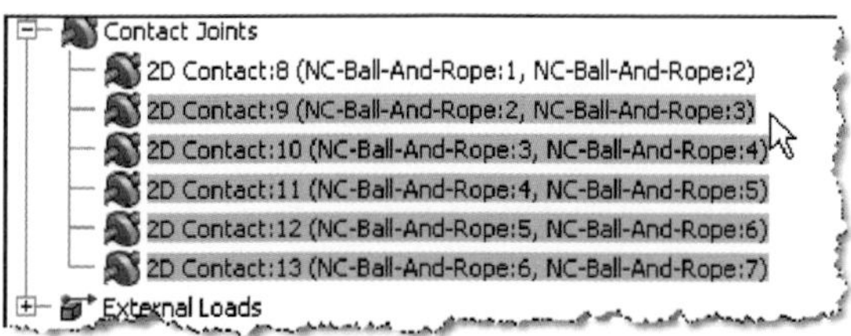

22. Change the **Restitution** value to 1 and the **Friction** value to 0. Click **OK**

23. Switch off the visibility of the projected sketches on the balls

Play – simulate Newton's cradle

24. Play the simulation

You will see that the balls in the middle remain static. The only balls moving should be the first and last. Now we offset the second ball and see how two balls behave when we rerun the simulation.

25. Select **Construction Mode** > Double click on the second spherical joint

26. Select **dof 3(R)** and change the starting position of the first ball to −40 deg

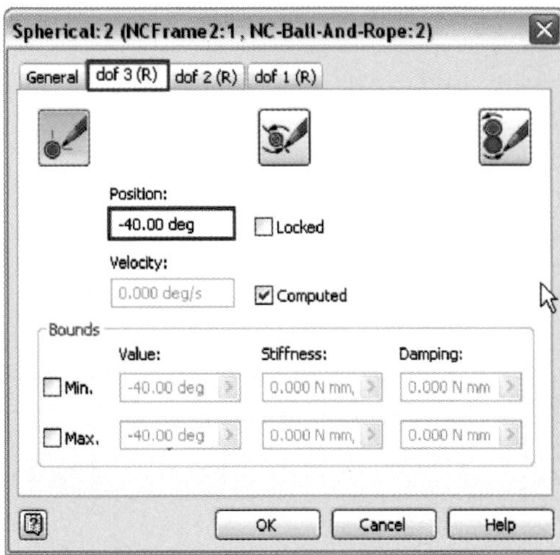

27. Click **OK**

28. Rerun the simulation

The three balls in the middle do not move and, thus, we have successfully simulated a Newton's cradle.

29. Close *Newton's–Cradle5.iam*

EXAMPLE 6

CAM Design – Imposed motion via the Input Grapher

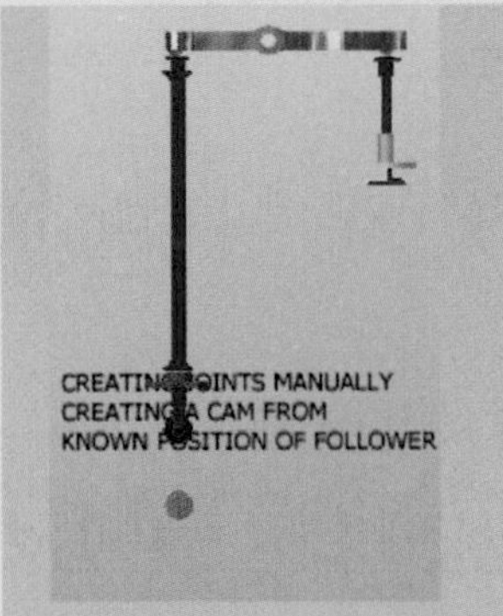

Workflow of Example 6

- Set gravity – external forces
- Define imposed motion using Input Grapher

Set gravity – external forces

1. Open *CAM2.iam*

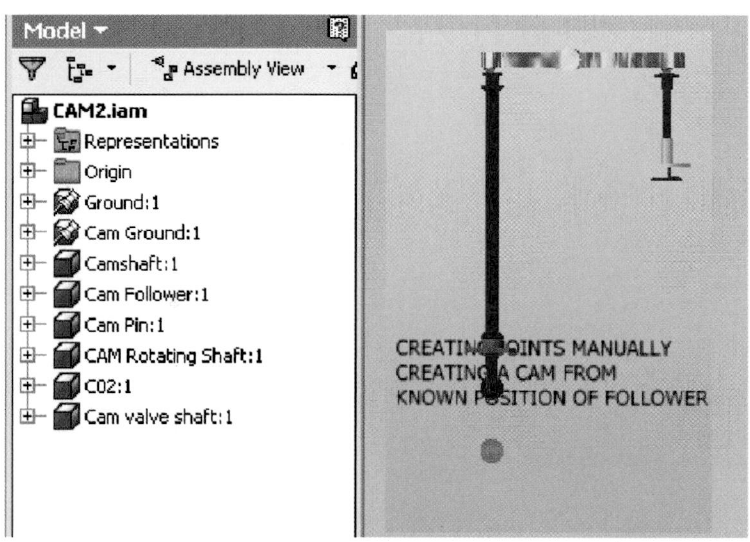

2. Select **Environments** tab > **Dynamic Simulation**

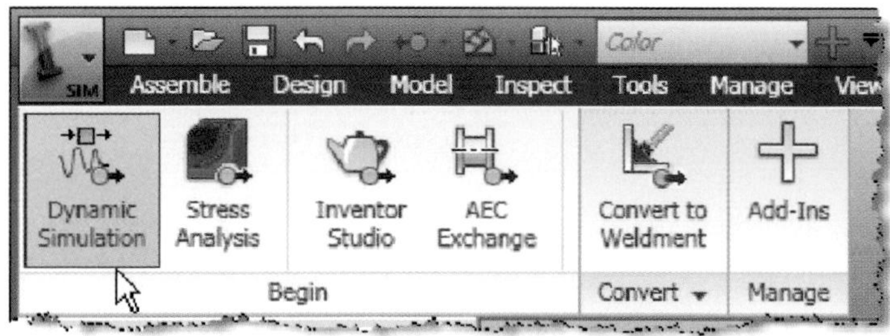

This will activate the Dynamic Simulation environment.

3. Right click **Gravity** under **External Loads** in the **Dynamic Simulation** browser > Select **Define gravity**

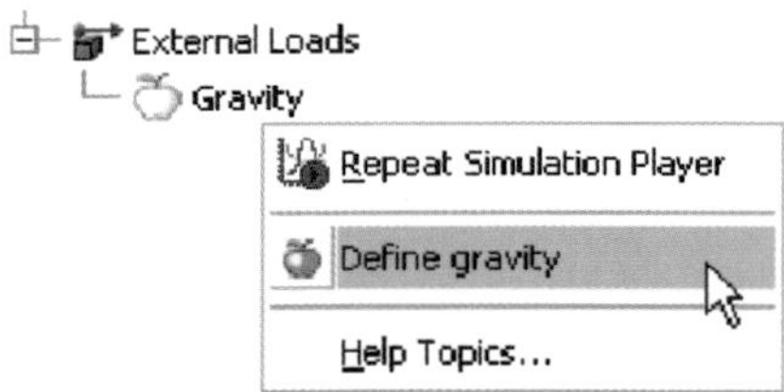

4. In the **Gravity** dialog box, define gravity by selecting the edge of ground component so that the gravity is pointing down as shown. Click **OK**

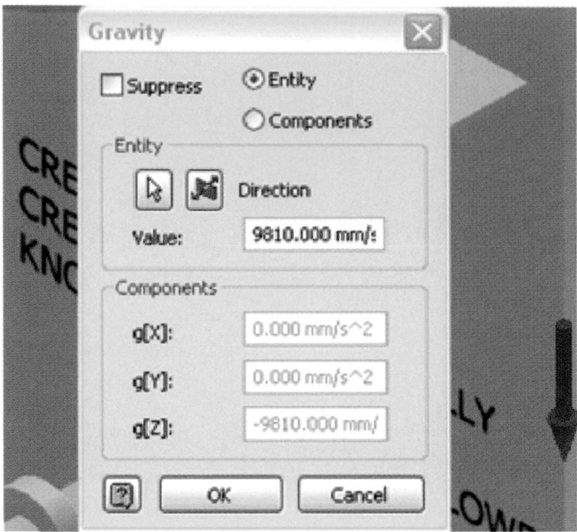

5. Play the simulation

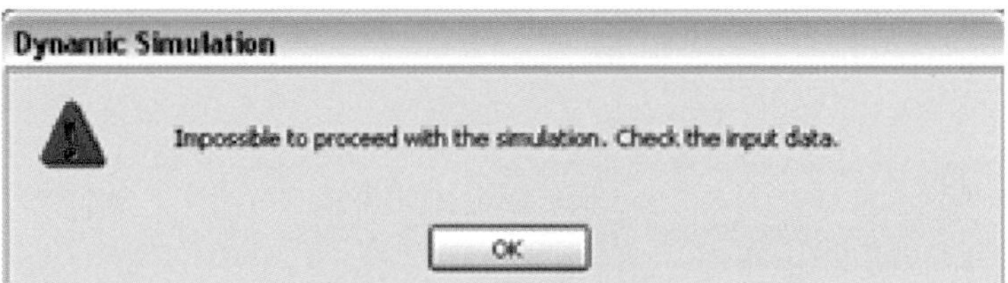

You will see that the components behave freely under gravity and after a couple of seconds the above warning may appear. This means that we need to properly define and restrict the motion of the assembly.

6. Select **Construction Mode** to redefine the simulation

Define imposed motion using Input Grapher

In this example, we will define the position of the valve using the Input Grapher functionality. The valve will open once for a very short time during one complete revolution of the unknown cam profile (to be determined).

1. Right click Cylindrical:3 (Cam valve shaft:1, Ground:1) and select **Properties**

2. Select the **dof 2(T)** tab > Right click in the **Position** value > Select **Set offset**

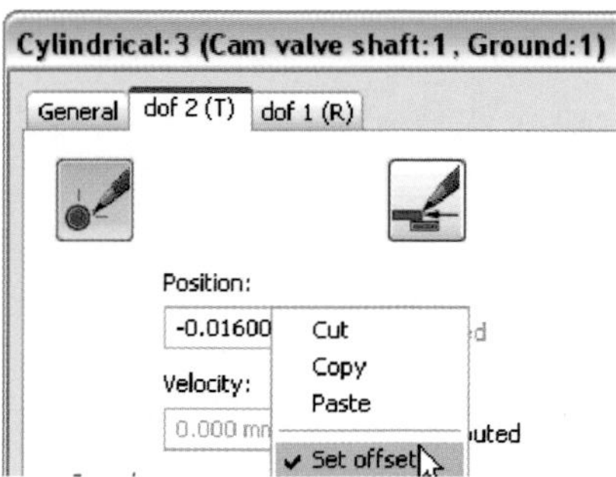

This will reset the current position of the valve to zero.

3. Click on the **Edit imposed motion** button

4. Select **Enable imposed motion** and change from **Velocity** to **Position**

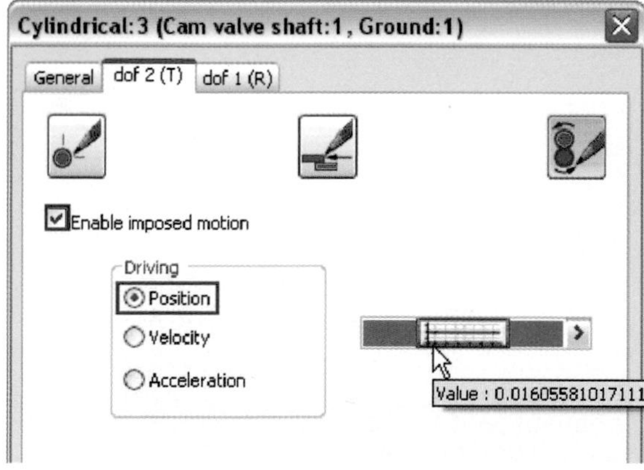

5. This allows us to set the positional parameter of the valve

6. Click in the value box to edit the Input Grapher values

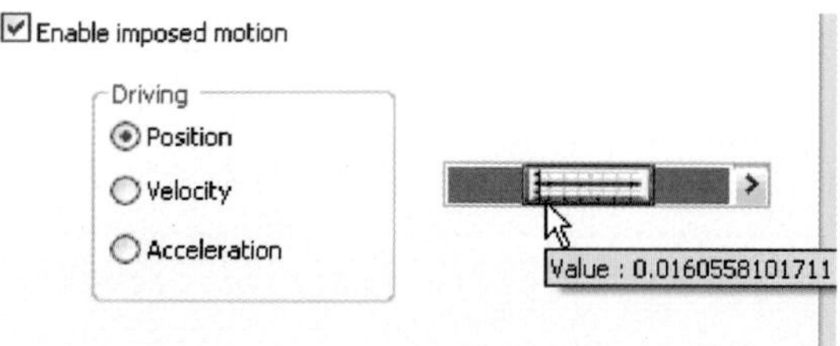

7. In the **Input Grapher** dialog box, change the values of Y1 to 0 and Y2 to 0

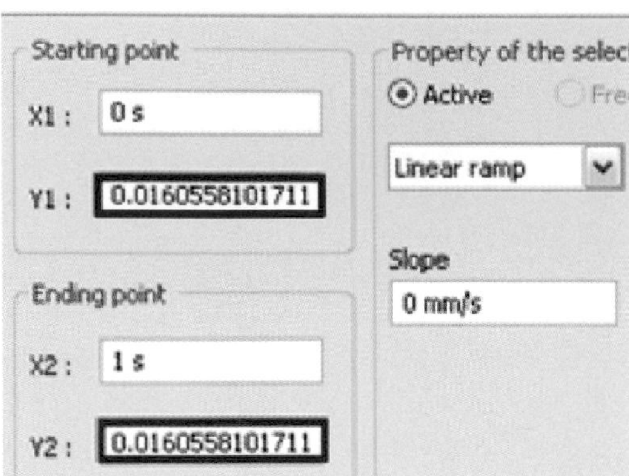

8. Double click four times in different positions along the X axis in the Input Grapher, as indicated. This will create four more points, as illustrated below:

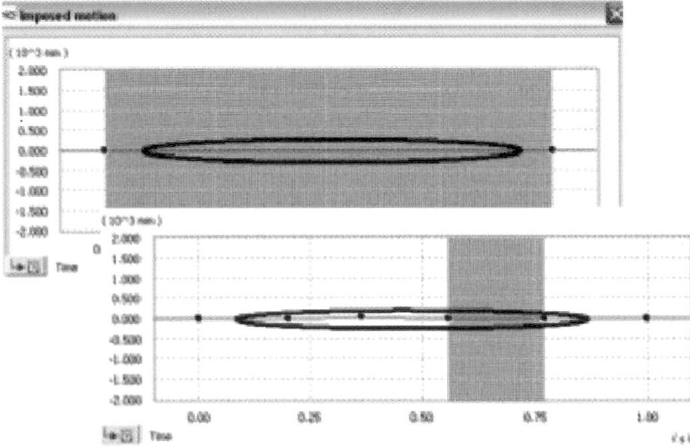

If you make a mistake, just right click any point and select **Remove Point**. The other alternative is to clear the whole definition using the **Clear curve definition** button, as illustrated below, and start again.

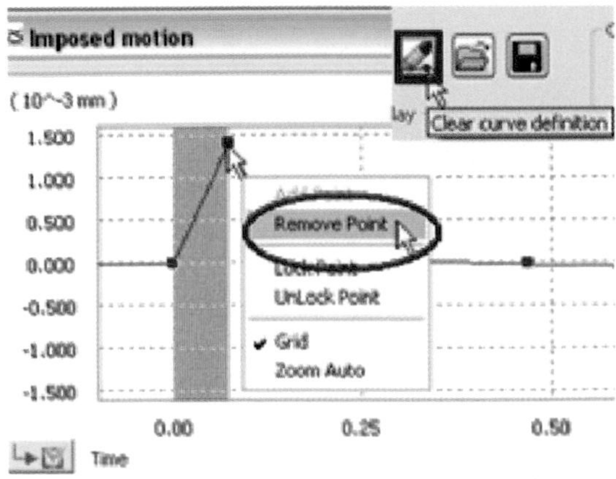

9. Click once anywhere between the first two points and specify the values as shown below

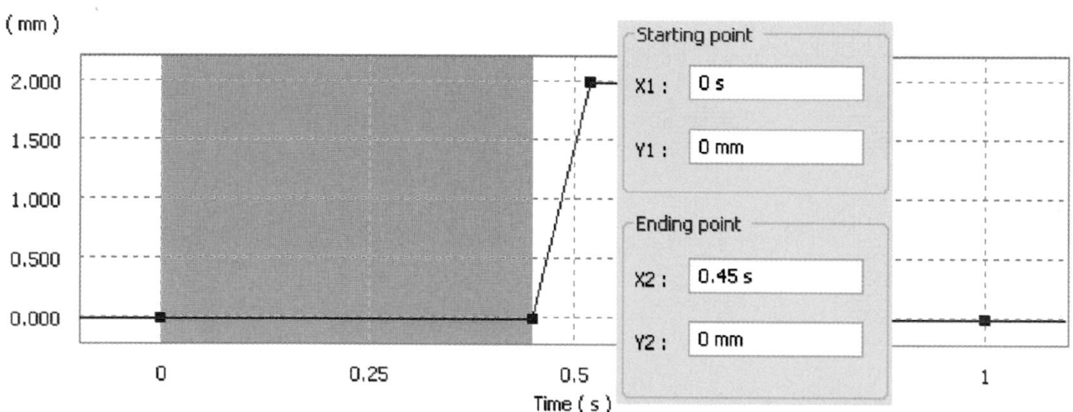

10. Click once anywhere between the second and third points and specify the values as shown below

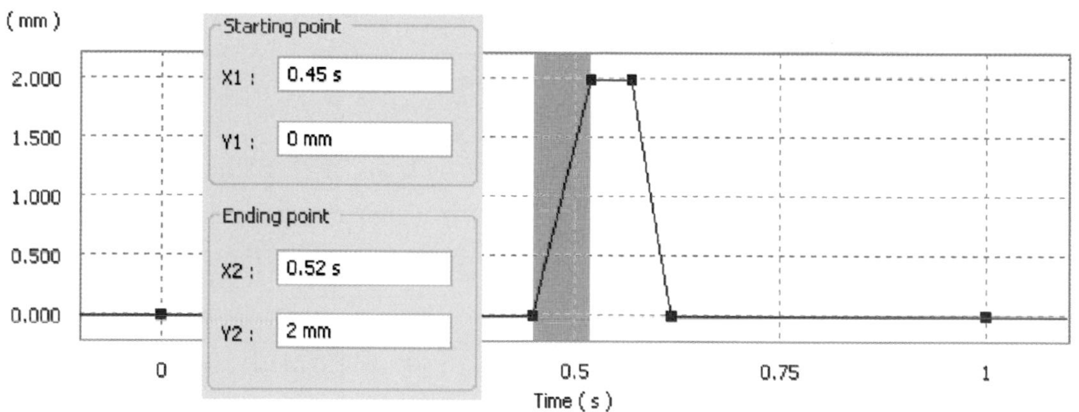

11. Click once anywhere between the third and fourth points and specify the values as shown below

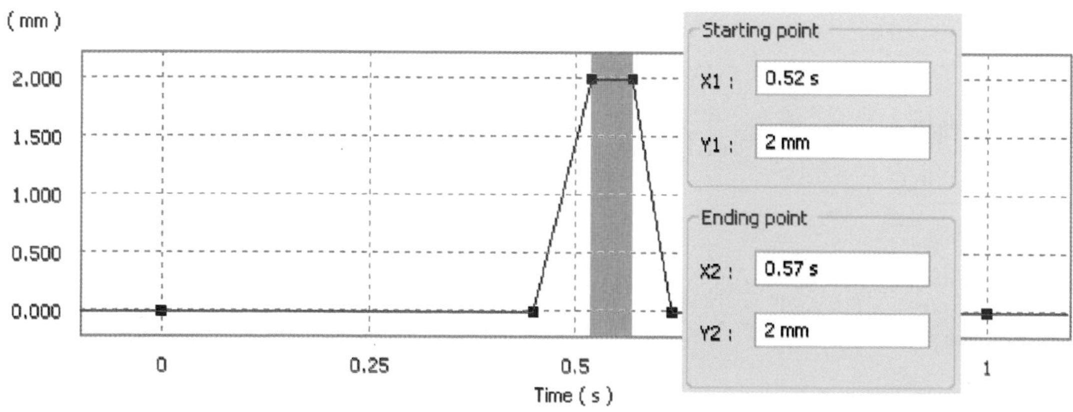

12. Click once anywhere between the fourth and fifth points and specify the values as shown below

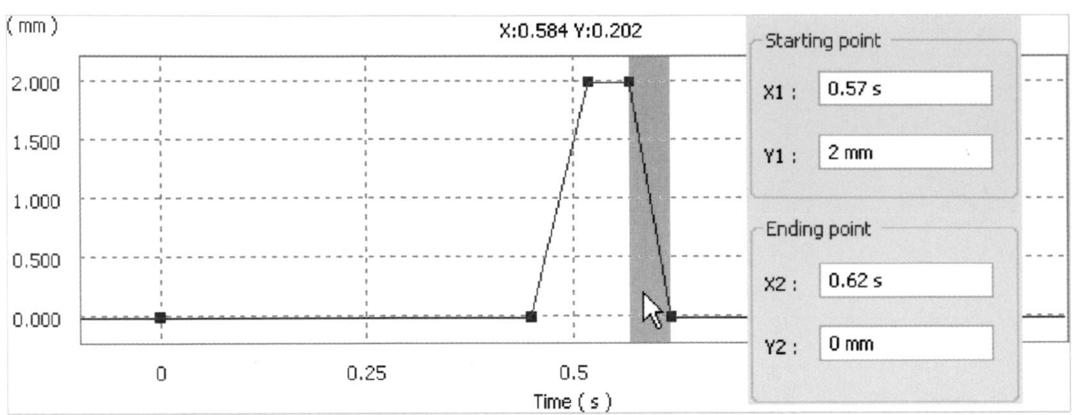

13. Click once anywhere between the fifth and last points and specify the values as shown below

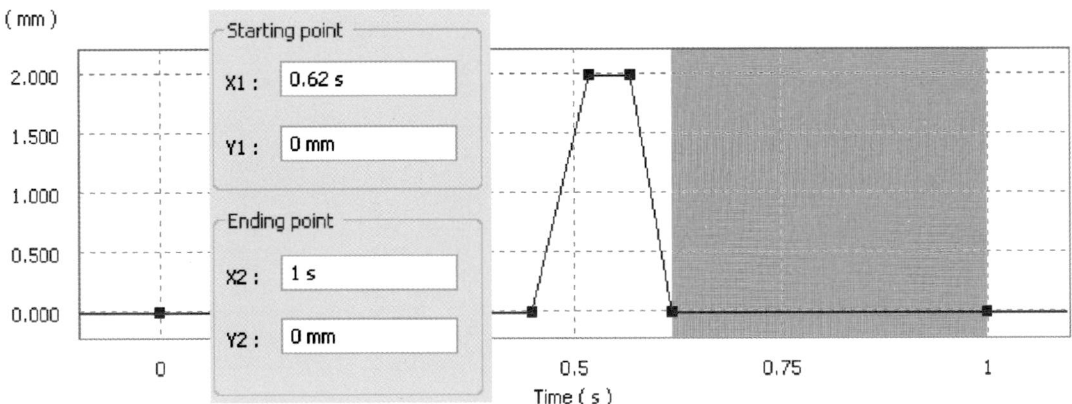

To obtain a smooth transition from the closed to open position, we will refine the graph between points two and three, and four and five by defining a cubic ramp between these points.

14. Click once anywhere between the second and third points > Select **Cubic ramp** from the list of available laws

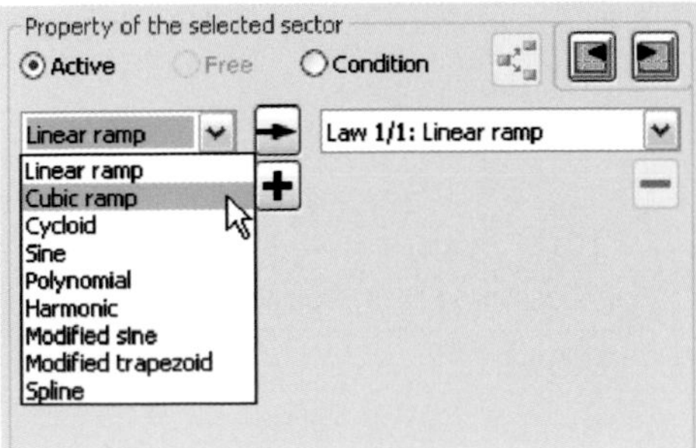

15. Click the **Replace the current law** button to change the existing linear ramp to cubic ramp

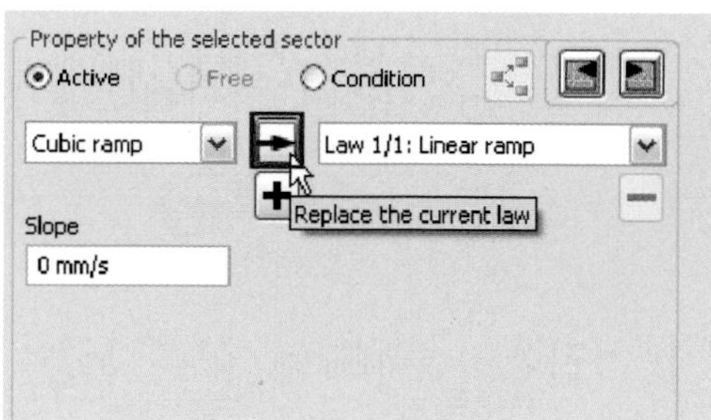

16. Repeat steps 14–15 to change the linear ramp between the fourth and fifth points to cubic ramp

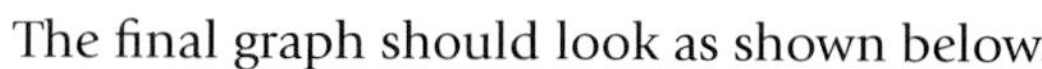

The final graph should look as shown below.

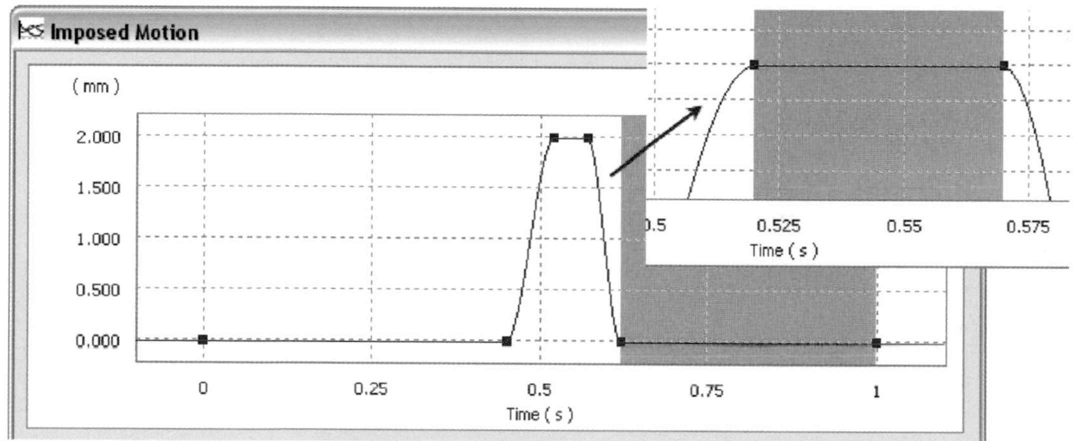

17. Click **OK** to exit the **Input Grapher** dialog box
18. Click **OK** again to exit the **Joint Properties** dialog box
19. Change the number of images in the simulation panel from 100 to 200, as shown

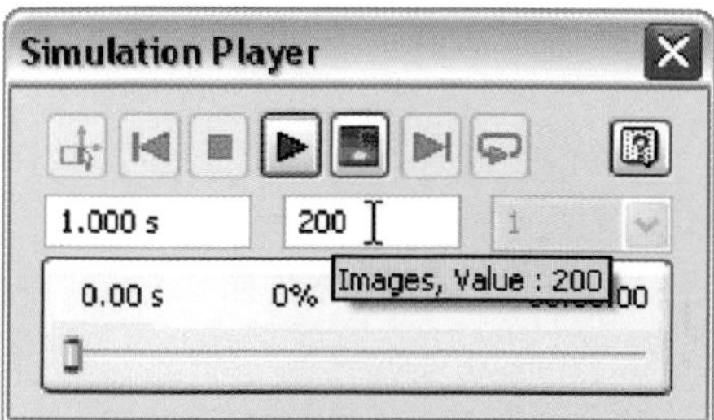

The higher the number of images, the more accurate the simulation will be, but the simulation will take longer to run and the file size will get larger.

20. Play the simulation to see the effect of the Input Grapher

You will see that the warning no longer appears.

You may need to zoom into the model to see the movement of the valve and the follower. Perhaps play the simulation with the **Continuous Loop** option enabled.

Here, I have shown you how you can create a specific law of motion. In reality, there will be many more points, in some cases in excess of 100, necessary to obtain a smooth operation of the cam and follower. As this can take a long time, another option would be to import this data via the spline law, if it exists. This will help to speed up the process of creating the laws of motion.

We will continue with this example in the next section, analyzing results, to create a cam from this known movement of the valve.

21. Close the file

EXAMPLE 7

Whitworth Quick Return Mechanism – Joint friction and imposed motion

Workflow of Example 7

- Set initial position
- Define imposed motion by applying constant velocity
- Define friction in joints

Set initial position

1. Open *Whitworth Quick Return2.iam*

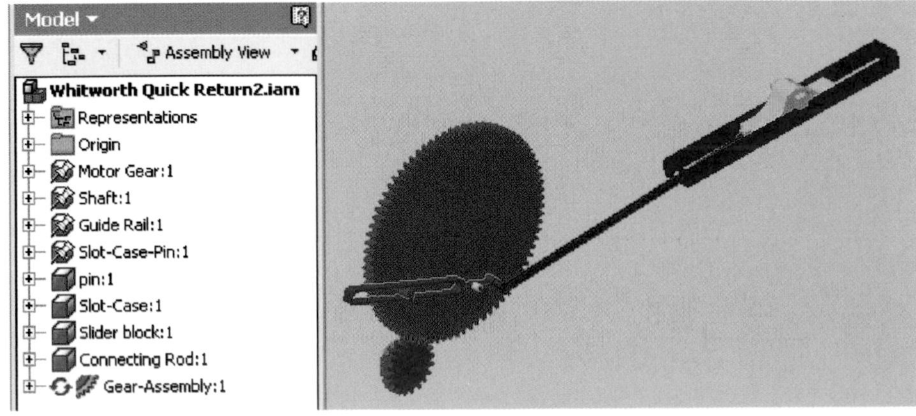

2. Select **Environments** tab > **Dynamic Simulation**
3. Double click on the Prismatic:2 joint
4. Select **dof 1 (T)** tab > Change the initial position to −6.660 mm

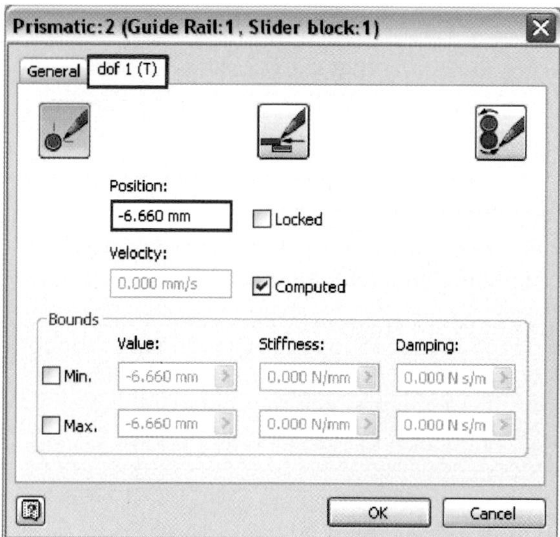

This is the furthest position to which the slider block can travel.

Define imposed motion by applying constant velocity

5. Click **OK** > Double click on the Revolution:3 joint

This is the gear that will be driven by a motor. Here, we will define a constant velocity and then determine the maximum torque required by the motor to maintain this constant velocity.

6. Select the **dof 1 (R)** tab > Check **Enable imposed motion.** Apply a constant velocity of −720 deg/s

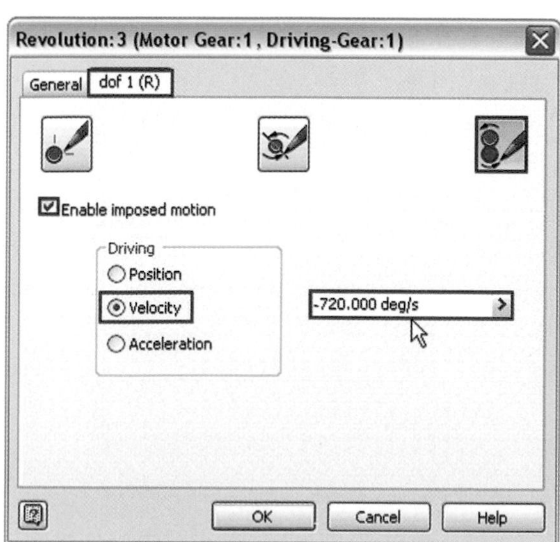

7. Click **OK**

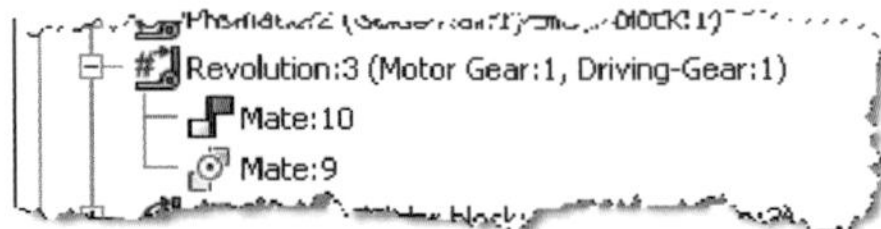

A # symbol appears in front of the revolution joint, implying that the joint has imposed motion applied to it.

8. Play the simulation

The slider block returns at higher velocity; this is a main function of the Whitworth Quick Return Mechanism and is used in various applications, including in shaping machines. This can also be verified by the Output Grapher by looking at the velocity of the slider.

9. Now select **Output Grapher** from the **Simulation** panel

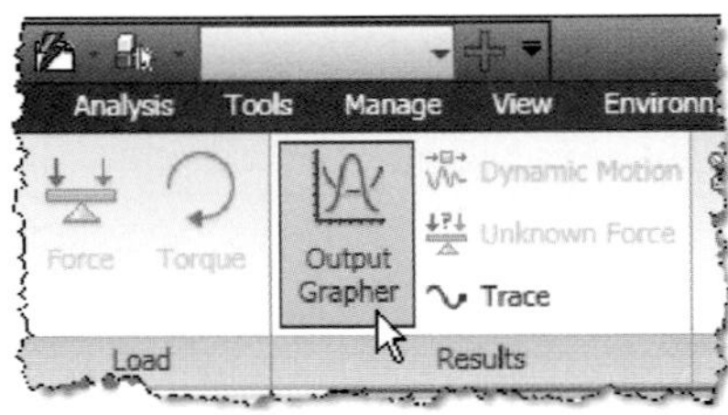

10. In the Output Grapher, select the velocity results of the Prismatic:2 joint, as shown below:

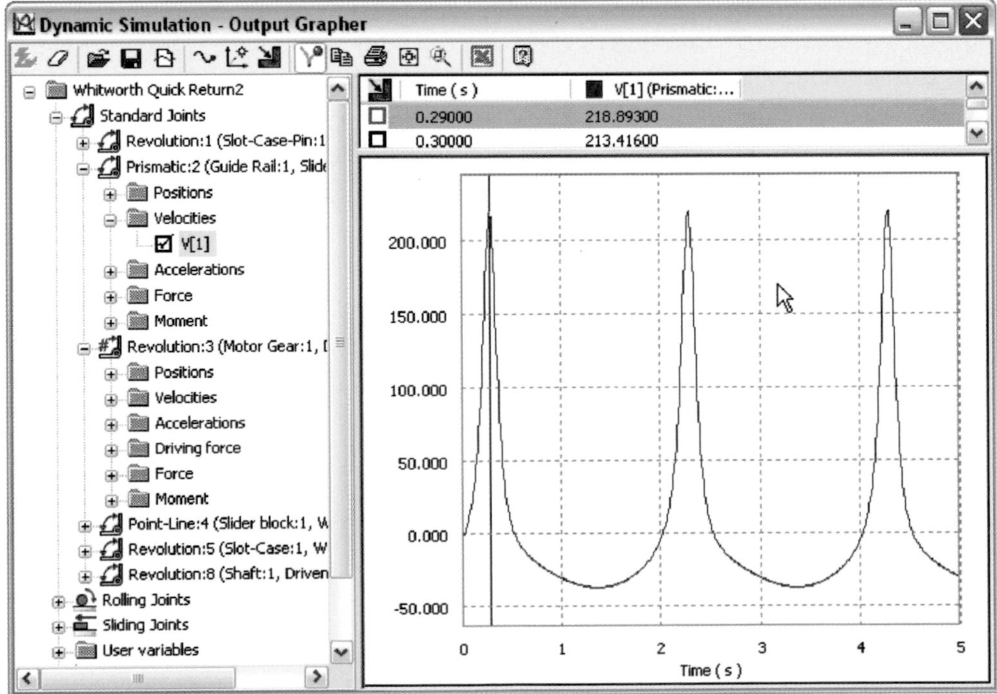

The maximum velocity on the return stroke exceeds 200 mm/s, whereas it never achieves more than 50 mm/s in the forward stroke, as expected – the actual value may slightly differ.

11. Now delete the curve by right clicking in the results column, as shown

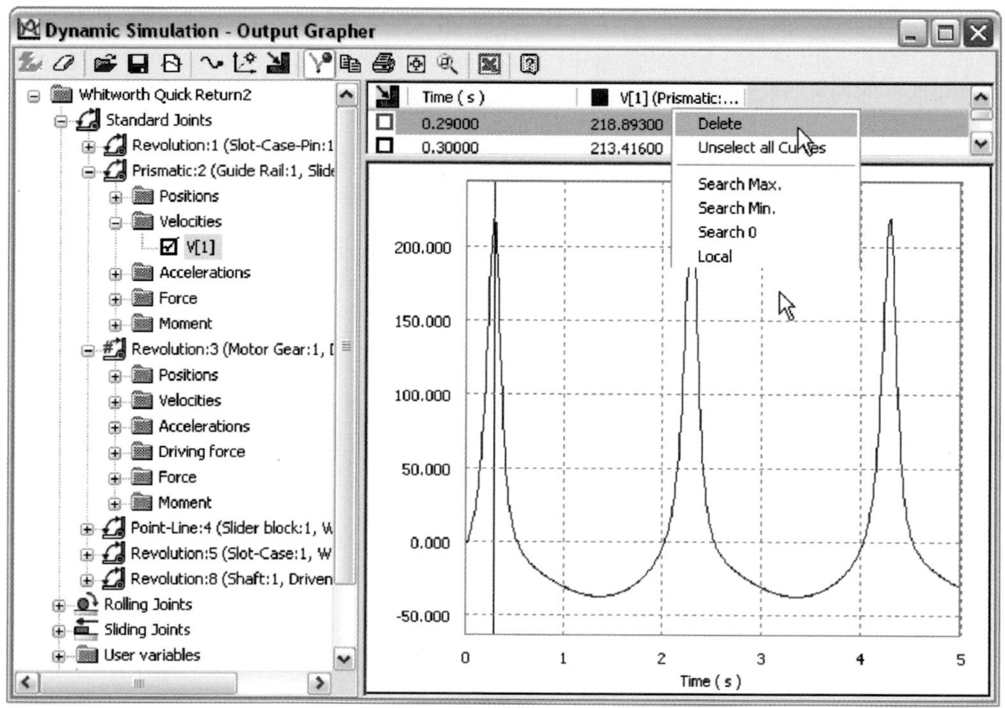

Next, we will check the torque required by the motor to maintain a constant velocity of 720 deg/s.

12. In the Output Grapher, select the driving force of the Revolution:3 joint, as shown below

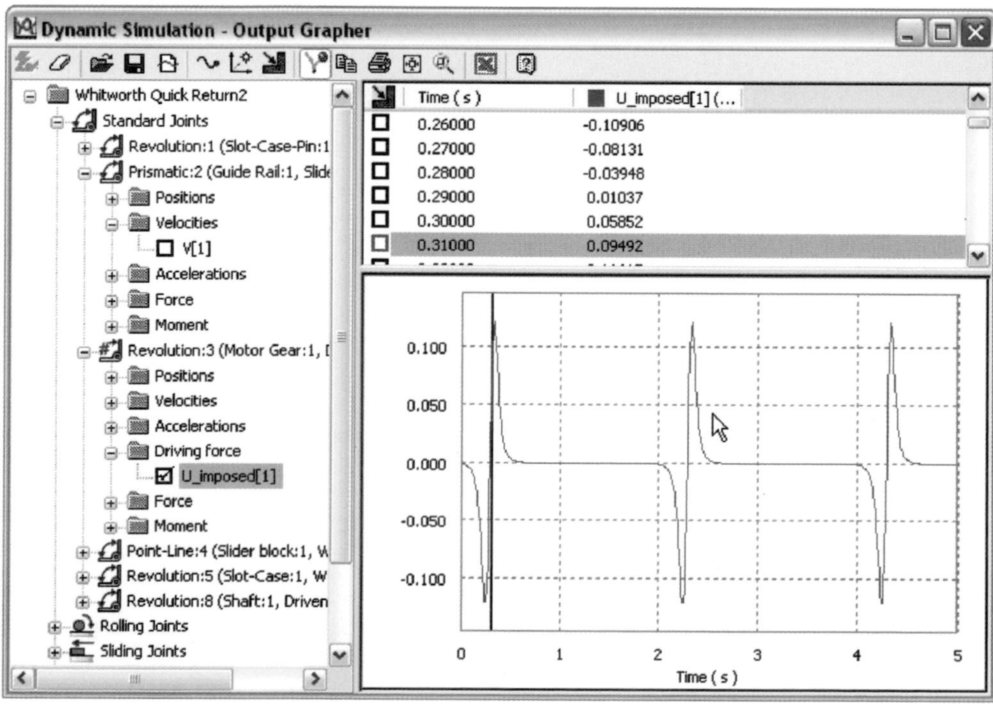

The torque is approximately 0 N/mm. The reason for this is that the joints are frictionless and, furthermore, there is no external force being applied to the assembly. Here, we will apply friction force between the slider block and guide rail.

Define friction in joints

13. Minimize the Output Grapher > Select **Construction Mode**
14. Double click on the Prismatic:2 joint > Select the **dof 1 (T)** tab > Click on **Enable joint force** > Specify 1 N force and 0.15 for friction > Click **OK**

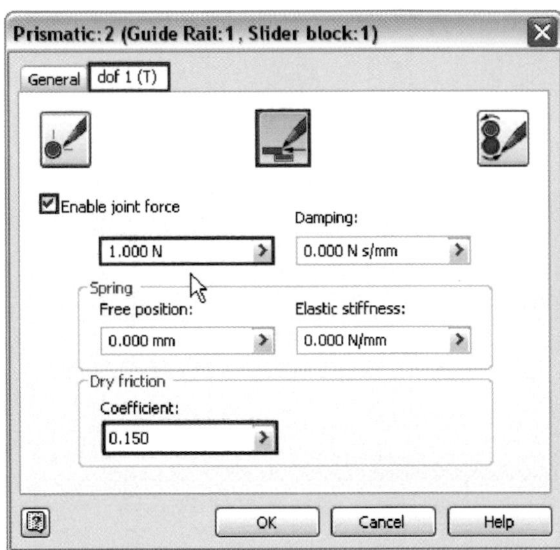

15. Play the simulation > Maximize the Output Grapher

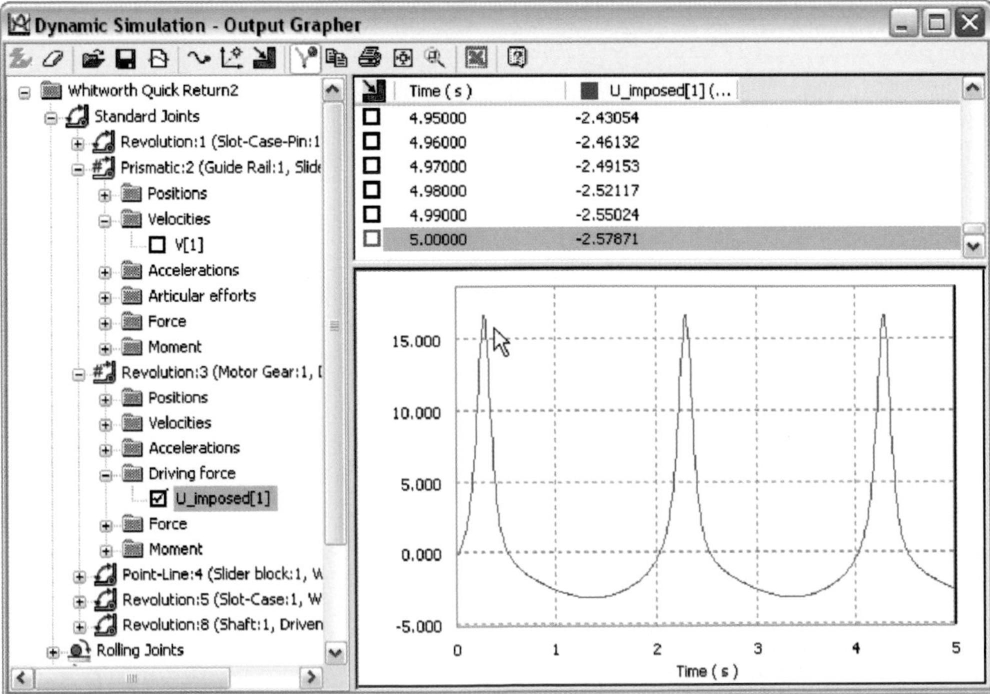

Torque has now increased to around 15 N/mm to maintain the constant velocity of 720 deg/s.

16. Right click on the data column > Select **Search Max.**

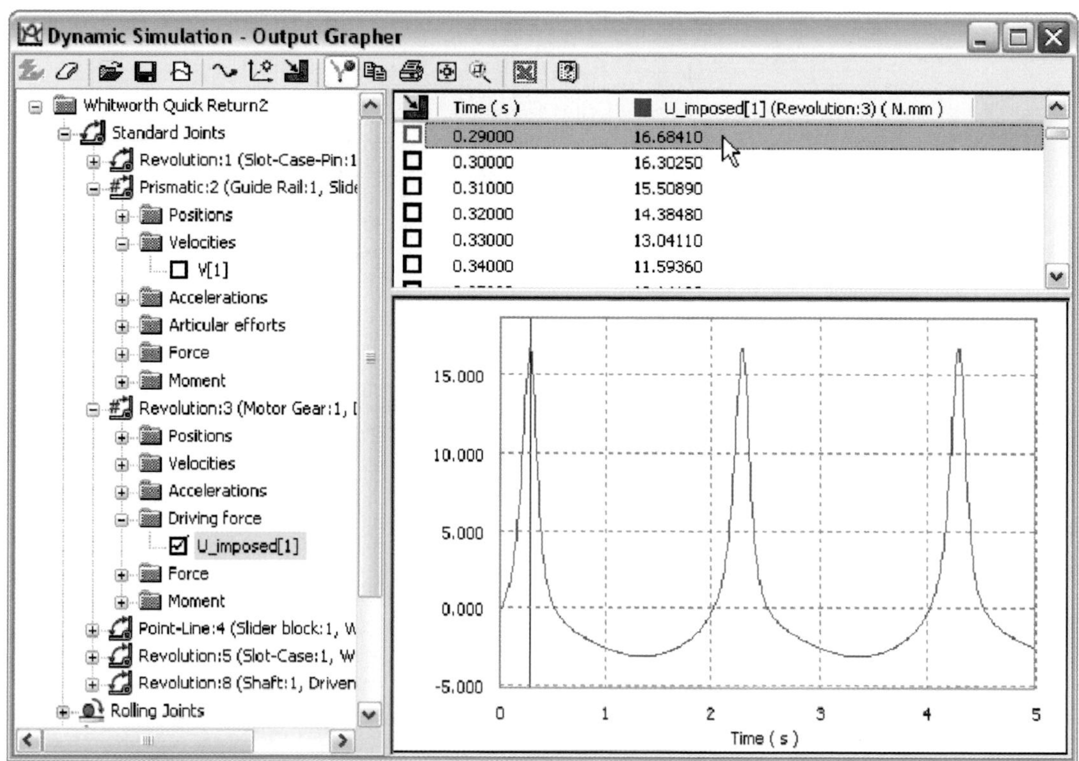

The maximum torque now required is 16.68 N/mm. The max value may differ slightly.

We will now discuss more of the Output Grapher functionality in detail in the following section.

17. Close *Whitworth Quick Return2.iam*

ANALYZING RESULTS

Within Dynamic Simulation, results can been seen and analyzed via the Output Grapher. The Output Grapher can be accessed from the **Simulation panel**.

Step 4

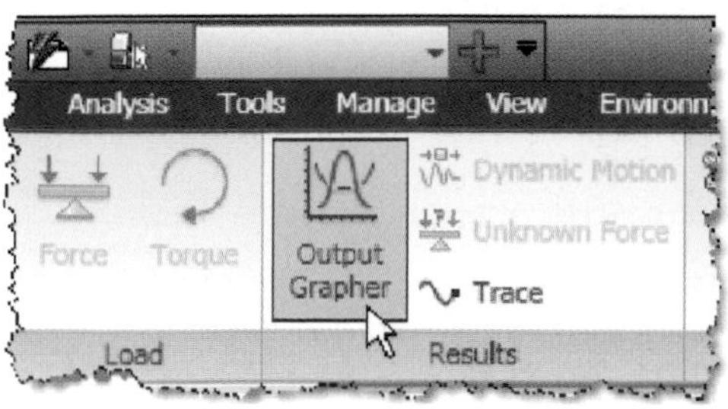

Output grapher

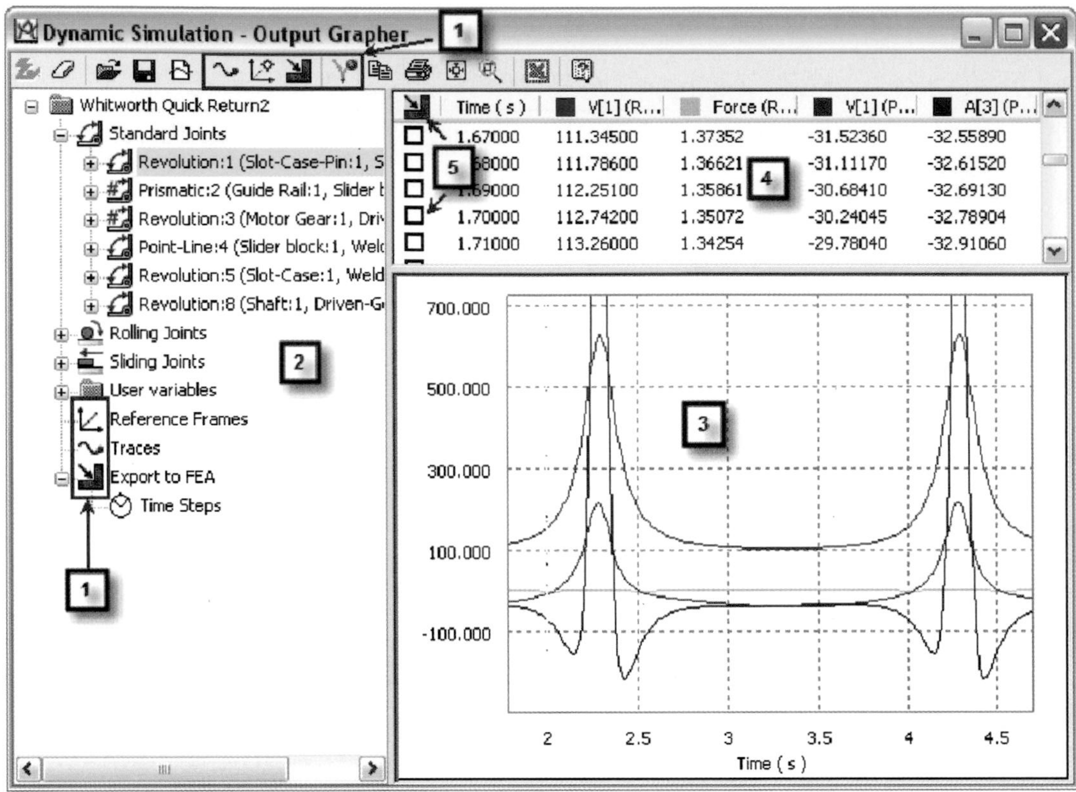

1. **Specialized Output Grapher Tools**

Here, there are four tools:

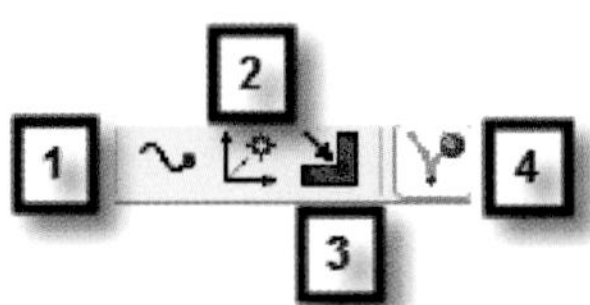

1. **Trace** – Allows creation of traces of components and joints, both visually and numerically (displaying data in Output Grapher).
2. **Reference Frame** – Gives the ability to examine results in reference to other components and user-defined origins.
3. **Export to FEA** – Here you specify the component to be analyzed within the Stress Analysis environment by identifying load-bearing faces.
4. **Precise Events** – Allows determination of the precise time of contact or impact events.

Precise information about contact events is now displayed as a specific time step in the Output Grapher.

2. Output Grapher tree
Here, you have access to all joints' data within the simulation. Joint data selected here have their values displayed in the graph and the time column.

3. Graphic area
The graphs of the selected joints' data are displayed here. Double click in this area to set a time in the graph that is synchronized to the time column and the display of the mechanism in the graphics window, and the simulation player. You also have the following capabilities within the graphic area:
- Scroll the mouse wheel to zoom in and out of the graph.
- Keep the mouse wheel pressed to pan around the graph.
- Double click the mouse wheel to reset the graph to its original state.

4. Time column
Contains columns for time steps and number of steps matching the time mode images in the simulation player. Each of the variables selected in the tree has a column.

5. Load transfer column
Here, you select single or multiple time steps to transfer loads to FEA.

In addition to simply displaying results of joints, as seen in the previous example, the Output Grapher has some advanced specialized tools, which will be explained next.

Output grapher environmental constraints – Snapshot of tools used throughout the book

	Environmental constraints	Examples using these environmental constraints	Design problems using these environmental constraints
1	Traces – Output Grapher data	8	3,4,5
2	Traces – Export	8	4,5
3	Precise events	9	
4	Export to FEA		8,9
5	New curves/user variables		4
6	Unknown force		4

Process of using the specialized tools within the output grapher

This section will explain the different tools available for creating environmental conditions using the same examples we used in the joints creation process.

- Example 8 – cam design
  - Create and export trace
  - Create cam shape based on trace
- Example 9 – Ball and staircase
  - Precise motion

The **Export to FEA** function is discussed in Chapter 9.

EXAMPLE 8

CAM design – Output trace

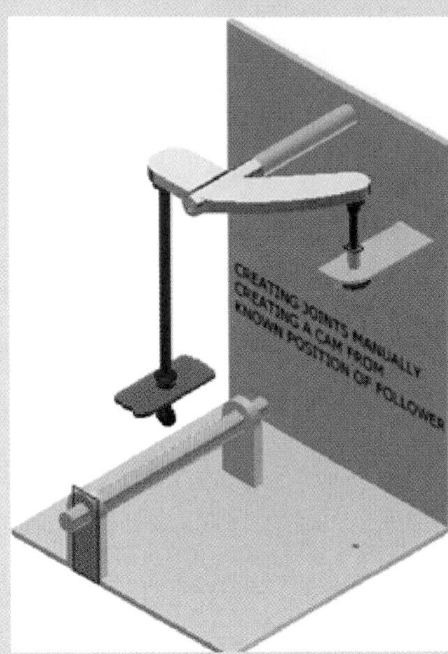

Workflow of Example 8

- Create trace
- Export trace
- Create cam from export trace
- Play simulation based on new cam

Extra Joints used in Example 8

- Sliding-cylinder curve

Create trace

1. Open *CAM3.iam*
2. Select **Environments** tab > **Dynamic Simulation**
3. Select **Output Grapher** from the **Simulation panel**
4. Select **Add Trace** from within the Output Grapher

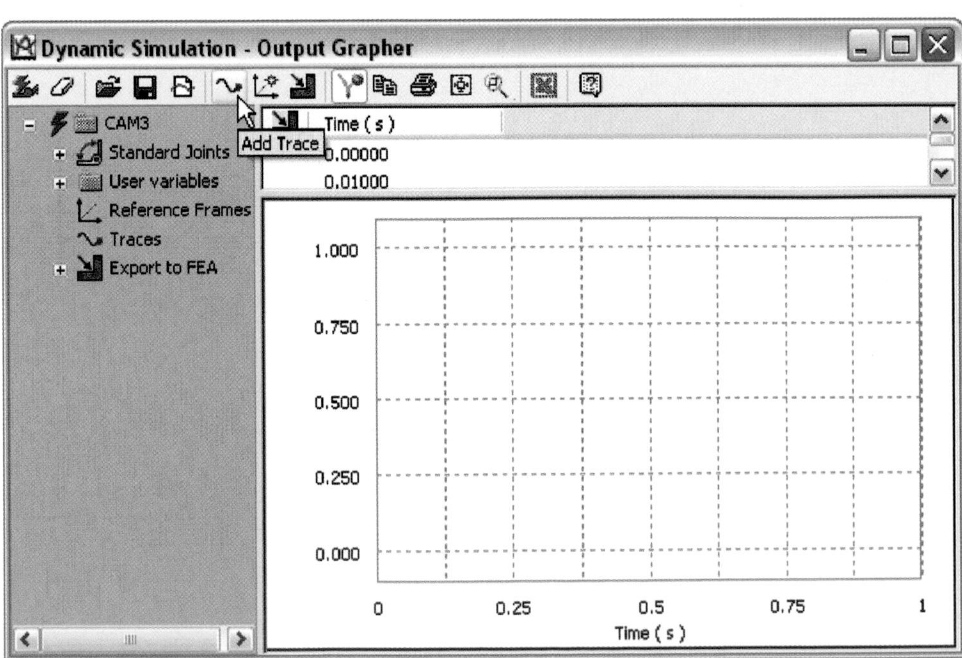

5. In the **Trace** dialog box, select the edge of the follower (to specify the origin of the trace) > Select CAM Rotating Shaft:1 as the reference geometry to be used to create the trace > Click **OK**

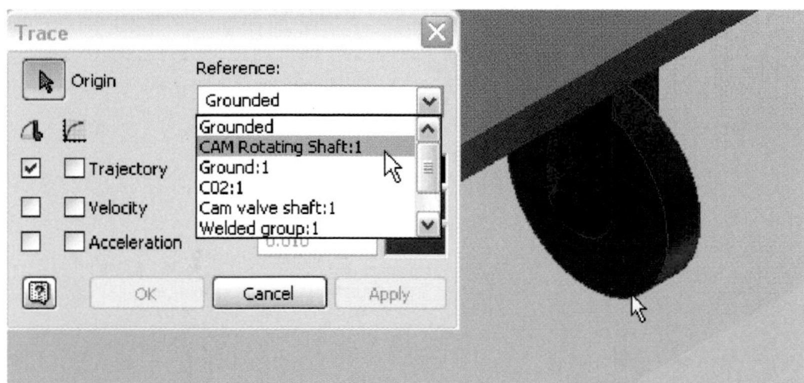

6. Minimize Output Grapher

7. Change the view of the assembly as shown

8. Play the simulation

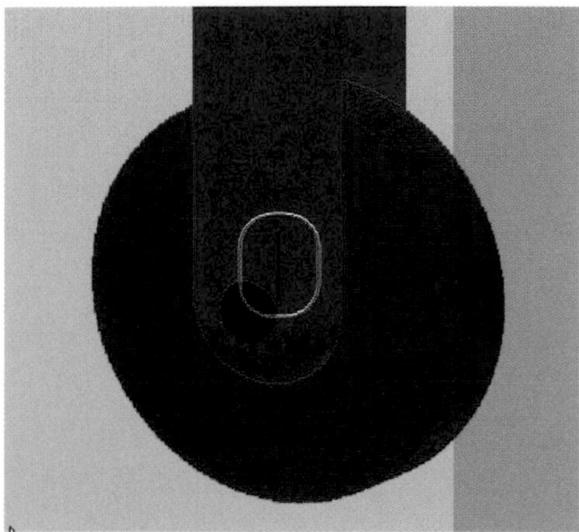

Note that the trace will go up and down. This is because we have not specified any rotation of the shaft.

9. Select **Construction Mode** > Right click the Revolution:1 joint > Select **Properties**

10. Select the **dof 1 (R)** tab > Check **Enable imposed motion** > Type 360 deg/s in the **Velocity** value box

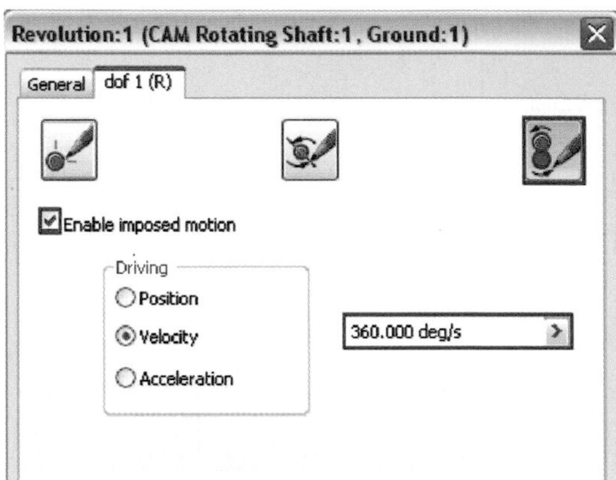

11. Click **OK** and change the view of the assembly as shown

Export trace

12. Play the simulation. You should now see the trace of the follower as shown

13. Maximize Output Grapher

14. Right click Trace:1 and select **Export to Sketch**

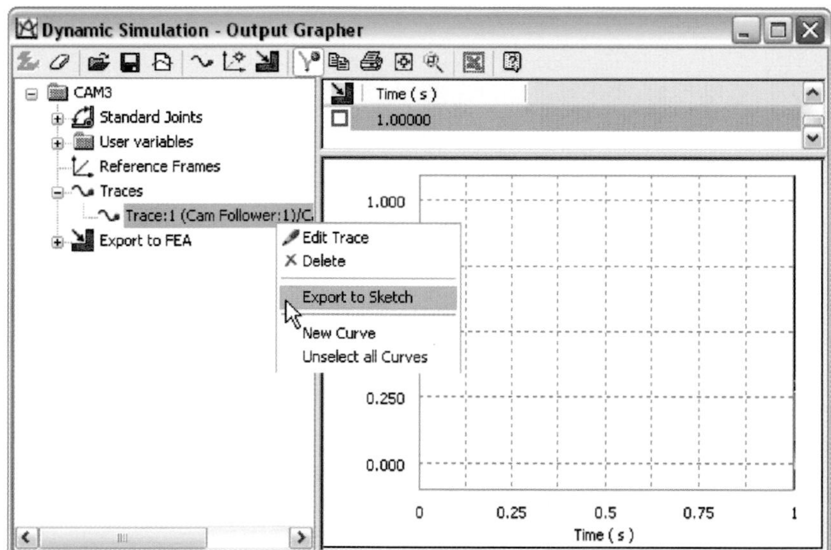

15. Select the cam rotating shaft as shown

The command prompts the user to select the component to which this sketch needs to be attached.

16. Close Output Grapher > Select **Construction Mode**

Create cam from export trace

17. Double click the cam rotating shaft. The Part environment will now be active

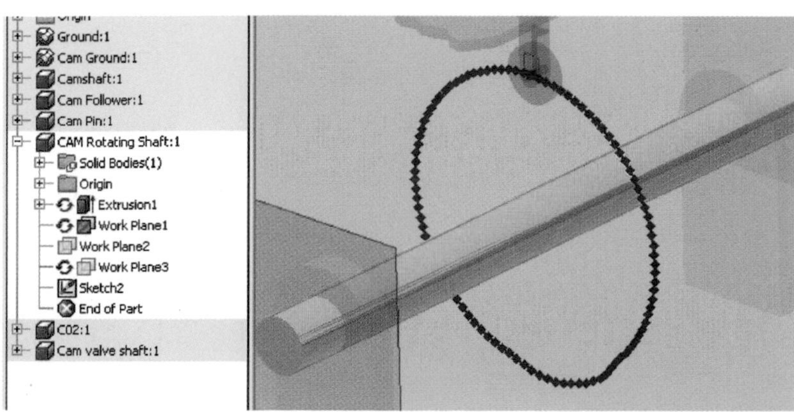

18. Right click Sketch 2 > Select **Edit Sketch** > Create an offset of 5 mm of the exported sketch > Select **Finish Sketch**

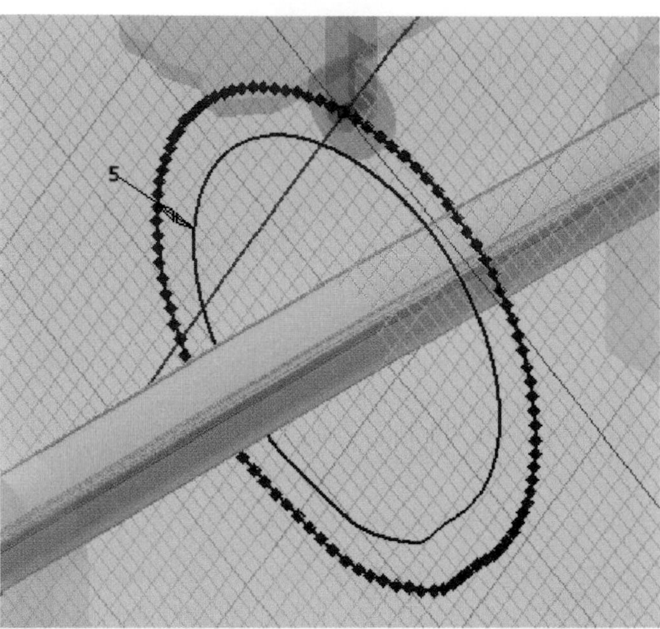

19. Extrude the sketch by 2 mm, as shown

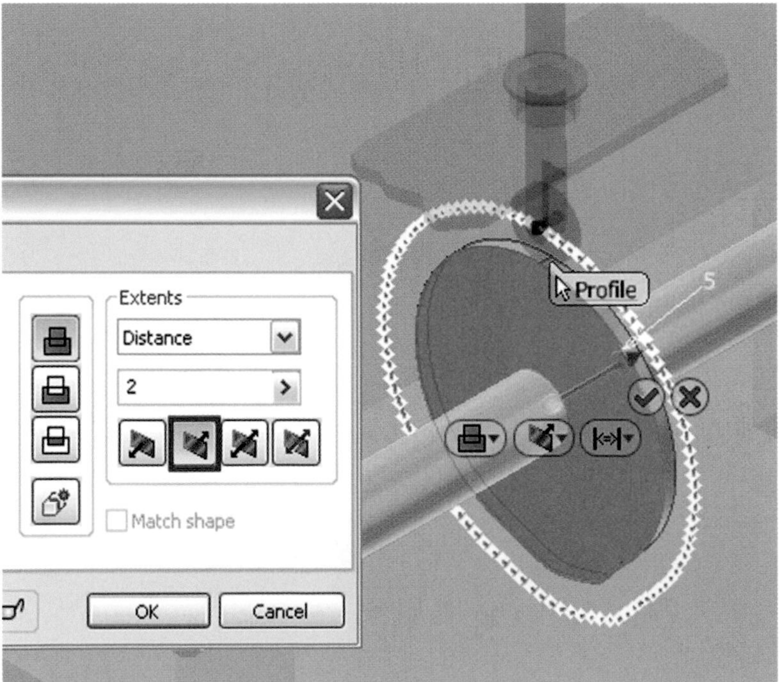

20. Click **Return** to go back to the Dynamic Simulation environment

21. Play the simulation > Select **Continuous Loop** to replay the simulation

You can further refine the Input Grapher by adding more points to define a more detailed graph, hence creating a smoother cam profile. Alternatively, you can import predefined data (in .txt format) using the spline law.

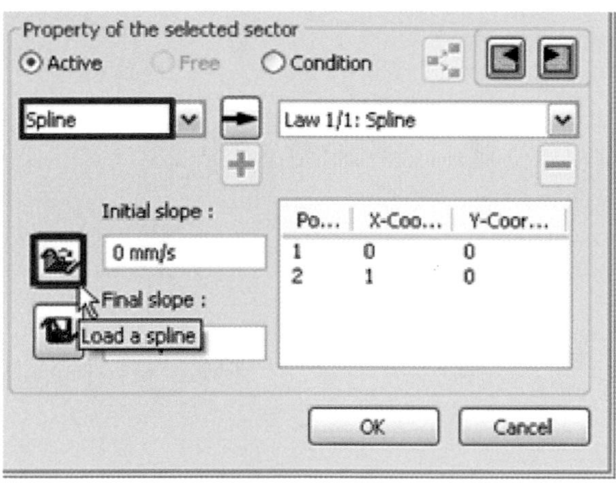

Now, as we have designed the cam we can use this new cam profile to drive the follower and hence open and close the valve accordingly.

Play simulation based on new cam

22. Select **Construction Mode**
23. Double click the Cylindrical:3 joint and deselect **Enable imposed motion**. Click **OK**

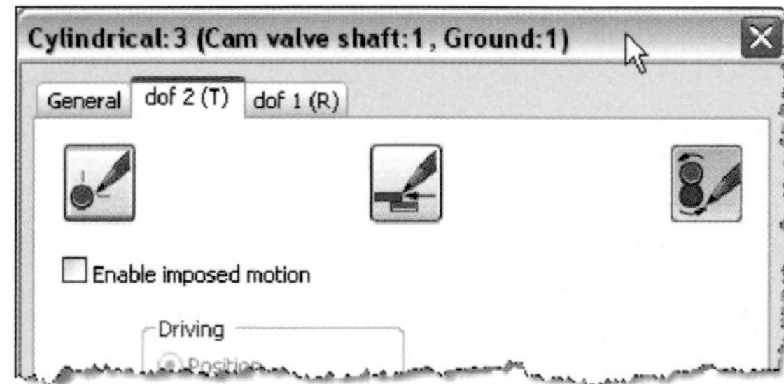

24. Select **Insert Joint** and select the Sliding: Cylinder Curve joint
25. For **Curve**, select the edge of the new cam, as shown

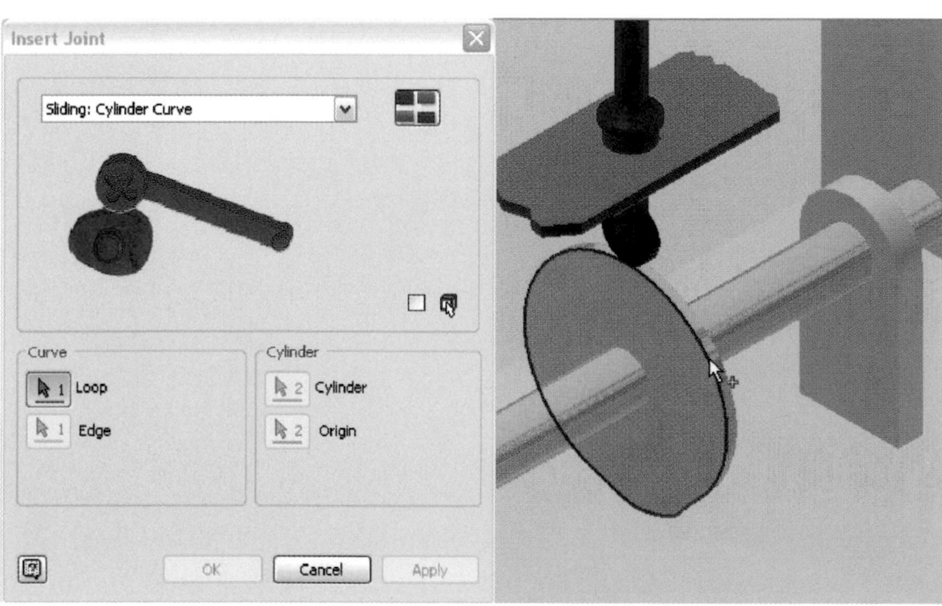

26. Select the **Cylinder** button

27. For **Cylinder**, select the edge of follower as shown > Click **OK**

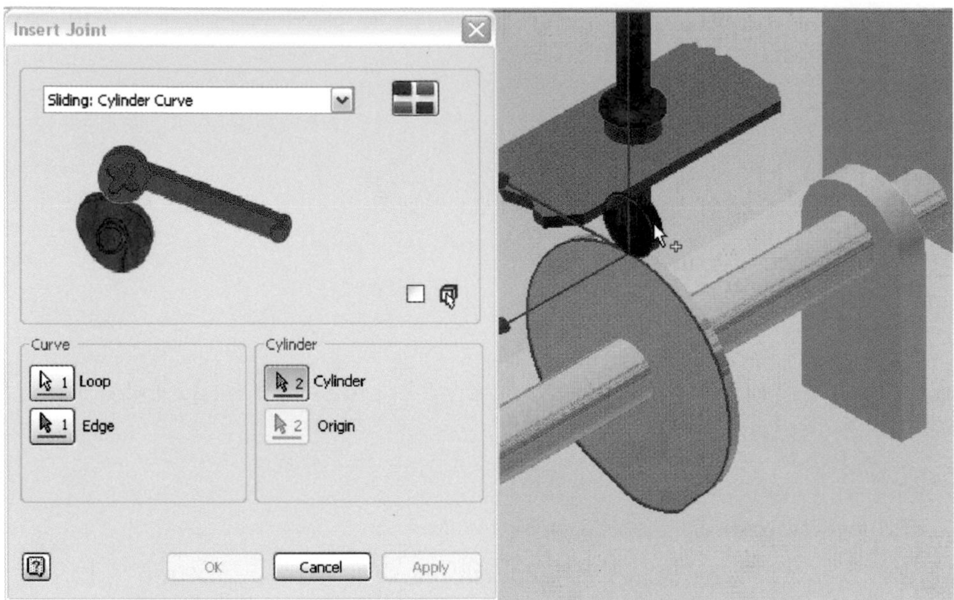

28. Play the simulation and, after a while, the following warning appears:

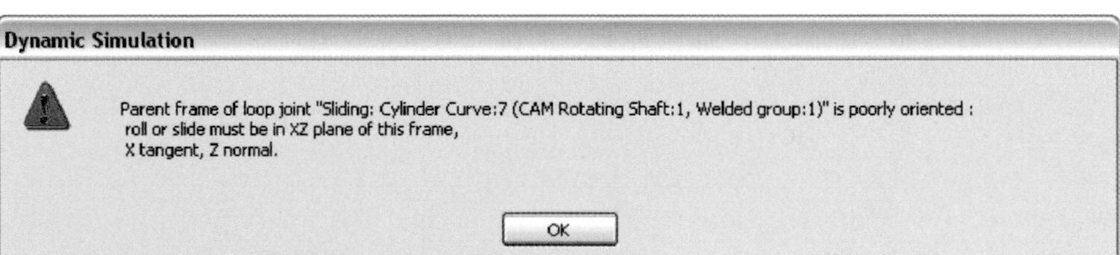

This warning appears because the follower is misaligned and rotated as shown below:

If we can stop the follower from rotating, we may be able to get the simulation to work.

29. Click **OK** > Select **Construction Mode**

30. Double click the Cylinder:4 joint and select **dof 1(R)**. Lock the position, as shown

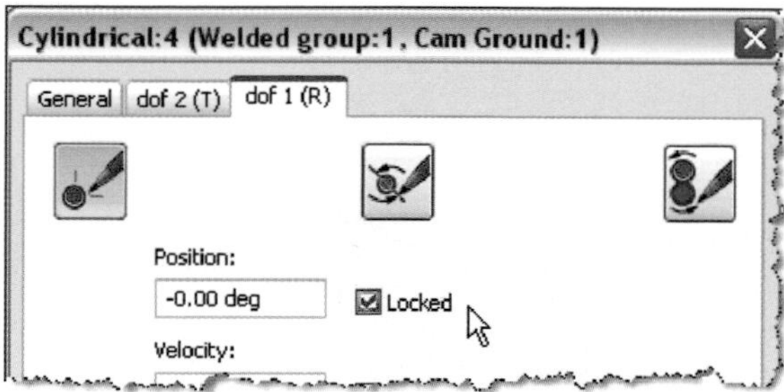

31. Click **OK** > Accept the warning message

32. Play Simulation

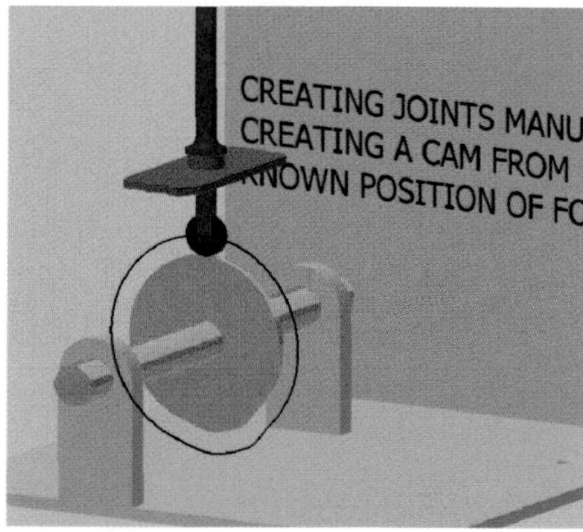

Note: Here, the cam now drives the opening and closing of the valve.

33. Close *CAM3.iam*

EXAMPLE 9

Ball and staircase – Precise events

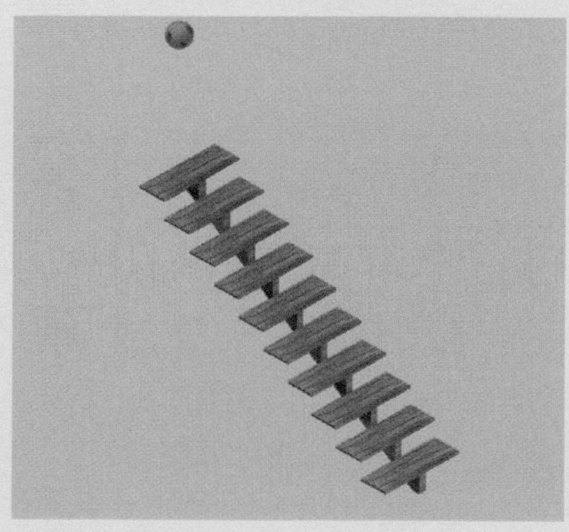

Workflow of Example 9

- View impact position of ball
- Set precise motion and view results

View impact position of ball

1. Open *ball-stair.iam*

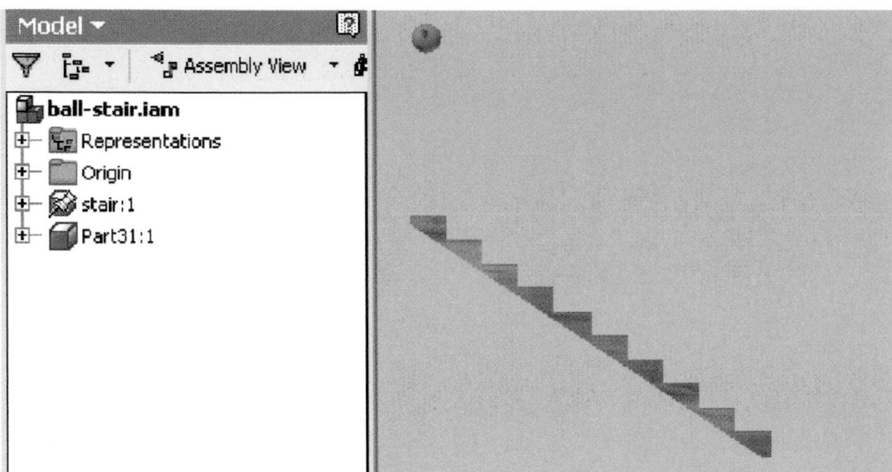

2. Select **Environments** tab > **Dynamic Simulation**
3. Play the simulation
4. Select Output Grapher

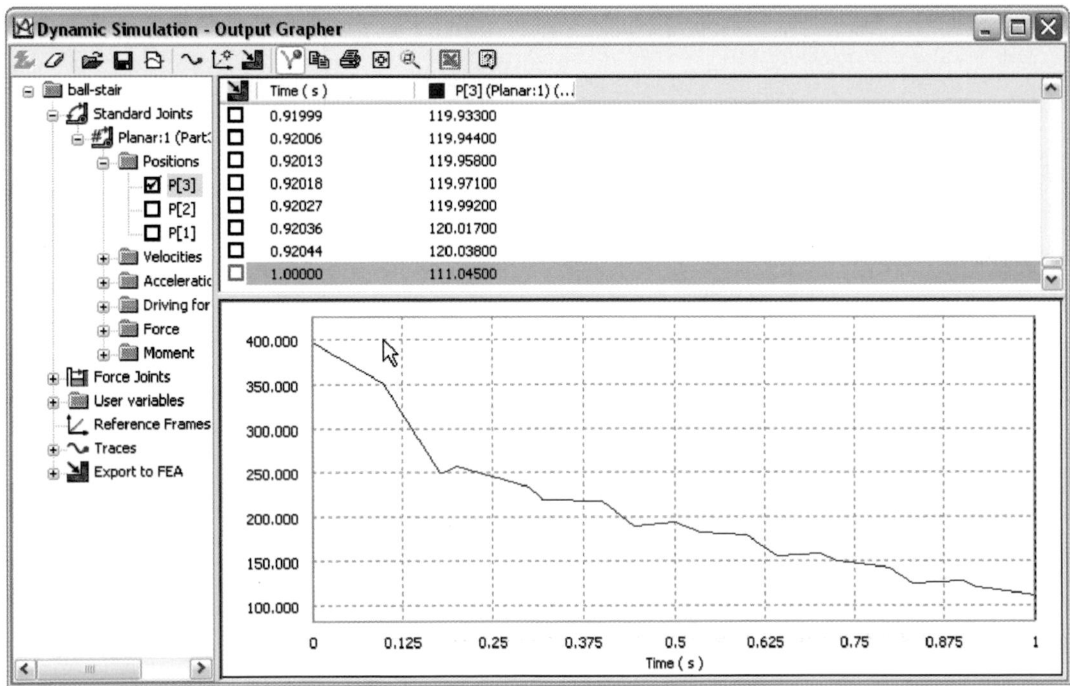

The **Precise Events** button is selected by default and we can see kinks in the graphs representing the precise motion at the points of contact. Also note that the time steps are not evenly calculated as the **Precise Events** button forces the software to calculate at key time events.

5. Deselect the **Precise Events** button and notice the difference in the curve

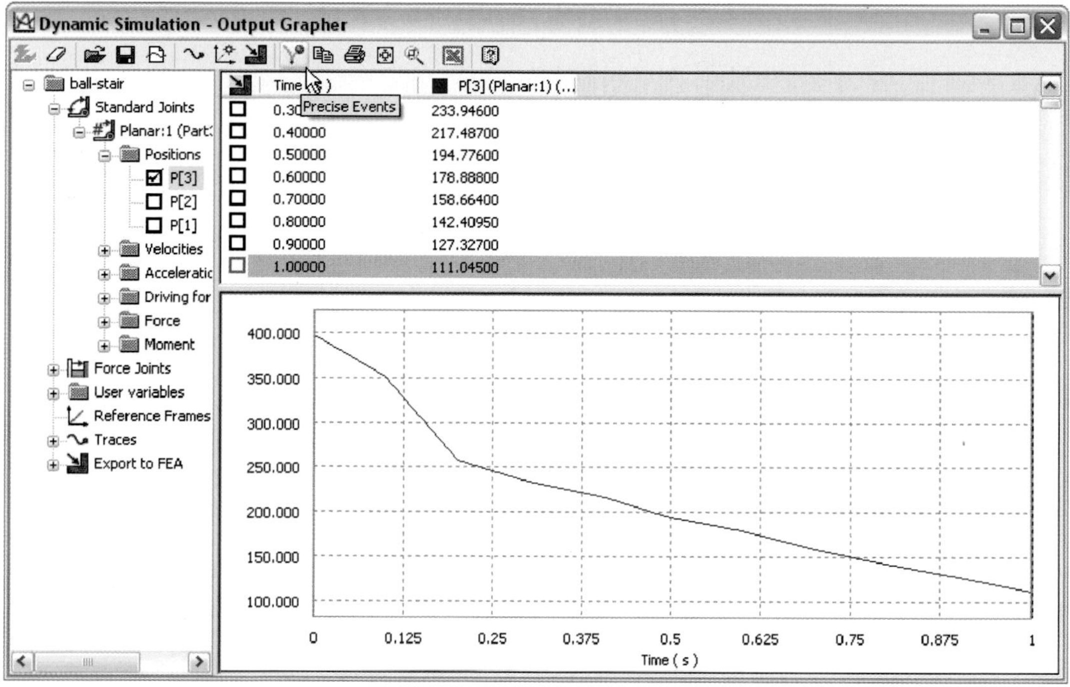

With the **Precise Events** button switched off, the time steps are now calculated with uniform time steps. However, this will not necessarily calculate and capture key contact events.

6. Now select the Force_max variable under Force Joints

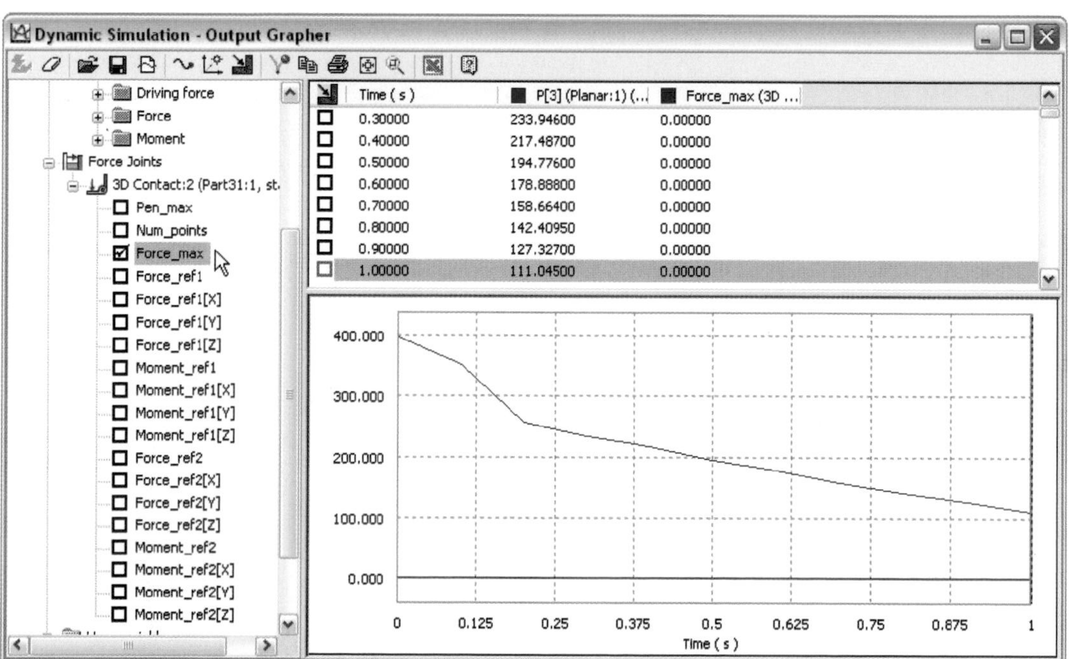

The graph will not display impact forces as the **Precise Events** button is deselected.

Set precise motion and view results

7. Select the **Precise Events** button and see how the graph now shows impact forces at each contact event between the ball and stair

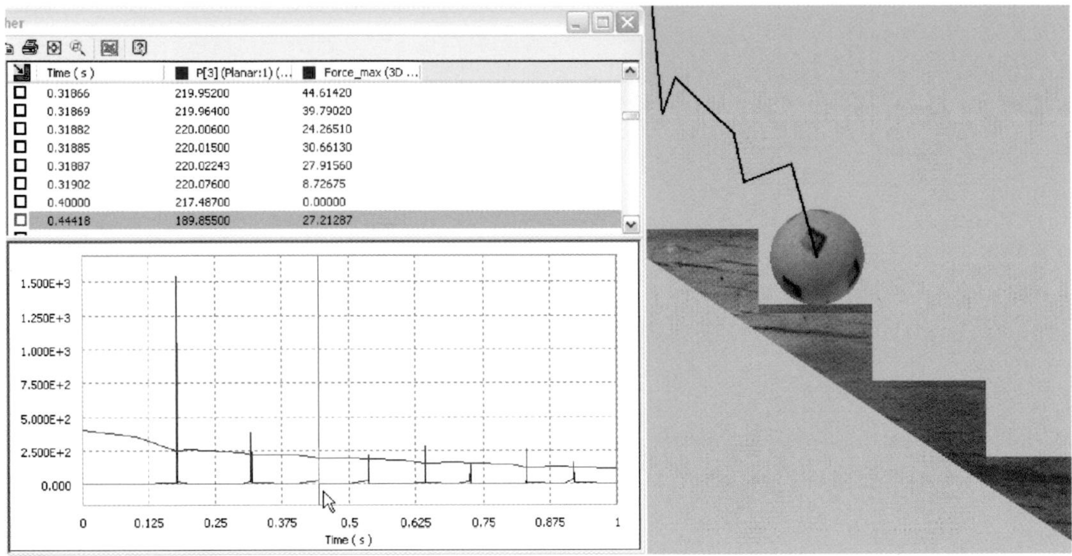

8. Close *ball-stair.iam*

CHAPTER 2

DP1 – Size a Motor

90 Degree M16 Bolt Removal Tool Design

(Design Problem courtesy of Hallimarine)

JOINTS INTRODUCED/COVERED IN THIS DESIGN PROBLEM

	Joints	Joint creation process
1	Revolution	Automatically converted
2	Rolling cylinder on cylinder	Automatically and manually created
3	Rolling cone on cone	Manually created

KEY FEATURES AND WORKFLOWS INTRODUCED IN THIS DESIGN PROBLEM

	Key features/workflows
1	Color Mobile Groups
2	Creating assembly constraints within the simulation environment
3	Imposed motion – constant translational velocity

INTRODUCTION

This tool is designed to remove eight M16 × 60 long-hex-head bolts from a hazardous environment that restricts human entry. The tool is deployed by a robotic manipulator. There are two input handles on either side of the unit; this is to cater for the restriction in reach by the manipulator. When the tool has removed the four lower M16 bolts, the manipulator rotates the tool 180 degrees to start removing the M16 bolts on the top flange. All nuts are welded in position and the bolts are pre-torqued to 95 N/m or 95 000 N/mm. The bolts are located on two flanges (top and bottom) of a pipe.

Up and Running with Autodesk Inventor Simulation 2011. ISBN: 978-0-12-382102-7
Copyright © 2010 Elsevier Inc. All rights of reproduction, in any form reserved.

NUTS - WELDED IN POSITION
4 OFF M16 BOLTS
4 OFF M16 BOLTS
NUTS - WELDED IN POSITION

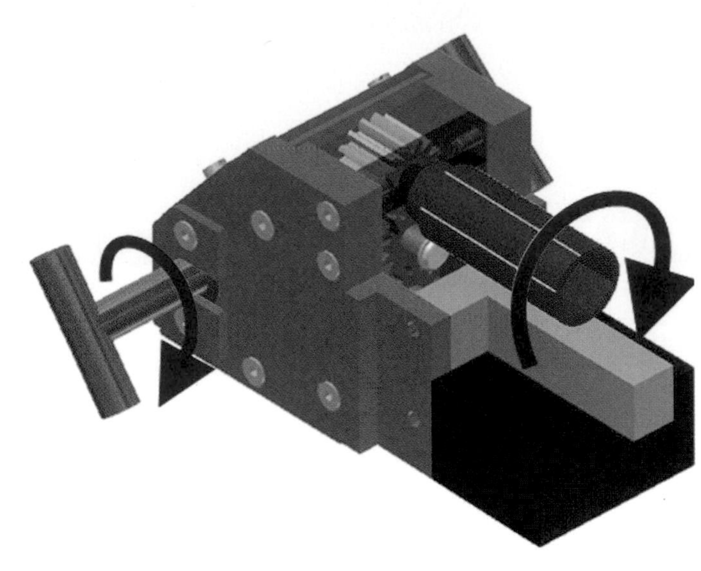

Hence, the main requirements of this design problem are to

- Determine the input torque (at the handle)
- Achieve a torque ratio of at least 1.8:

$$\text{Torque ratio} = \frac{\text{Output torque}}{\text{Input torque}} \geq 1.8$$

WORKFLOW OF DESIGN PROBLEM 1

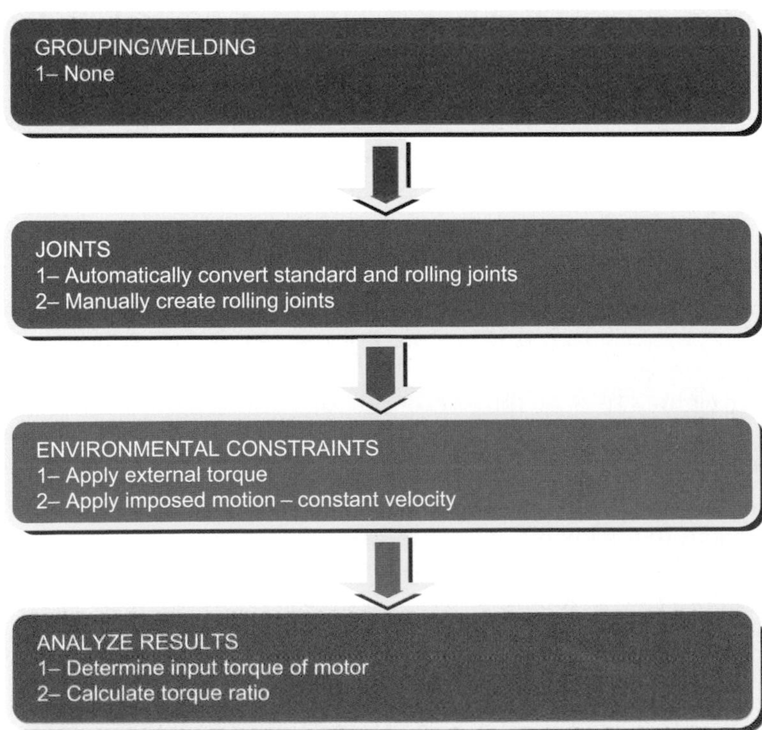

Joints

In this design problem, the primary goal is to determine the input torque of the gears. Therefore, we may concentrate only on the motion between the gears, the handle, and the removal tool. The rest of the components can be treated as grounded, and therefore there is no need to group or weld these components together as no joints will be created between the grounded components.

AUTOMATICALLY CONVERT STANDARD AND ROLLING JOINTS

1. Open *Bolt-Removal.iam*

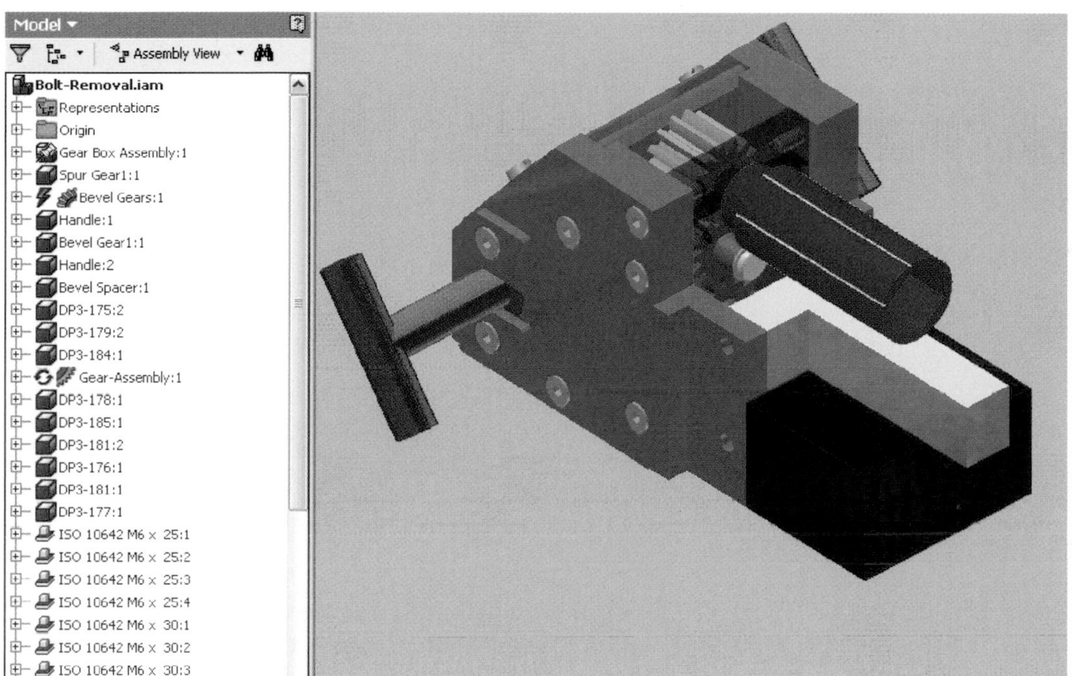

2. Select **Environments** tab > **Dynamic Simulation**

3. Select **Simulation Settings**

4. Select **Automatically Convert Constraints to Standard Joints** within the **Dynamic Simulation Settings** dialog box

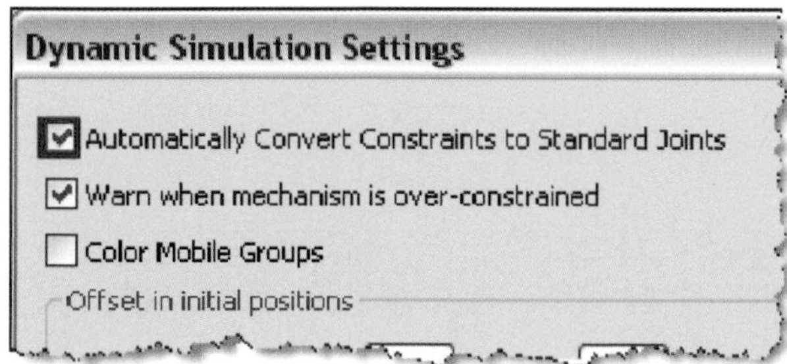

5. Click **OK**

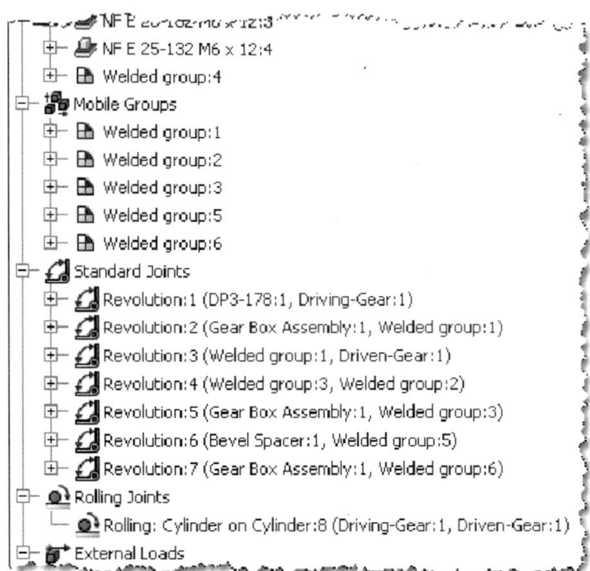

Dynamic Simulation will create rolling joints automatically if the motion constraints have been created by Design Accelerator.

Based on predefined assembly constraints, Dynamic Simulation has welded some components together in addition to creating revolution joints.

Any motion constraints created manually will not be converted automatically as this functionality is not currently supported.

Design Accelerator only allows the creation of a maximum of two gears at a time. Therefore, you will always have to create extra motion constraints manually between gears, if more are required, within the assembly, and therefore you will need to create rolling joints manually within the Simulation environment.

As it is not clear which components are mobile as all are part of a welded group, we will use the **Color Mobile Group** tool.

6. Select **Simulation Settings**

7. Select **Color Mobile Groups** > Click **OK**

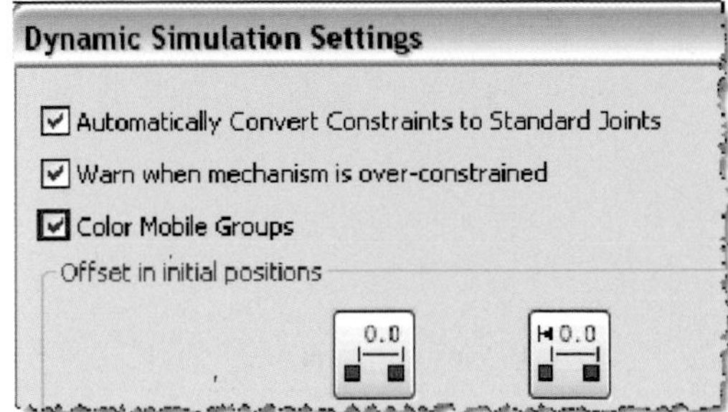

Mobile Color assignments may differ slightly.

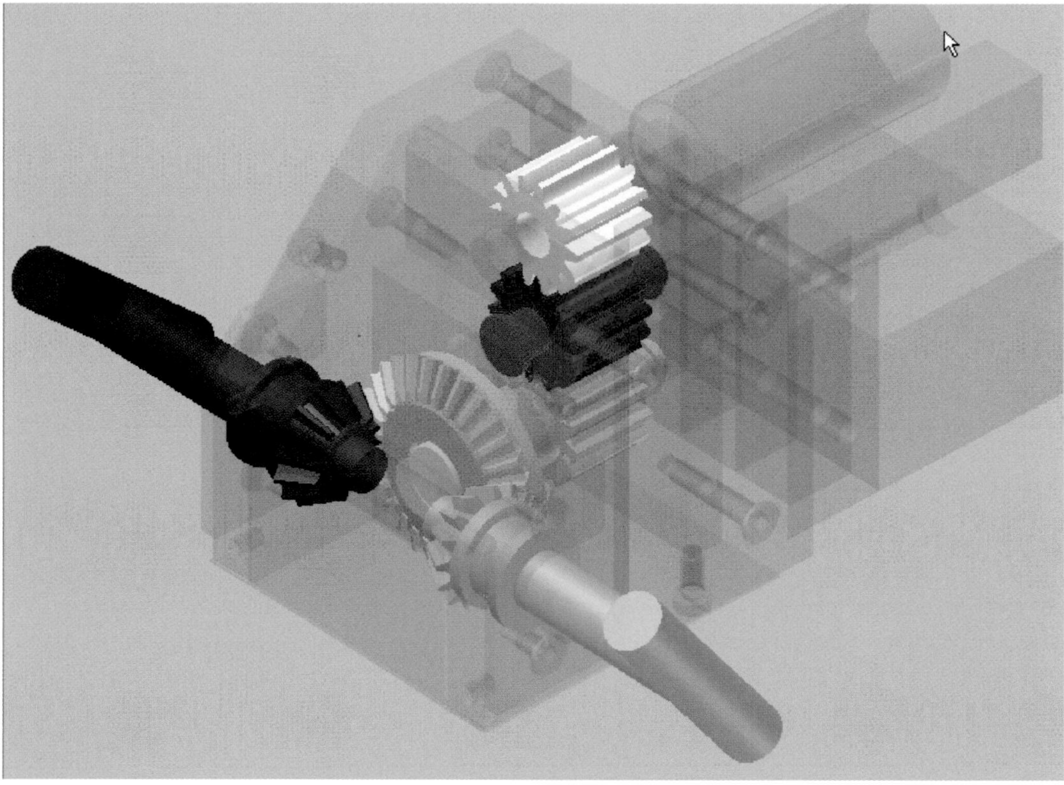

By selecting **Color Mobile Groups**, all components that have no motion or are grounded will have color glass applied to them, aiding the analysis and visualization of the assembly.

The removal tool is shown as grounded and is not attached to the white spur gear; hence, motion will not be transferred. So, first we need to attach the spur gear to the output tool. We will do this by applying more assembly constraints between the two components to remove any degrees of freedom. Another reason that this has happened is that we did not pay enough attention, initially, to the Assembly Constraints.

8. Press **C** to activate the **Place Constraint** dialog box within the Simulation environment

9. Apply a mate constraint between the gear and removal tools > Click **OK**

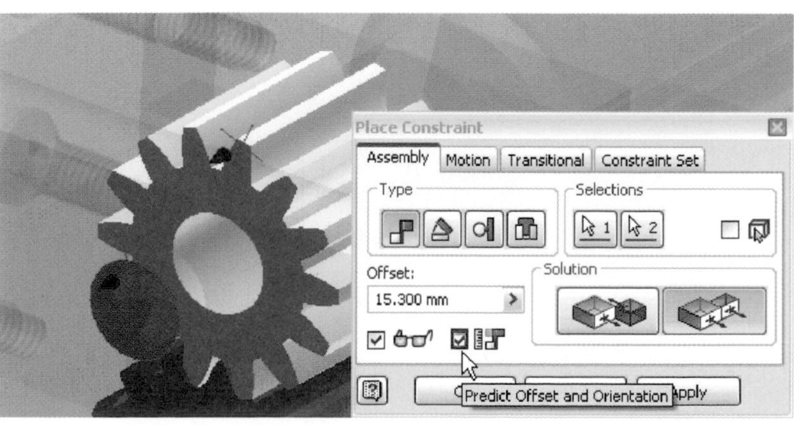

Make sure the **Predict Offset and Orientation** tool is selected to avoid misplacing components.

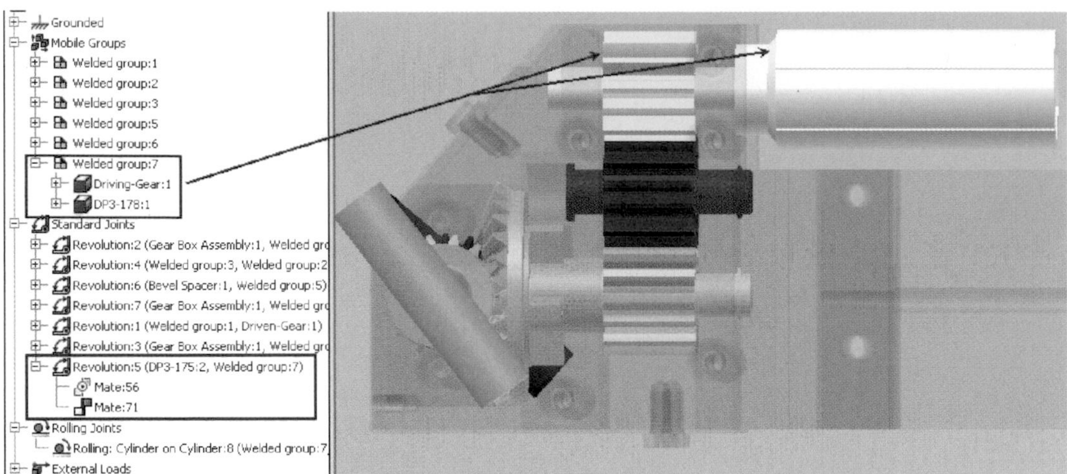

We now have a new welded group between the two components, as illustrated by the white color, and a new revolution joint between the welded group and a component from ground.

We could have created a subassembly of all the fasteners and all the components that have no motion.

MANUALLY CREATE ROLLING JOINTS

The next step is to create rolling joints for the rest of the spur and bevel gears. Switch visibility off for the gear box assembly to aid in the selection of construction surfaces.

The construction surfaces of the gears have already been made visible from within the Part environment; by default they are invisible.

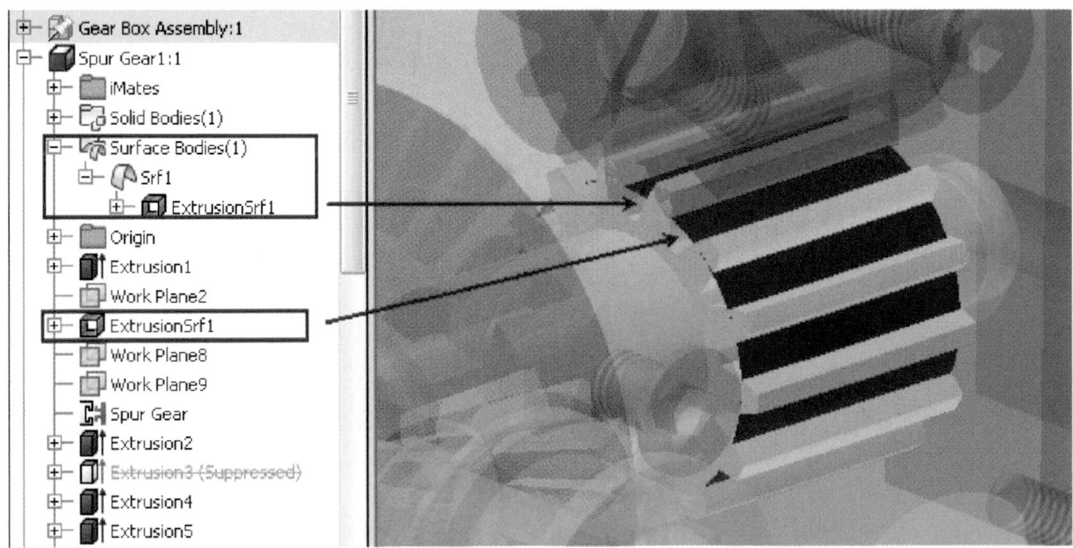

10. Select **Insert Joint** > Select the **Rolling: Cylinder on Cylinder** joint from the joints list

Make sure you select the **1 Constraint: Rolling** option, as this will simulate reality accurately, before selecting the first cylinder.

11. For Component 1, select the Driven-Gear:1 construction surface

12. For Component 2, select the construction surface of Spur Gear1

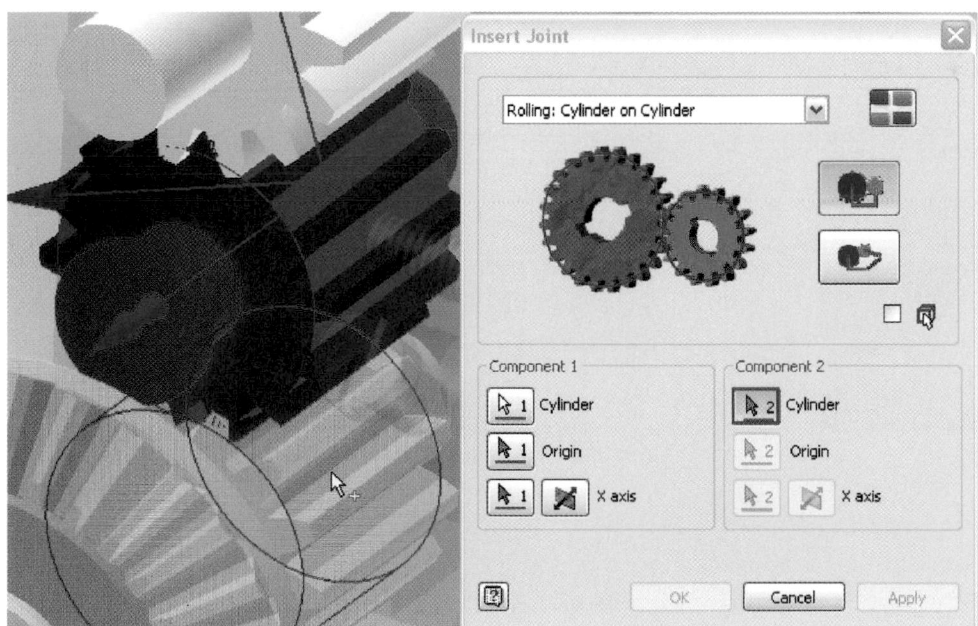

13. Click **OK**

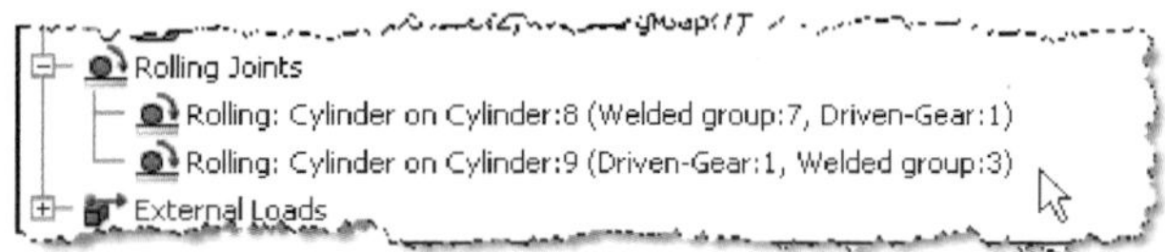

14. Select **Insert Joint** > **Rolling: Cone on Cone**

15. For Component 1, select the construction surfaces of the Bevel Gear1

16. For Component 2, select the construction surfaces of Bevel Gear2

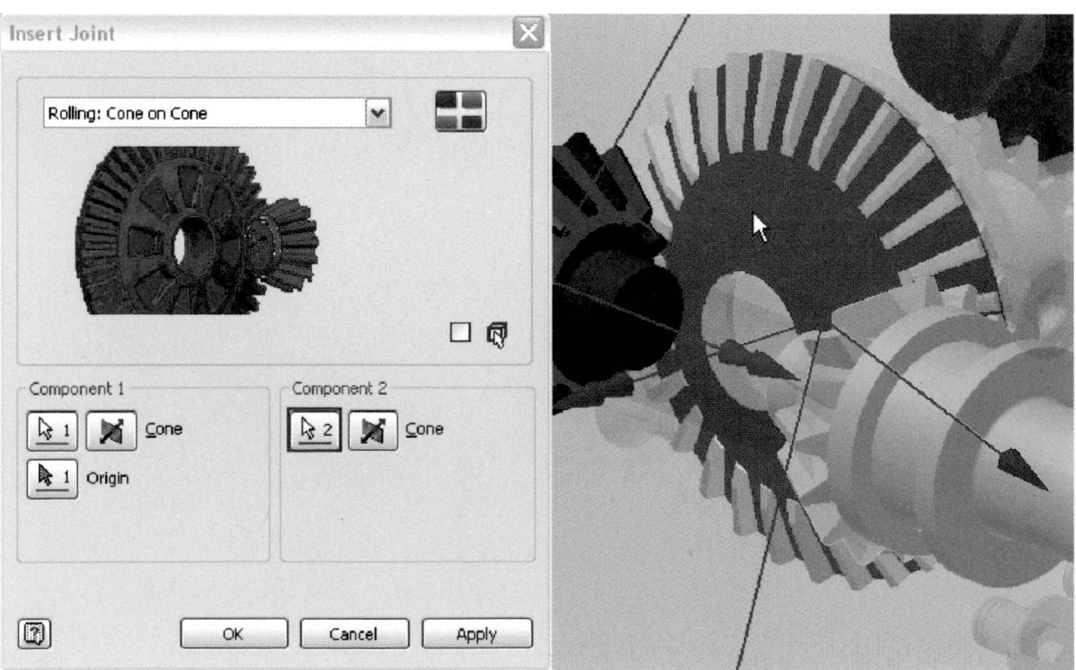

17. Click **Apply**, so we can continue creating another joint

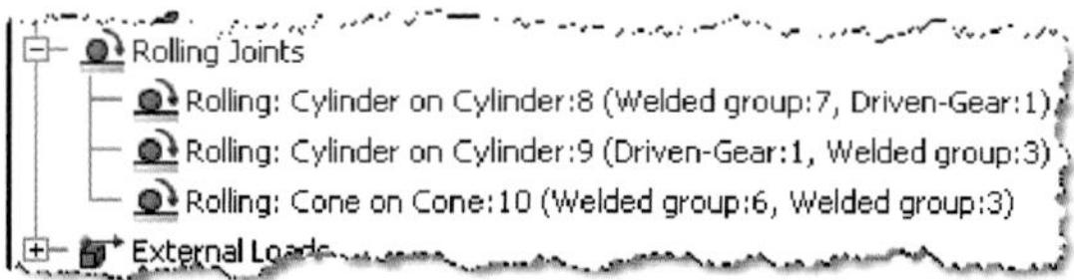

18. For Component 1, select the construction surfaces of Bevel Gear1

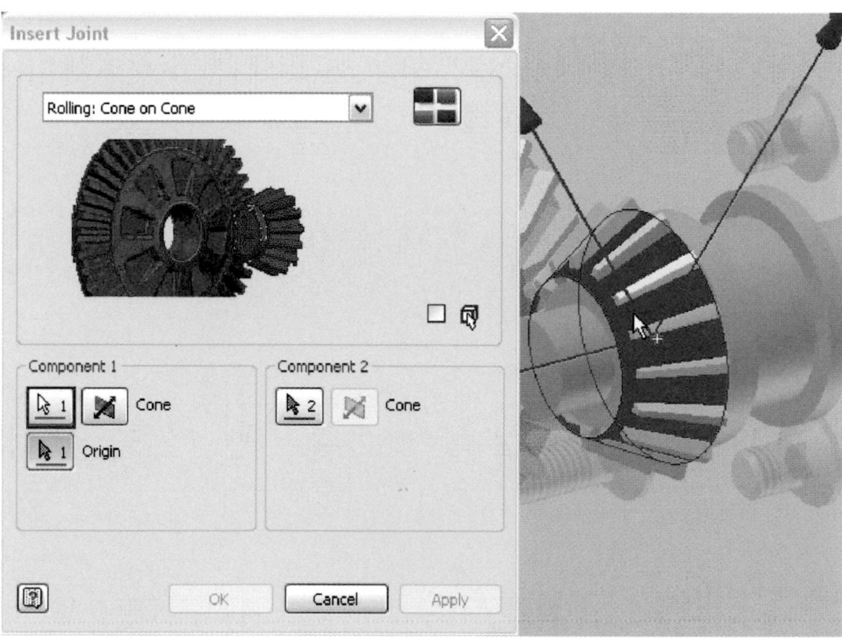

This bevel gear is another instance of the first bevel gear.

19. For Component 2, select the construction surfaces of Bevel Gear2

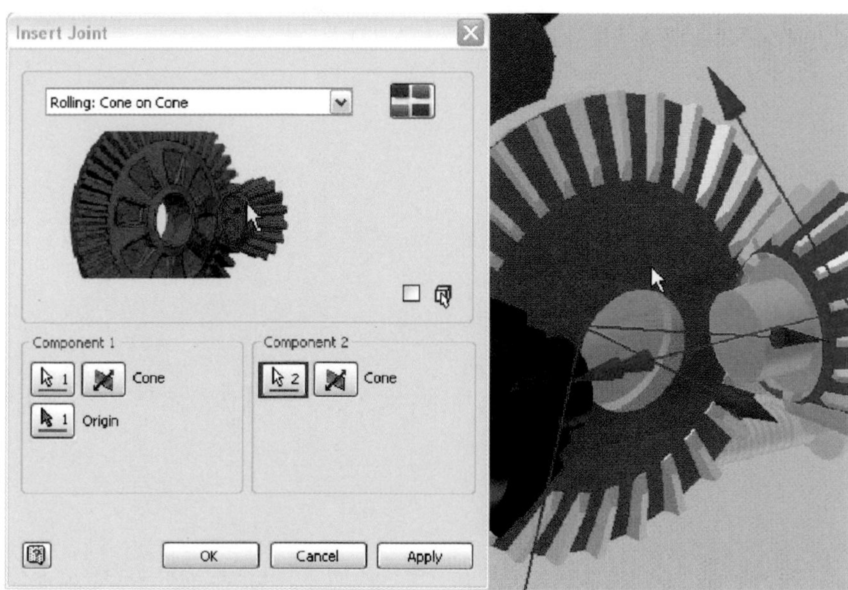

20. Click **OK**

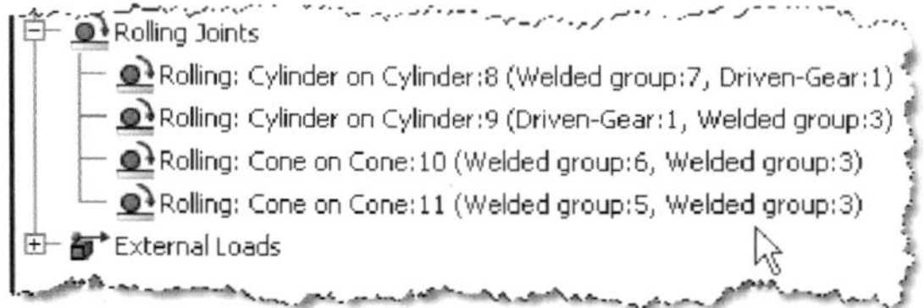

Now we have all the necessary joints created. The next step is to create environmental constraints.

21. Now make the gear box assembly housing visible

Environmental constraints

Here we need to predefine the output torque on the removal tool to simulate the pretightening of the bolts to 95 N/m.

22. Select **Torque** from the **Dynamic Simulation panel**

23. For location, select the edge of the torque tool

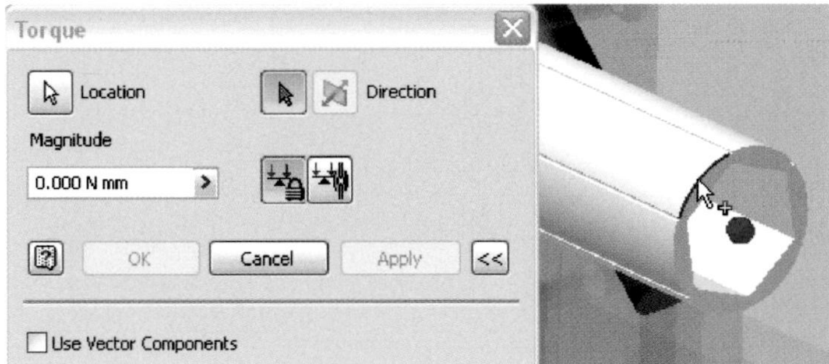

24. For direction, select the edge along the torque tool

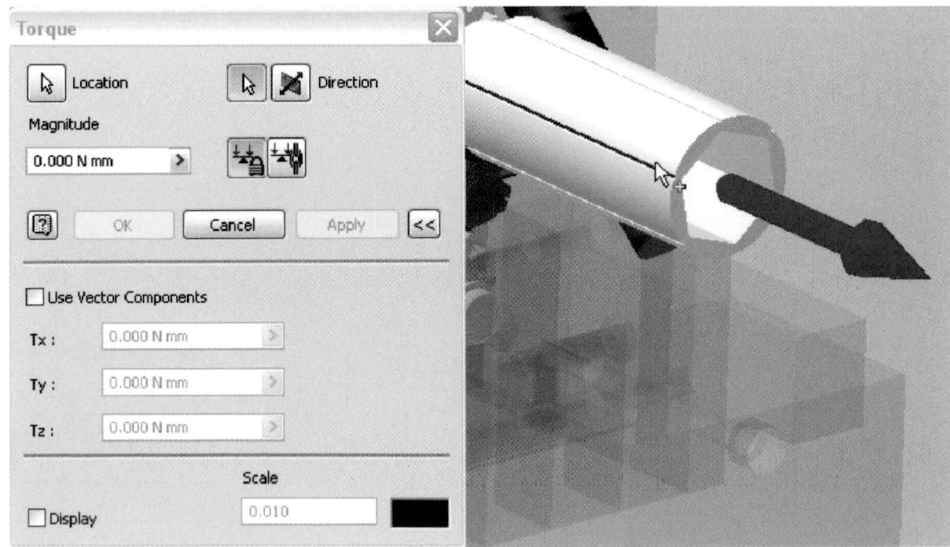

25. Specify 95 N/m in the **Magnitude** box > Click **OK**

Specifying N/m after the value will convert the value into N/mm (95 000 N/mm).

Finally, we need to specify the speed (360 deg/s) of the manipulator required to untighten the bolts using either handle.

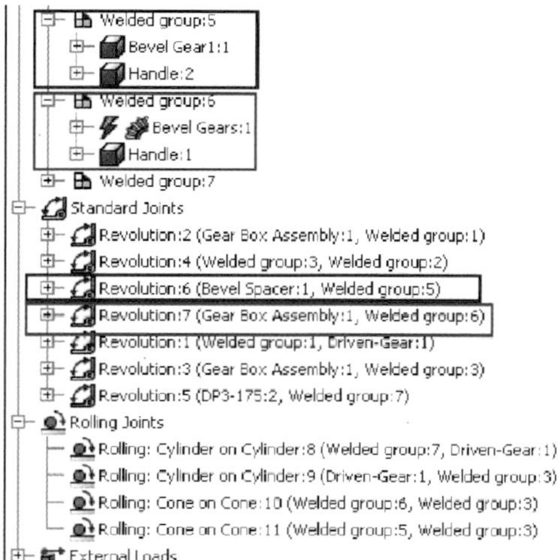

You can pick either the Revolution:6 or Revolution:7 joint as both are related to each handle. We will use the former joint.

26. Double click on the Revolution:6 joint

27. Select the **dof 1 (R)** tab in the Revolution:6 dialog box

28. Select **Imposed Motion** > Select **Enable imposed motion** > Apply a velocity of 360 deg/s

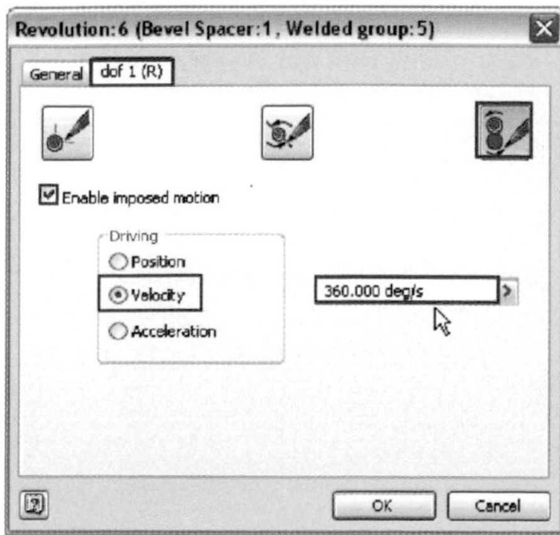

29. Click **OK** > Play the simulation

Analyze results

30. Select the Output Grapher > Select the Driving force value under the Revolution:6 joint node

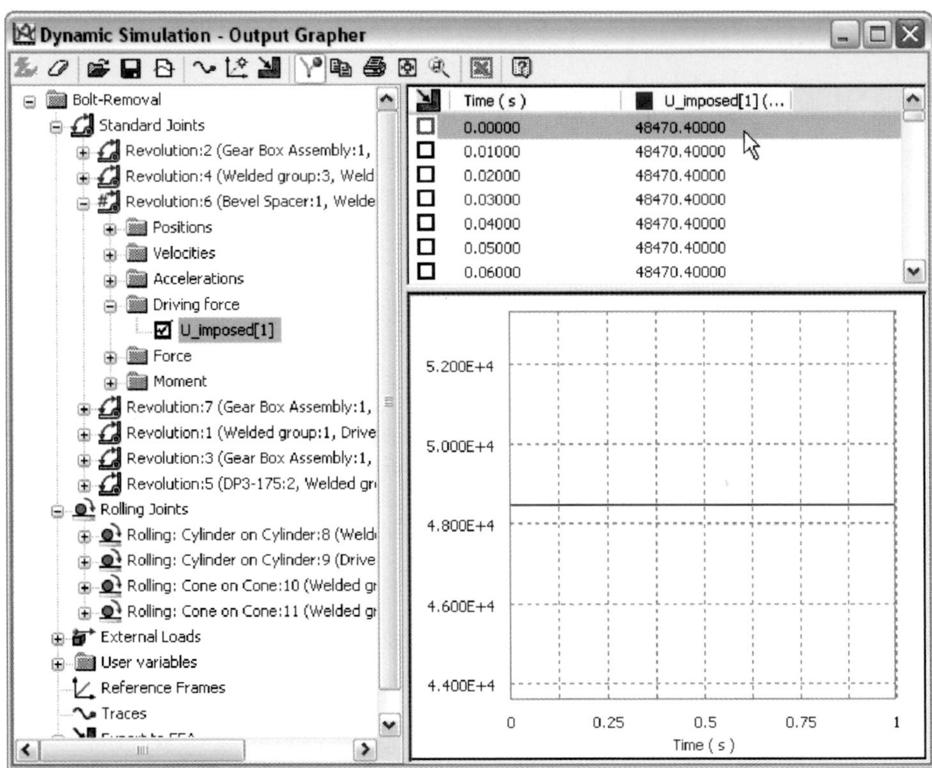

The graph and the time column indicate that a maximum torque of 48 470 N/mm is required to maintain a constant velocity of 360 deg/s for the handles.

$$\text{Torque ratio} = \frac{95000}{48470} = 1.96$$

Since the calculated torque ratio of 1.96 is higher than 1.8, the proposed design is acceptable. However, by altering some of the gear ratios/modules, you can do additional analysis to see whether the torque ratio can be further improved without altering the overall size and shape of the main body.

CHAPTER 3

DP2 – Size a Jack

New Mechanism Design for an Existing Bridge

(Design Problem courtesy of British Waterways)

JOINTS INTRODUCED/COVERED IN THIS DESIGN PROBLEM

	Joints	Joint creation process
1	Revolution	Manually converted
2	Cylindrical	Manually converted
3	Rolling: cylinder on plane	Manually created
4	2D Contact	Manually created

KEY FEATURES AND WORKFLOWS INTRODUCED IN THIS DESIGN PROBLEM

	Key features/workflows
1	Mechanism Status and degree of mobility
2	Locking degrees of freedom
3	Imposed motion – constant translational velocity

INTRODUCTION

The types and sizes of bridge on Britain's canals are almost as numerous as the bridges themselves, and many are manually operated. A typical example shown below is opened by pulling the chain attached to the bridge arm to raise the deck, which can weigh many tons. The bridge is finely balanced using counter weights in the arms and opened by a

Up and Running with Autodesk Inventor Simulation 2011. ISBN: 978-0-12-382102-7
Copyright © 2010 Elsevier Inc. All rights of reproduction, in any form reserved.

single operator pulling a chain attached to the end of the balance beam. Once the bridge is fully opened, the operator secures the bridge by anchoring the chain to the ground.

As part of an ongoing program of safety improvements, these bridges are being modified to allow them to be operated from either side of the canal and to provide restraint to a large moving mass.

The visual impact of any modifications must be minimized due to the heritage value of the canals, which is closely regulated.

The solution is required to lift and lower the bridge in a controlled manner while allowing operation from either side of the canal. The method chosen for opening/closing the bridge is a hydraulic cylinder (jack), which is to be placed underneath the bridge deck.

As part of the proposed design, including the jack, two user pedestals will be provided, one on each side of the canal, so that the bridge can be operated by a single boater. The cylinder will then be operated via a power pack located in the nearest user pedestal.

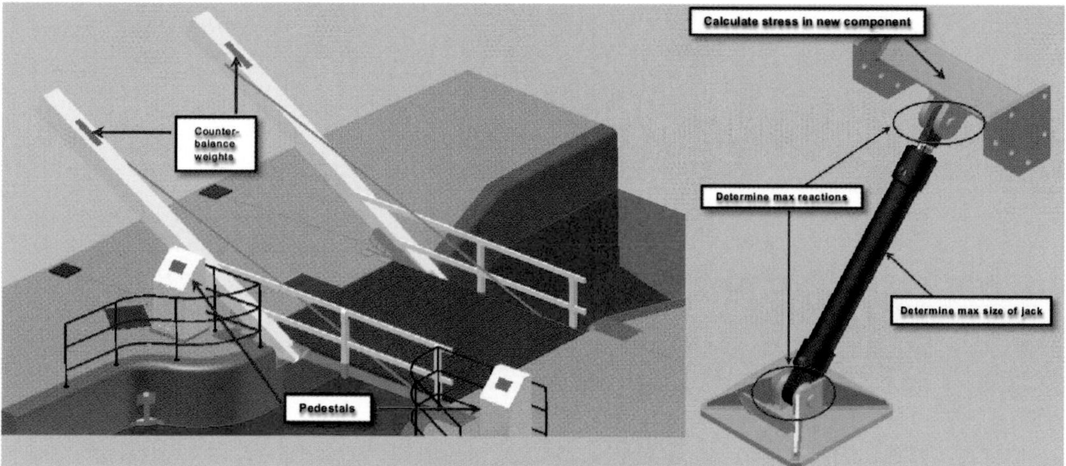

Hence, the main requirements for the proposed design are to determine:

- Maximum loads induced by the jack into the modified structure
- Maximum force required by the jack to fully open the bridge

In addition to the main requirements, the following criteria will be taken into account:

- The counterbalance weights will be removed as these would simulate the worst-case scenario. In other words, the maximum jack force calculated is greater than that if the weights were to remain present. This makes sense, as up to now, a single operator has been expected to open the bridge without too much effort.
- The pump required to operate the jack is to be housed within the pedestals, which means the pump size will be limited. With this limitation in mind, the velocity rate of cylinder operation can be calculated from various data.

The stroke of the cylinder is 482 mm and the bore is 63 mm. This will give:

Volume of 1503 ml = 1.5 liters approximately

The pump chosen will deliver 3 liters/min at a pressure up to 50 bar. This means:

Time to raise bridge = 30 seconds

Therefore, the cylinder operates at a rate of $482 \div 30 = 16$ mm/s.

WORKFLOW OF DESIGN PROBLEM 2

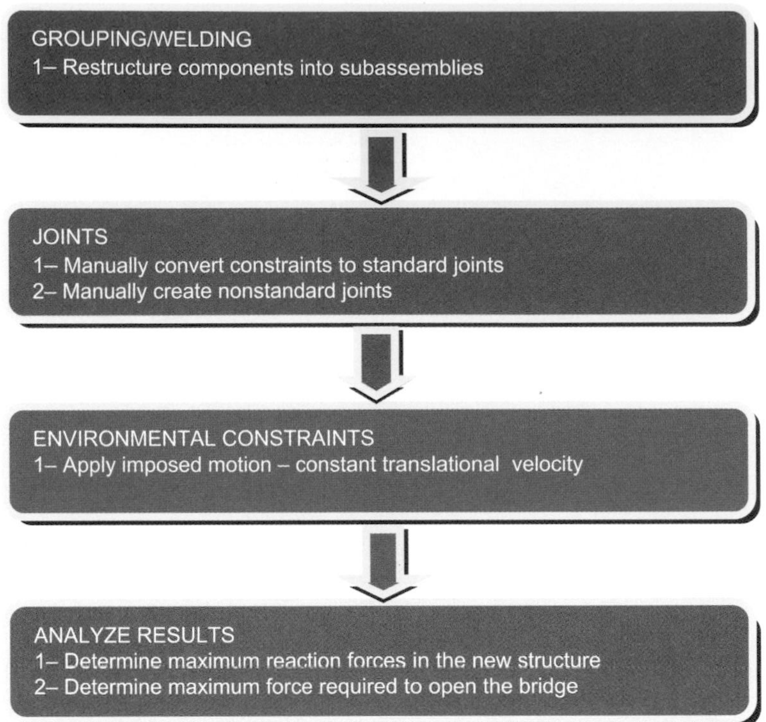

Grouping/welding

As the main purpose of this design problem is to determine the maximum force of the new jack, most of the components within the bridge assembly are to be grouped and welded together. This will reduce the number of joints created and needed and, hence, will make the process of simulating and analyzing easier and much more efficient.

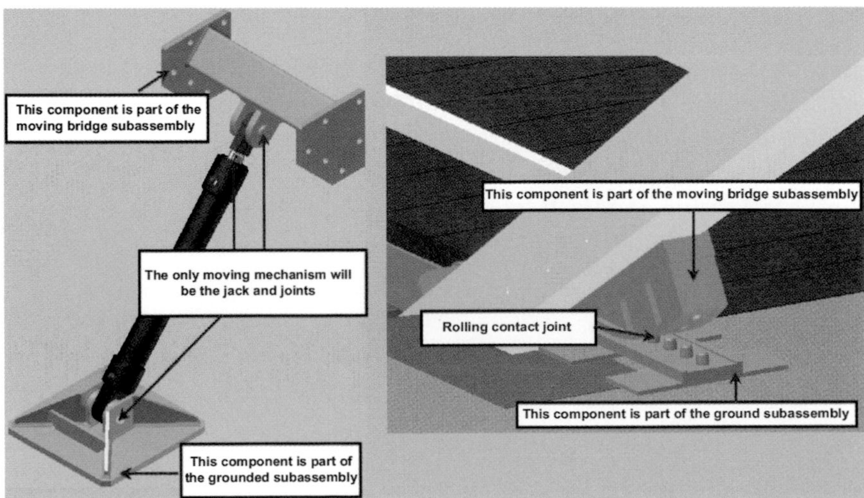

Before actually creating joints, it helps to visualize the problem and the possible joints required. This will enable you to decide which components can be grouped together, helping to reduce the number of joints needed.

1. Open *Bridge.iam*

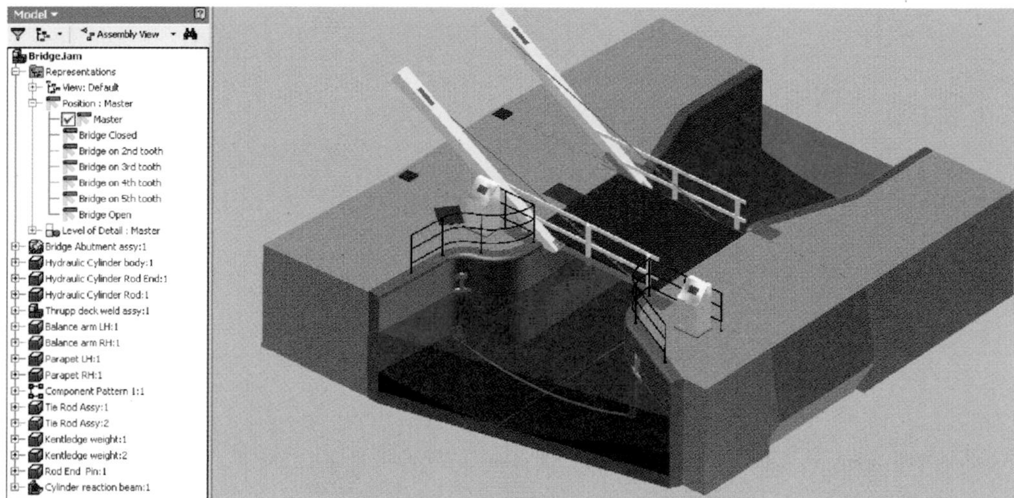

2. Select all the following components > Right click > **Component** > **Demote**

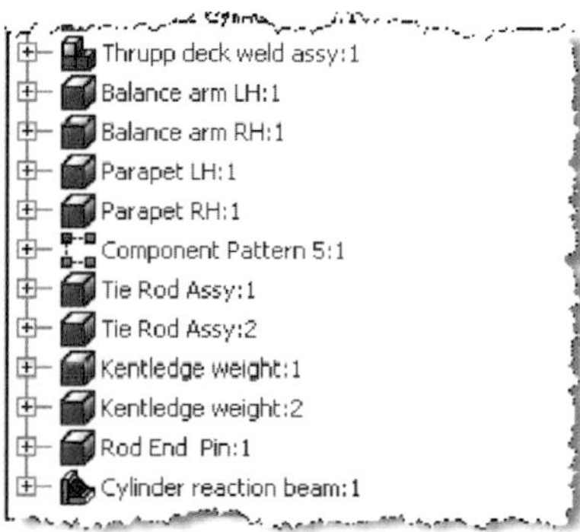

All these components have no relative motion between them. Additionally, we are not interested in determining the strength of these components as they are not to be altered, as mentioned earlier. For these reasons we will group them together.

The mass properties of the bridge need to be accurate as this will determine on the size of the jack required; thus we need to include all components of the assembly in the simulation.

3. Name the Subassembly 'Moving bridge' > Click **OK**

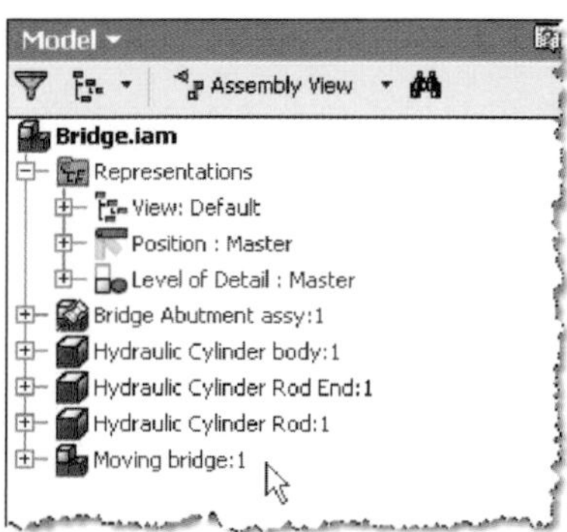

We need to suppress the counterbalance weights (*Kentledge weight.ipt*) as we will not be including these in the simulation study, as mentioned in the introduction.

4. Suppress the following two components within the new Moving-Bridge subassembly:

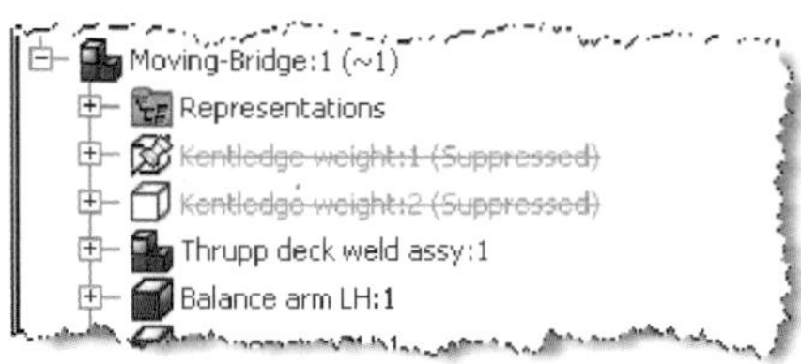

5. Select **Environments** tab > **Dynamic Simulation**

IMPORTANT—If components appear twice or are missing within the simulation browser, you will need to rebuild all, without finishing dynamic simulation, using the following approach:

Select the Manage tab, select **Rebuild All**, then reselect the **Dynamic Simulation** tab.

Components should now appear similar to the assembly browser. Continue to the next step.

6. Select the following two components > Right click > Select **Weld parts**

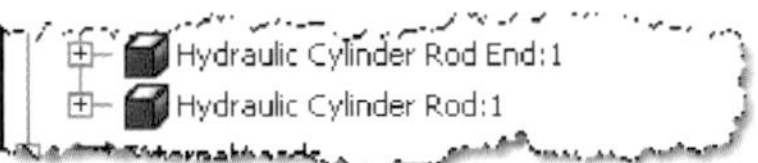

These jack components move together and welding them together eliminates the need to create an extra joint between them.

7. Select View tab and make all work features visible

8. Select **Half section view** and then select the work plane through the middle of the assembly to create the half section

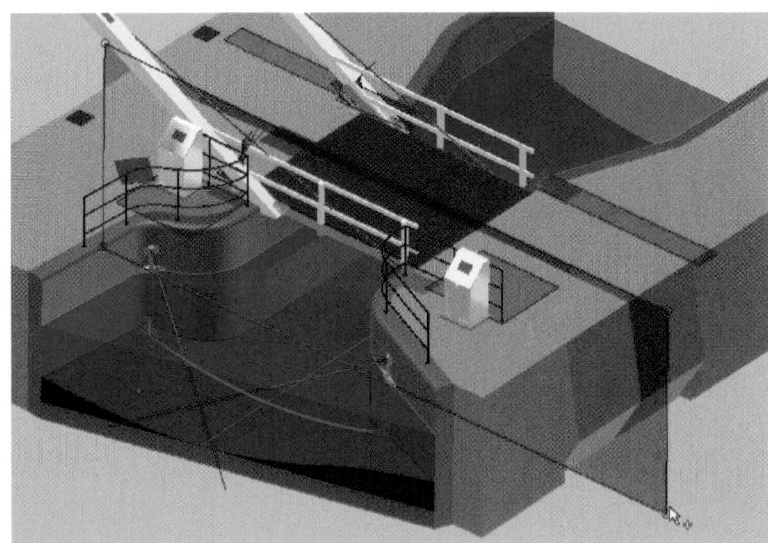

This will help to make the selection of the jack components easier.

9. Now switch off the visibility of all work features

10. Select the **Dynamic Simulation** tab to go back to the Simulation environment

Joints

Initially, we will create standard joints converting existing constraints and then all the remaining joints required will be created manually.

MANUALLY CONVERT CONSTRAINTS TO STANDARD JOINTS

11. Click on **Convert Constraints** in the **Dynamic Simulation** panel

12. In the **Convert Assembly Constraints** dialog box, select the components Hydraulic Cylinder body:1 and Bridge Abutment assy:1

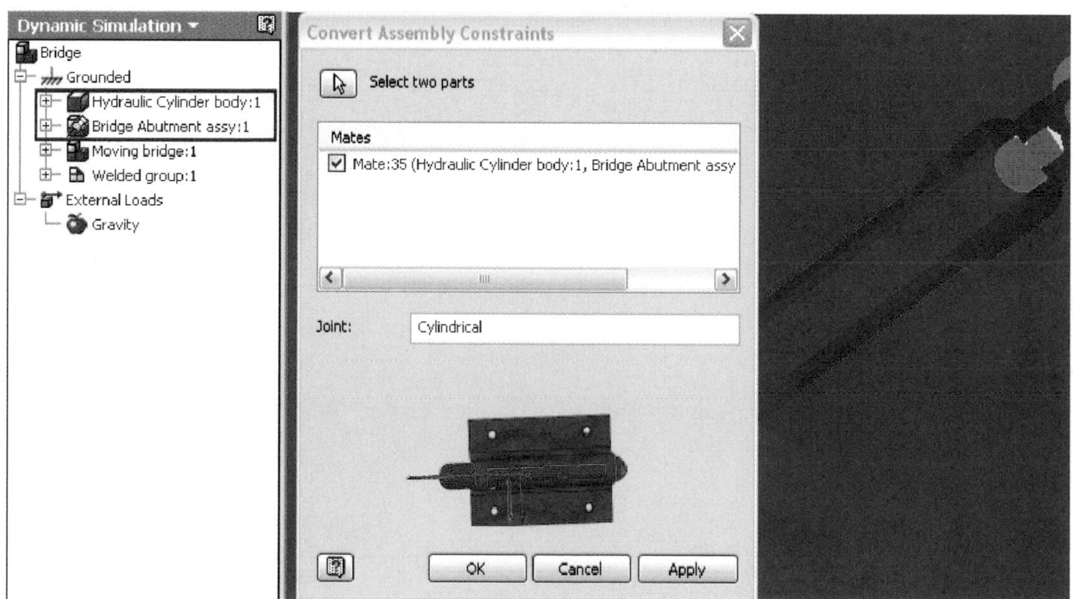

It may be easier to select components within the **Dynamic Simulation** browser rather than in the graphic window.

13. Click **Apply**

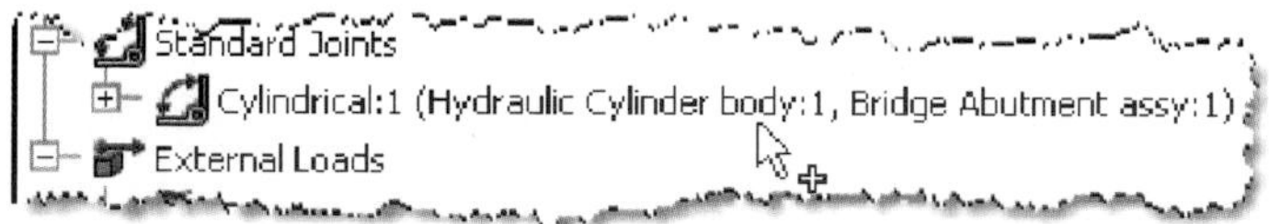

14. For the next two components, select Hydraulic Cylinder body:1 and Hydraulic Cylinder Rod:1

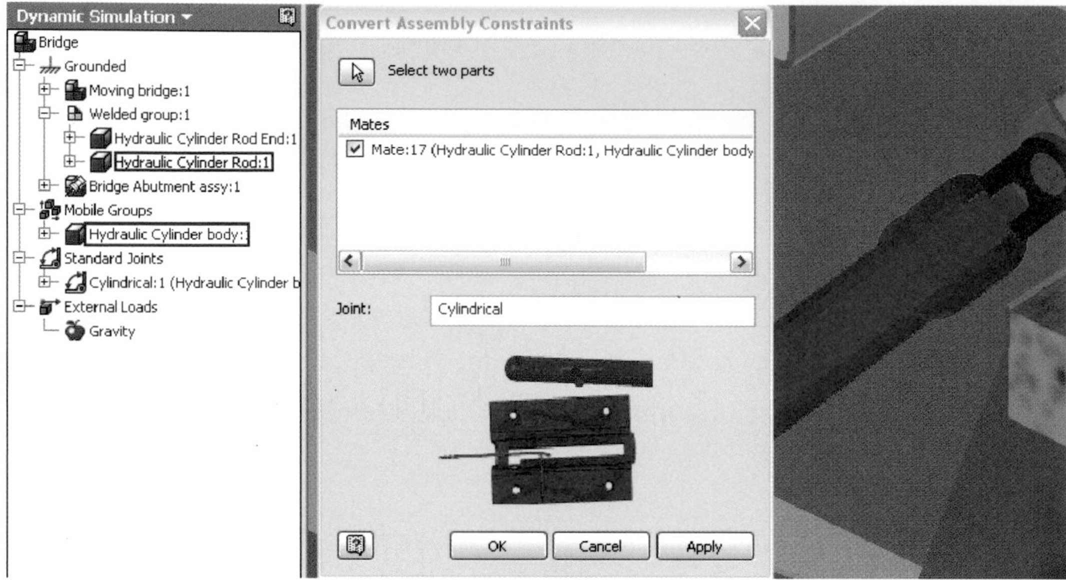

15. Click **Apply**

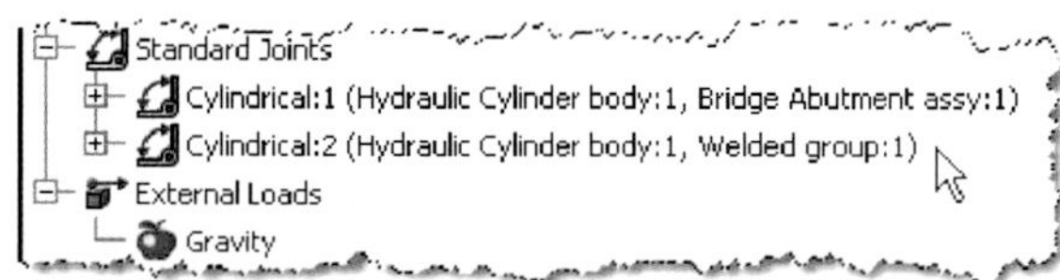

16. For the final two components, select Hydraulic Cylinder Rod End:1 and the Moving bridge:1 assembly

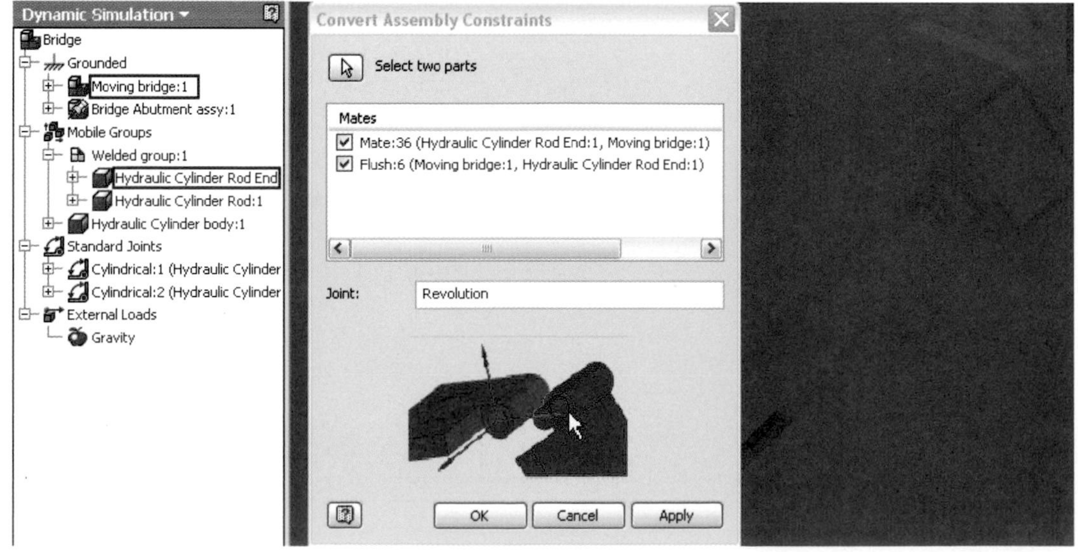

17. Click **OK**

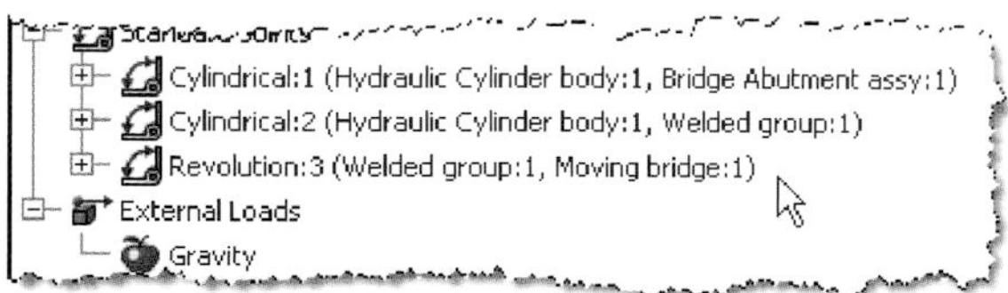

MANUALLY CREATE NONSTANDARD JOINTS

18. Select **Insert Joint** > **Rolling: Cylinder on Plane** from the pulldown menu

19. For Plane 1, select the face of component, as shown

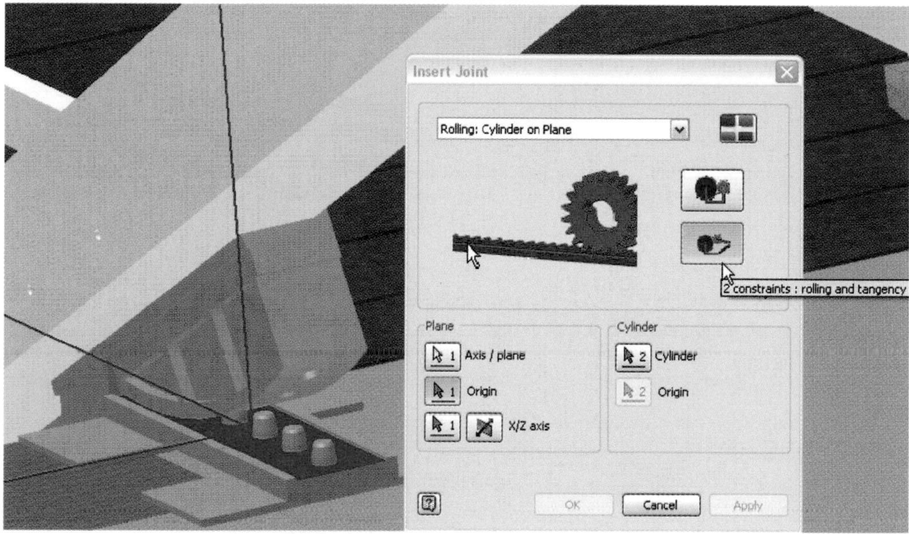

Make sure you select 2 Constraints: Rolling and Tangency option as this simulates the bridge movement more closely (as illustrated in the above image).

20. For Cylinder 2, select the face of the other component, as shown

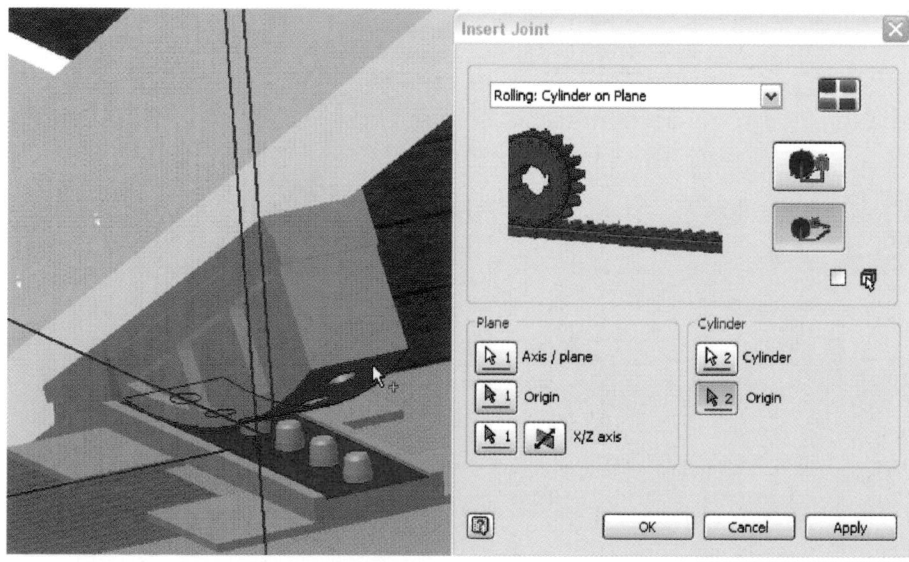

21. Click **OK**

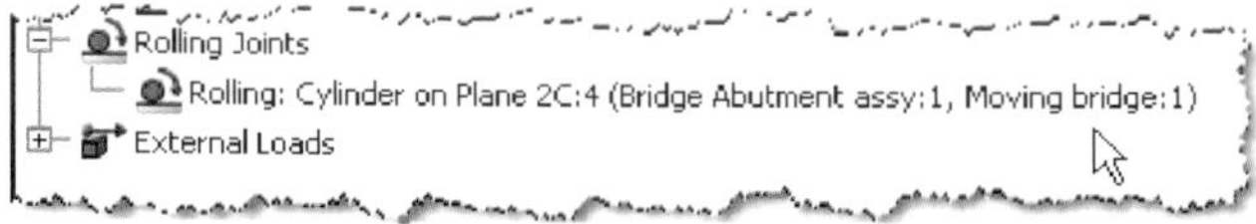

22. Select **Mechanism Status** to determine the degree of mobility of the assembly

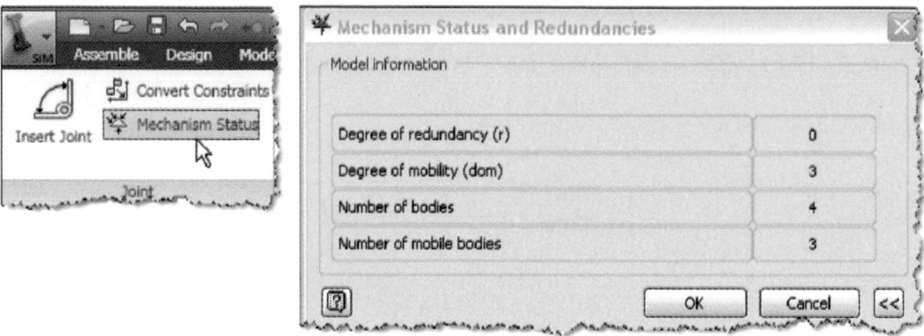

It is advisable to reduce the degree of mobility to 1. In reality, the bridge will have only one degree of mobility, which allows the bridge to open and close.

23. Click **OK**

24. Double click on the Cylindrical:1 joint to alter the properties > Select the **dof 2 (T)** tab > Lock the position > Click **OK**

Locking the translational degree of freedom makes this joint behave like a revolution joint

25. Double click on the Cylindrical:2 joint > Select the **dof 1 (R)** tab > Lock the position > Click **OK**

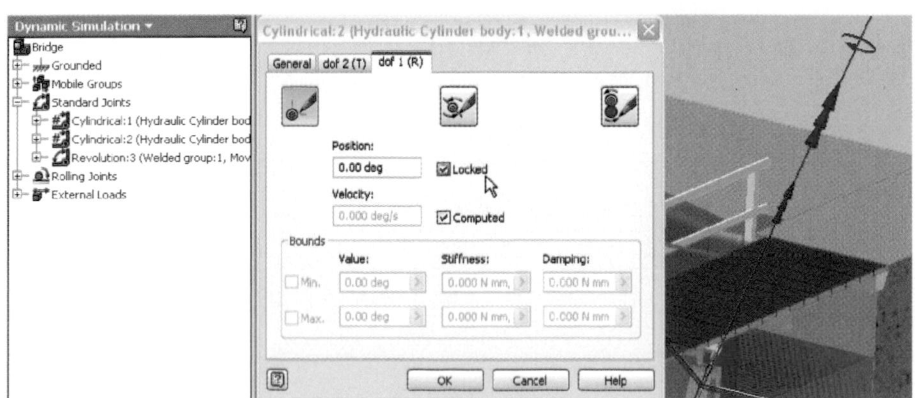

Locking the rotational degree of the cylindrical joint makes this joint behave like a prismatic joint.

Creating another constraint between the two components or subassemblies (face–face) would have resulted in a prismatic joint using the **Convert Constraints** tool.

26. Select **Mechanism Status**

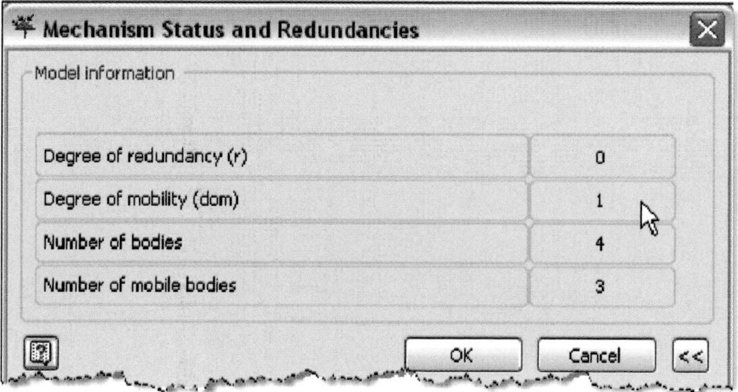

The degree of mobility is now 1, as desired.

27. Select the View tab > End the section view > Reselect the **Dynamic Simulation** tab

Next we will create a 2D contact joint between the stopper on the ground and the bridge arm.

28. Select **Insert Joint** > Select **2D Contact**

29. For Curve 1, select the face as shown

30. For Curve 2, select the face as shown

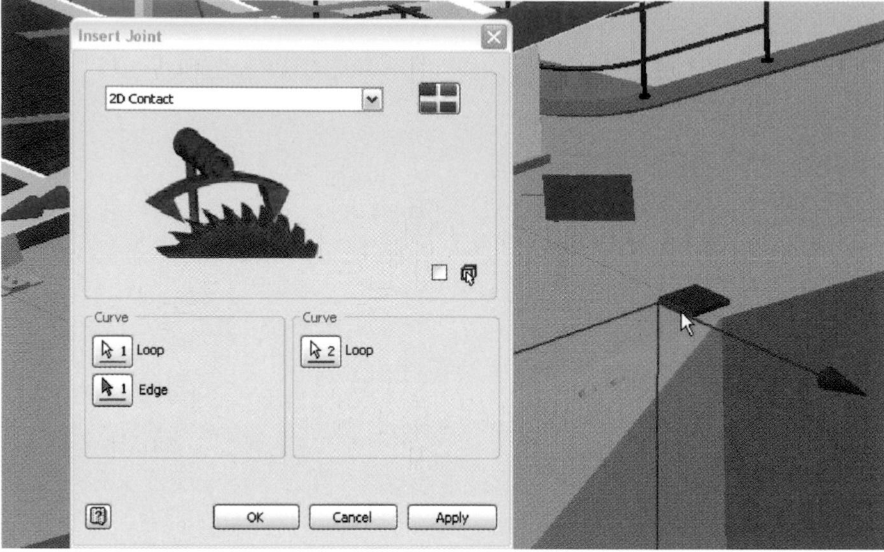

Select the inside of the bridge arms as the contact will not work otherwise. This is because the outside faces are tapered and therefore not in the same plane of the face of the stopper.

31. Click **OK**

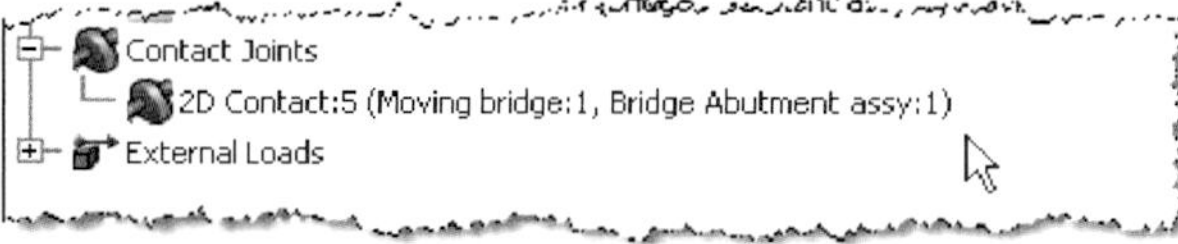

Environmental constraints

Now we are ready to simulate the environmental conditions and, as mentioned before, we need to specify that the jack operates at 16 mm/s and thus that it will take the bridge 30 seconds to fully open.

32. Double click the Cylinder:2 joint to edit the properties > Select **dof 2 (T)** > Check **Enable imposed motion** > Specify 16 mm/s for **Velocity** > Click **OK**

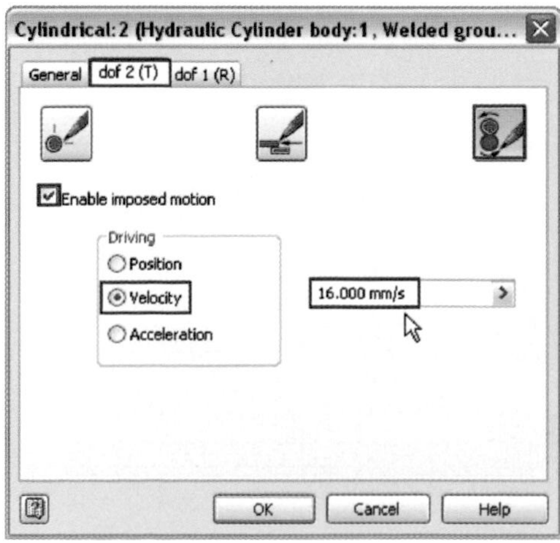

33. Set the simulation time to 30 seconds and the image value to 300 > Play the simulation

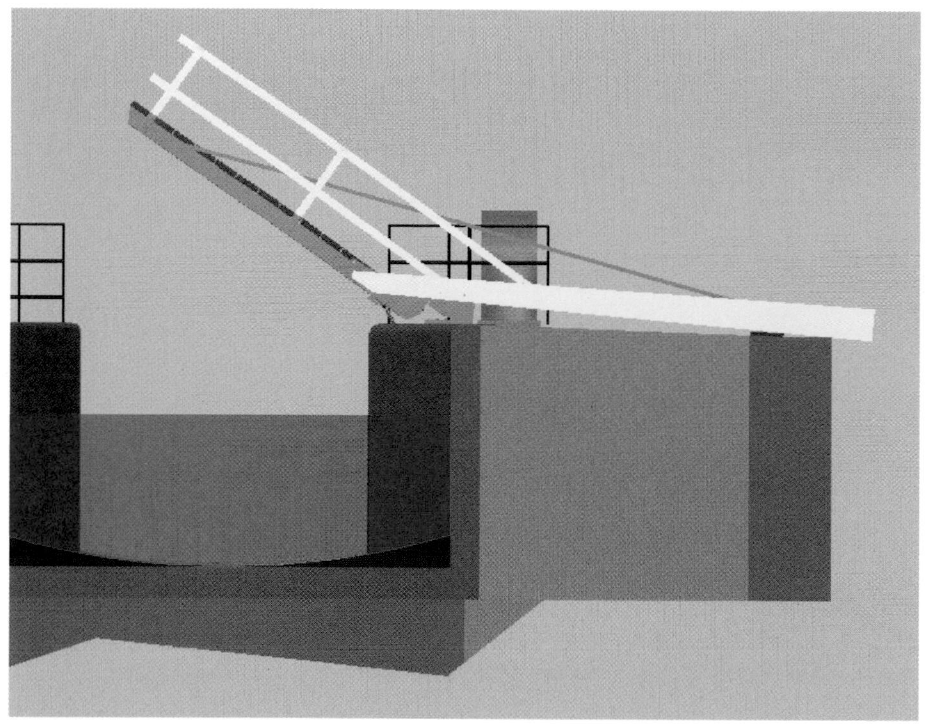

Accept the warning to continue to play the simulation. This warning appears as we have locked the degrees of freedom of some joints.

34. Click **OK** to the warning

The warning basically indicates that the bridge arm is now in contact with the stopper and therefore cannot open further.

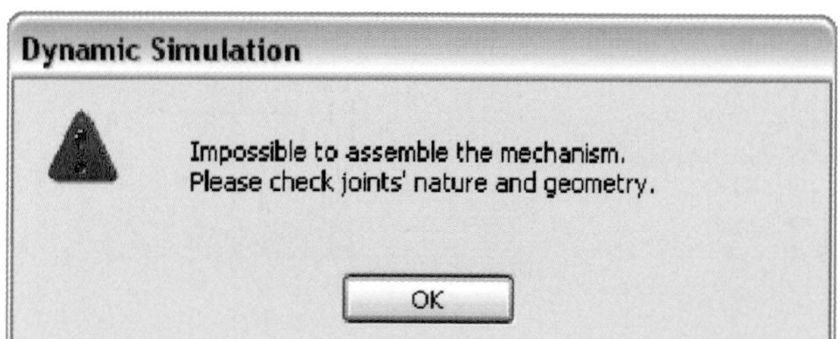

The bridge fully opens in just under 30 seconds, as predicted by calculations done by hand.

Analyze results

Using the Output Grapher, we can determine the maximum force required by the jack and the maximum force acting on the new structure.

35. Select the Output Grapher and select the Driving Force U_imposed[2] value under the Cylindrical:2 joint node

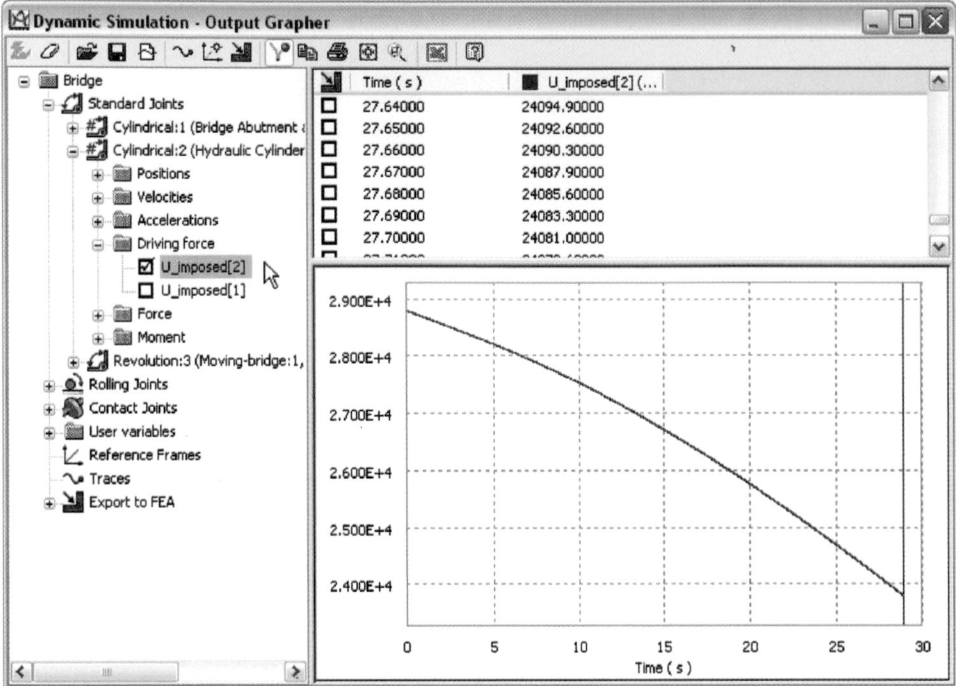

This driving force is the force required to maintain 16 mm/s velocity of the intended jack.

The Output Grapher also indicates that the jack requires the maximum force when it starts to lift the bridge; this requirement then steadily reduces. This is due to the shift in position of the center of gravity of the moving bridge as it opens.

36. Right click anywhere on the U_imposed[2] column > Select **Search Max**

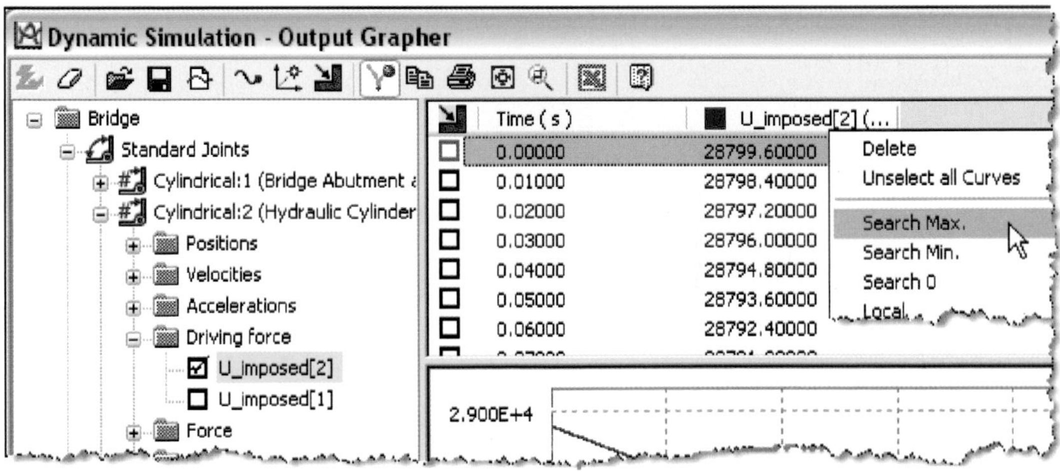

The maximum force required by the jack to open the bridge is 28 799 N, though the value may differ slightly.

Due to the position and leverage of the jack mechanism, the maximum force required by the jack is less than the overall weight of the bridge (approximately 38 000 N).

Hence, by simulating the bridge, we have determined the exact size of the force, which is also less than the total weight of the bridge, indicating the need for a smaller power pack to drive the jack.

37. Deselect the U_imposed[2] value > Select the force under the Revolution:3 node

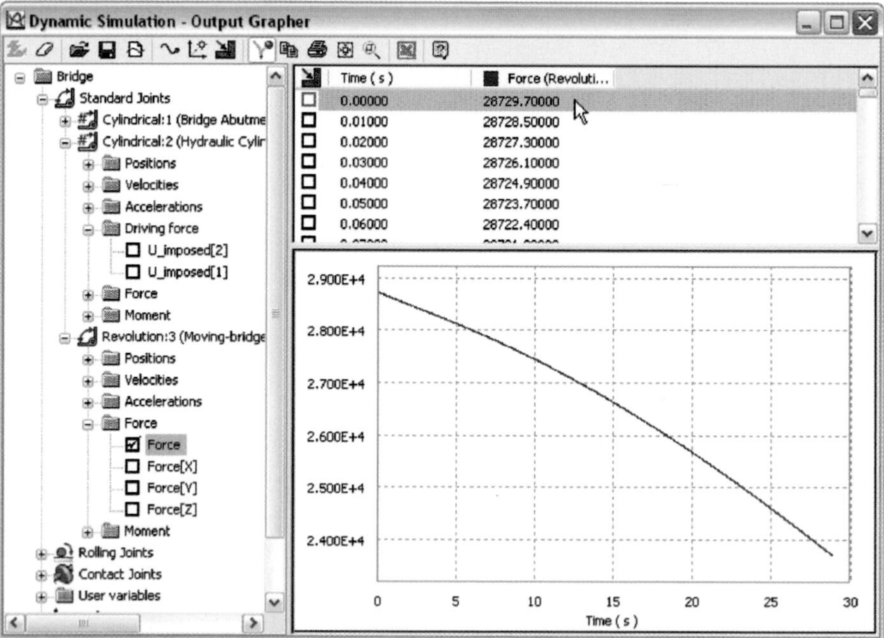

The maximum force is at the start, as indicated.

The maximum reaction force is 28 729 N, though the value may differ slightly. This is the force acting on the new structure, as illustrated below.

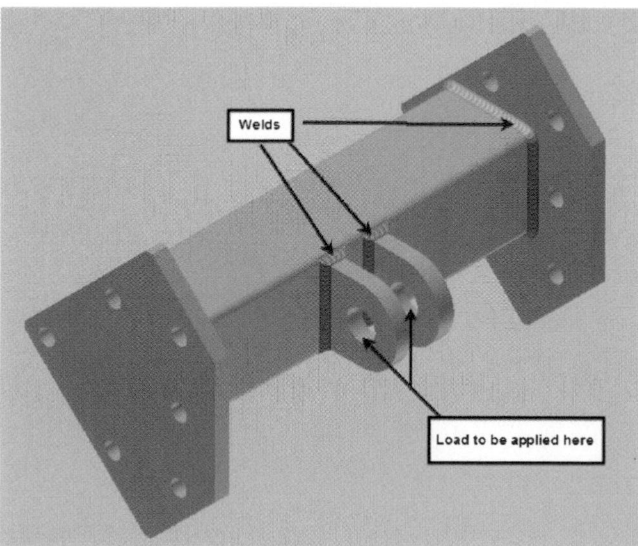

We will perform an assembly analysis on the above modified structure, using the calculated load, in Chapter 13.

CHAPTER 4

DP3 – Size Multiple Actuating Jacks

New Mechanism Design for a Transporter Ramp

(Design Problem courtesy of In-CAD Services Ltd)

JOINTS INTRODUCED/COVERED IN THIS DESIGN PROBLEM

	Joints	Joint creation process
1	Spherical	Automatically created
2	Point-line	Automatically created
3	Cylindrical	Automatically created
4	Prismatic	Automatically created

KEY FEATURES AND WORKFLOWS INTRODUCED IN THIS DESIGN PROBLEM

	Key features/workflows
1	Repairing redundant joints by altering assembly constraints
2	Mechanism Status and degree of mobility
3	Locking degrees of freedom
4	Initializing position of joints
5	Imposed motion – Input Grapher

INTRODUCTION

In-CAD Services, a specialist consultancy provider, was tasked to design a twin transporter ramp to take a maximum load of 5 tons equally spread over both ramps. This transporter ramp is to be secured to the chassis of the associated trucks and lorries, via four mounting lugs.

Up and Running with Autodesk Inventor Simulation 2011. ISBN: 978-0-12-382102-7
Copyright © 2010 Elsevier Inc. All rights of reproduction, in any form reserved.

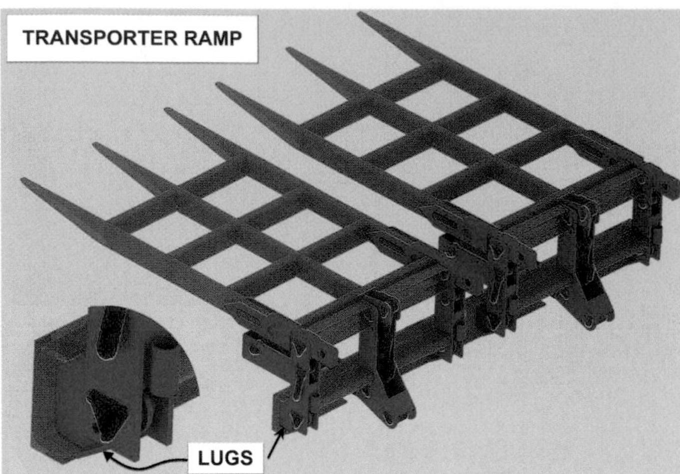

In addition to designing the transporter ramp, the necessary size of both jacks to operate the ramp's dead load (ramp weight) needs to be calculated. Further, there is a need to investigate the maximum reaction loads of the mounting lugs to help us understand whether they can withstand the design load. As both ramps are identical, one ramp will be used for the purposes of simulation and analysis.

Hence, the main requirements for this design problem are to determine:

- Minimum force of the jacks necessary to unfold the ramp.
- Maximum reaction forces on the mounting lugs.

In addition to the main requirements, the following criteria will be taken into account:

- Jack 1 will take 15 seconds to unfold the first stage of the ramp.
- A minimum of 10 seconds is required to place locking pins.
- Jack 2 will take 15 seconds to unfold the ramp so that it is ready to be loaded.

The calculated force on the mounting lug will be much higher, as the assembly is acting as a cantilever, with the load being transferred to the two lugs. In reality, the ramp would be placed on the road at the other end, as shown below.

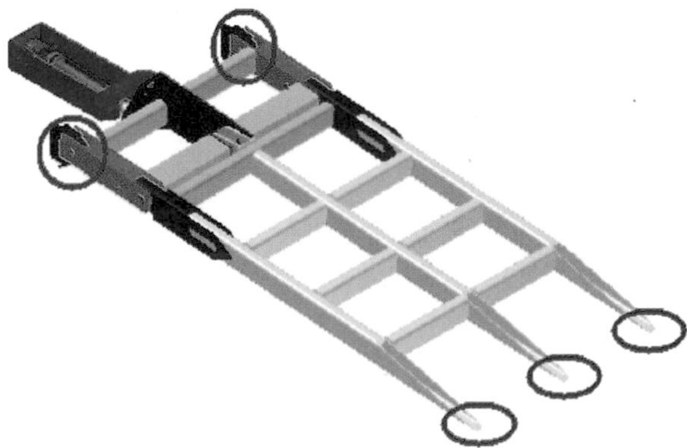

To obtain the true reactions within the lugs, we can either include geometry representing the road, so that we can simulate contact between the road and the ramp, or we can split the loads proportionally.

We will use the second example and this is explained below.

The above diagram shows five points at which there will be reactions. We already know that the reactions between the lugs are identical. We will assume that the sum of the reactions at the road end will be the same as the sum of the reactions in the lugs. This assumption is based on the fact that the load will be uniformly spread on the ramp.

Sum of reactions on lugs = Sum of reactions on other side (road)

Hence, the reaction load in the lugs will be halved for the purposes of determining the structural integrity of the mounting lugs.

WORKFLOW OF DESIGN PROBLEM 3

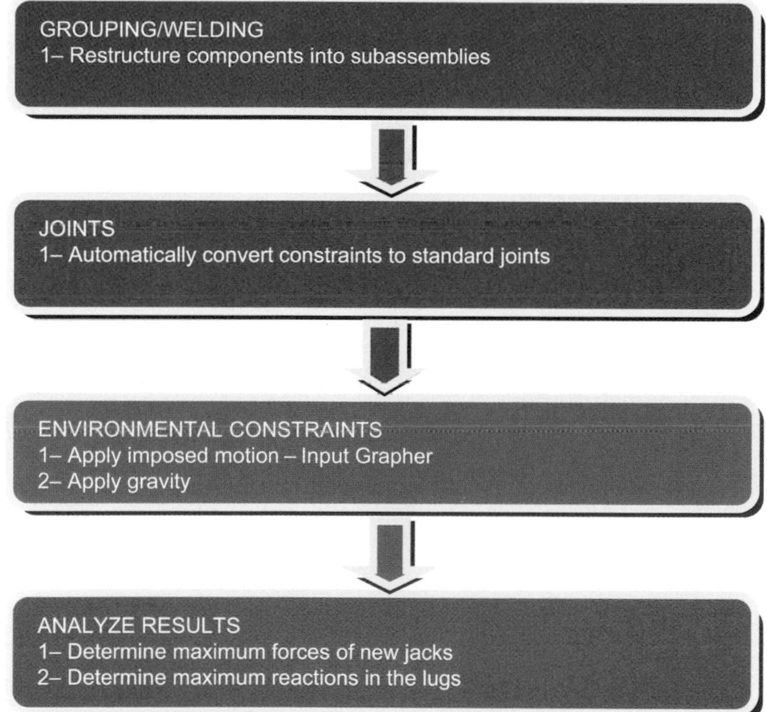

Grouping/welding

In this design problem, the main aim is to determine the size of the jack required so that the mechanism functions satisfactorily within the field. In addition to the jacks, we need to determine whether the brackets are strong enough when the ramp is fully loaded. As we have already analyzed the main pins, fasteners, and bolts, we are not interested in determining the reactions and strength of these individual parts. However, we will include all these items, as the jack also needs to overcome their total mass. To simplify the analysis, we will combine the bulk of the fasteners and pins together into a subassembly, as individually they have no relative motion between them. Further, this will help to significantly reduce the number of joints required.

Restructure components into subassemblies

1. Open *Ramp.iam*

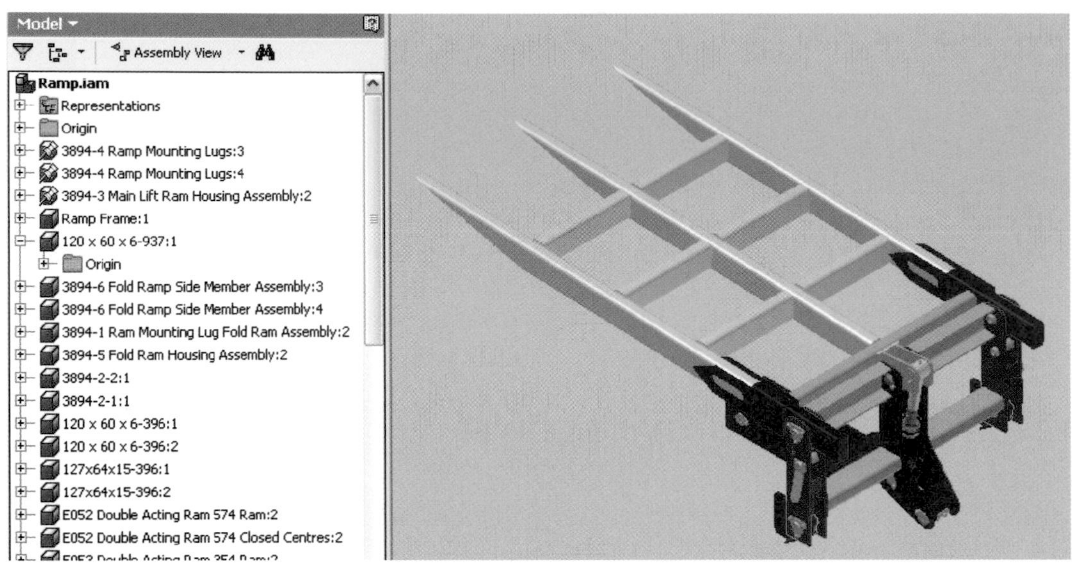

2. Select all the following components > Right click > **Component** > **Demote**

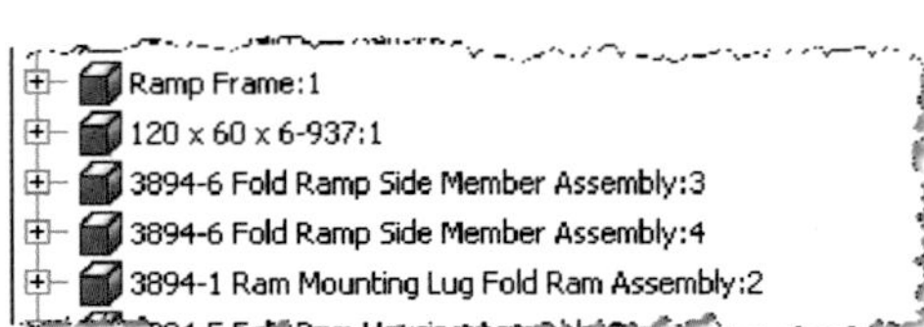

These make up the first stage of the ramp unfolding mechanism and are normally welded together, and hence can be treated as one component for simulation purposes.

3. Name the subassembly 'Frame-Main' > Click **OK** > Accept the warning

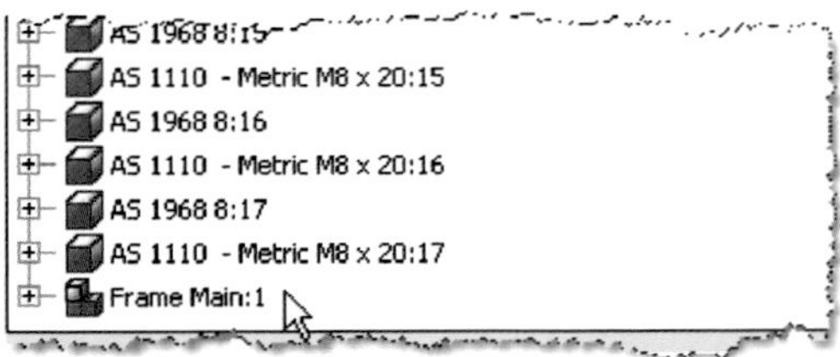

The warning message refers to the fact that assembly constraints may be lost between components that are not part of the new subassembly.

4. Select all of the following highlighted components > Right click > **Component** > **Demote**

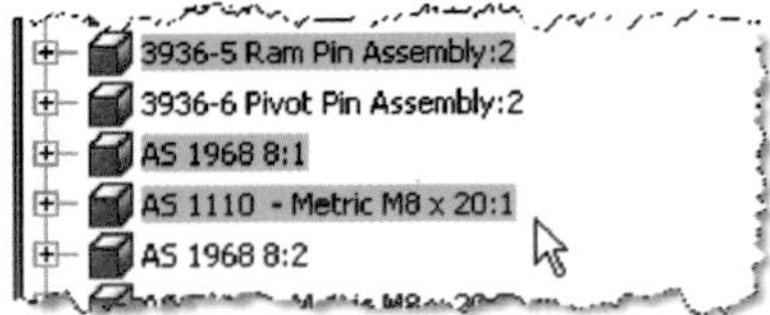

These are the fasteners securing the second jack and thus are not associated with the main ramp.

5. Name the subassembly 'Fasteners-2nd Jack' > Click **OK** > Accept the warning

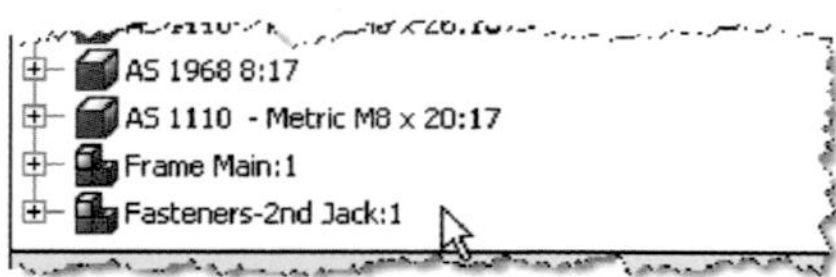

6. Select all of the following components > Right click > Select **Component** > Select **Demote**

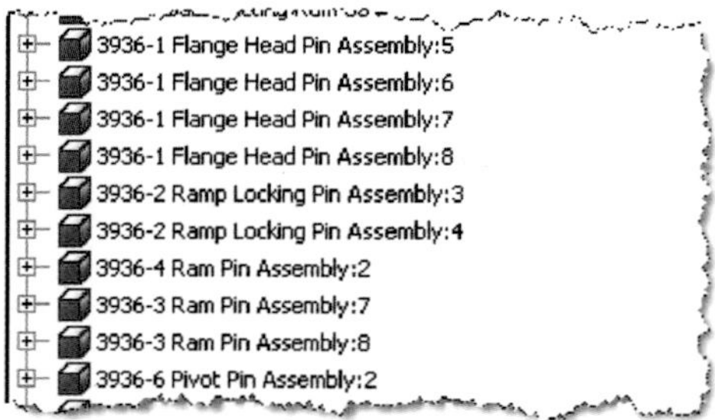

These components are comprised of the main locking pins for the frame-folding assembly and the first jack.

7. Name the subassembly 'Frame-Folding-Pins' > Click **OK**

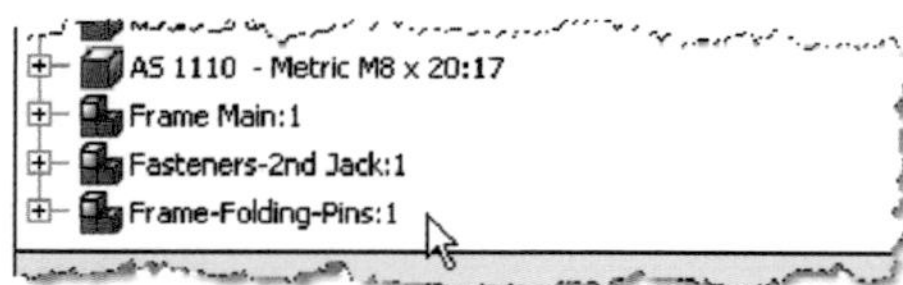

8. Select the following components > Right click > **Component** > **Demote**

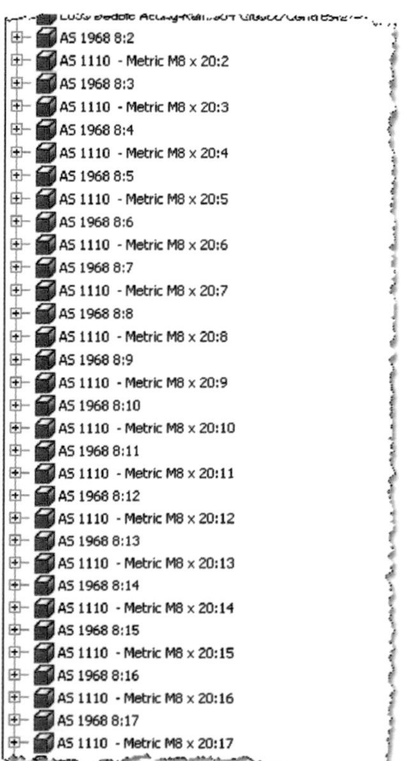

These components comprise all the fasteners for the frame-folding assembly and first jack and their mass will have an impact on the size of jack required.

9. Name the subassembly 'Frame-Folding-Fasteners' > Click **OK** > Accept the warning

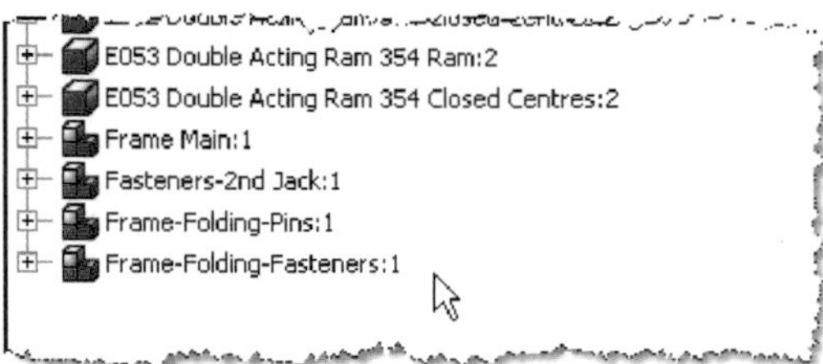

10. Finally, select the following components > Right click > Select **Component** > **Demote**

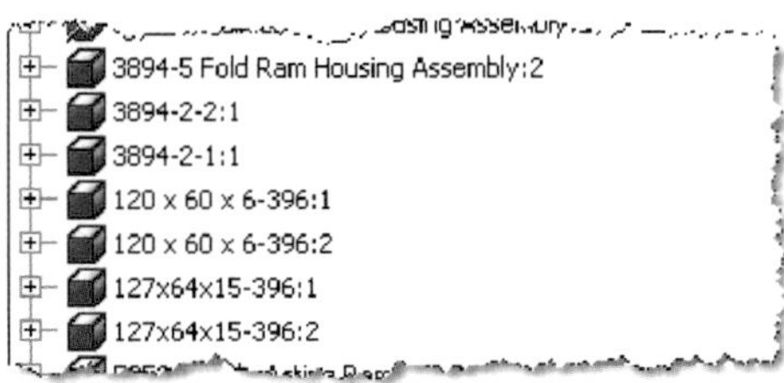

11. Name the subassembly 'Frame-Folding' > Click **OK** > Accept the warning

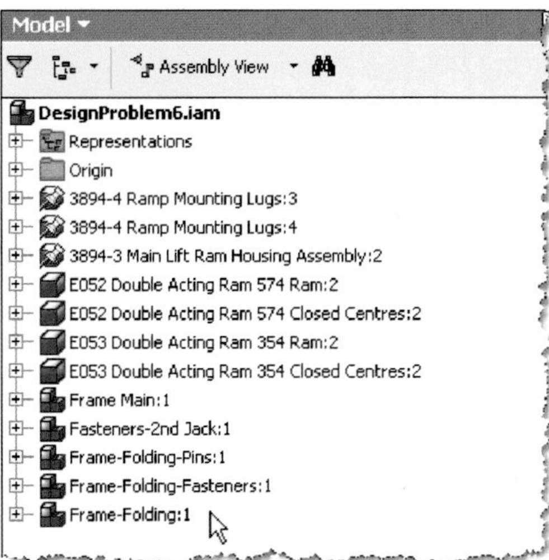

As a result of creating these new subassemblies, we have lost the constraints between adjacent components and subassemblies. At this stage, we can either create constraints or create joints manually within the Dynamic Simulation environment. The decision is entirely based on personal preferences.

Joints

Here, we will start by creating some constraints within the Assembly environment and then continue the process within the Simulation environment. This way we will immediately be able to see which joints are created from applying the constraints and can modify them if necessary.

To make selecting some features easier, it may be helpful to activate **Pick Part** first, within the **Place Constraint** dialog box, as this will ensure you are selecting the correct feature associated with the component or subassembly.

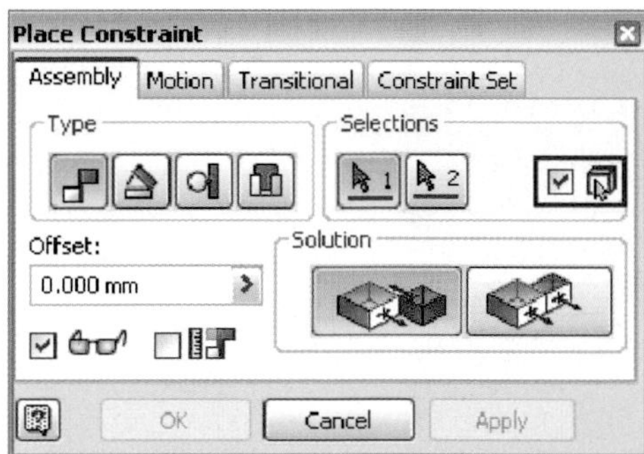

12. Select **View** tab > **Object Visibility** > Select **All Workfeatures**

This will make some of the key work features visible that will be used to create constraints. These work features have been created beforehand (and made invisible) with the view that they will be needed to create joints.

13. Select the following components > Right click > **Visibility**

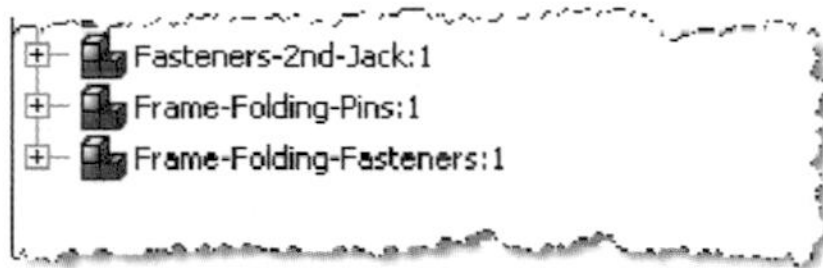

This will make the subassemblies invisible and help to make selecting work features easily when creating constraints.

14. Press **C** on the keyboard to activate the **Place Constraint** dialog box

15. For the first selection, select a point of the subassembly Frame-Main:1, as shown below

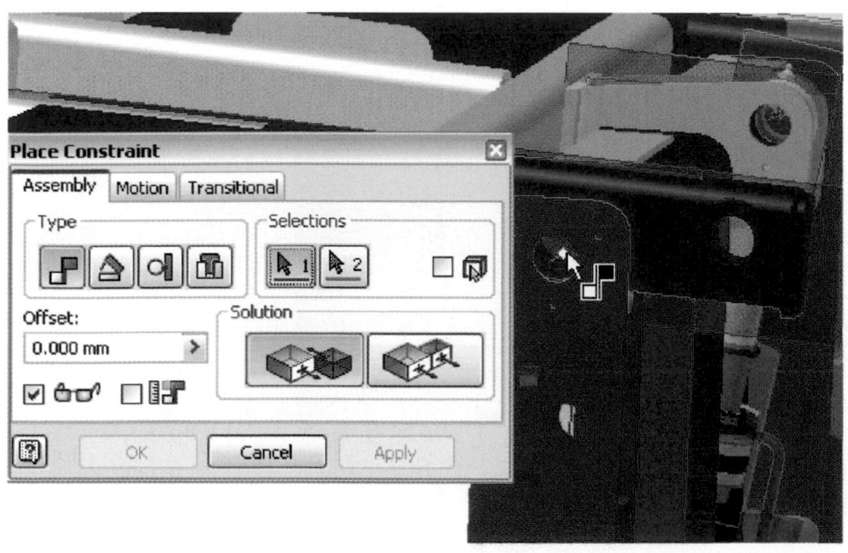

16. For the second selection, select the point of the subassembly Frame-Folding:1 in the same location as shown below

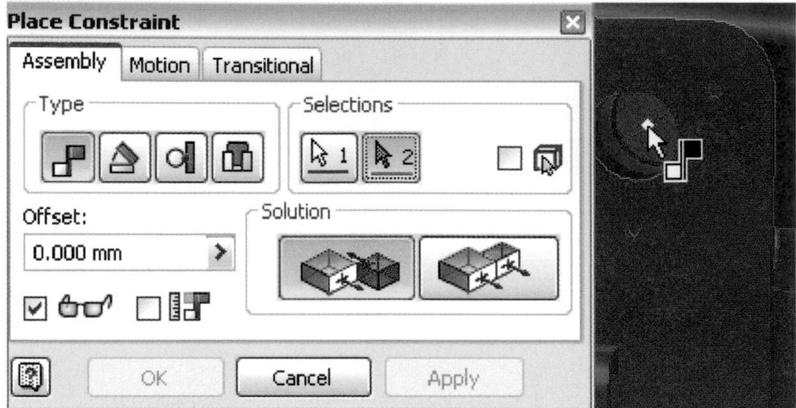

The workpoints from both components are in the same location

17. Click **Apply** so we can continue applying a similar constraint on the other side

18. For the first selection, select the point of the subassembly Frame-Main:1 as shown below

19. For the second selection, select the axis of the subassembly Frame-Folding:1 in the same location as shown below

20. Click **OK.** The following constraints will appear under the Frame-Main:1 subassembly (and also under Frame-Folding:1)

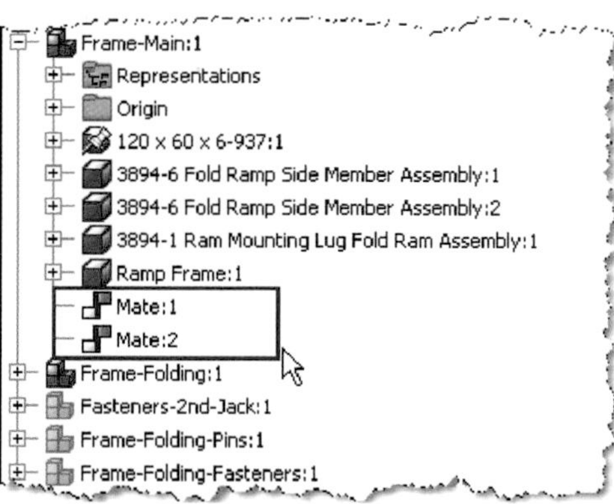

Additional constraints information, including for Mate:1 and Mate:2, can be activated by selecting **Tools > Application Options > Assembly > Display command names after constraint name.** This will make it easier to identify constraints.

We have not used axis-axis + plane-plane constraints (equivalent to a revolution joint) because Frame-Main:1 and Frame-Folding:1 have three pivot points (three joint connections), as shown below.

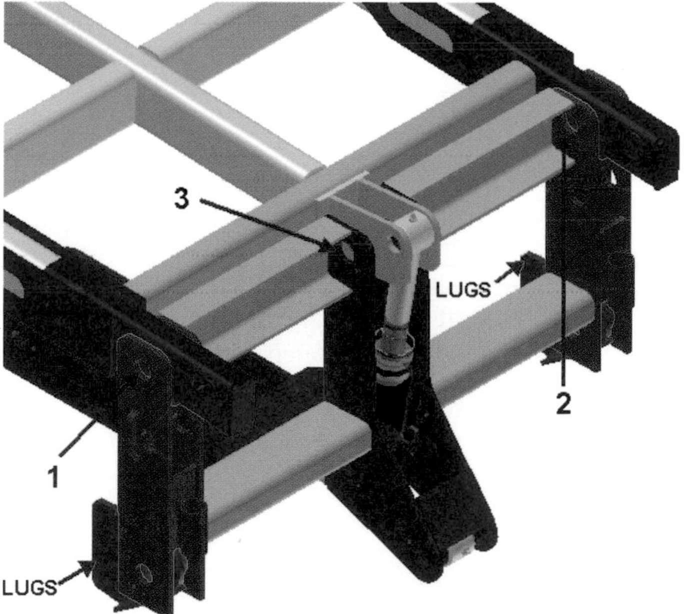

In this design problem, we have ignored the third joint and only applied constraints at joints 1 and 2 (refer to the redundancy section for further explanation). This will mean that the total reaction between these two subassemblies will be split between these two joints. The reactions will therefore be higher than normal, and we will use this as a built-in factor of safety when we analyze the lugs for strength, as there are two of them on either side.

One point-point constraint and one point-axis constraint have been applied. This is because, when transferred to Dynamic Simulation, the following two joints will be created with no redundancies:

Point-point constraint – spherical joint – three degrees of freedom restricted

Point-line constraint – point-line joint – two degrees of freedom restricted

If an insert constraint was created on either side, redundancies in the model would have been created within the Simulation environment.

The rest of the constraints will be created within the Dynamic Simulation environment.

21. Select **Environments** tab > **Dynamic Simulation**

> **IMPORTANT** – If components appear twice or are missing within the simulation browser, you will need to rebuild all, without finishing Dynamic Simulation, using the following approach:
>
> Select the Manage tab, select **Rebuild All,** then reselect the **Dynamic Simulation** tab.
>
> Components should now appear similar to the assembly browser. Continue to the next step.

AUTOMATICALLY CONVERT CONSTRAINTS TO STANDARD JOINTS

22. Select **Simulation Settings**
23. Select **Automatically Convert Constraints to Standard Joints** within the **Dynamic Simulation Settings** dialog box
24. Click **OK**

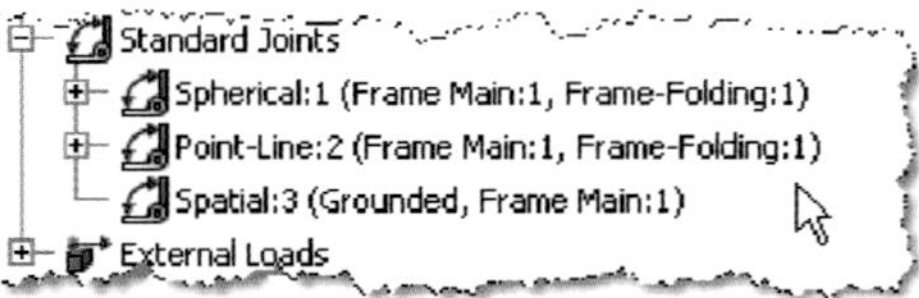

An extra spatial joint is also created, which will in turn create an extra six degrees of freedom. As we start creating the rest of the joints, this spatial joint will change or become obsolete.

In the following steps, we will create joints automatically by using assembly constraints, in the Dynamic Simulation environment, between the lugs and the Frame-Folding:1 subassembly. Also note that the frame has two connection joints, and hence we will not be able to use a revolution joint as this creates redundancies if applied to both connection joints. Therefore, we will use point-line and spherical joints.

25. Press **C** on the keyboard to activate the **Place Constraint** dialog box
26. For the first selection, select a point of the Frame-Folding:1 subassembly, as shown below

27. For the second selection, select the point of component 3894-4 Ramp Mounting Lugs:3 at the location shown below

28. Click **Apply**

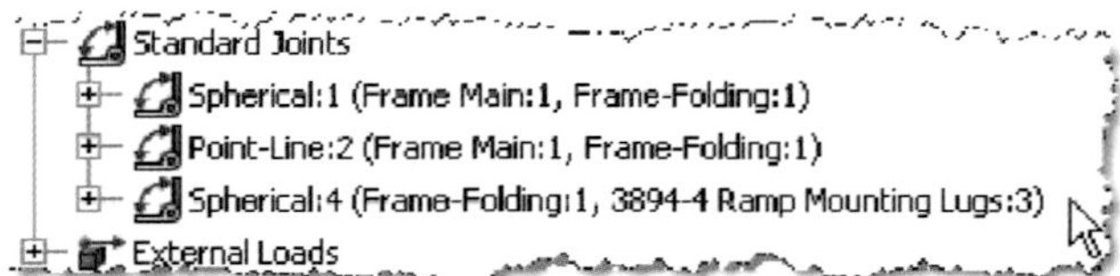

Next we will create a constraint on the other side, between the other lug and the folding frame.

29. For the first selection, select the point of the subassembly Frame-Folding:1 shown below

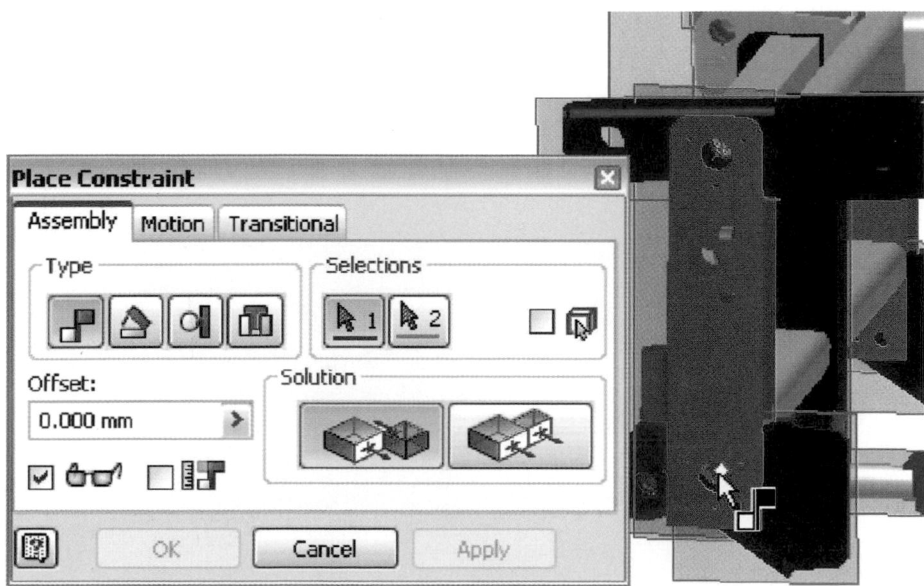

30. For the second selection, select the axis of component 3894-4 Ramp Mounting Lugs:4, as shown below

31. Click **Apply**

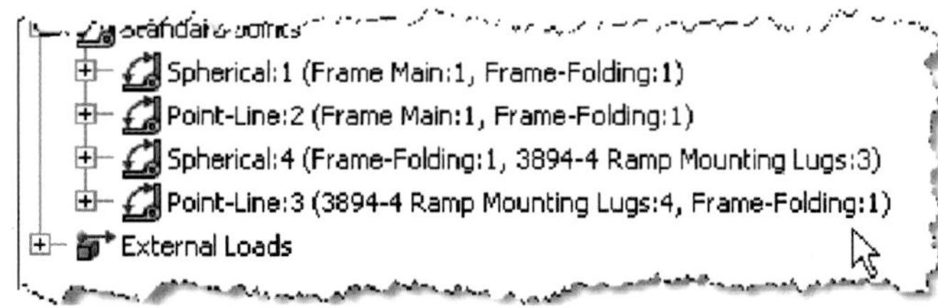

In the following steps, we will create constraints so that the first jack is constrained and attached to Frame Main:1 and Frame-Folding:1.

32. Select the cylindrical axis of the Frame Main:1 assembly, as shown

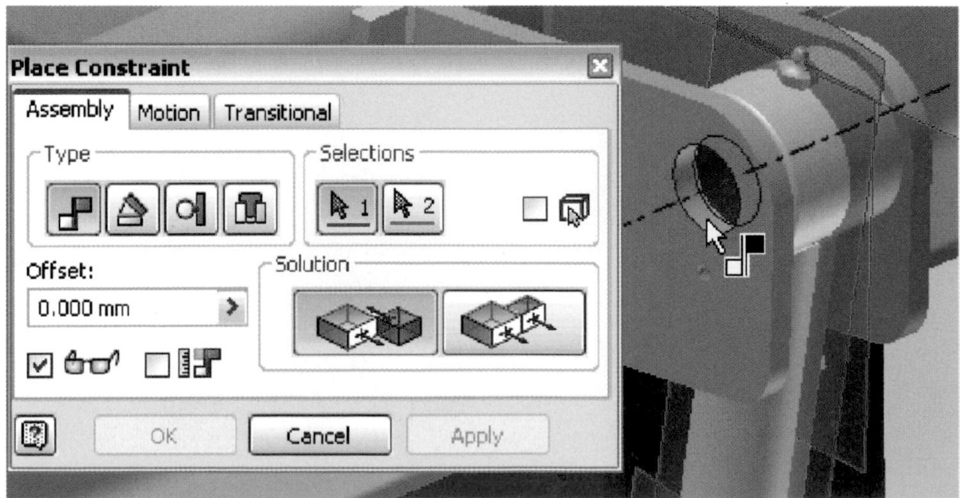

33. For the second selection, select the cylindrical axis of component E053 Double Acting Ram 354 Ram:2, as shown below

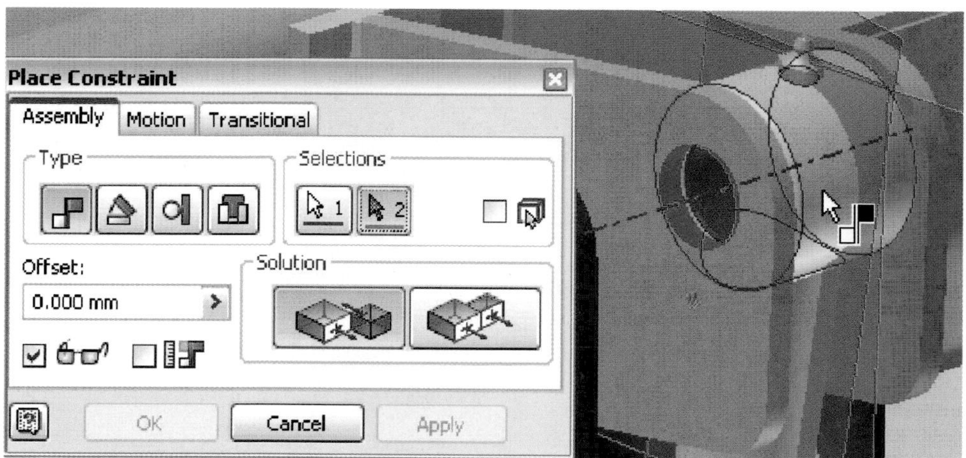

34. Click **Apply**

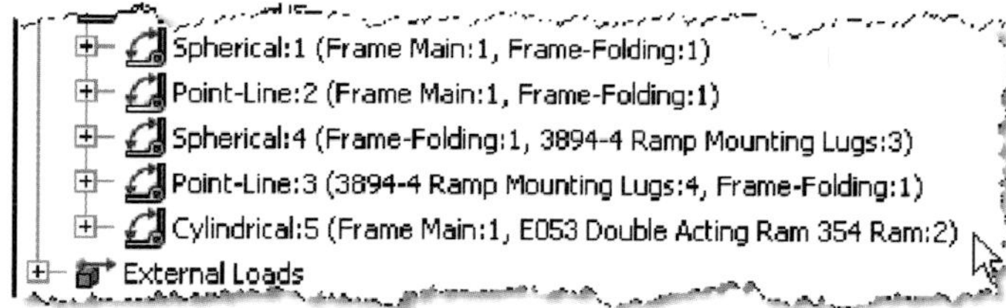

35. For the first selection, select the cylindrical axis of component E053 Double Acting Ram 354 Ram:2, as shown below

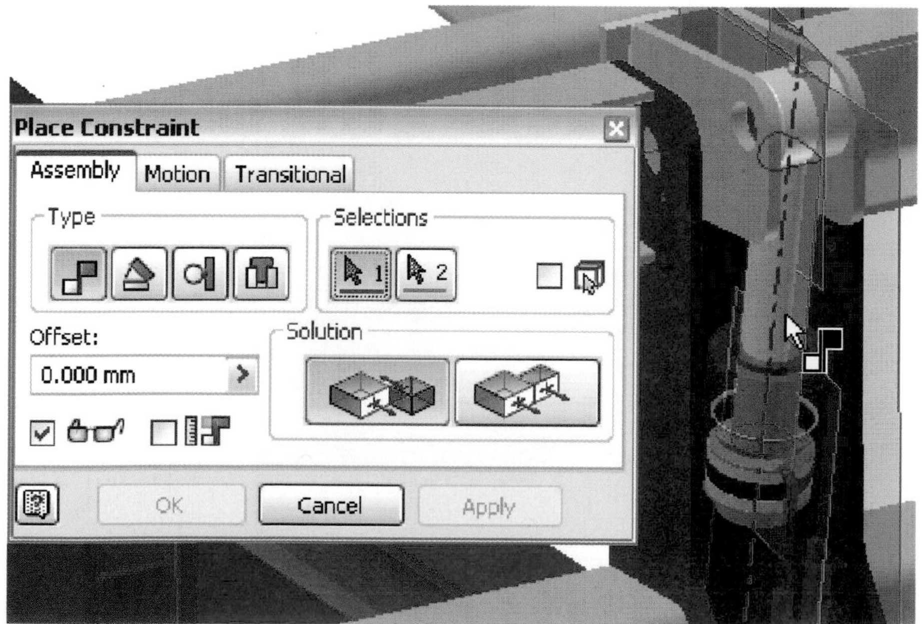

36. For the second selection, select the cylindrical axis of component E053 Double Acting Ram 354 Closed Centers:2, as shown below

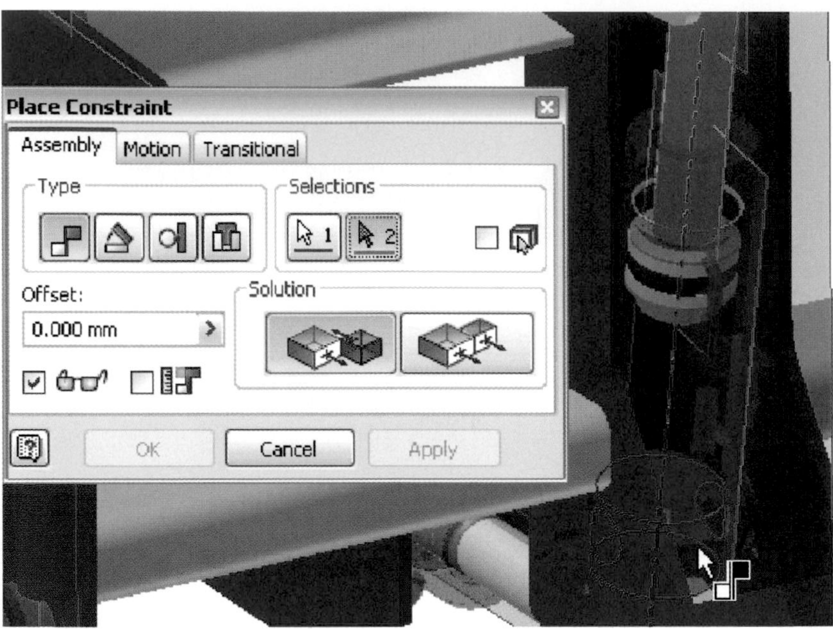

37. Click **Apply**

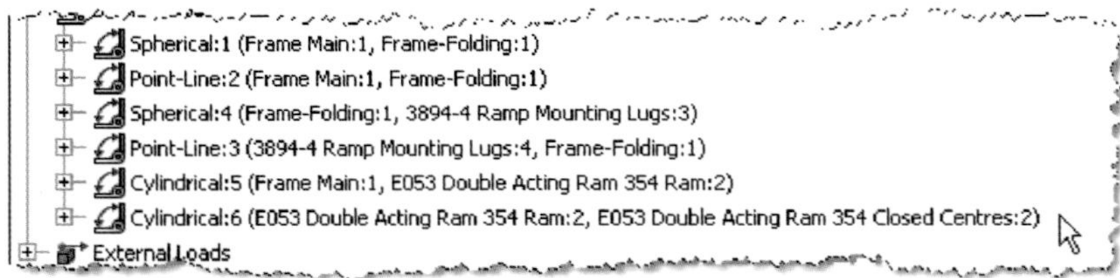

38. For the first selection, select the cylindrical axis of the subassembly Frame-Folding:1, as shown below

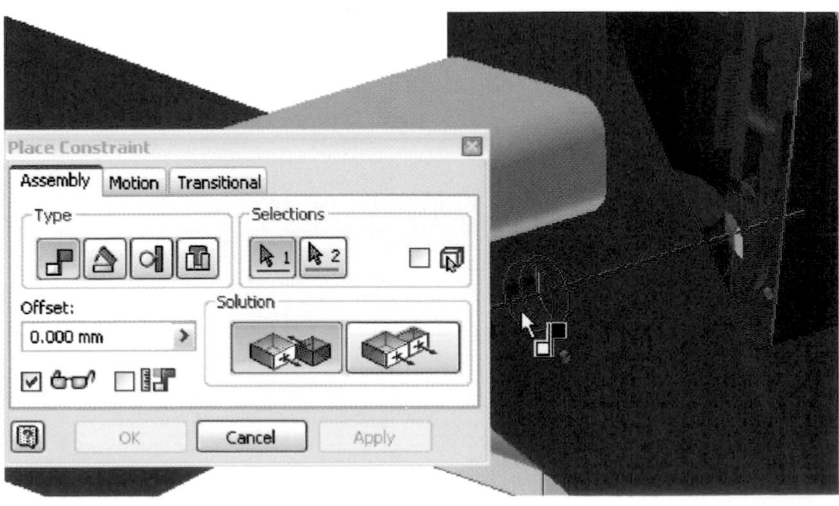

39. For the second selection, select the cylindrical axis of component E053 Double Acting Ram 354 Closed Centers:2, as shown below

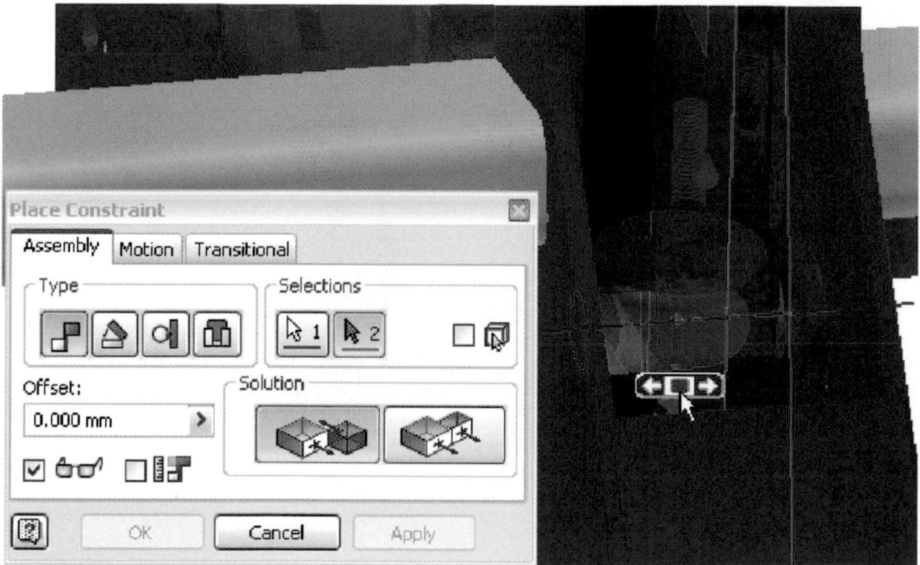

You may need to use **Select Other**, as shown above.

40. Click **OK**. Accept the redundancy warning

41. Click **OK**

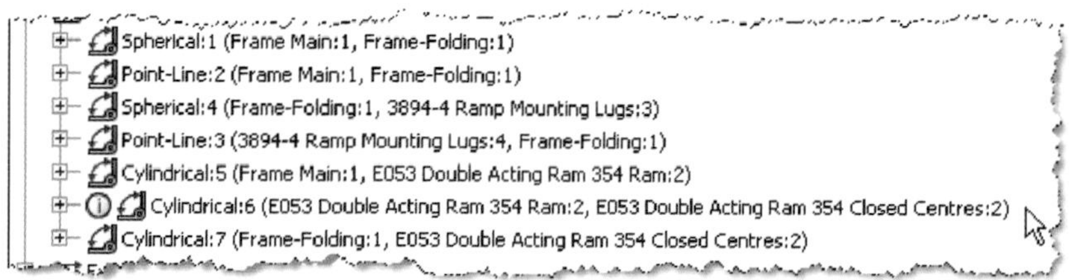

It is not always necessary to alter the redundant joint suggested by Dynamic Simulation. Altering another joint can remove redundancy within the model.

Cylindrical:6 becomes redundant. As we have created the joints using assembly constraints, we cannot repair it. The only alternative is to alter the constraint within the Assembly environment. We will alter the Cylindrical:5 joint to a point-line joint by altering the constraint from axis-axis to axis-point.

42. Expand the Cylindrical:5 joint in the simulation browser > Right click Mate:5 > Select **Edit**

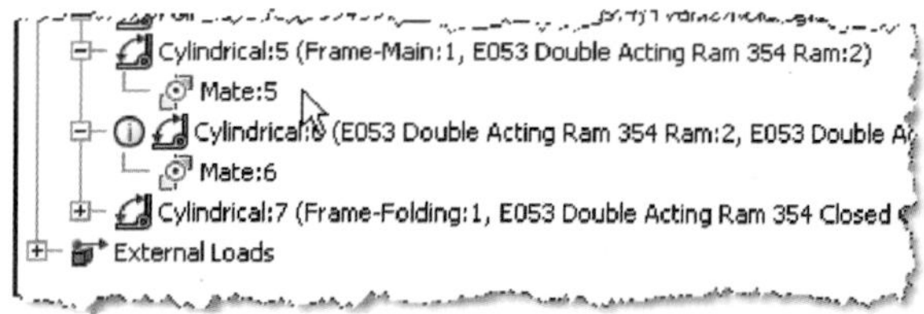

43. Reselect component 2 and select the point as shown. Click **OK**

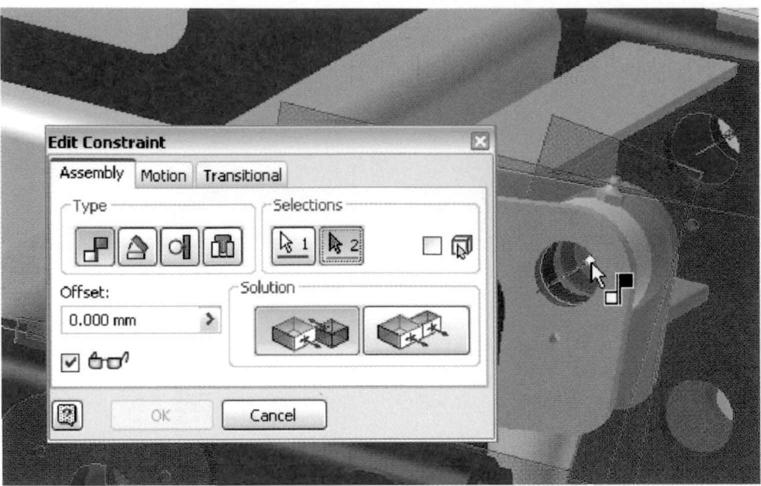

No redundancy warnings appear and the 'Cylindrical:5' joint now becomes 'Point-Line:5'.

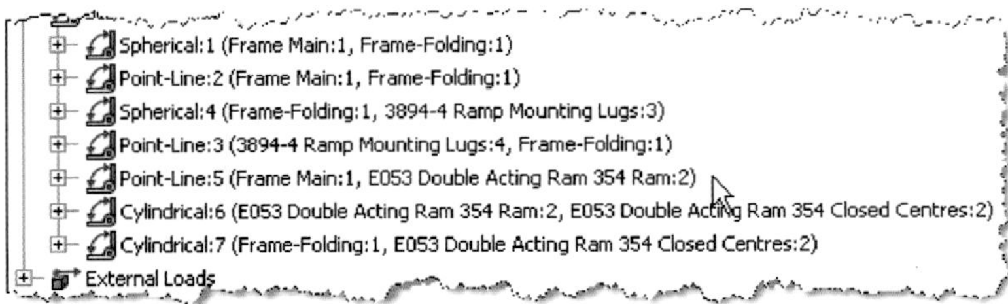

In the remaining steps, we are going to constrain jack 2 to Frame-Folding:1 and the ground.

44. Press **C** to activate the **Place Constraint** dialog box

45. For the first selection, select the cylindrical axis of the subassembly Frame-Folding:1, as shown below

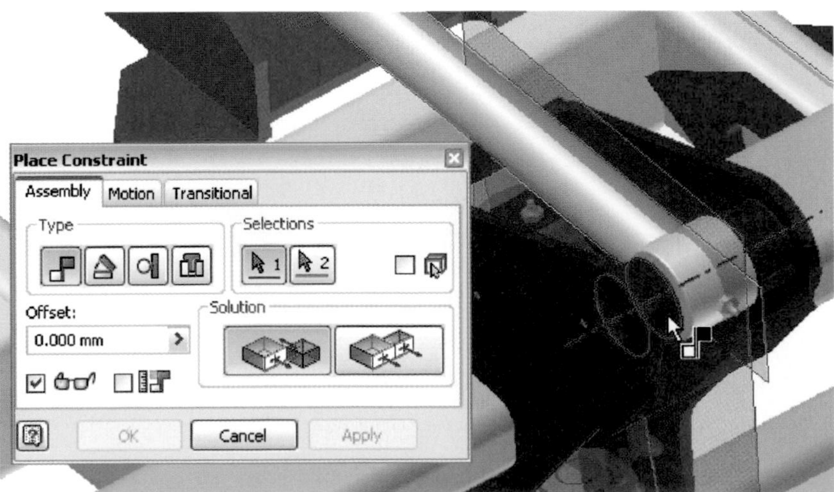

46. For the second selection, select the point of component E052 Double Acting Ram 574 Ram:2, shown below

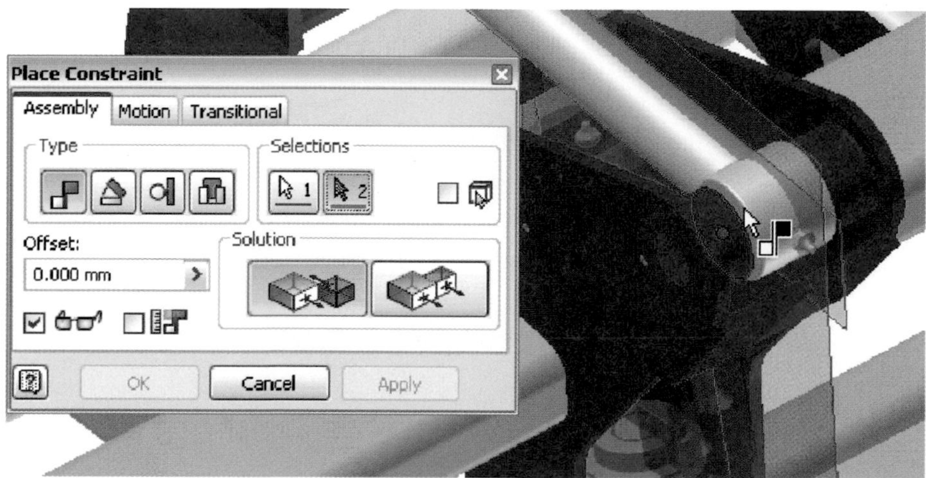

47. Click **Apply**

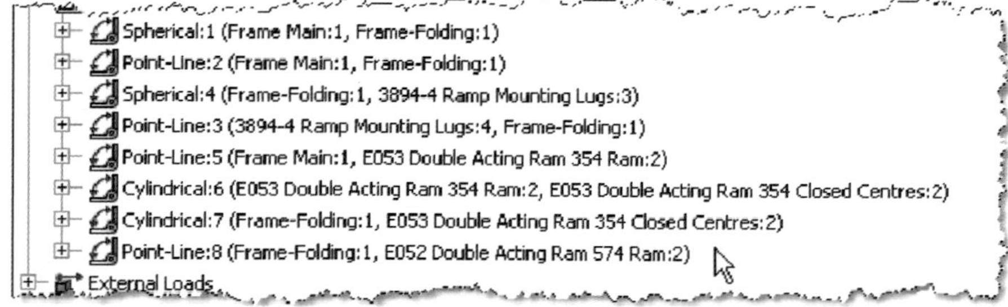

48. For the first selection, select the cylindrical axis of component E052 Double Acting Ram 574 Ram:2, as shown below

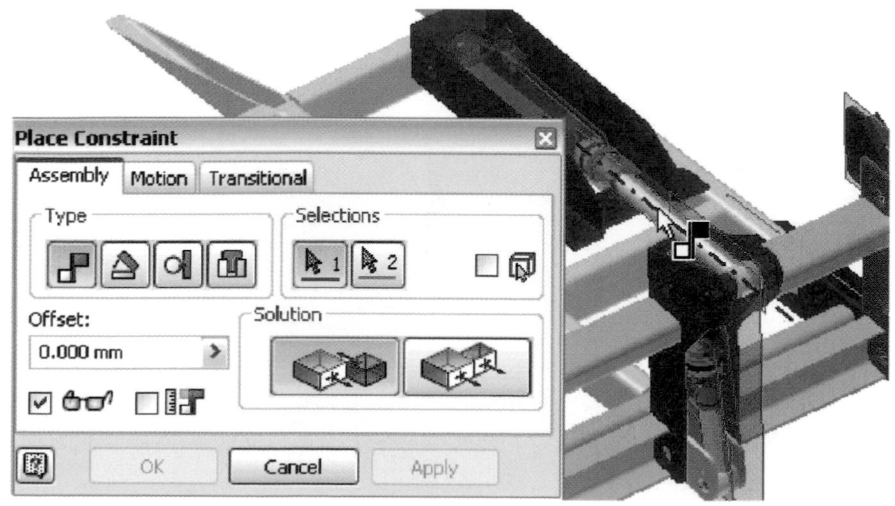

49. For the second selection, select the cylindrical axis of component E052 Double Acting Ram 574 Closed Centers:2, as shown below

50. Click **Apply**

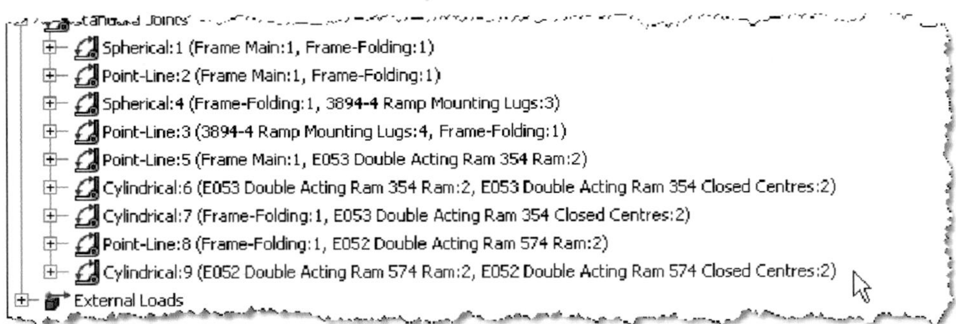

51. For the first selection, select the cylindrical axis of component 3894-3 Main Lift Ram Housing Assembly:2, as shown below

52. For the second selection, select the cylindrical axis of component E052 Double Acting Ram 574 Closed Centers:2, as shown below

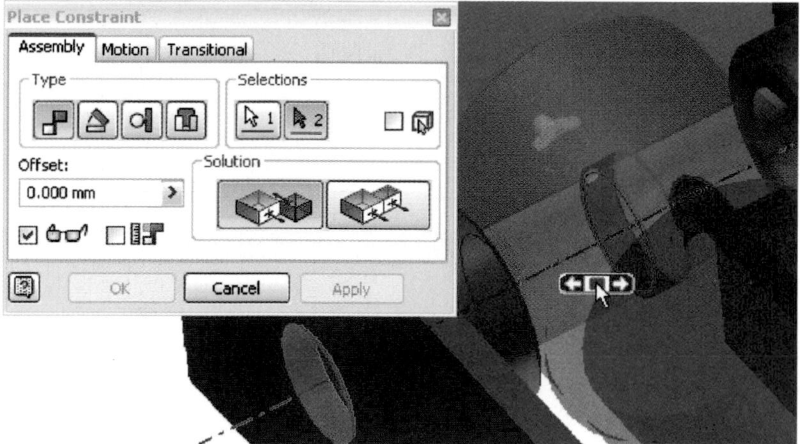

You may need to click **Select Other** to cycle through to select the axis.

53. Click **OK**

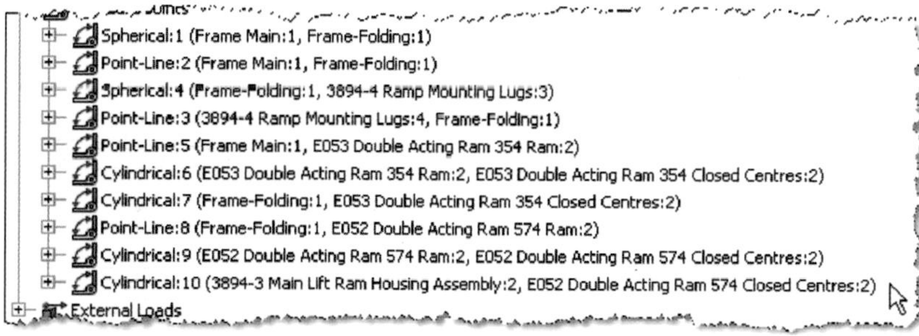

54. Press **F6**

55. Select the **View** tab > **Object Visibility** > Deselect **All Workfeatures**

In the next step, we will check the mechanism status of the assembly.

56. Select the **Dynamic Simulation** tab

57. Select **Mechanism Status** to determine the degree of mobility of the assembly

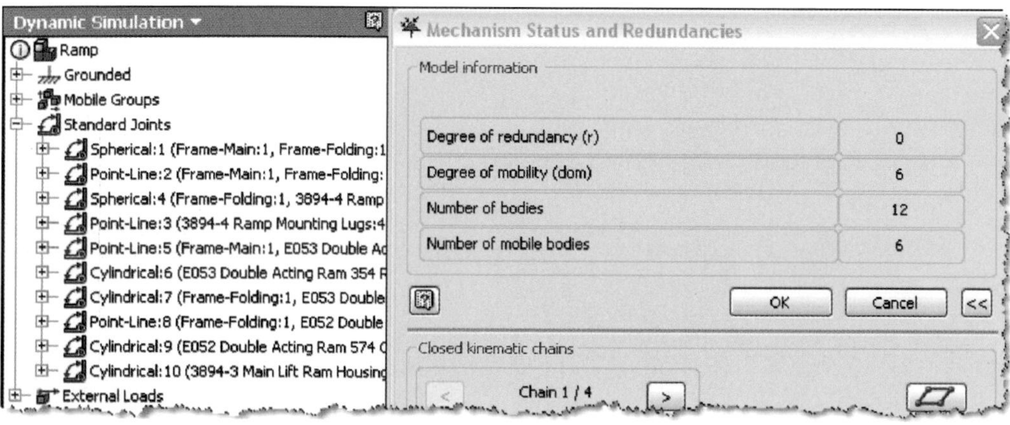

In the **Mechanism Status and Redundancies** dialog box there are six degrees of mobility. This means that there are five components that need to have their degrees of freedom restricted. This can be achieved by either creating more assembly constraints or modifying joints by locking some degrees of freedom. In this design problem, we will choose to lock degrees of freedom. The target is to reduce the degree of mobility to at least two, preferably to one.

58. Click **OK** to exit the dialog box

The first place to look at which joint to lock, in this example, would be the cylindrical and point-line joints, as they have several degrees of mobility.

59. Double click the Cylindrical:6 joint to alter its properties > Select the **dof 1 (R)** tab > Lock **Position** > Click **OK**

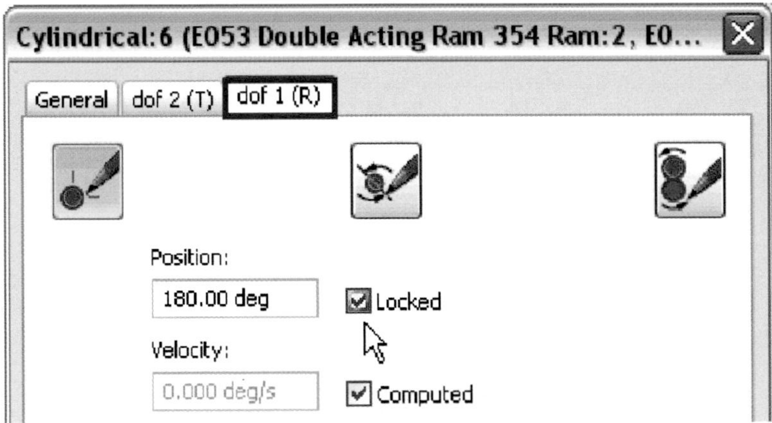

60. Double click the Cylindrical:7 joint to alter its properties > Select the **dof 2 (T)** tab > Lock **Position** > Click **OK**

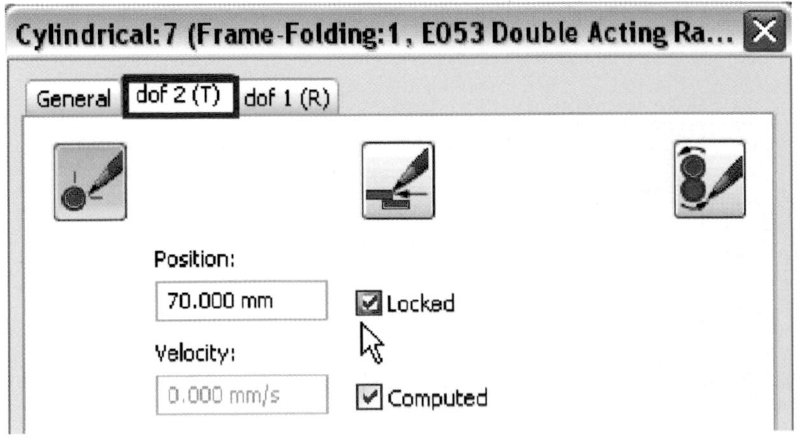

61. Double click the Cylindrical:9 joint to alter its properties > Select the **dof 1 (R)** tab > Lock **Position** > Click **OK**

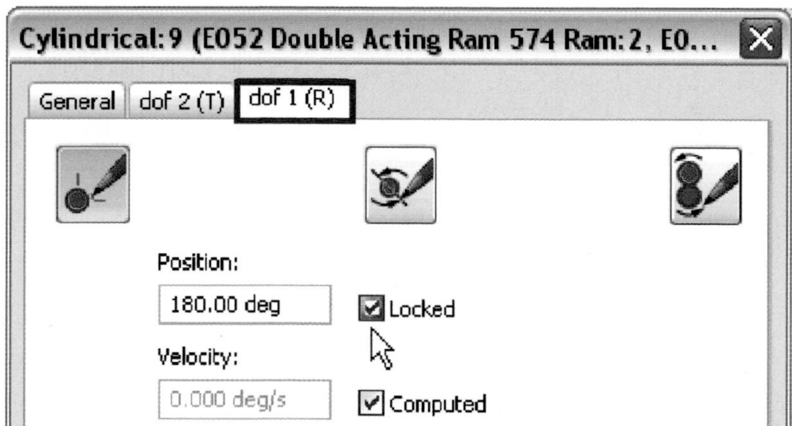

62. Double click the Cylindrical:10 joint to alter its properties > Select the **dof 2 (T)** tab > Lock **Position** > Click **OK**

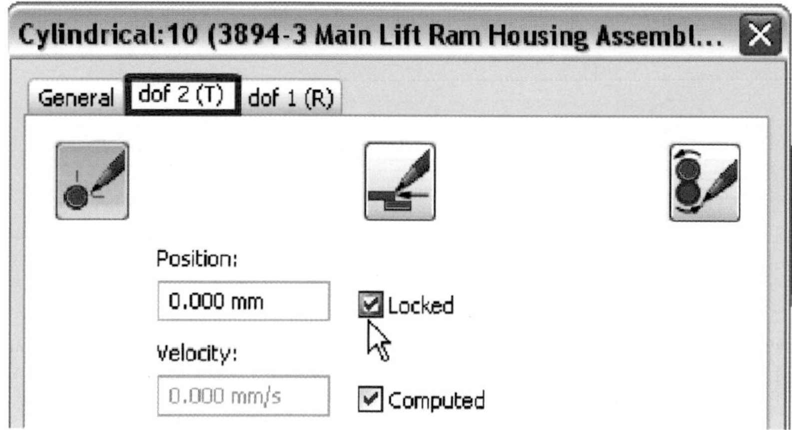

Next we will check the degree of mobility status again.

63. Select **Mechanism Status**

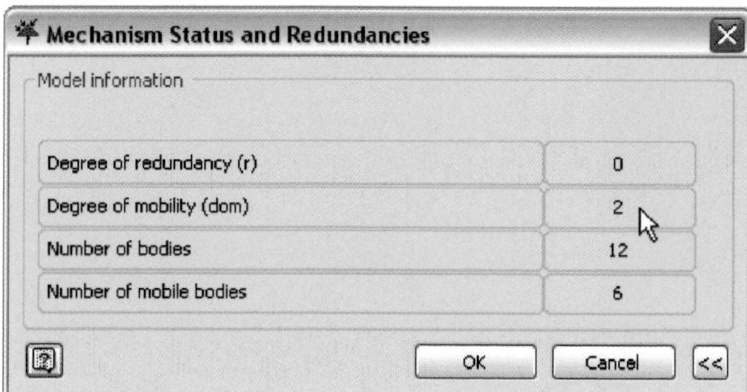

In the **Mechanism Status and Redundancies** dialog box, we now have two degrees of mobility.

64. Click **OK**

We could go further and start to restrict the degree of mobility to one by going through the remaining point-line joints. However, we should be okay to continue without restricting the extra degree of freedom.

You cannot run the tool **Unknown Force** unless you have one degree of mobility.

Environmental constraints

In this section, we need to determine the amount of travel required by the two jacks to completely unfold the ramp. We will first determine the amount of travel for the first jack.

We know that Frame-Main:1 needs to rotate until the holes shown below are aligned so that the locking pin can be inserted.

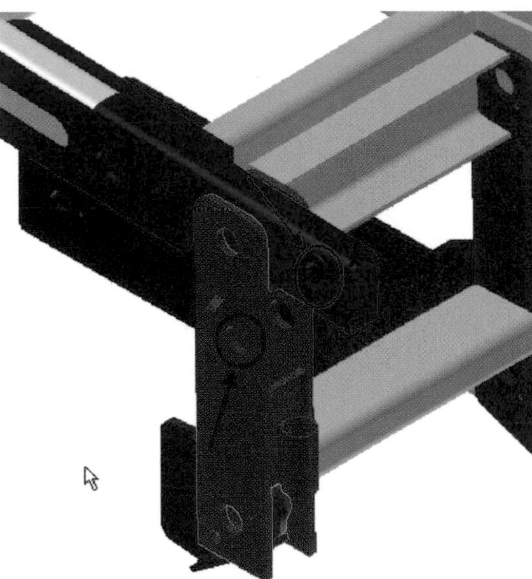

For the holes to align, Frame-Main:1 should be touching Frame-Folding:1 as shown.

Before we can simulate the travel, we need to set the initial positions of both jacks.

It is easier in this example to create more angle constraints within the Assembly environment in order to set the initial position of the ramp. We will then suppress the constraints so that

Dynamic Simulation does not convert them to more joints, as the **Automatically Convert Constraints to Standard Joints** option is activated.

65. Select **Model** from the **Dynamic Simulation** browser

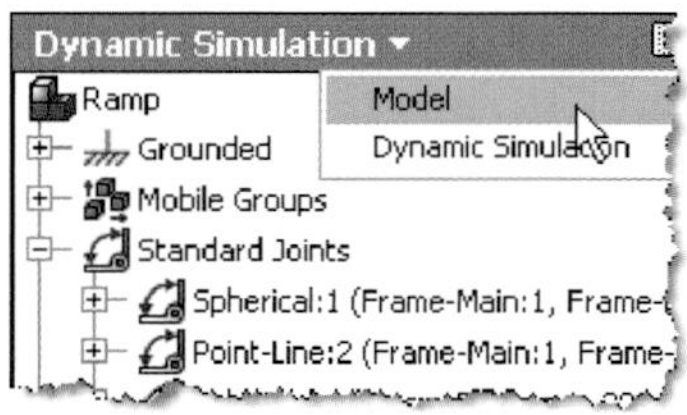

66. Press **C** to activate the **Place Constraint** dialog box

67. Select the angle-type constraint and, for the first component, select Frame-Main:1 as shown

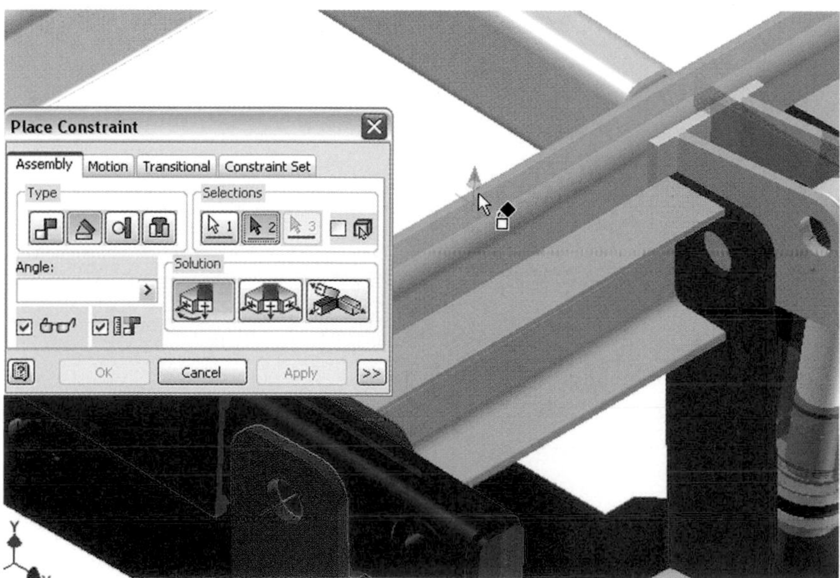

68. For the second component, select 3894-3 Main Lift Ram Housing Assembly:2. Make sure the angle is set to zero

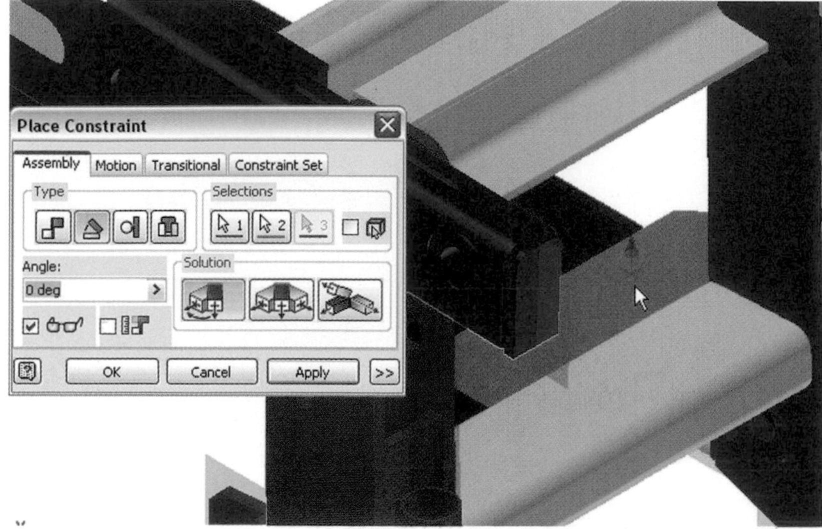

69. Click **Apply**, so we can continue creating more constraints

70. For the first component, select Frame-Folding:1 as shown

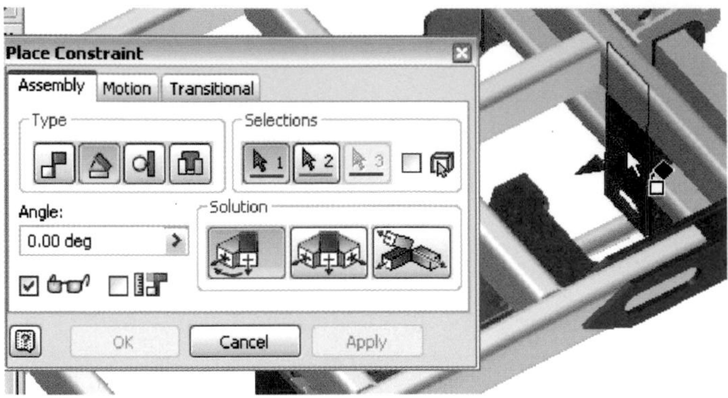

71. For the second component, select 3894-3 Main Lift Ram Housing Assembly:2 and type in 90 deg to set the position of the angle as shown

72. Click **OK**

73. Expand the Frame-Main:1 subassembly. Right click Angle:1 (0.00 deg) and select **Suppress**

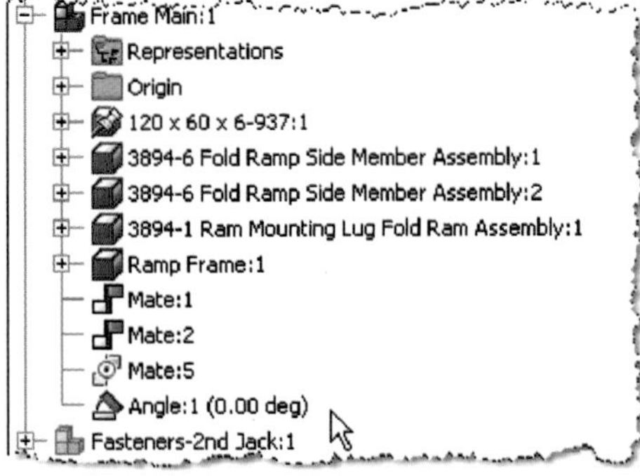

74. Expand the Frame-Folding:1 subassembly. Right click Angle:2 (90.00 deg) and select **Suppress**

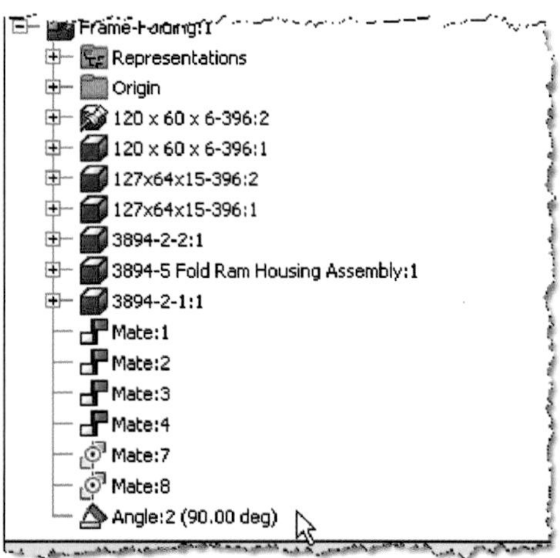

75. Select **F6**

76. Now select **Dynamic Simulation** from the **Assembly** browser

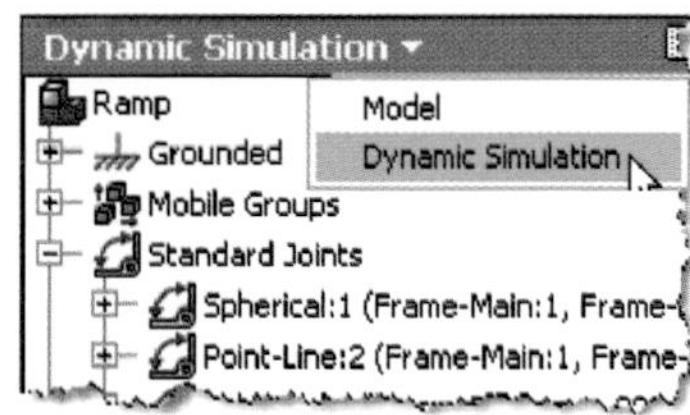

Initially, we need to unfold the Frame-Main:1 ramp by 90 degrees so that we can place the locking pins. We can use either the Spherical:1 or the Point-Line:2 joint; we will use the latter.

77. Double click the Point-Line:2 joint

78. In the Point-Line:2 joint dialog box, select the **dof 3 (R)** tab, as shown

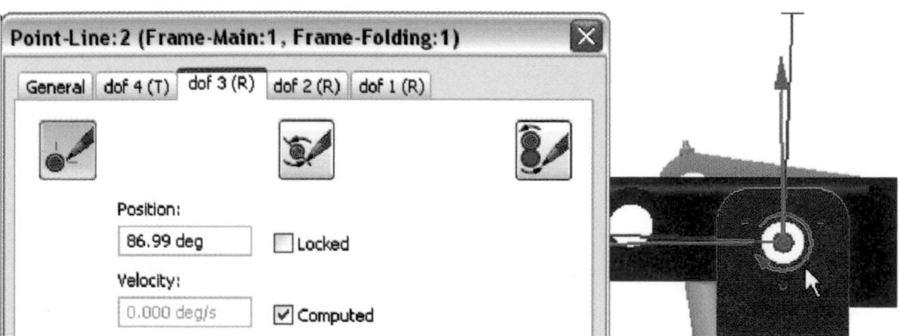

We need to change the position by 90 degrees to represent the final position. We can either add 86.99 + 90 in the **Position** value or set the original to 0 and then simply add in 90. We will use the latter option.

79. Right click in the **Position** magnitude > Select **Set offset**

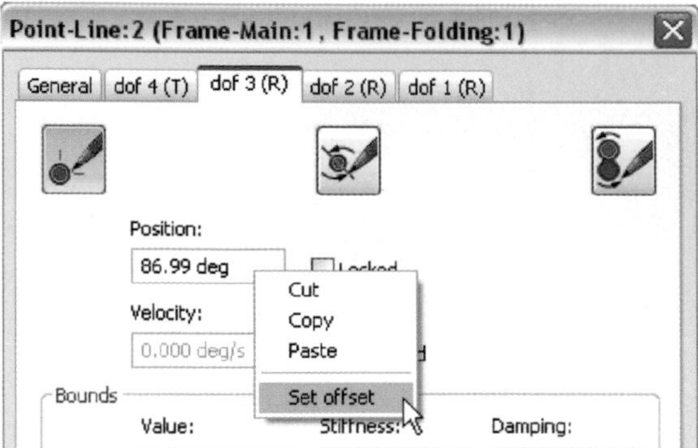

This will reset the value to 0 so that we can simply add in 90 rather than working out 86.99 + 90.

80. Type in 90 > Click **OK**

IMPORTANT—If ramp unfolds the other way then enter –90 rather than 90.

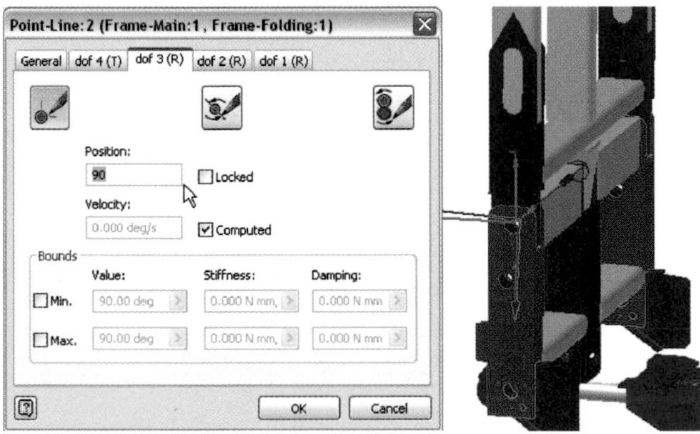

Finally, we need to unfold the folding ramp by 95 degrees so we can start loading. We can either use the Spherical:4 or the Point-Line:3 joint; we will use the former.

IMPORTANT—As a result of suppressing angle constraints in the earlier step, the Point-Line:3 joint may be renamed as Point-Line:4. Therefore, be careful to use the correct joint for the remaining steps.

81. Double click the Point-Line:3 joint

82. In the Point-Line:3 joint dialog box, select the **dof 3 (R)** tab, as shown. Type in –95 after the initial value of 89.3

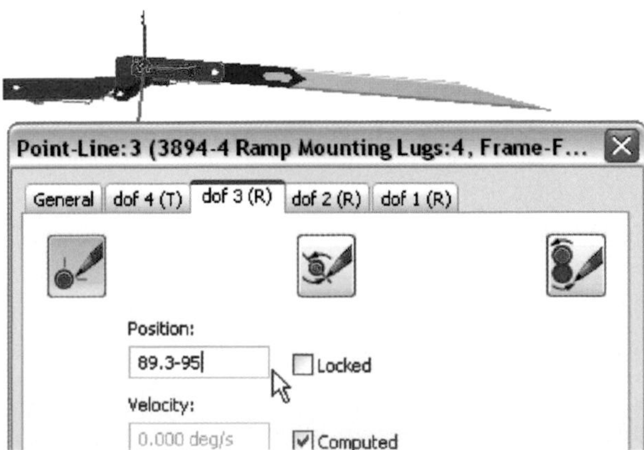

This will rotate the ramp by another 95 degrees clockwise (final position).

83. Click **OK**

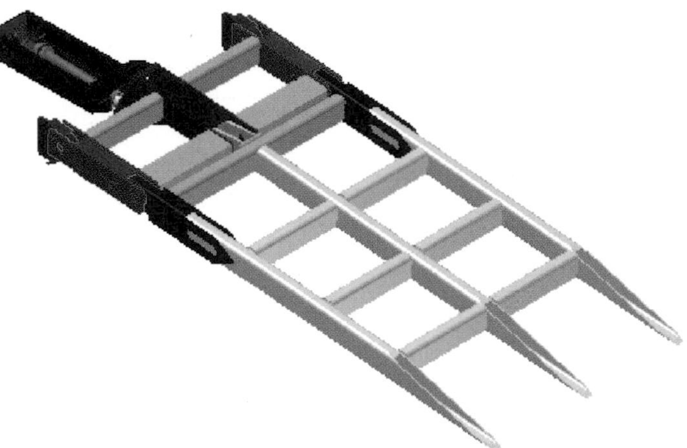

Now we will measure the final travel distance of both jacks represented by the Cylindrical:6 joint (jack 1) and the Cylindrical:9 joint (jack 2).

84. Open Output Grapher from the Dynamic Simulation panel

85. Expand the Output Grapher nodes to select position P[2] for the Cylindrical:6 and Cylindrical:9 joints

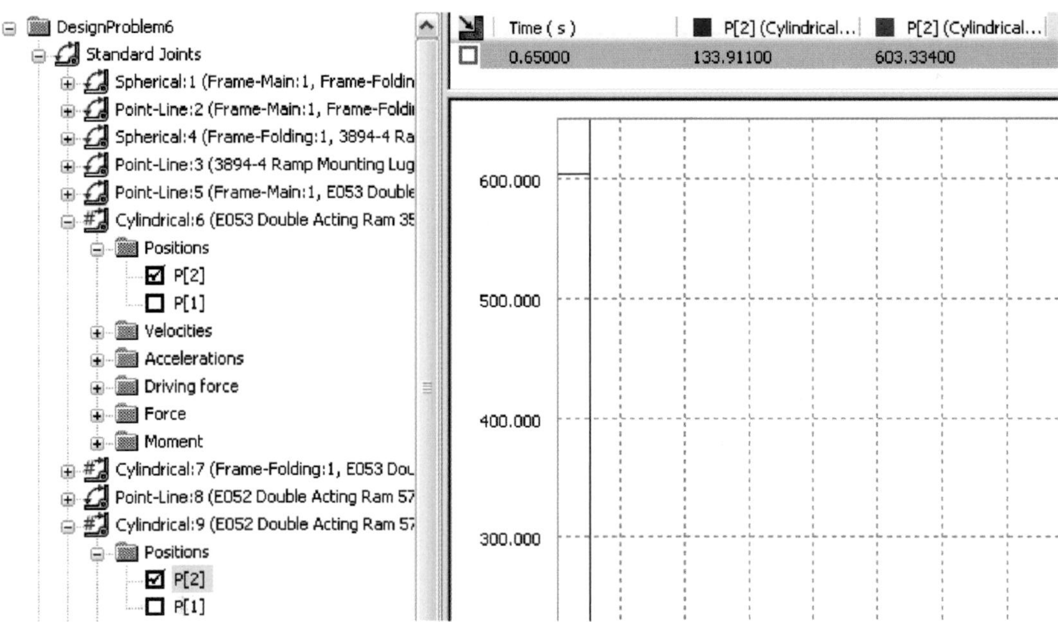

86. Play the simulation for about 0.5 seconds

87. Stop the simulation and note the following values:

Cylindrical:6 133.91 mm
Cylindrical:9 603.33 mm

We will use these values to size the jacks in the next section.

88. Close Output Grapher

89. Select **Construction Mode**

90. Double click the Point-Line:2 joint

91. Change the **Position** value back to 0 > Click **OK**

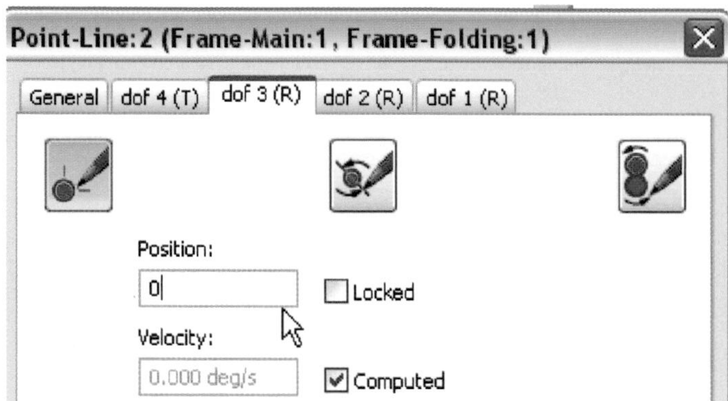

92. Double click the Point-Line:3 joint (or the Point-Line:4 joint)

93. Change the **Position** value back to the folded position. Type in −5.7 + 95

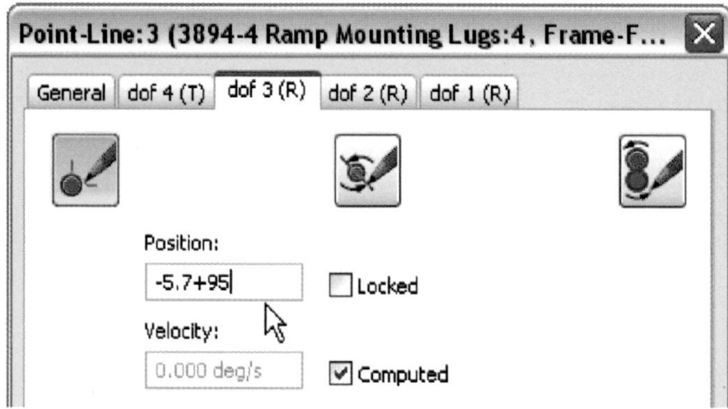

94. Click **OK**

The assembly should be back to its folded/unloaded position.

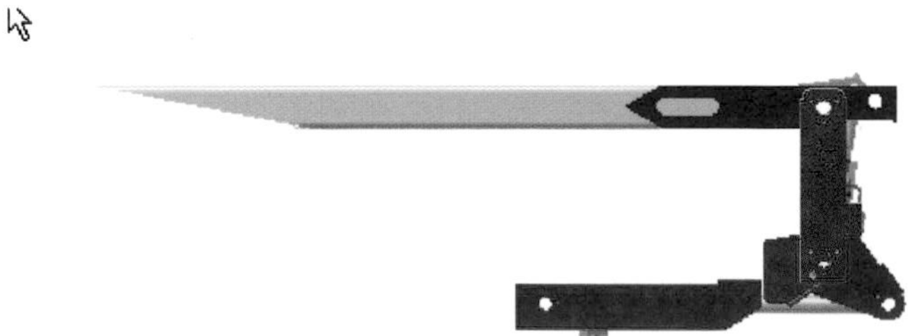

APPLY IMPOSED MOTION – INPUT GRAPHER

95. Double click the Cylindrical:6 joint > Select the **dof 2 (T)** tab > Select **Enable Imposed Motion** > Select **Position** > Select the **Input Grapher**

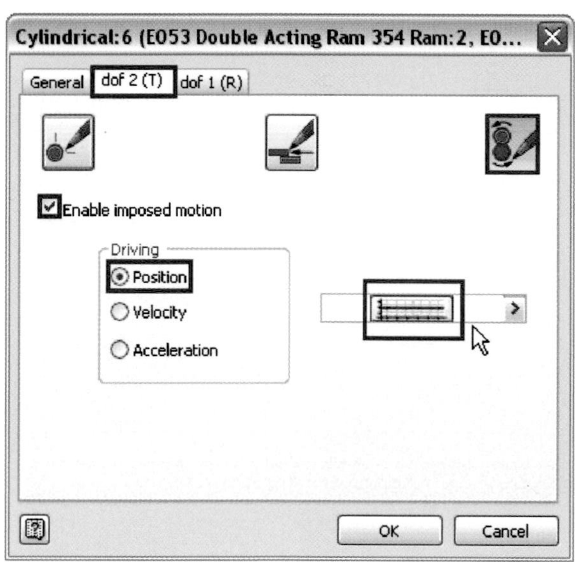

96. In the **Imposed Motion** dialog box, change the following values:

X1 = 0 seconds Y1 = 0 mm

X2 = 15 seconds Y2 = 133.91 mm

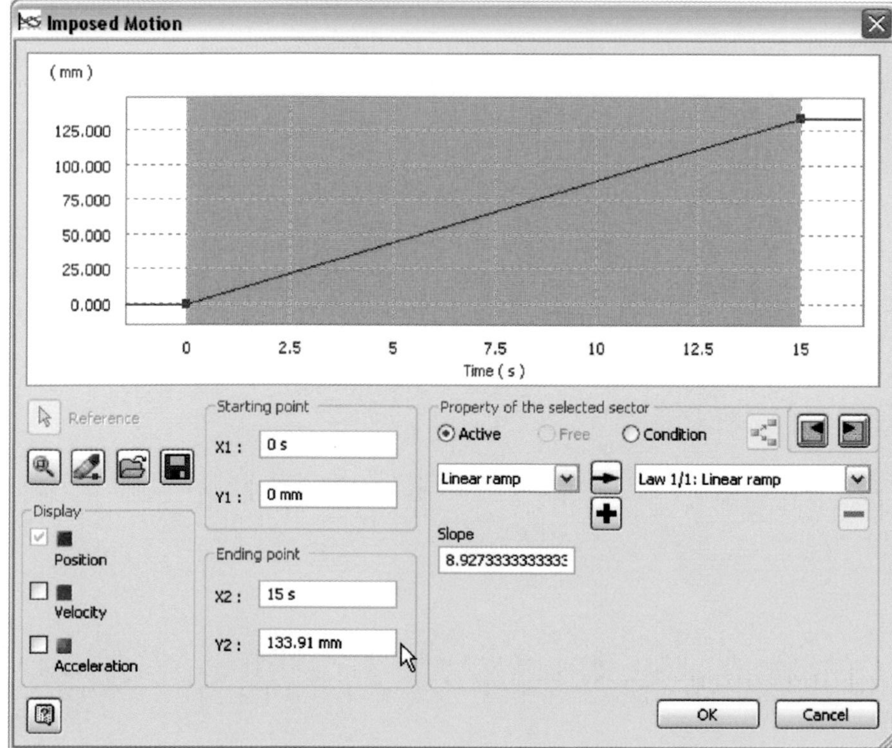

This allows the first jack to unfold the Frame-Main:1 ramp within 15 seconds. After this, a locking pin is placed for safety reasons and also stops the load being transferred to the first jack.

97. Click **OK** twice

98. Double click the Cylindrical:9 joint > Select the **dof 2 (T)** tab > Select **Enable Imposed Motion** > Select **Position** > **Input Grapher**

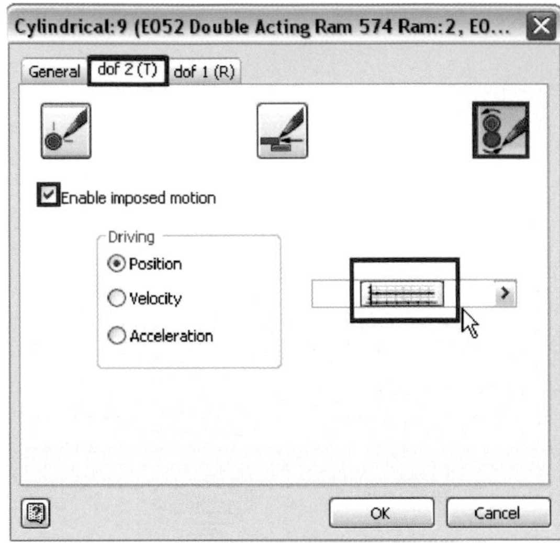

99. In the **Imposed Motion** dialog box, change the following values:

X1 = 25 seconds   Y1 = 291.6 mm (initial position)

X2 = 40 seconds   Y2 = 603.33 mm (final position)

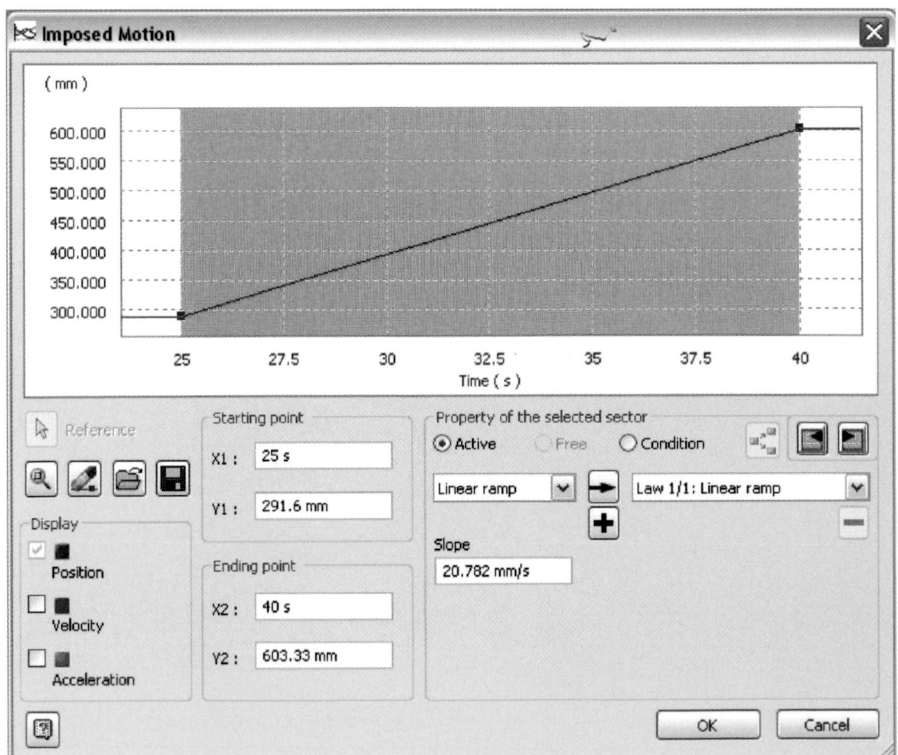

The graph allows a 10 second delay so that the locking pin can be placed before the second jack is activated. The second jack takes 15 seconds to unfold the ramp to its loading position.

100. Click **OK** twice

APPLY GRAVITY

In the next steps, we will run the simulation to determine the maximum forces required by the jacks.

101. Double click **Gravity**

102. Untick the **Suppress** button to activate the gravity. Note that gravity is acting down

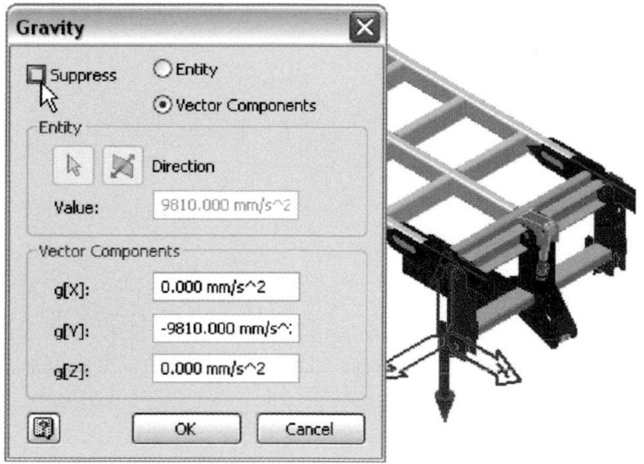

103. Click **OK**

Analyze results

104. Change the final time in the **Simulation panel** to 40 seconds and the **Images** value to 1200 > Play the simulation > Accept the warning

105. Select the **Output Grapher**

106. Deselect the P[2] values for the Cylindrical:6 and Cylindrical:9 joints

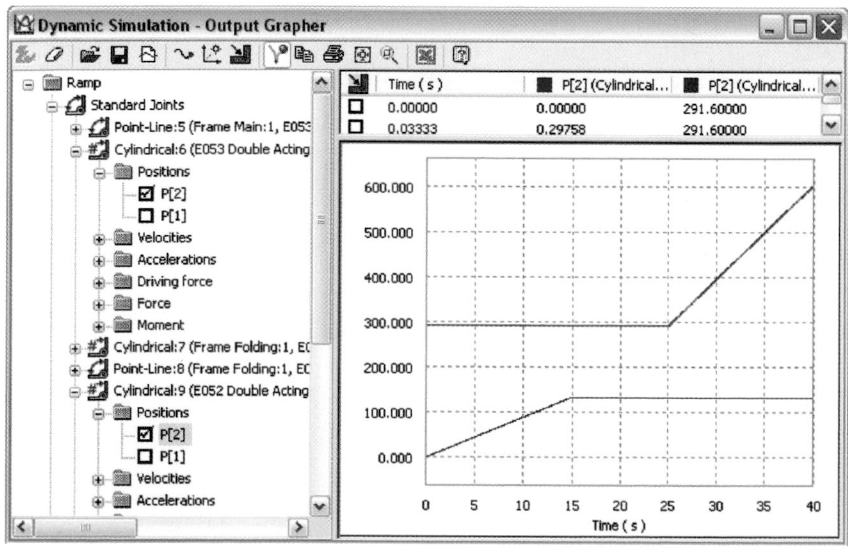

107. Select the U-imposed [2] Driving force values for the Cylindrical:6 and Cylindrical:9 joints, as shown

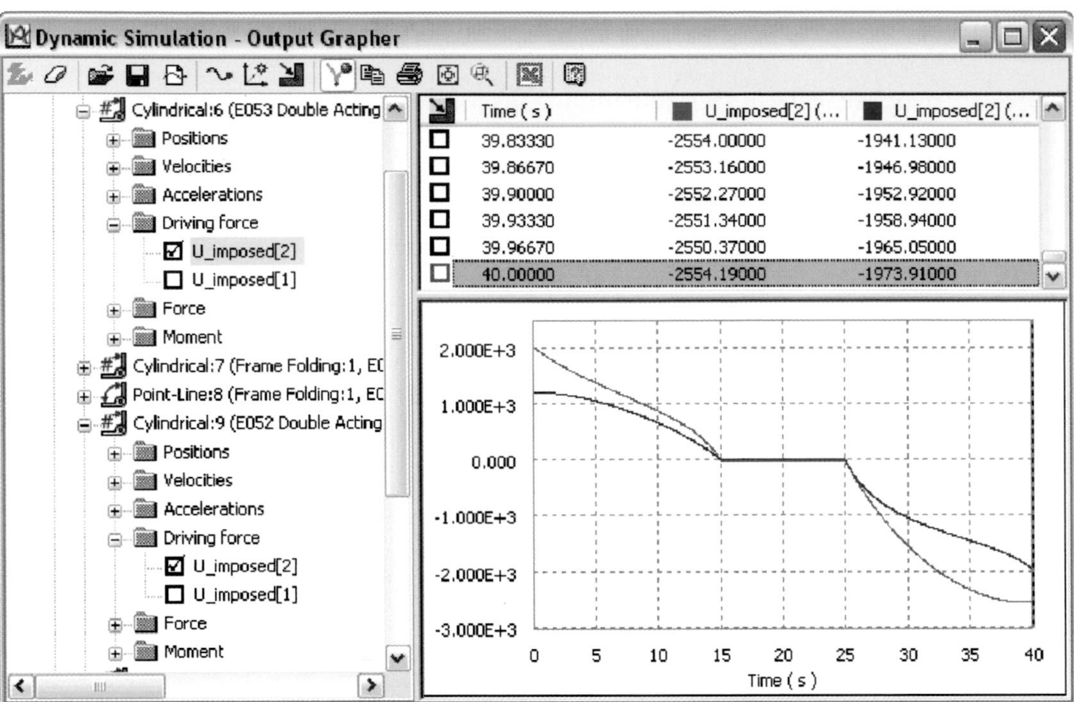

Negative denotes that the jack is in compression and positive that it is in tension.

We get maximum and negative forces because the locking pin has not been included in the simulation and, as a result, both jacks are active during the complete simulation.

Below is the ramp ready to be loaded.

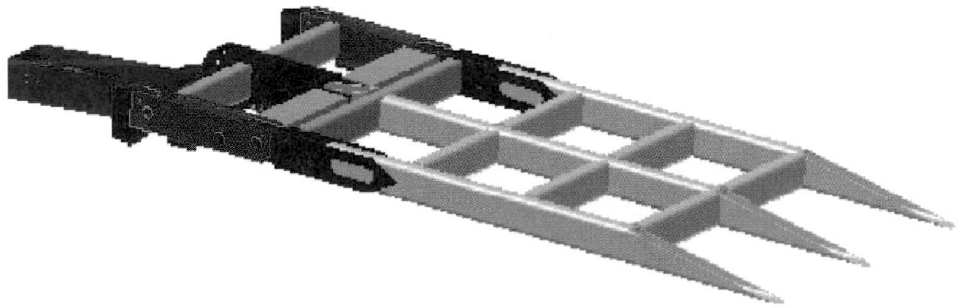

The maximum force of both jacks is around 2000 N (use **Search Max.** to determine the exact value).

108. Close the **Output Grapher** > Click on **Construction Mode**

Determine the maximum reactions in the lugs.

The ramp is designed to take loads of 2.5 ton (25 000 N). So, in this section, we will apply a force to simulate loading conditions on the ramp.

109. Select **Force** > Select the corner of Main-Frame:1 as shown for the location of the joint

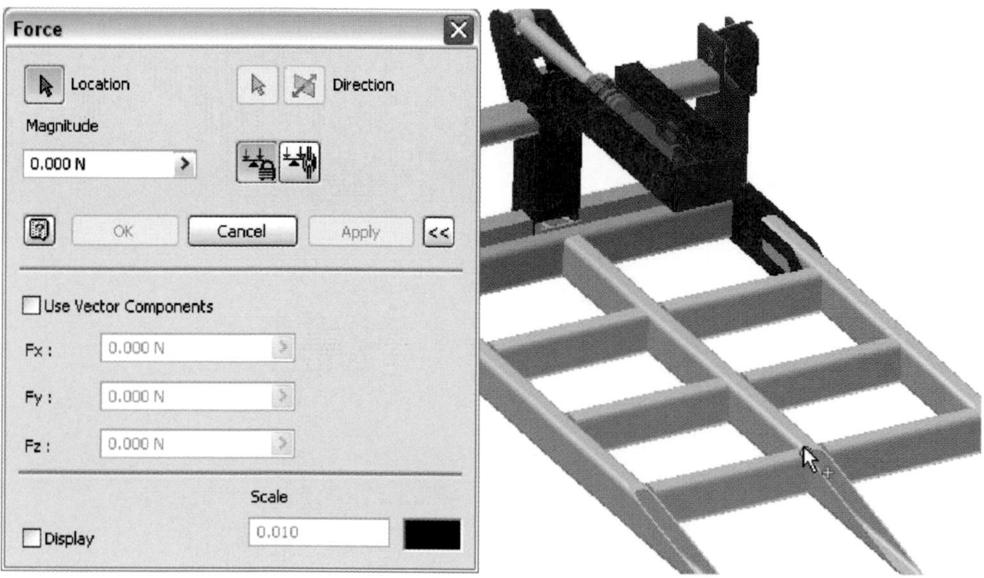

110. For the direction of the force, select the edge of Main-Frame:1, as shown

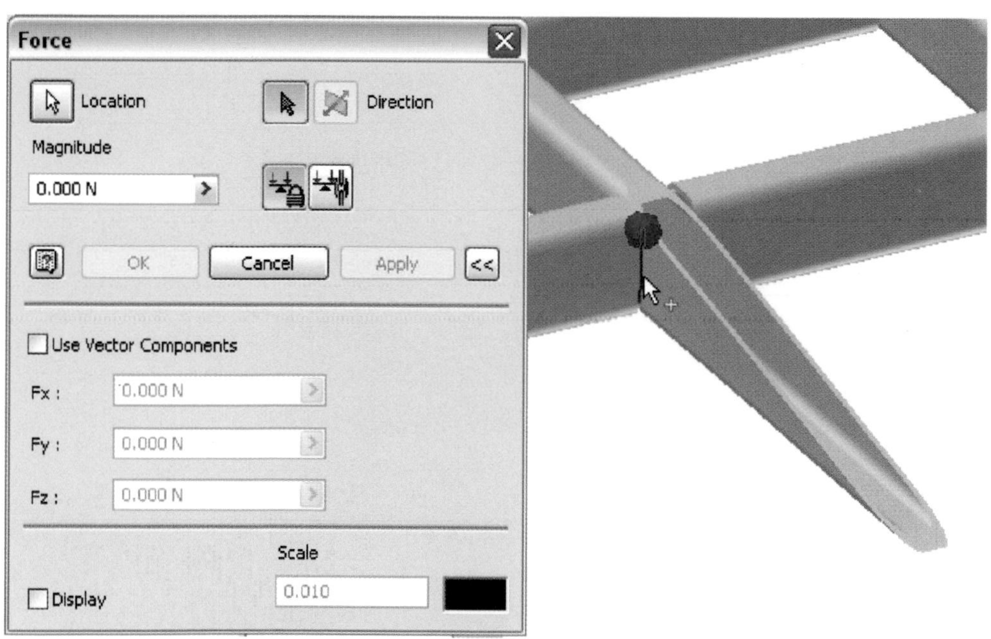

111. Select **Associative Load Direction**

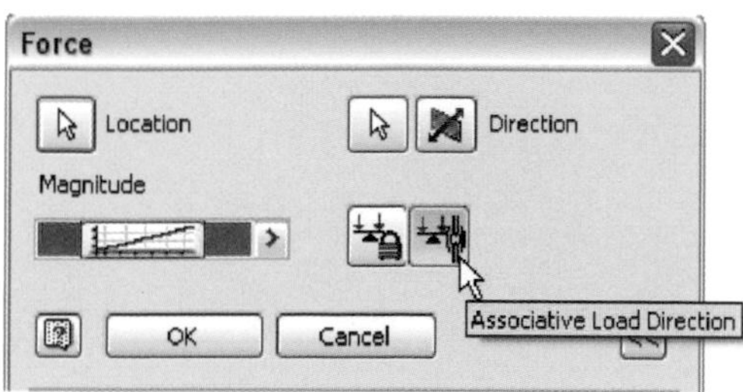

112. For **Magnitude**, select **Input grapher**

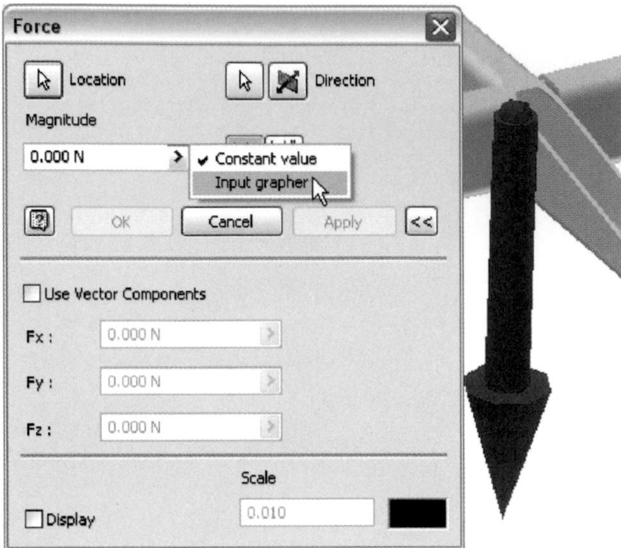

113. In the **Magnitude** Input Grapher dialog box, change the following values and click **OK**

X1 = 45 seconds Y1 = 0 N

X2 = 46 seconds Y2 = 25 000 N (this represents 2.5 ton)

114. Select **Display** in the **Force** dialog box > Change the scale to 0.0001 > Click **OK**

115. Change the final time to 50 seconds and **Images** to 1500 > Play the simulation > Accept the warning

116. Open the **Output Grapher** once the simulation has been completed > Right click anywhere on any of the Force column data results > Select **Unselect all Curves**

117. Expand and select Force for both Spherical:3 and Point-Line:4, as shown in the Output Grapher

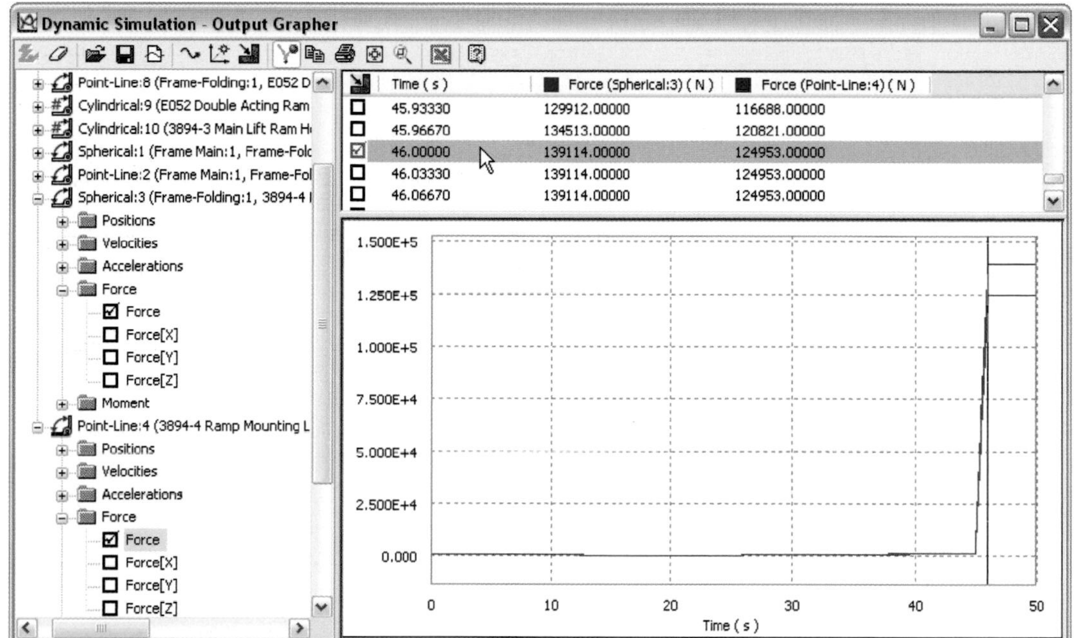

The maximum force is slightly different in both lugs; this is mainly due to the different joints created to avoid redundant joints. In reality, we would expect identical results, and, for this reason, the lugs designed are identical. For stress analysis purposes, we will use the slightly higher force of 139 114 N in the spherical joint. The values may slightly differ.

So, according to the assumption notes at the beginning of this chapter, the actual load transferred to the lug will be half of 139 114 N, which is **69 557 N**, and will be used to test the structural integrity of the lug later, in Chapter 10.

118. Close the **Output Grapher** > Click **Construction Mode** > Close the file

CHAPTER 5

DP4 – Advanced Simulation Settings

Racing Car Engine Connecting Rod

(Design Problem courtesy of Triple Eight Engineering Ltd)

JOINTS INTRODUCED/COVERED IN THIS DESIGN PROBLEM

	Joints	Joint creation process
1	Revolution	Automatically converted
2	Spherical	Automatically converted
3	Point-line	Automatically converted
4	Prismatic	Automatically converted

KEY FEATURES AND WORKFLOWS INTRODUCED IN THIS DESIGN PROBLEM

	Key features/workflows
1	Repairing redundant joints by altering assembly constraints
2	Input Grapher – conditions – logic equations
3	Initializing all joints to 0
4	Friction in joints

INTRODUCTION

Triple Eight Race Engineering designs, manufactures, and races Vauxhalls in the British Touring Car Championship (BTCC) on behalf of General Motors. Since 2001, Triple Eight has won six Drivers championships, five Team Championships, and seven Manufacturer Championships in the BTCC.

Up and Running with Autodesk Inventor Simulation 2011. ISBN: 978-0-12-382102-7
Copyright © 2010 Elsevier Inc. All rights of reproduction, in any form reserved.

In developing race-winning cars, Triple Eight utilizes Autodesk Inventor and Dynamic Simulation. One of the critical design issues in developing race-winning cars is weight, as this has a considerable impact on the performance of the cars. In this design problem, we will look at the key components of the engine and how we make effective use of Dynamic Simulation to simulate the explosion of gases on the piston-crank assembly.

Therefore, in this design problem, we will determine:

- The time taken for the engine speed to reach 7000 rpm.
- The engine torque with friction not taken into account.
- The engine torque with friction taken into account.
- The reaction forces acting on the connecting rod.

As the engine comprises four connecting rods that are identical, we will only analyze one connecting rod to simplify the simulation and analysis.

In addition to the main requirements, the following criteria will be taken into account:

- Max. combustion pressure = 90 bar or 9 N/mm^2 (1 bar = 1 KPa = 0.1 MPa = 0.1 N/mm^2).
- The cross-sectional area of the piston head is approximately 6000 mm^2.

In addition, we need to convert maximum combustion pressure to force, where

Maximum force = Cross-sectional area of piston head × Pressure

Maximum force = 6000 × 9 = 54 000 N = 54 KN

In normal operating conditions, the pressure will rise and fall; this maximum force will not be always active, the mean pressure will be lower, and for the purposes of this simulation we will use 10 KN and 25 KN.

For friction analysis, the following will be taken into consideration:

Friction in the key component, including bearings = **0.2**. (This data is also available in the *Mechanical Engineer's Handbook*.)

The torque required to overcome this friction is calculated using

$$\text{Torque} = -(v) \times \text{sqrt}(\text{sqr})Fr[\text{X}] + \text{sqr}(Fr[\text{Y}])) \times cf \times R$$

where *Fr* is the force in DOF of the joint.

Damping coefficient (D) = **0.05**. (This data is also available in the *Mechanical Engineer's Handbook*.)

The damping coefficient is dependent on a variety of factors including the viscosity of the lubricant material, speed, temperature, etc.

The damping coefficient creates torque opposite to the velocity of the joint and, thus, the torque required to overcome this is calculated by

$$\text{Torque} = -D \times v$$

WORKFLOW OF DESIGN PROBLEM 4

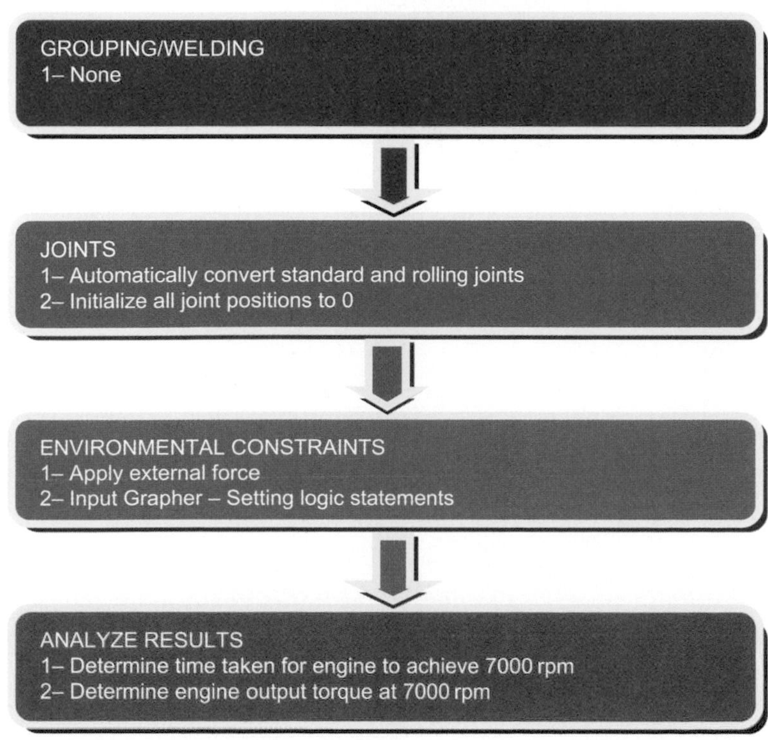

As there are only eight components, of which three are grounded, there is no need to restructure them into subassemblies.

Joints

1. Open *Piston.iam*

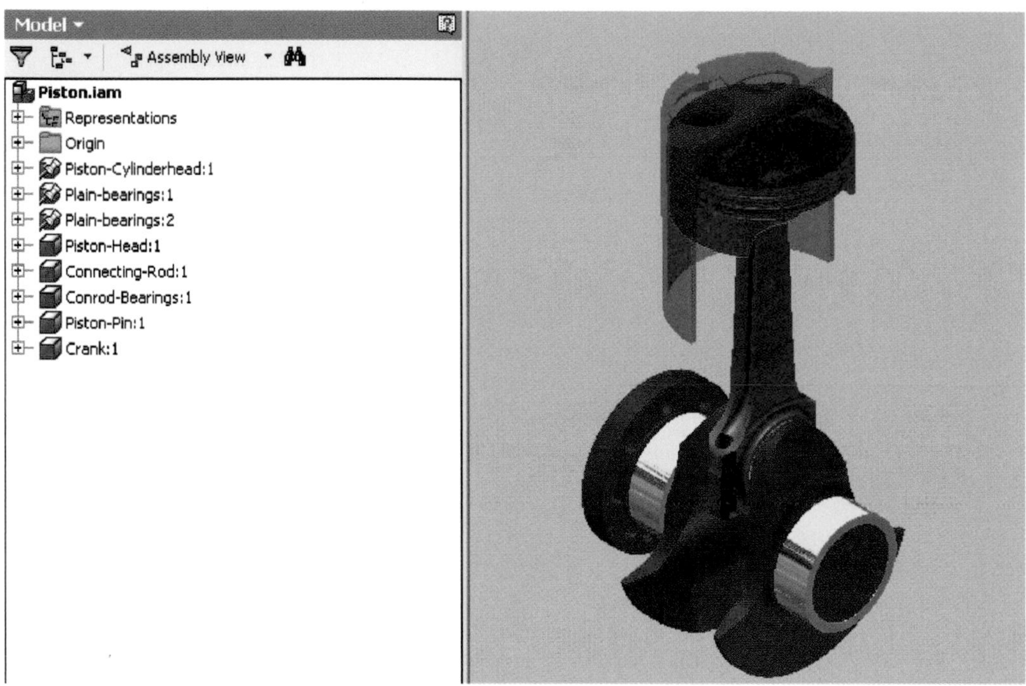

2. Select **Environments** tab > **Dynamic Simulation**

3. Select **Simulation Settings**

4. Select **Automatically Convert Constraints to Standard Joints** > Select **All initial positions at 0.0**

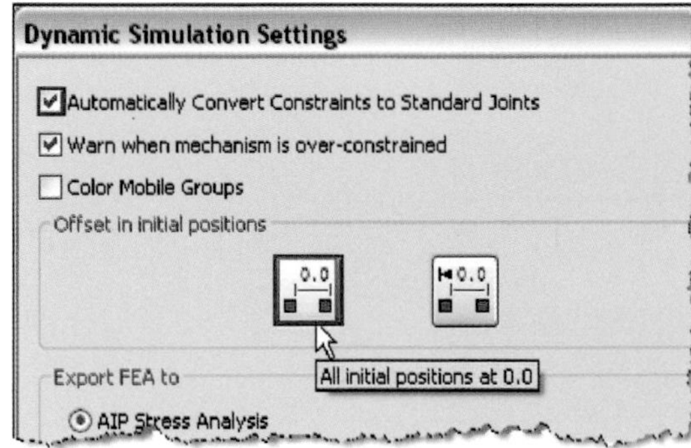

Selecting **Offset** in all initial positions will make all values in the **Joints Position** tab return to zero, making it easier to set the initial positions of joints.

5. Select the **Expand** button >Select **Input angular velocity in revolutions per minute (rpm)**

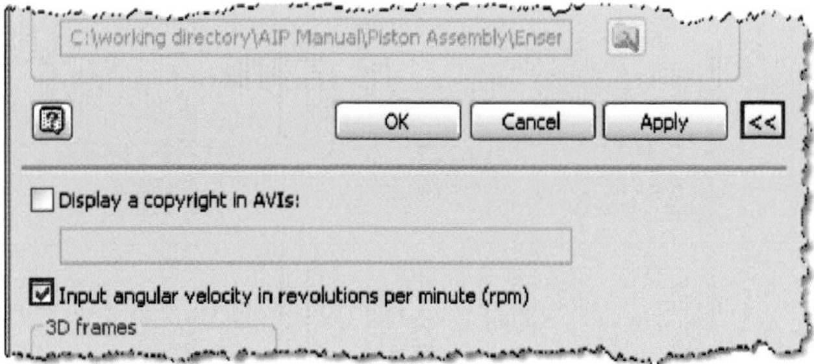

This will allow you to specify rotational velocity in revolutions per minute rather than degrees per second.

6. Click **OK**. Accept the warning regarding over-constraint

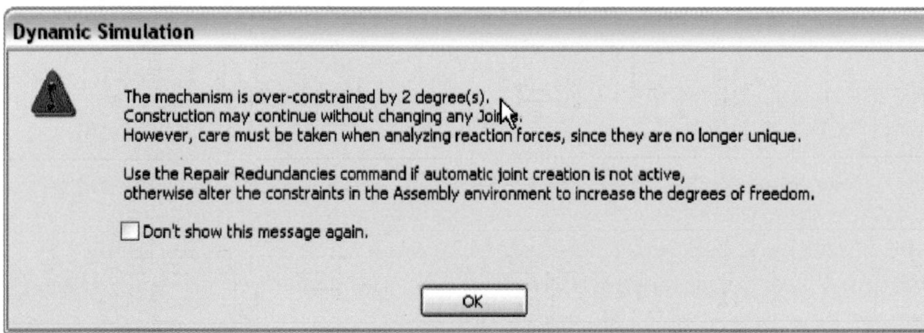

Based on predefined assembly constraints, Dynamic Simulation has attempted to create joints and, as a result, has over–constrained the model by two degrees of freedom, resulting in redundant joints.

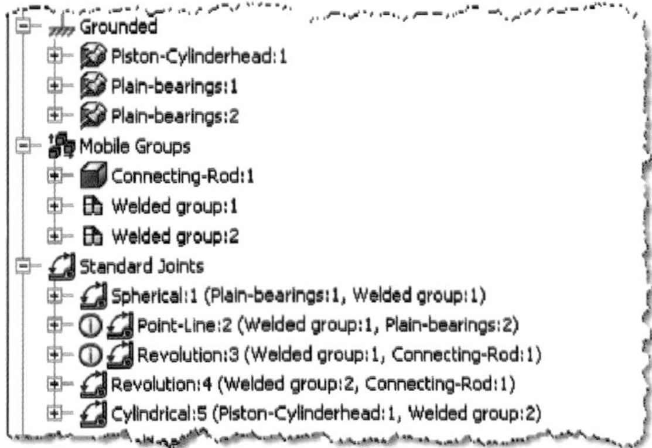

It is not necessary to fix the redundant joints suggested by Dynamic Simulation in order to remove redundancy from the model. Altering other nonredundant joints can remove redundancy.

Based on predefined assembly constraints, Dynamic Simulation has welded some components together in addition to creating revolution joints.

Before we can continue, we need to remove redundancy from the model.

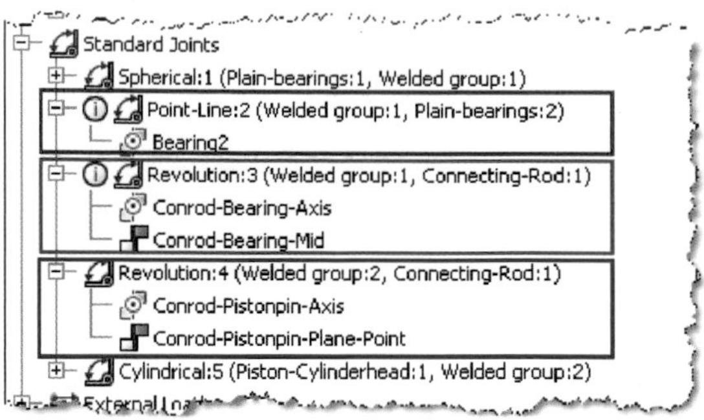

We cannot further release two degrees of freedom by altering the Point-Line:2 joint as it is the least restricted joint, by two degrees of freedom. Removing these two degrees of freedom would result in this joint becoming spatial. Hence, the components will not be connected.

We could suppress one of the constraints in the Revolution:3 redundant joint to release some degrees of freedom. The first Conrod-Bearing-Axis constraint aligns the two components about the central axis, and suppressing this constraint would disconnect the components. The second constraint aligns the two components symmetrically, about the axis. Suppressing this constraint would only release one degree of freedom. Note that suppressing this constraint would not necessarily misalign the position of the components as other joints would restrict their axial position.

In the Revolution:4 joint, we could suppress the Conrod-Pistonpin-Plane-Point constraint only, as the other aligns the components centrally about their axes. Suppressing the Conrod-Pistonpin-Plane-Point constraint would release two degrees of freedom, and hence would remove redundancy from the model completely.

Another option would be to modify the constraints within Dynamic Simulation or the Assembly environment.

7. Expand the Revolution:4 joint > Right click the Conrod-Pistonpin-Plane-Point constraint > Select **Suppress**

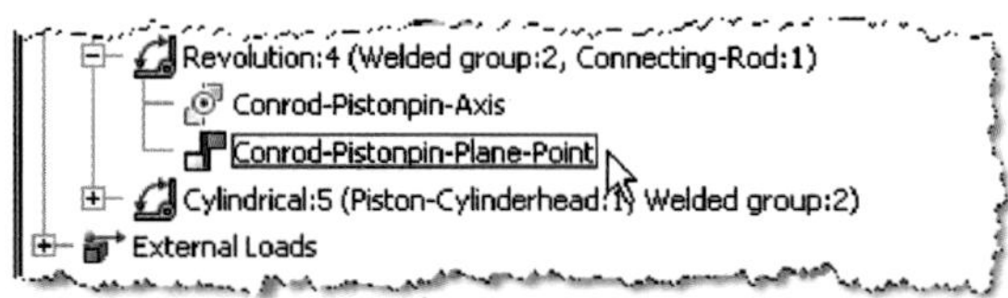

This will remove redundancy from the model, which you can check by selecting the **Mechanism Status** tool.

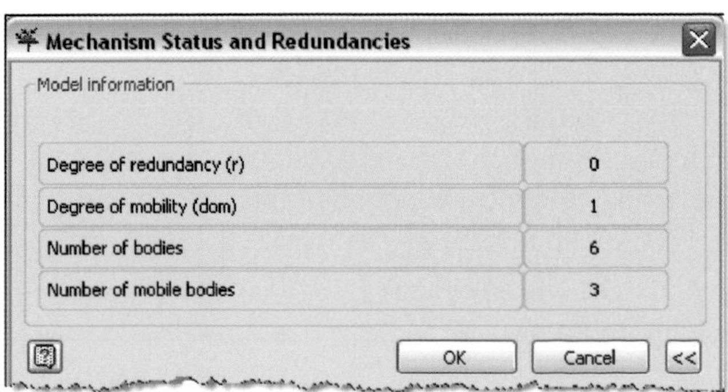

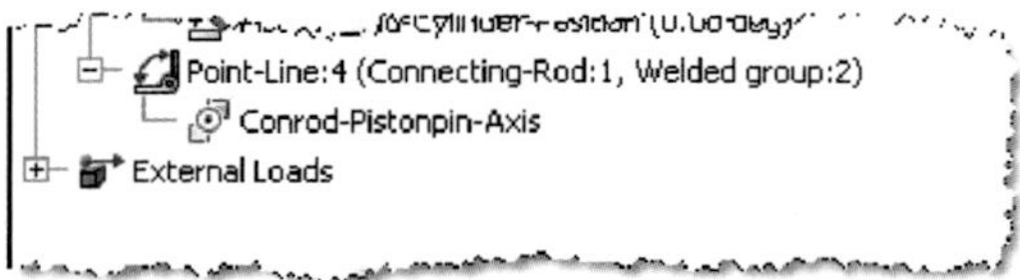

We expected a cylindrical joint as the only constraint left is an axis–axis constraint. To avoid redundancy in the model, Dynamic Simulation automatically alters the joints.

Now we have all the necessary joints created. The next step is to create environmental constraints.

Environmental constraints

Here we need to apply force on the piston head to simulate combustion forces and to control the force (combustion) until the engine speed achieves 7000 rpm (42 000 deg/s).

8. Select **Force** from the **Simulation** panel

9. For **Location**, select a point on the piston head

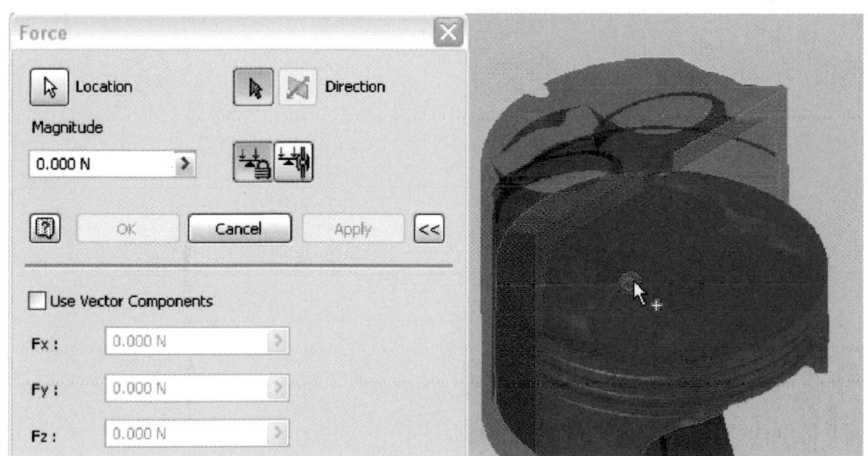

10. For **Direction**, select the edge of the cylinder

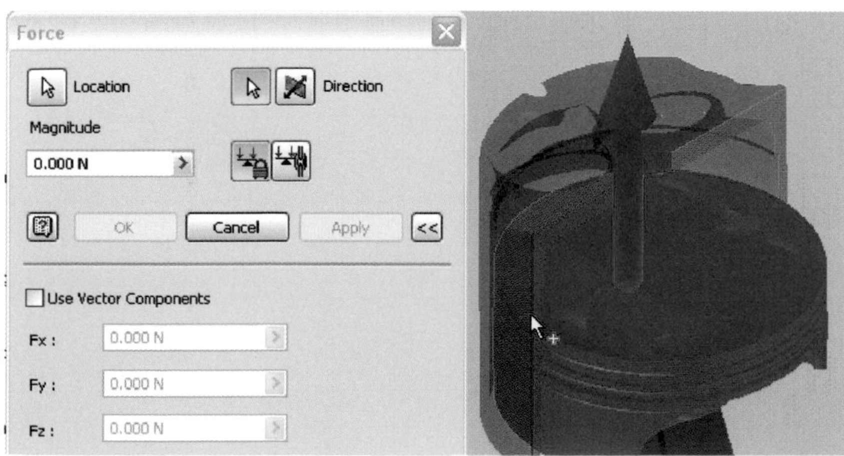

11. Change the direction so that the arrow points downward > Specify 10 000 N

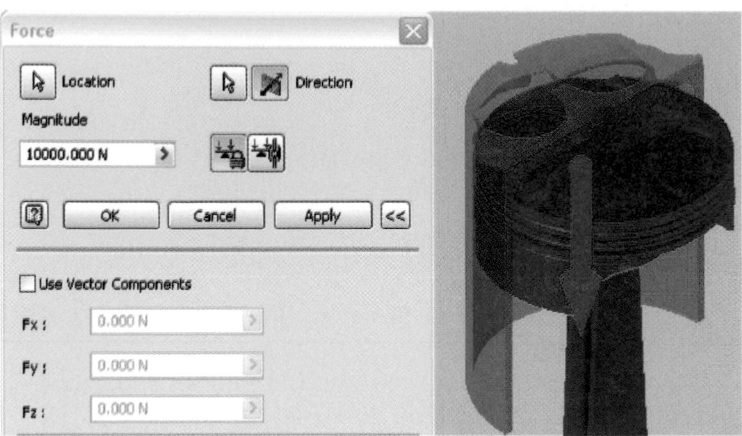

12. Select **Display** > Specify 0.000 01 for **Display Scale** > Click **OK**

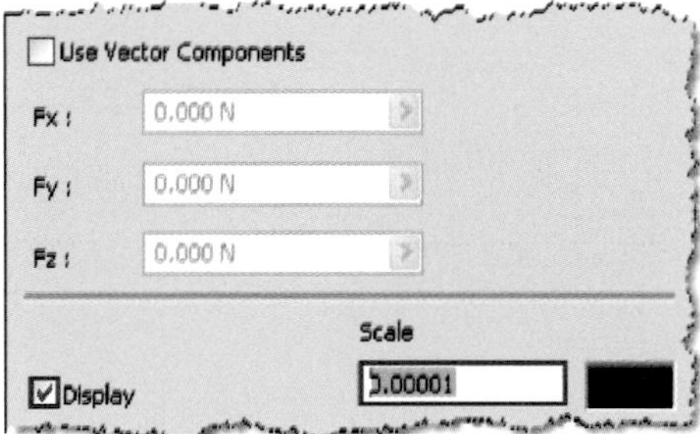

Next we are going to define the starting position of the piston.

IMPORTANT—If the position is not set to zero, either initialize all positions to zero again, within the simulation settings, or right click and select **Set offset,** within the magnitude value, before specifying 90 deg.

13. Double click the Spherical:1 joint > Select the **dof 1(R)** tab > Select 90 deg for **Position** > Click **OK**

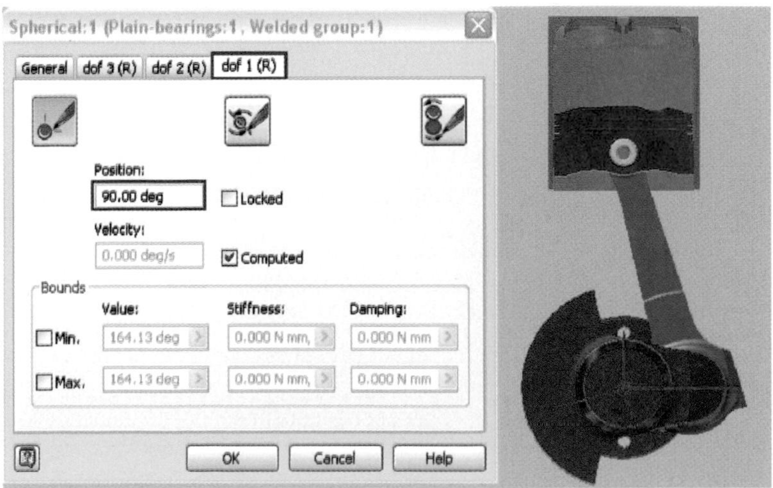

14. Set the final time to 0.02 seconds > Set the **Image** value to 200 > Play the simulation

Analyze results

15. Select Output Grapher > Select the Force value under the External Loads node

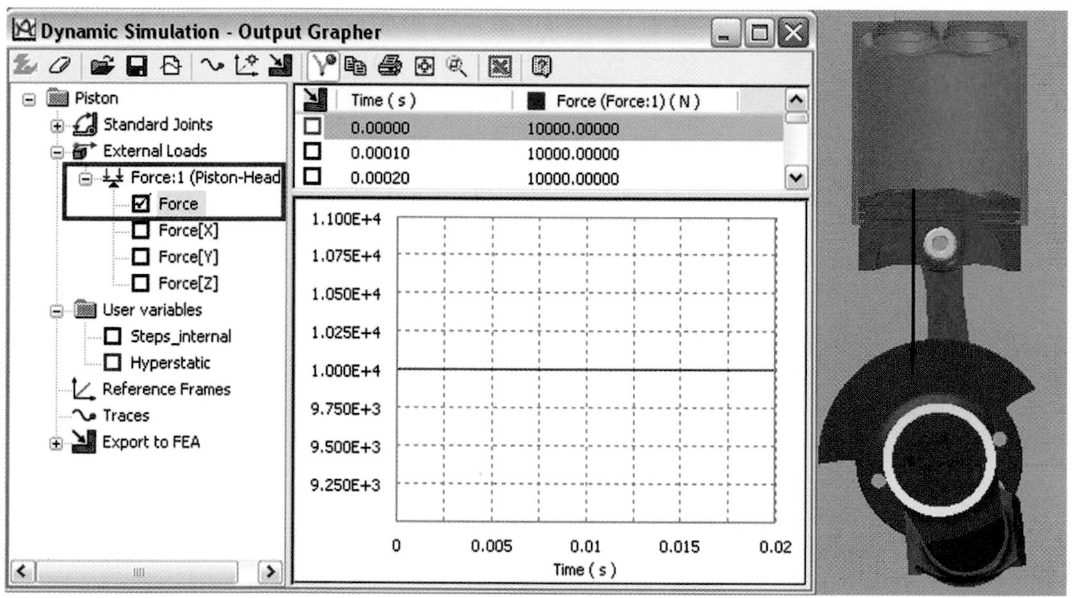

16. Now select the Velocity value for the Prismatic:5 joint

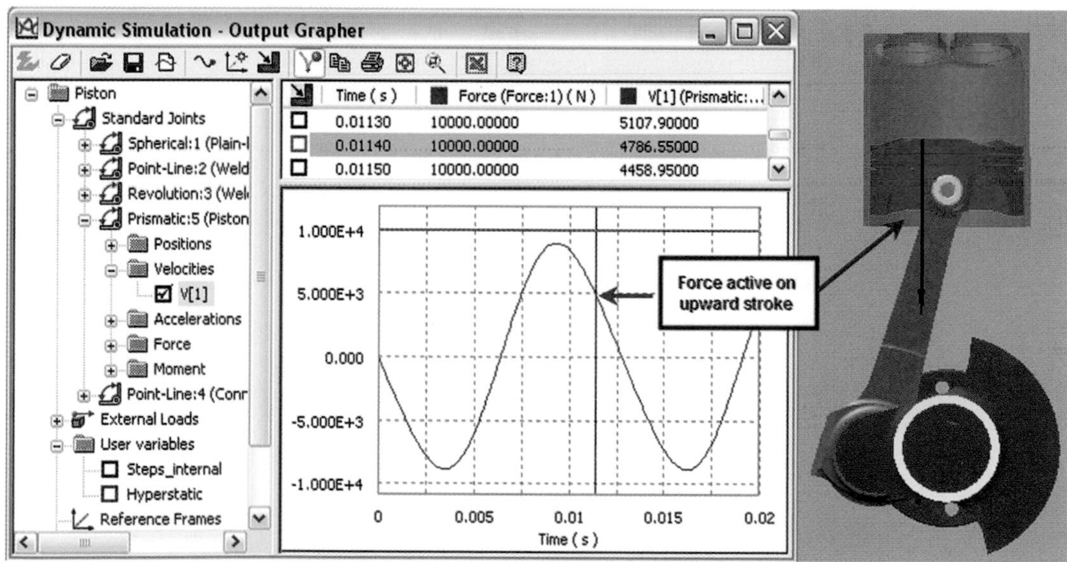

This shows us that the force is active on the upward stroke of the piston. We need to switch the force off on the upward stroke of the piston.

Use the mouse wheel to pan and zoom in the Output Grapher to help pinpoint the exact location of interest.

In the following steps, we will control the force so that it switches off on the upward stroke.

17. Minimize **Output Grapher** > Select **Construction Mode**

18. Double click Force 1 > Select **Input grapher**

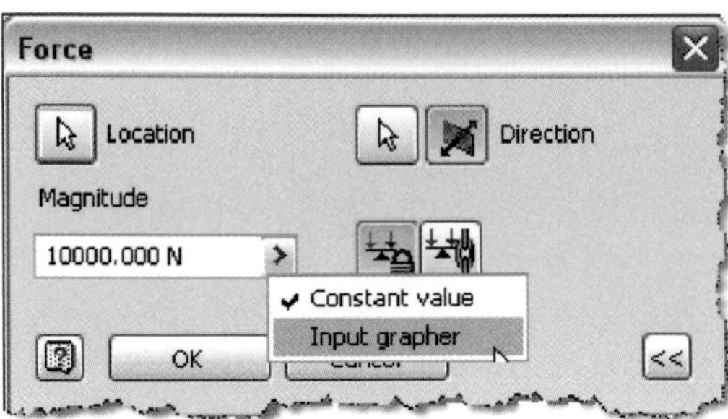

19. Type 0s for *X*1 > 10 000 N for *Y*1 > 1s for *X*2, and 10 000 N for *Y*2

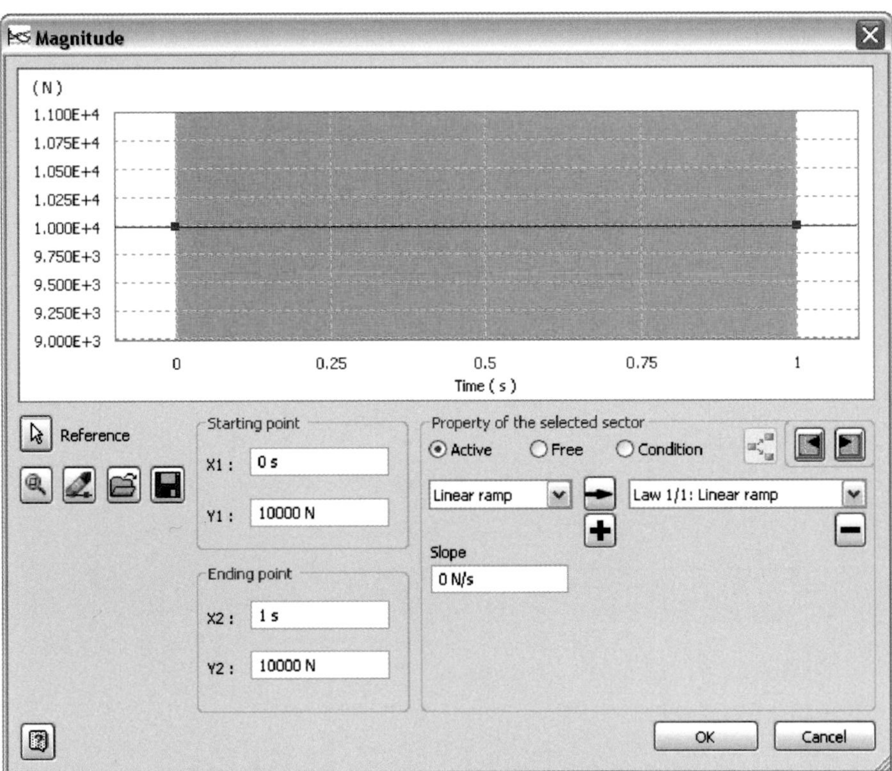

20. Select the **Condition radio** button in the Input Grapher

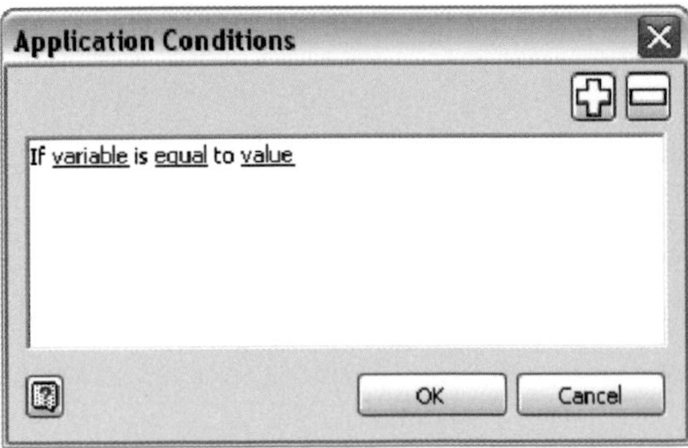

21. Select variable in the Application Conditions > Select V[1] for the Prismatic:5 joint

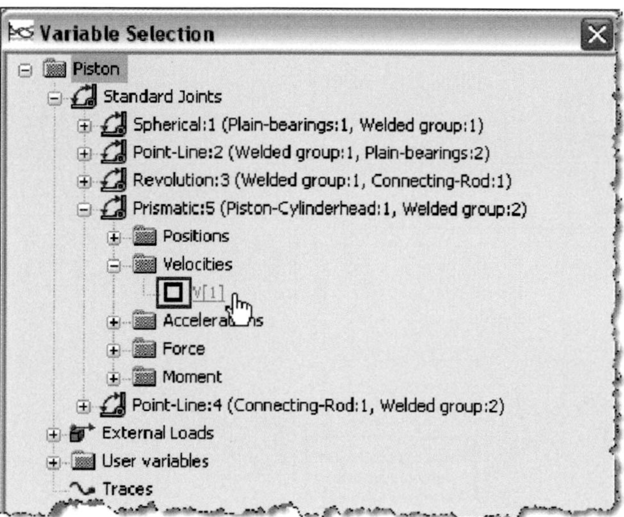

This is the value we are controlling against the force.

22. Select **equal logic** > **less than or equal**

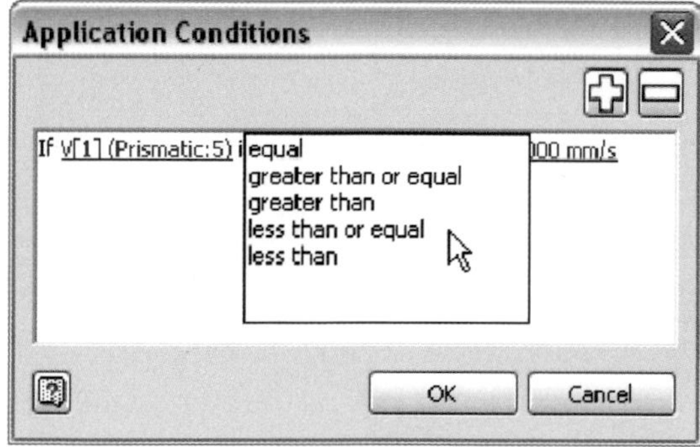

23. Specify a value of 0 mm/s > Click **OK** three times

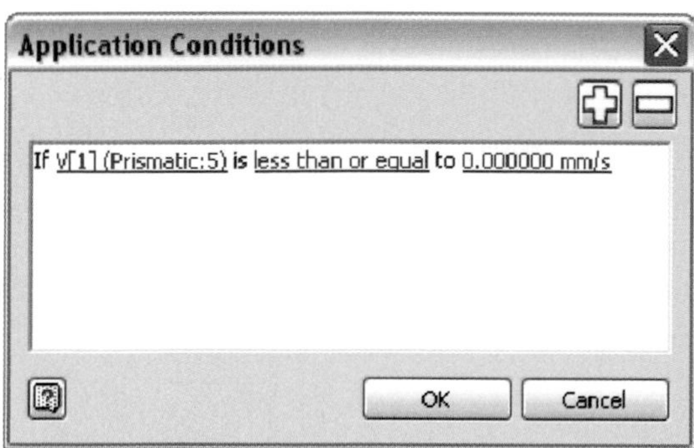

24. Set the final time to 0.04 seconds > Play the simulation

25. Maximize the Output Grapher > Zoom in on the graph to see the value of the force on the upward stroke

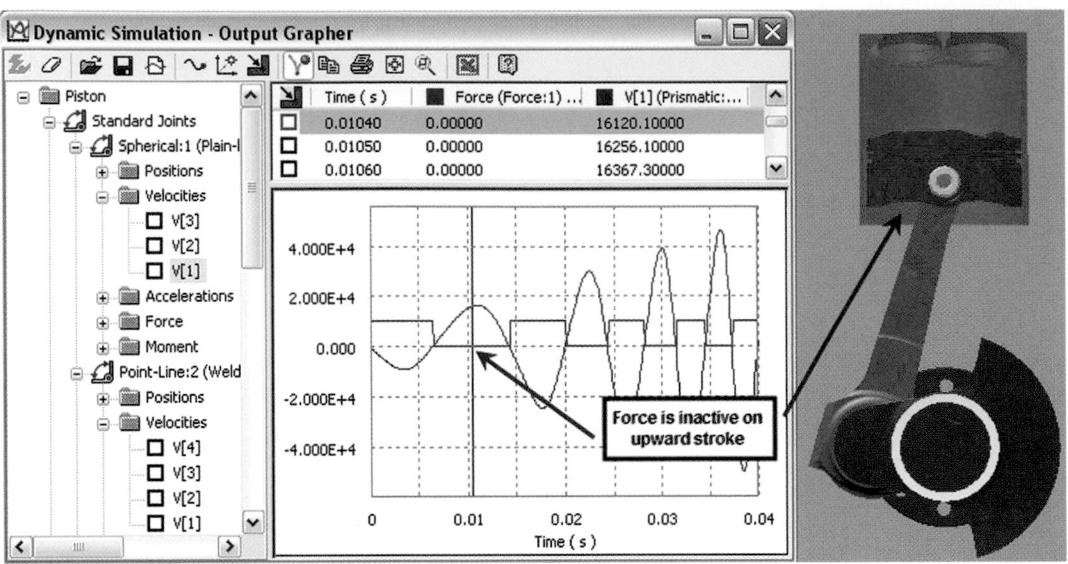

26. Select V[1] under the Spherical:1 joint to determine the velocity of the crank/engine

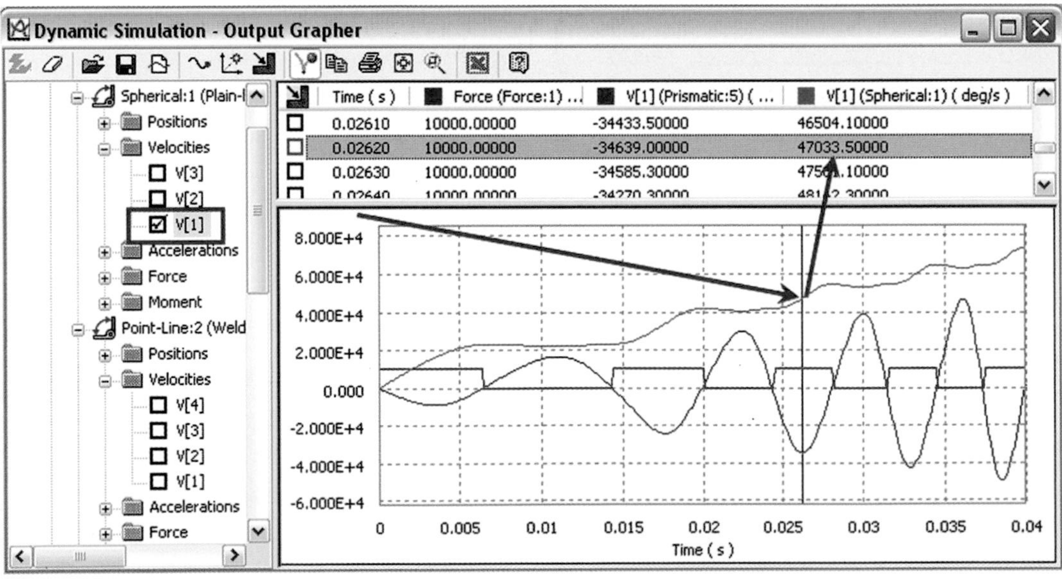

The crank velocity keeps increasing as a result of force being applied continually. We now need to stop applying the force once the engine (crank) has achieved a velocity of 7000 rpm, and determine the time taken to achieve the desired velocity.

27. Minimize the Output Grapher > Set **Construction Mode**

28. Repeat Steps 17–20 > Select **+** to add more conditions

29. Select variable in the Application Conditions > Select V[1] for the Spherical:1 joint

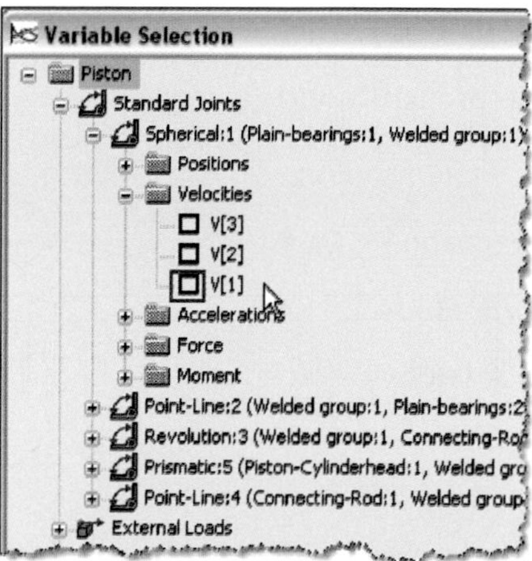

30. Select **equal logic** > **less than or equal**

31. Specify a value of 7000 rpm

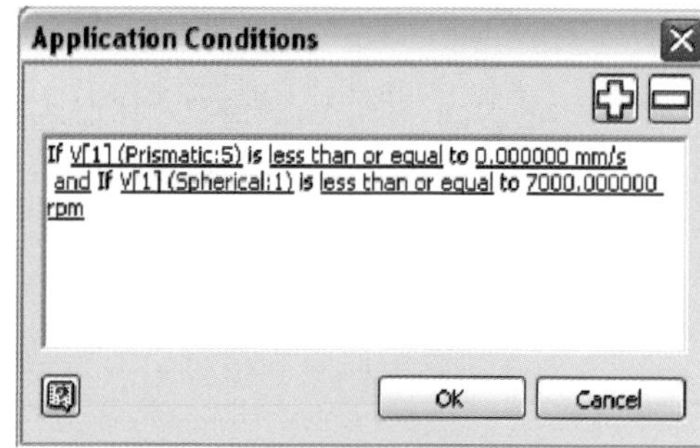

32. Click **OK** three times

33. Play the simulation

34. Maximize the Output Grapher

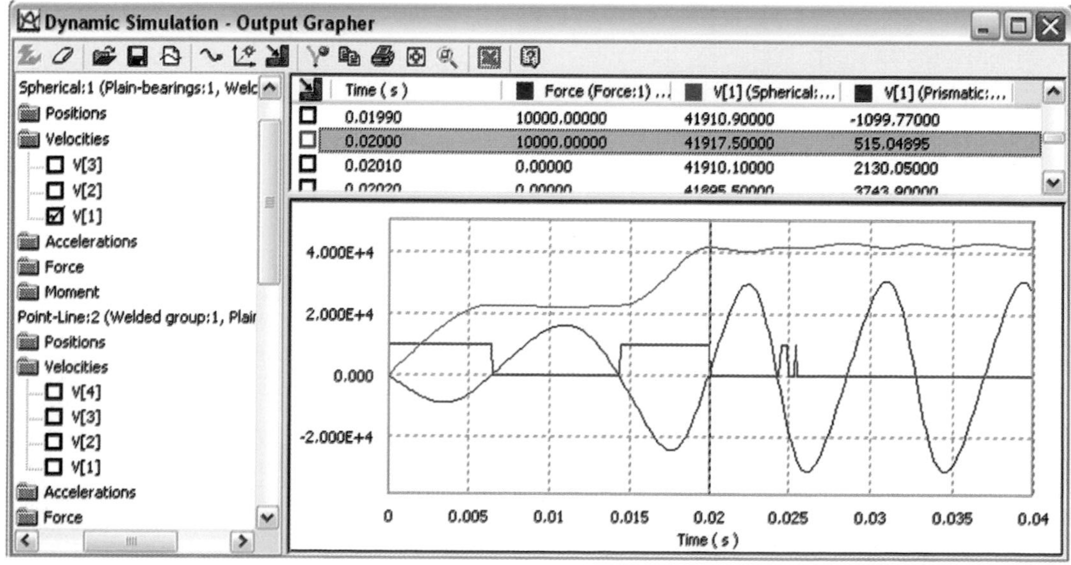

From the graph, we can see that the engine reaches the target engine speed within 0.02 seconds.

The reason why the velocity is not constant is due to the inertia and dynamics of the model. It is possible to further tune the results by altering the mass properties and geometry of the model. The velocity of the crank can be further refined by trying to simulate all four pistons and by introducing a flywheel.

Time taken to reach 7000 rpm (42 000 deg/s) = 20 ms (for a 10 000 N force).

35. Minimize the Output Grapher > Click **Construction Mode**

36. Change the force from 10 000 N to 25 000 N > Click **OK** twice

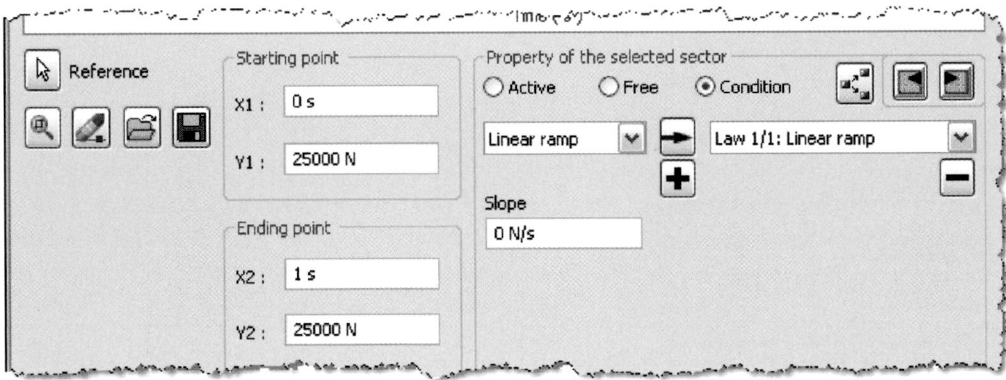

37. Play the simulation

38. Maximize the Output Grapher

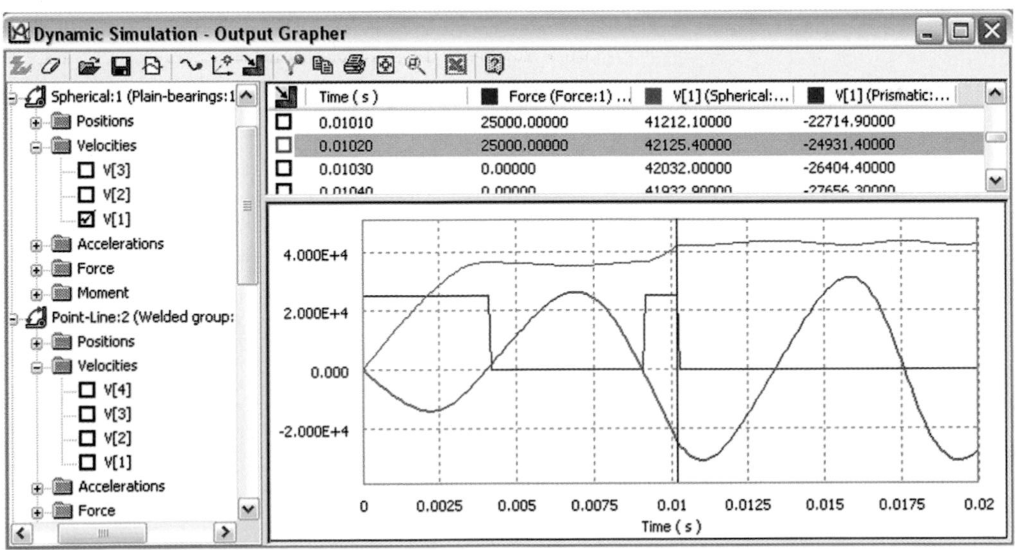

From the graph, we can see that the engine reaches the target engine speed within 0.0102 seconds.

Time taken to reach 7000 rpm = 10.2 ms (for a 25 000 N force).

It is also important to note that the simulation analyzed upto this point has not taken friction into consideration. In the following section, we will determine the torque of the crank at 7000 rpm with and without friction in the joints.

39. Minimize the Output Grapher > Select **Construction Mode**

40. Right click Force:1 > Select **Suppress**

41. Double click the Spherical:1 joint > Select the **dof 1(R)** tab > Select **Imposed Motion** > Select **Enable imposed motion** > Specify 7000 rpm for **Velocity** > Click **OK**

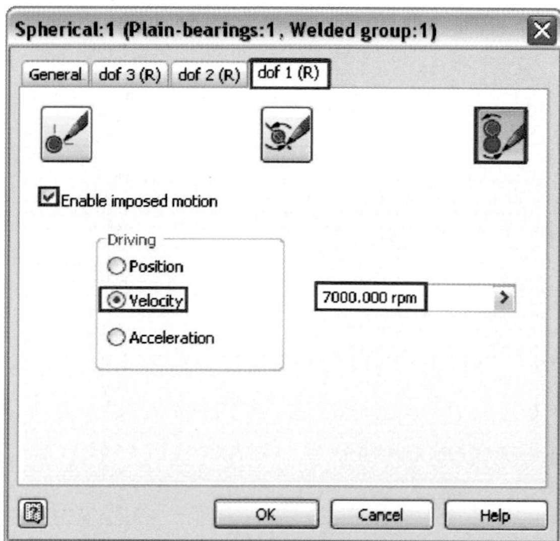

42. Set the final time to 0.02 s > Play the simulation

43. Maximize the Output Grapher > Select **Unselect all Curves** > Select the driving force for the Spherical:1 joint

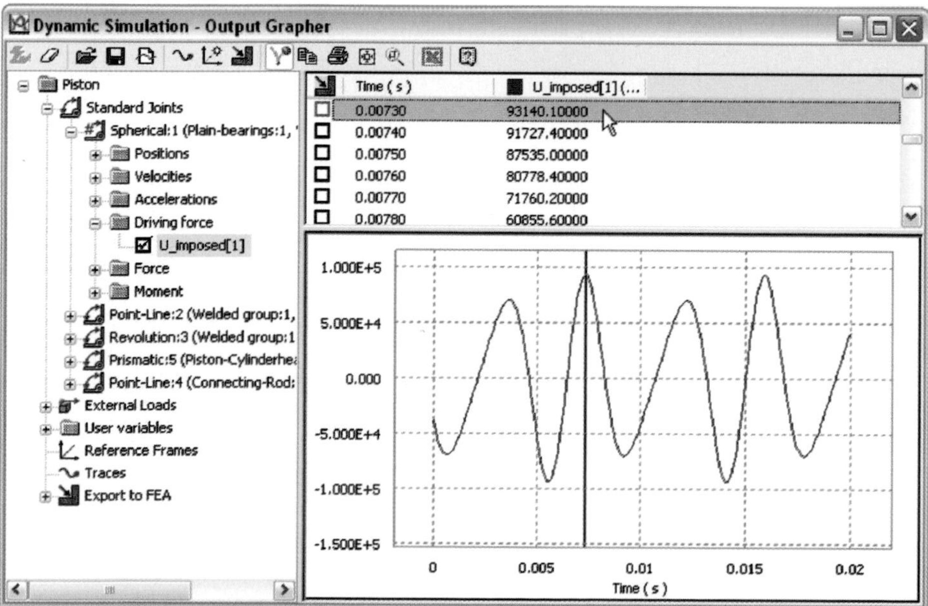

The torque is cyclic between 93 140 Nmm (93.14 Nm) and −93 416 Nmm (−93.41 Nm)

Now we will introduce friction in the joints.

44. Minimize the Output Grapher > Select **Construction Mode**

45. Double click the Spherical:1 joint > Select the **dof 1(R)** tab > Select **Enable joint torque** > Specify 0.05 for **Damping** > Specify 0.2 for **Coefficient** > Specify 25 mm for **Radius** > Click **OK**

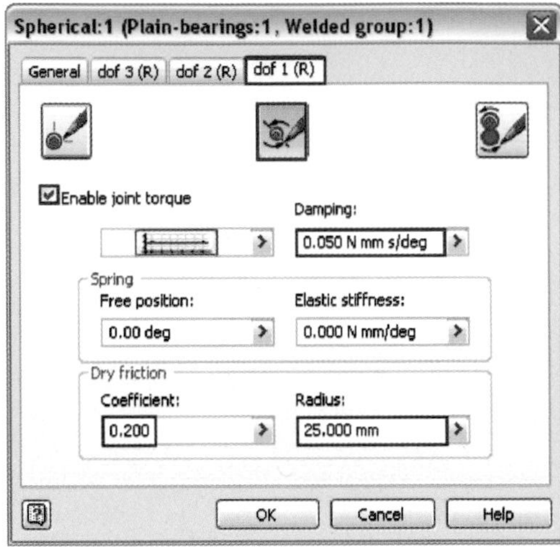

46. Repeat Step 45 for the following joints and specifications:

Joint	dof tab	Damping value	Friction coefficient	Radius
Point-Line 2	dof 3(R)	0.05	0.2	25 mm
Revolution 3	dof 1(R)	0.05	0.2	17.5 mm
Prismatic 5	dof 1(T)	0.05	0.2	–
Point-Line 4	dof 3(R)	0.05	0.2	9 mm

47. Play the simulation

48. Maximize the Output Grapher > Select the Driving force for the Spherical:1 joint

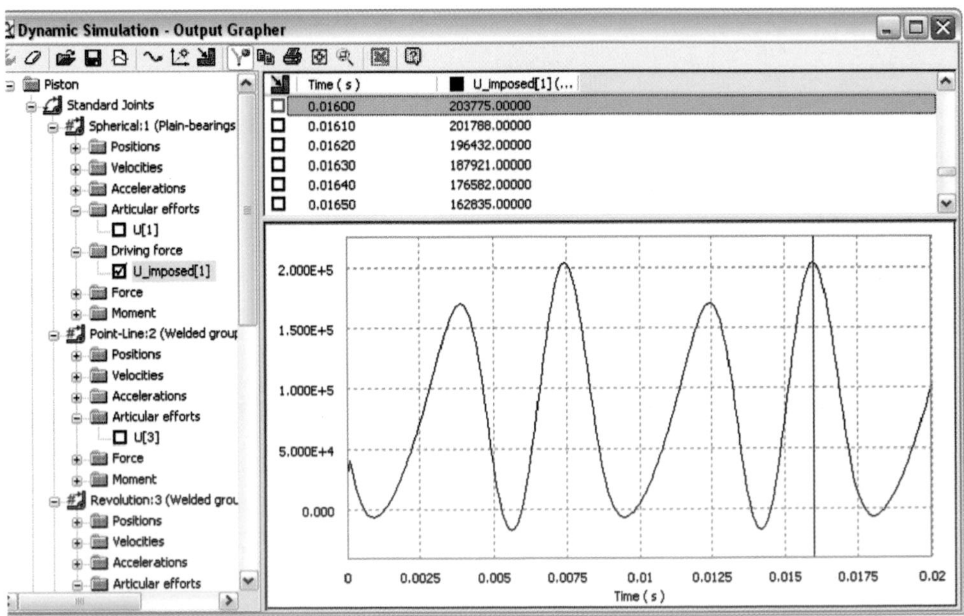

The torque is cyclic between 203.77 Nm and −31 Nm. Thus, maximum torque has increased.

49. Deselect U-imposed[1]

50. Select U[dof] under Articular efforts for all joints

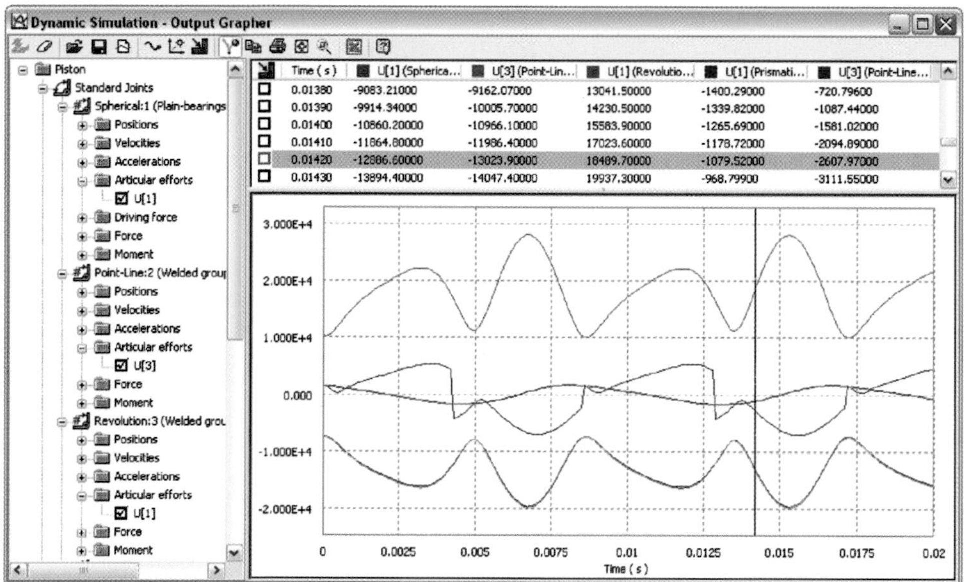

These are the torques required (loss) to overcome the friction.

We will now determine the reaction loads acting on the connecting rod.

51. Right click anywhere in data column > Select **Unselect all Curves**

52. Select the Force variables for both the Spherical:1 and the Point-Line:2 joints

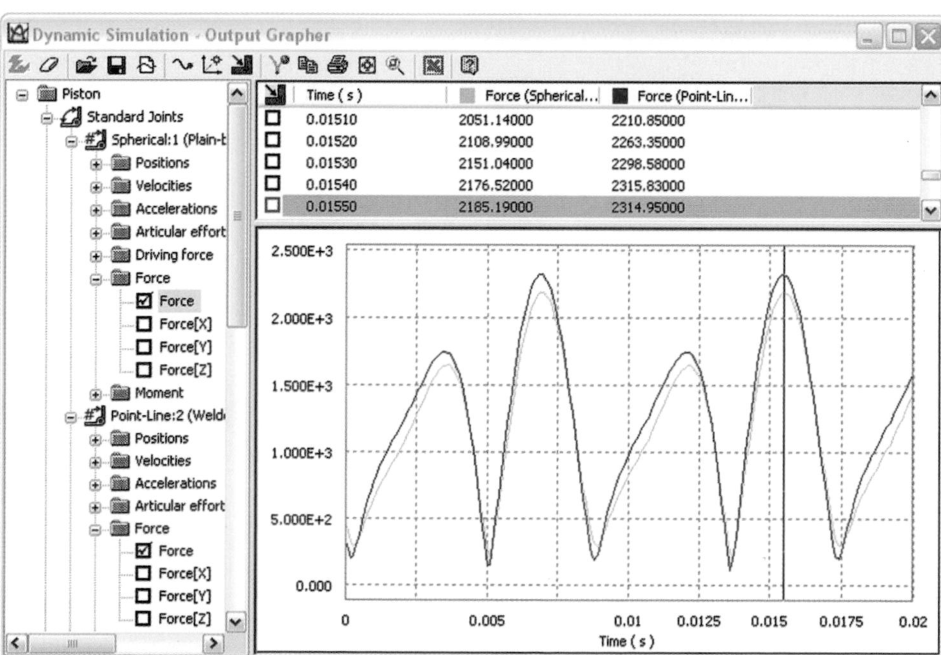

In Chapter 11, we will export these reaction loads acting on the connecting rod to the Stress Analysis environment to validate the structural integrity of the connecting rod operating at 7000 rpm.

CHAPTER 6

DP5 – Size a Spring

Agricultural Spring Mechanism Design

(Design Problem courtesy of Simba Ltd)

JOINTS INTRODUCED/COVERED IN THIS DESIGN PROBLEM

	Joints	Joint creation process
1	Revolution	Automatically converted
2	Spherical	Automatically converted
3	Cylindrical	Automatically converted
4	Prismatic	Automatically converted
5	Spring	Manually created

KEY FEATURES AND WORKFLOWS INTRODUCED IN THIS DESIGN PROBLEM

	Key features/workflows
1	Input Grapher – change reference axis – define force
2	Force – Export trajectory to Output Grapher
3	Output Grapher – create new curve – **Set as Reference**
4	**Unknown Force**

INTRODUCTION

Simba has a reputation for developing and testing innovative products within the agricultural and cultivating sector and has manufacturing facilities in the UK, Hungary, and Poland. A typical product is a ground-engaging tine, which is attached to an agricultural cultivator chassis via a spring-cushioned clamp assembly.

Up and Running with Autodesk Inventor Simulation 2011. ISBN: 978-0-12-382102-7
Copyright © 2010 Elsevier Inc. All rights of reproduction, in any form reserved.

The tine tip must near-constantly be maintained in a position where it is capable of penetrating hard soil, while at the same time having the capability to ride upward over obstructions such as stones present in the soil profile. Thus, excessive tip forces generated when hitting immovable objects and that could otherwise damage the machine chassis must be avoided.

The requirements of this design problem are to:

- Design a spring for the clamp assembly. This needs to provide sufficient tip force to penetrate hard soil, and allow for a limited amount of tip movement in the (mainly) horizontal plane within the working range of tip loads. This will provide for shattering hard soil as the tip travels at a consistent working depth. The loads used have been determined by field trials of the tine across a range of soil conditions.

- Confirm the total tip movement within the safe working range of the spring. The clamp assembly is designed to allow for a high degree of vertical tip movement later in the tripping sequence, so that the tip can ride upward over obstructions present. The model is used to determine the maximum tip force generated through the tripping range. This, in turn, can then be applied to the chassis (either by transferring the force directly into FEA or by normal methods of stress calculation) to confirm that the chassis design is adequate for the application.

WORKFLOW OF DESIGN PROBLEM 5

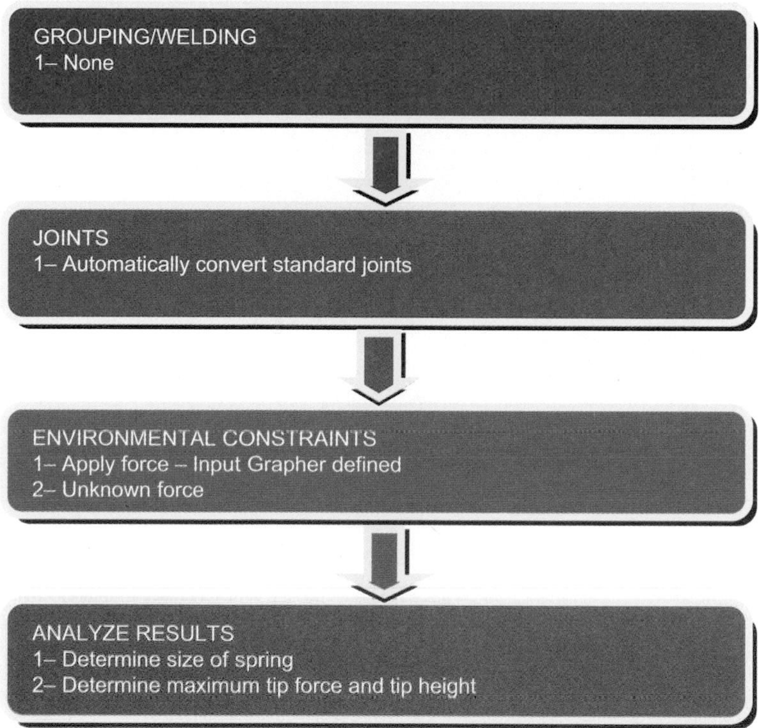

Grouping/welding

In this design problem, the main requirement is to determine the size of the spring; with this in mind, all the components have already been grouped together into subassemblies, keeping in mind that the joints needed to be created.

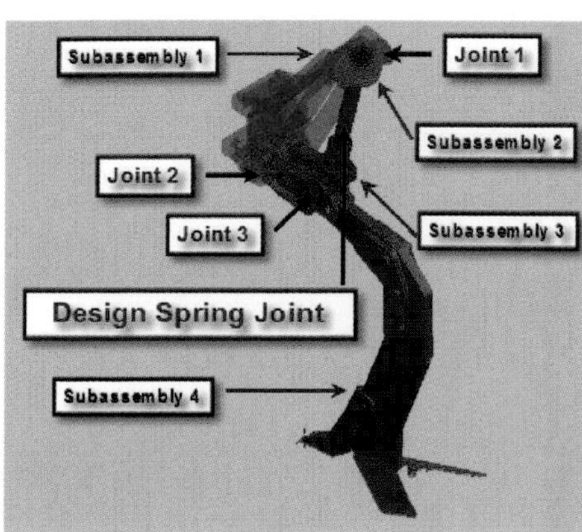

Joints

1. Open *Spring-Mechanism.iam*

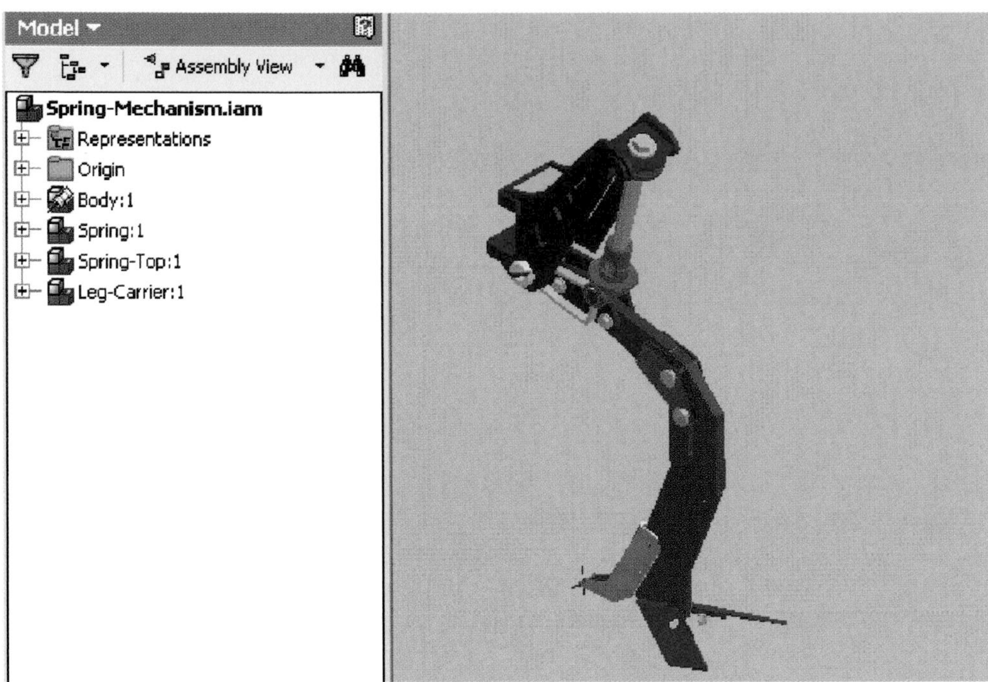

2. Select **Environments** tab > **Dynamic Simulation**

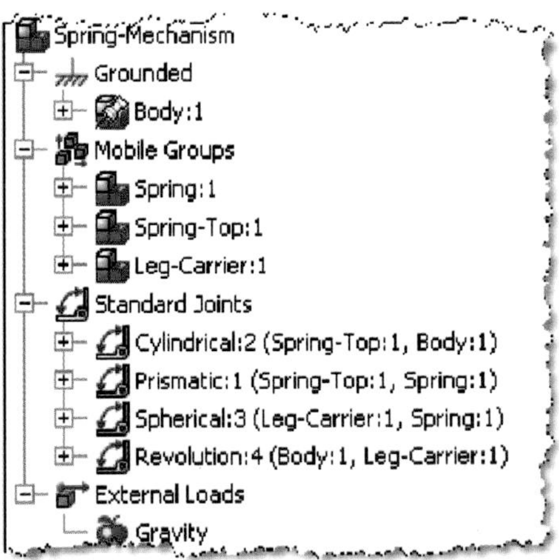

Environmental constraints

Based on significant engineering test data, the following conditions are to be used for determining the size of the spring:

- Maximum deflection of tip to be less than 100 mm.
- Force range of 1500–1750 N to simulate normal working conditions.

These conditions are graphically illustrated below.

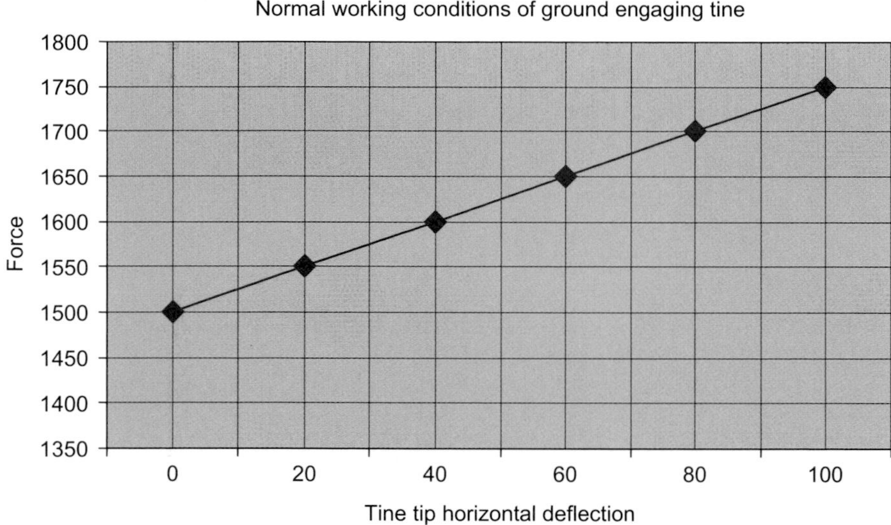

Finally, these conditions are illustrated below in terms of the product.

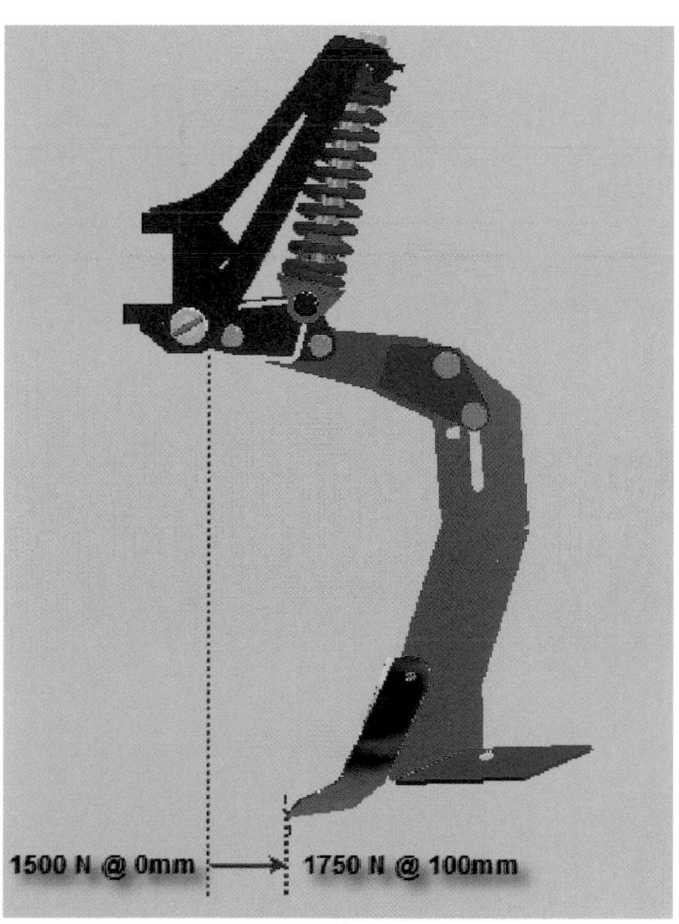

Before we proceed to determine the size of the spring, it would be beneficial to determine some basic spring behavior. Usually, compression springs have a constant stiffness and behave linearly as indicated in the graph below.

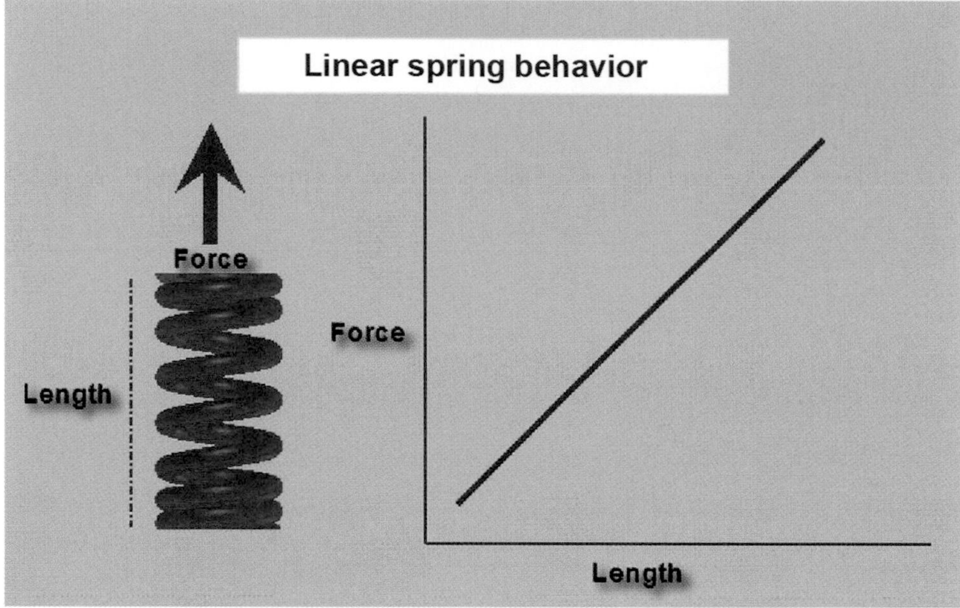

Where a nonlinear characteristic is needed (as in this case, once the tip has moved out of its working range to lift over an obstruction), this can be achieved by the geometry of the mechanism or assembly, as illustrated below. Here, the tip force can increase at a reducing rate once it is above its working range, so less stress is imposed on the assembly. This will become clear as the design work proceeds.

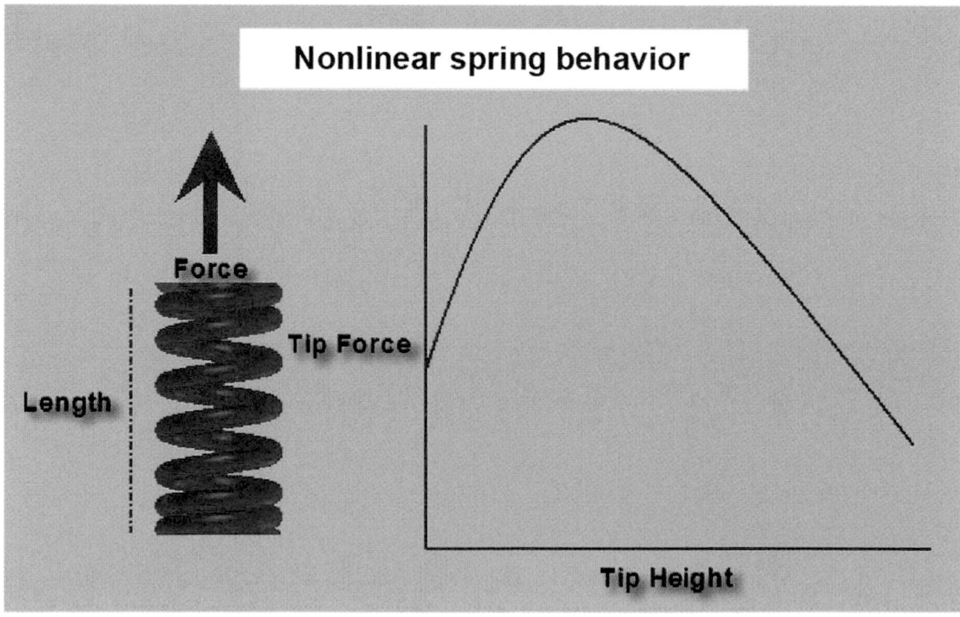

For the purposes of this design problem, we will assume a linear behavior of the spring for the normal working conditions of the tine, as this will help us to determine the spring stiffness from the gradient of the curve, as illustrated below.

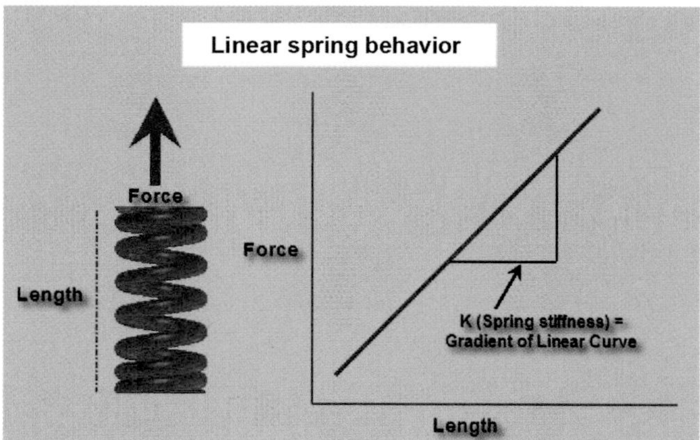

We now need to simulate these conditions, including

- Defining the horizontal displacement of the tine.
- Defining the force as a function of the horizontal displacement of the tine.

DEFINING THE HORIZONTAL DISPLACEMENT OF THE TINE

3. Press **Alt** . on the keyboard. > Zoom into the model as shown

This will make the user workpoints visible.

4. Select the Output Grapher > Select **Add Trace**

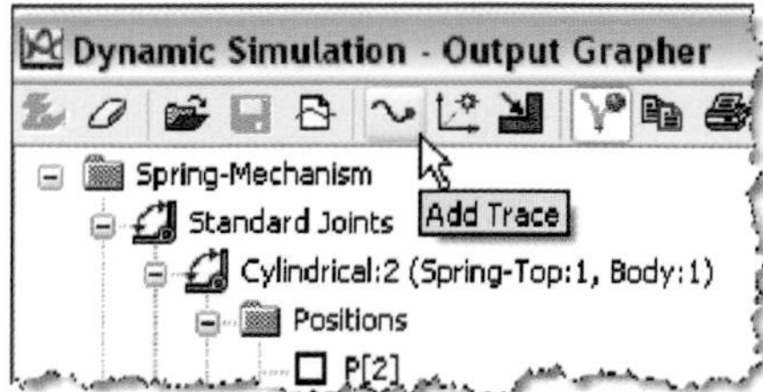

5. Select the workpoint on the tine tip > Select **Display Trace** > **Output Trace**

6. Click **OK**

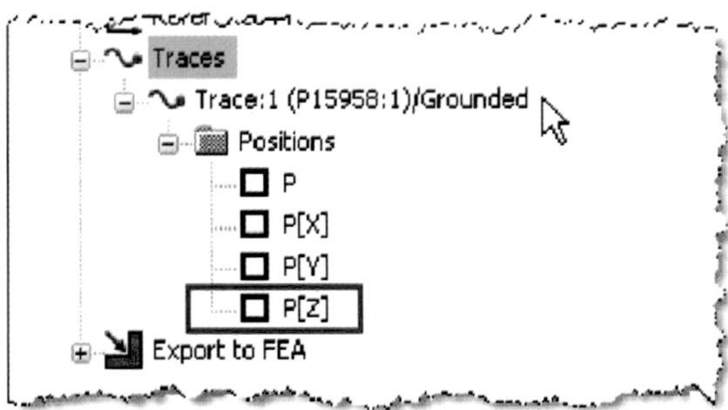

Selecting the **Output Trace** value has added the X, Y, and Z positions of the tine tip into the Output Grapher. We will use the Z (horizontal) position to define the force function in the next step.

DEFINING THE FORCE AS A FUNCTION OF THE HORIZONTAL DISPLACEMENT OF THE TINE

7. Close the Output Grapher > Select **Force** from the **Dynamic Simulation panel**

8. For the **Location** of **Force**, select the workpoint as shown

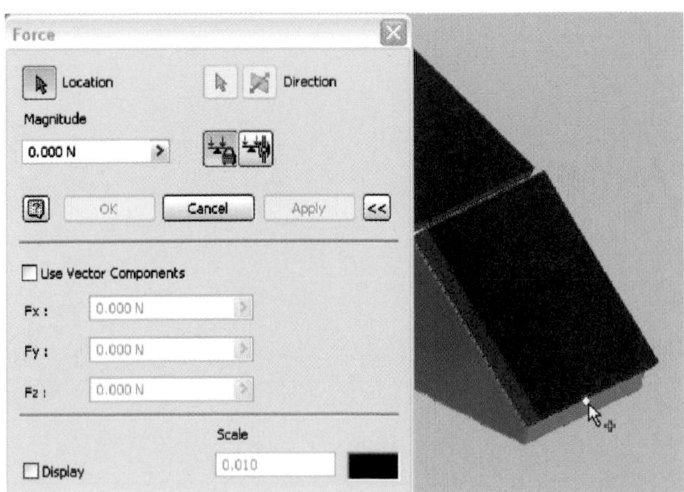

9. For **Direction**, select the edge of the body as shown

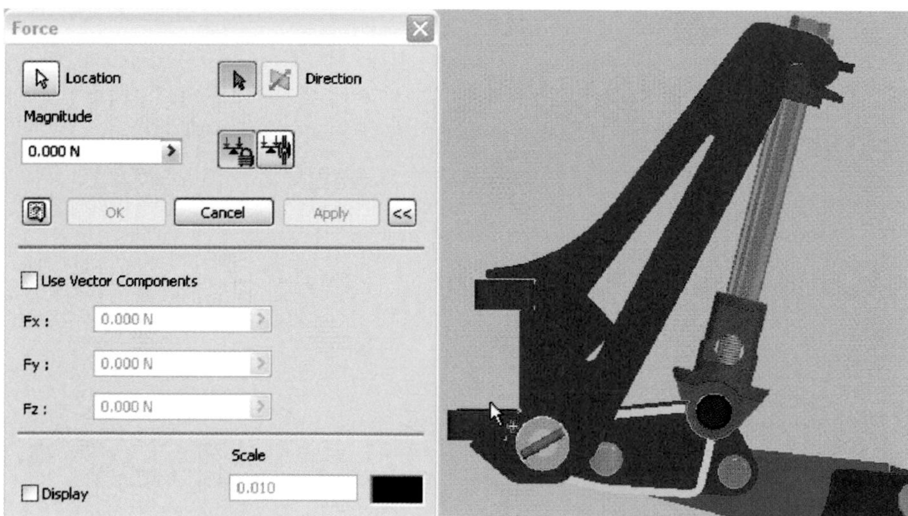

10. Select the Input Grapher to define the magnitude of **Force**

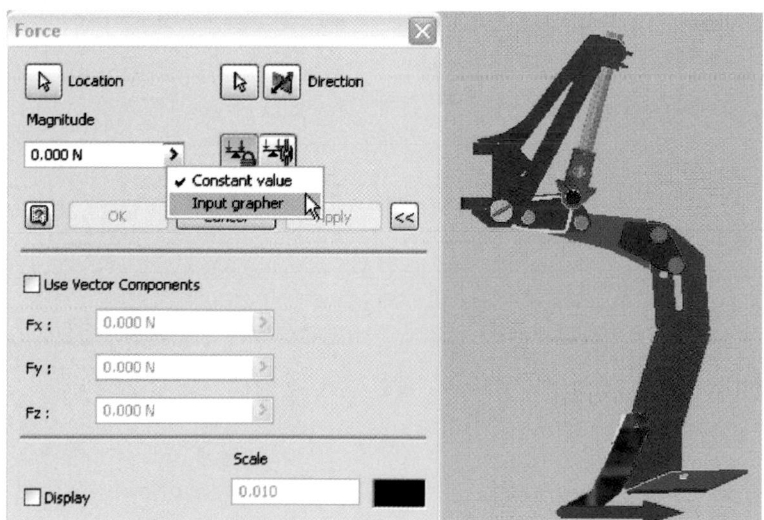

11. In **Select Reference** P[Z] under Trace:1, select and click **OK**

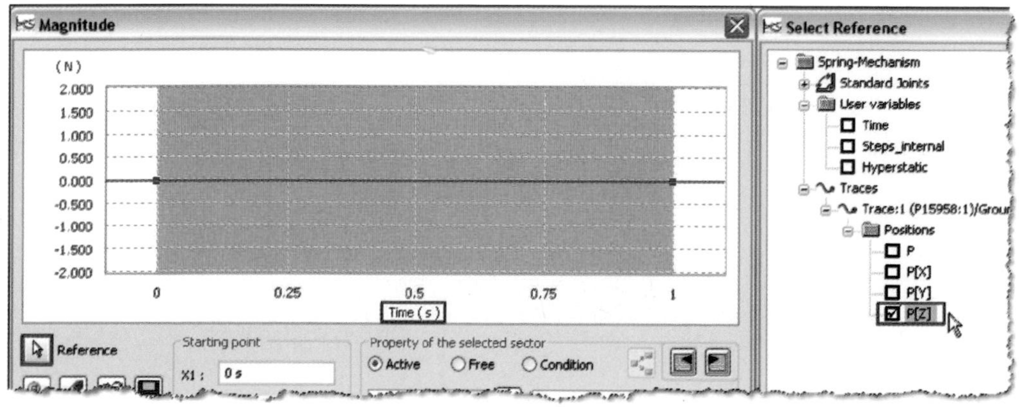

This allows us to define **Force** as a function of the horizontal tine position, as required.

12. Type 0 mm for X1, 1500 N for Y1, 100 mm for X2, and 1750 N for Y2

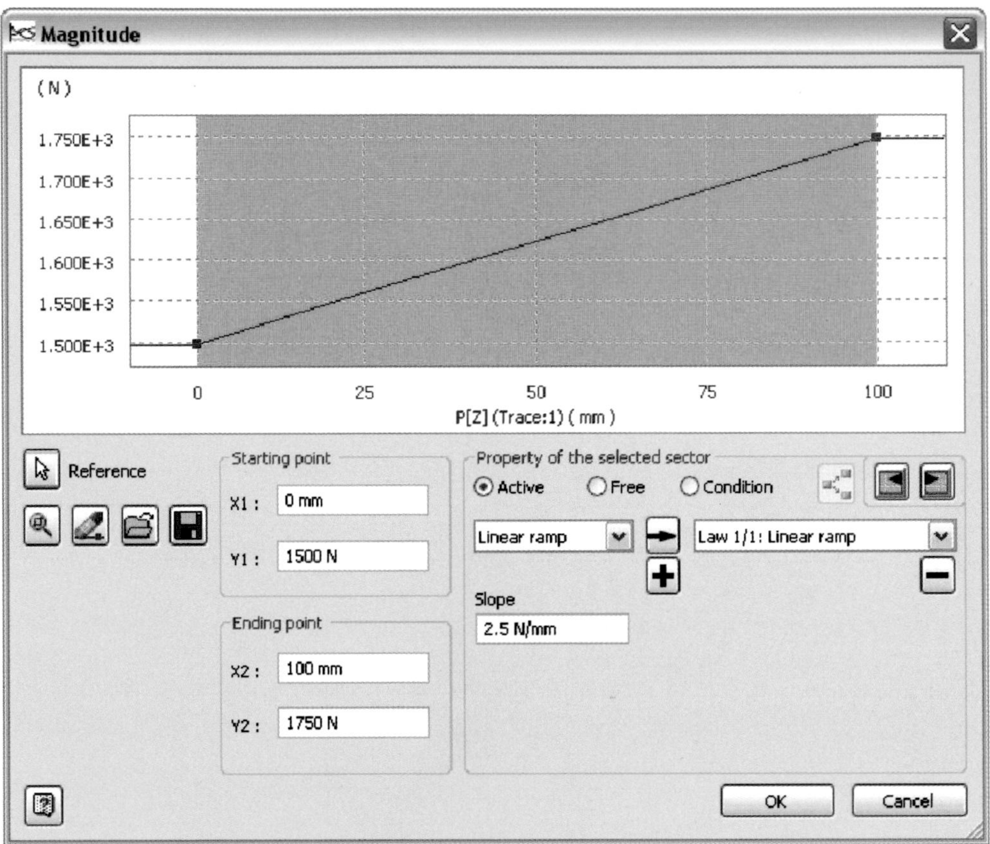

The force function is defined linearly between 0 and 100 mm displacement of the tip. Beyond this range, the graph assumes a constant force function, which is not the case in reality as we would expect the force to increase linearly with displacement. Therefore, we will modify the force function.

13. Select **Next sector**

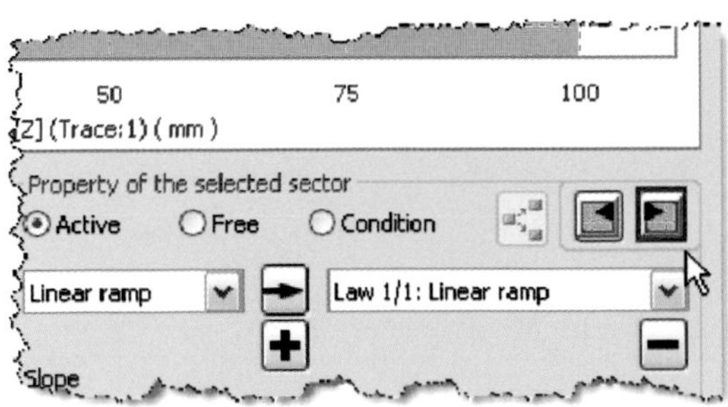

14. Select **Constant slope**

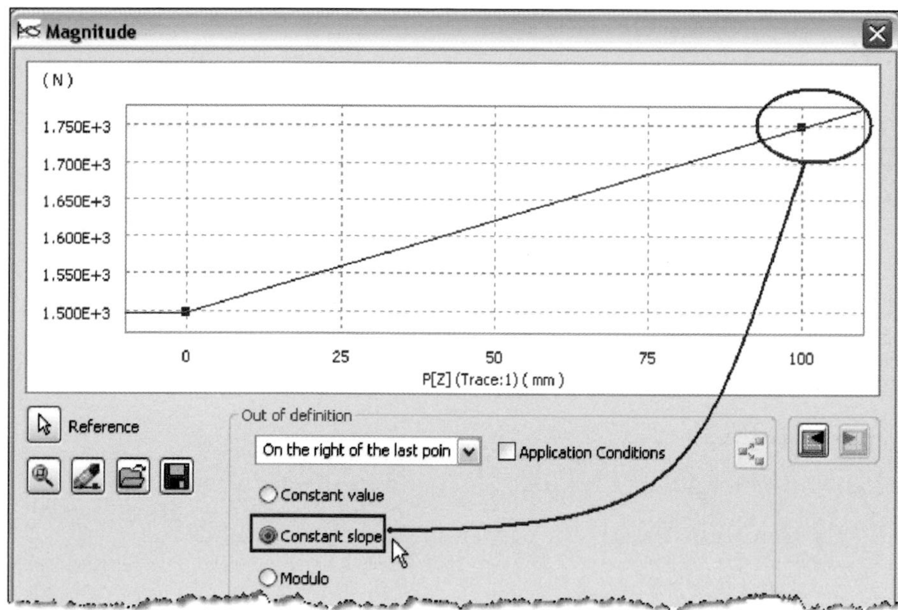

15. Select **Previous sector** twice

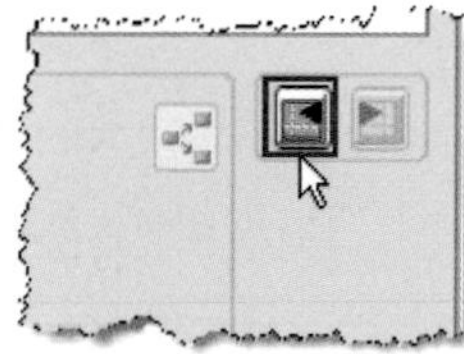

16. Select **Constant slope** > Click **OK**

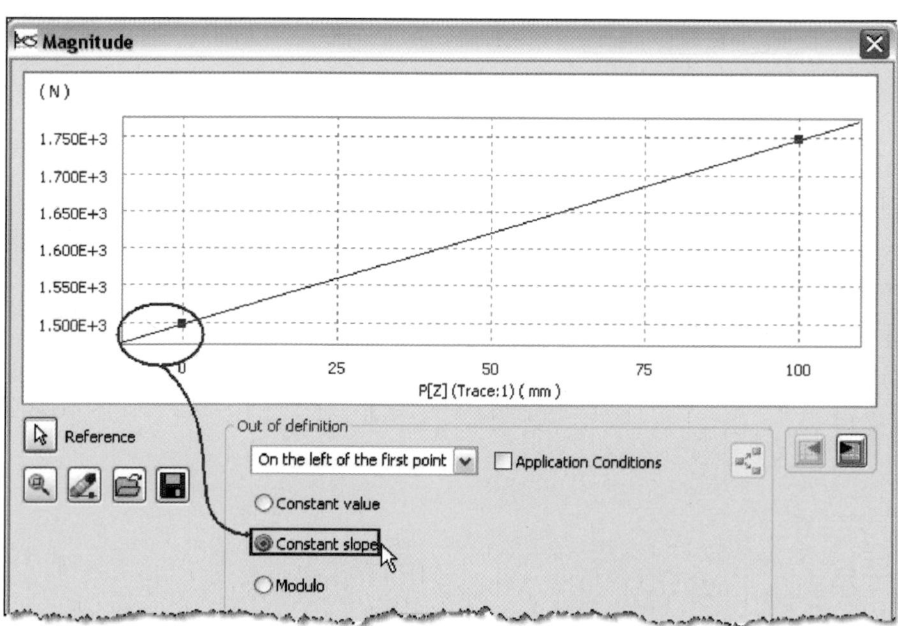

17. Select **Display** and enter 0.0001 for the **Scale** value. Click **OK**

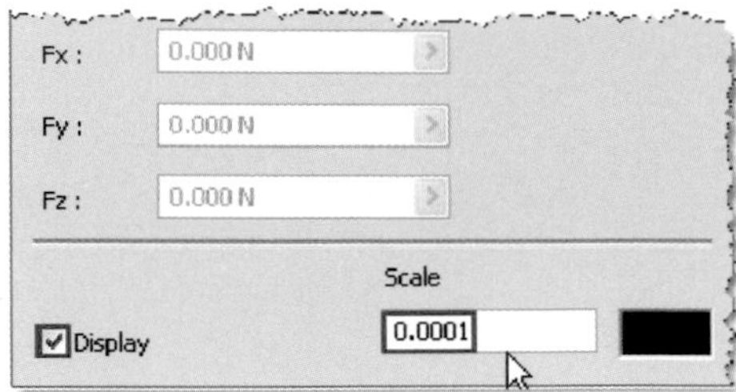

Analyze results

Since the force is now defined, we are ready to determine the size of the spring required to operate in normal working conditions.

DETERMINE SIZE OF SPRING

18. Select **Unknown Force** from the **Dynamic Simulation** panel

19. Select **Jack** > For **Location 1**, select the edge of the spring top, as shown

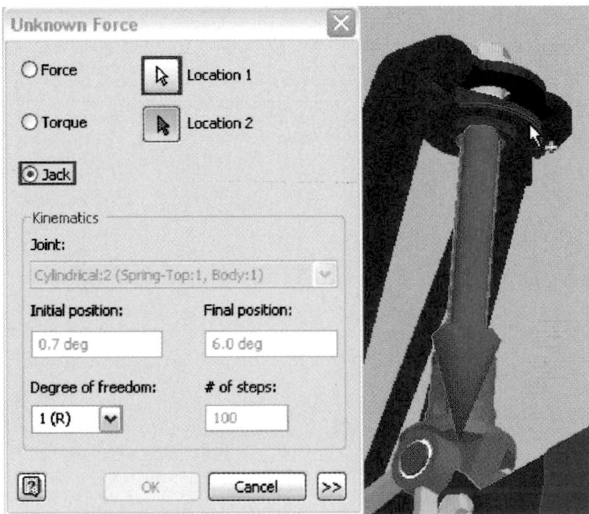

20. For **Location 2**, select the edge of the spring, as shown

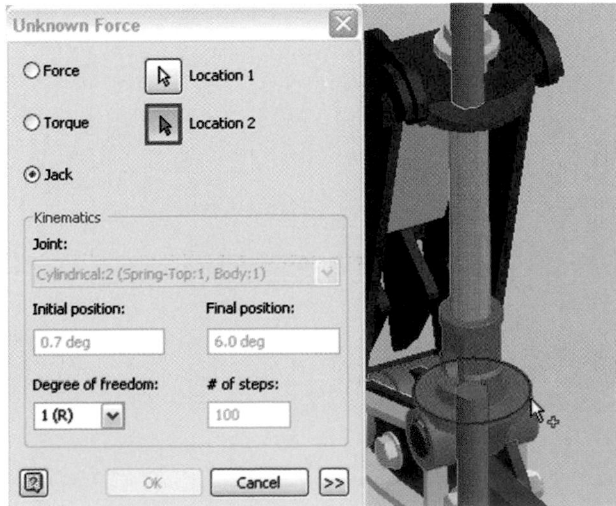

21. Select the Revolution:4 joint > Specify 7.9 deg (a further seven degrees) as the **Final position**

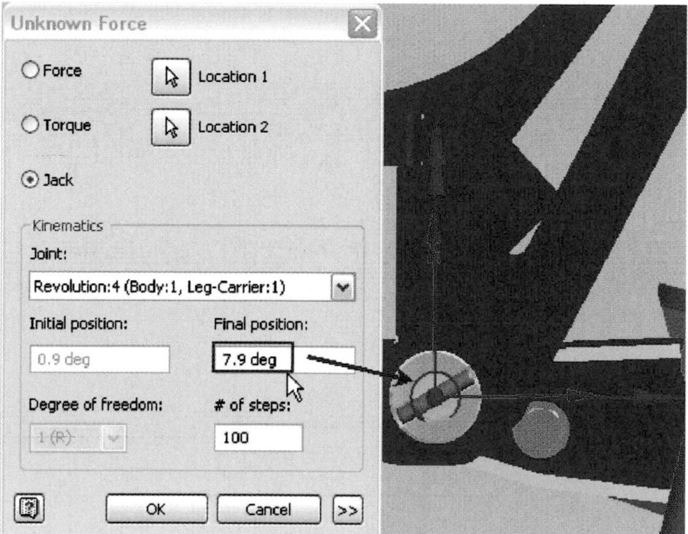

This is the allowable rotation for the tip to operate in normal conditions.

22. Click **OK**

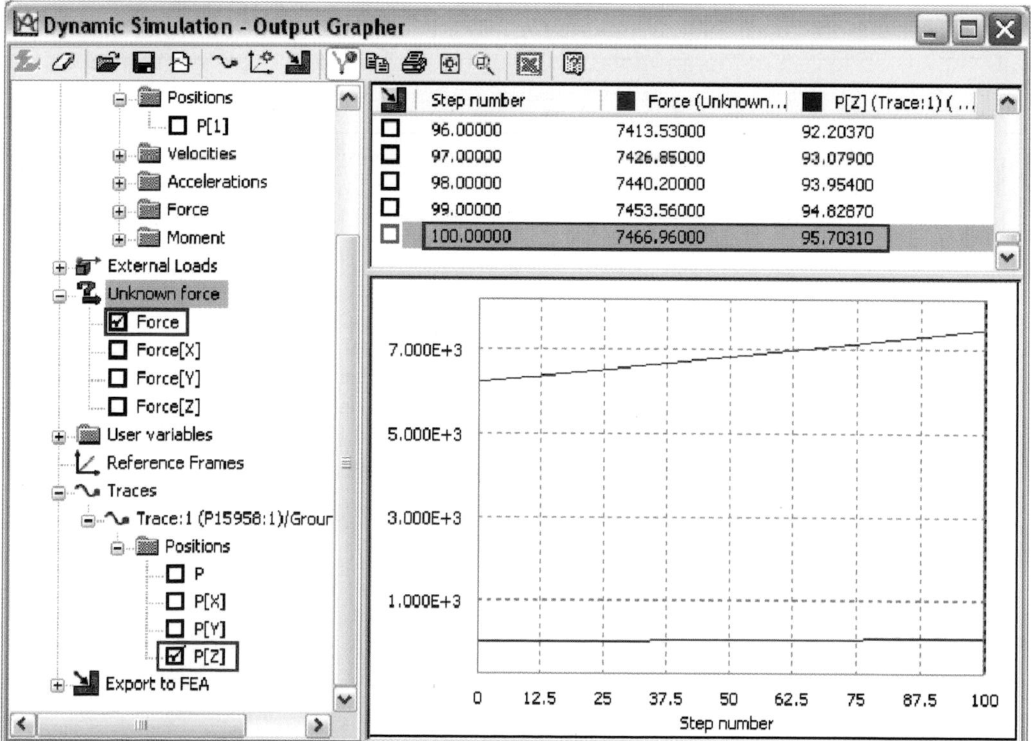

Seven degrees of rotation cause the tip to move approximately **95.7 mm** in the horizontal plane, and the force at this position that the spring needs to react is **7466.96 N**. You may need to select P[Z].

23. Delete all data curves except Force (Unknown force) (N)

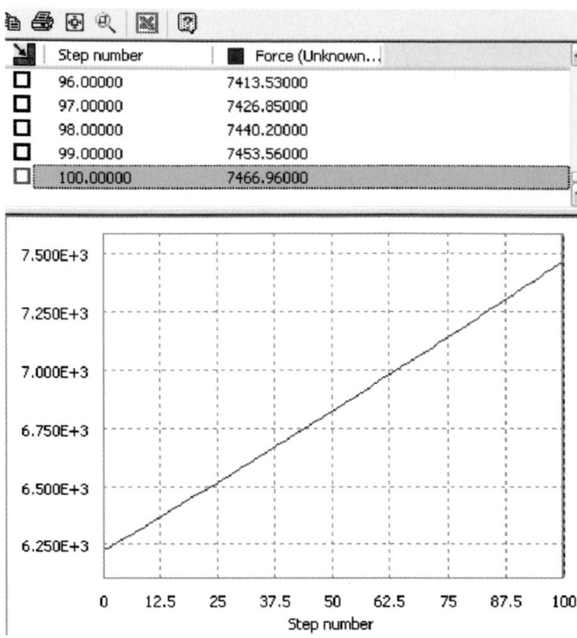

Step number	Force (Unknown...
96.00000	7413.53000
97.00000	7426.85000
98.00000	7440.20000
99.00000	7453.56000
100.00000	7466.96000

Ideally, we need to be able to define the force as a function of the spring length so we can determine the stiffness of the spring. We can manually determine the spring length at the minimum and maximum force values and then calculate stiffness. However, we are going to calculate the spring length automatically.

To create the spring length, we need to define the extremities of the spring; hence, we will need to **Output Trace** these positions and then define an equation that calculates the length of the spring.

24. Right click **Trace** in the Output Grapher and select **Add Trace**

25. For **Origin**, select the edge of the spring top > **Display** > **Output Trace** > **Apply**

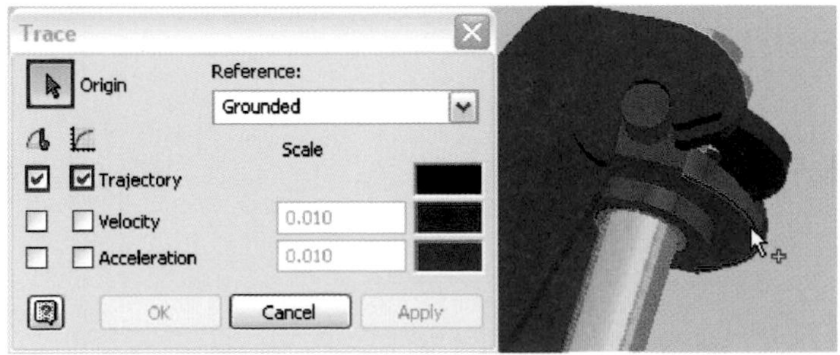

Now we need to select the other end of the spring.

26. For **Origin**, select the edge of the spring > **Display** > **Output Trace**

27. Click **OK**

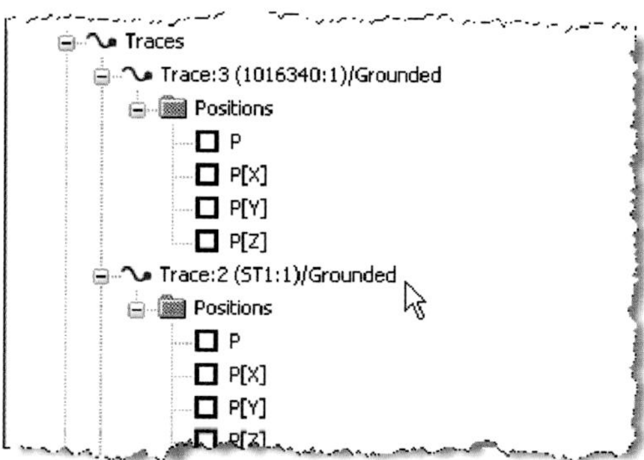

The end positions of the spring are now available to measure and analyze via the Output Grapher.

Now we will use these trace values to automatically determine the spring length.

28. Select **New Curve** from the Output Grapher

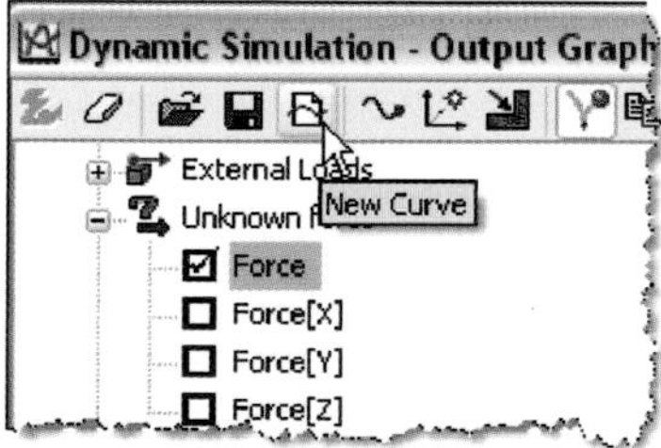

29. For **Name**, type in 'Spring Length'. Type in the following equation:

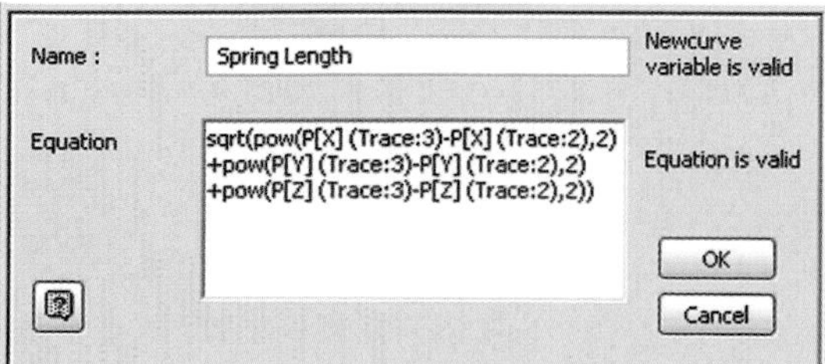

Click on the P[ ] values beneath the appropriate Trace nodes in the Output Grapher to avoid typing in the position text.

The equation represents the distance of a straight line

$$ab = \sqrt{(Xa - Xb)^2 + (Ya - Yb)^2 + (Za - Zb)^2}$$

where *ab* is the linear distance between two points.

30. Click **OK**

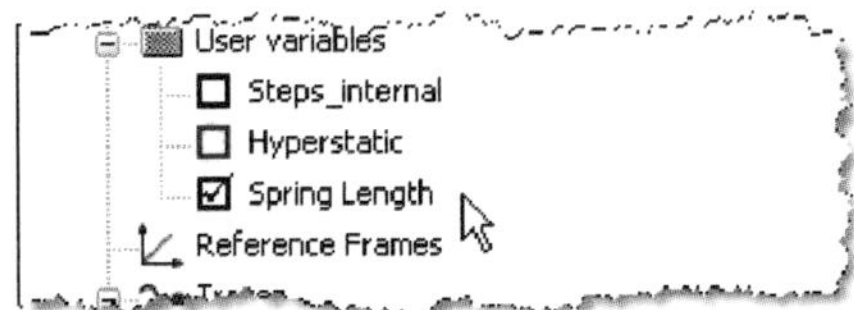

Notice that Spring Length appears under the User variables node.

31. Right click the Spring Length variable > Select **Set as Reference**

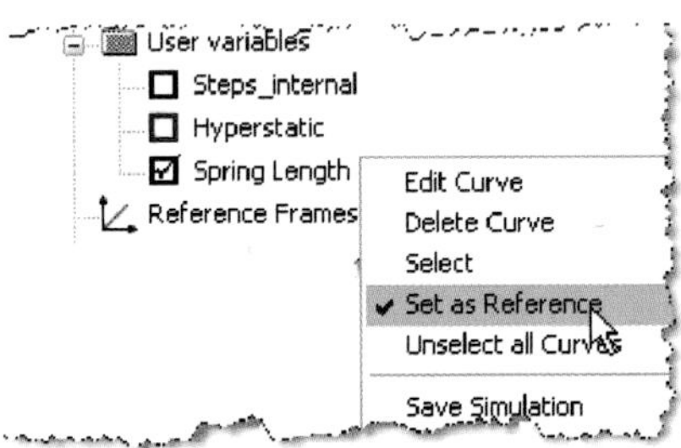

Notice now that the curve is plotted against Spring Length instead of Step number, showing a maximum force value 7466.96 N at a spring length of 290.65 mm.

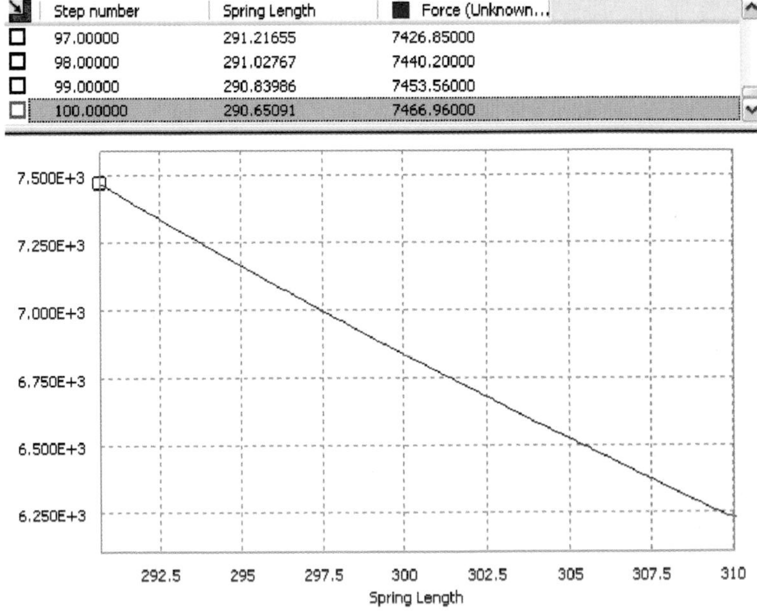

Step number	Spring Length	Force (Unknown...
97.00000	291.21655	7426.85000
98.00000	291.02767	7440.20000
99.00000	290.83986	7453.56000
100.00000	290.65091	7466.96000

32. Right click in the Force column > Select **Search Min**

Step number	Spring Length	Force (Unknown...
0.00000	310.05562	6231.64000
1.00000	309.85400	6233.28000
2.00000	309.65343	6244.97000
3.00000	309.45216	6256.68000

The graph shows a minimum force value 6231.64 N at a spring length of 310 mm.

From the graph, we can now calculate the required stiffness and preload of the required spring.

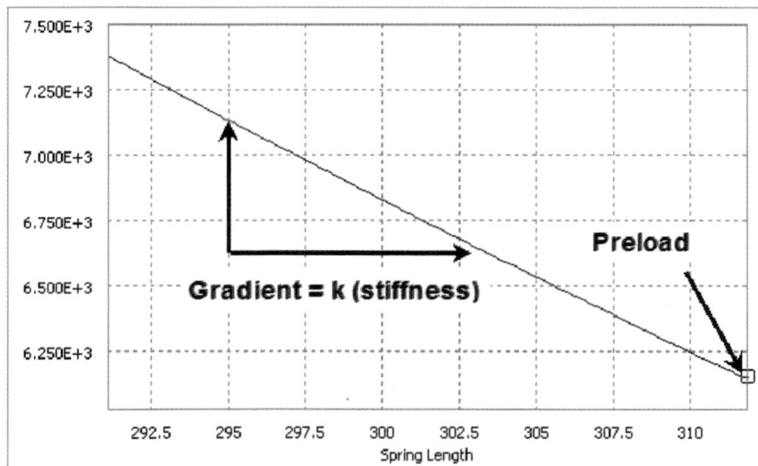

Preload = Minimum value

= 6231.64 N

Stiffness = (Maximum value–Minimum value)/(Change in spring length)

= (7466.96–6231.64)/(310–290.65)

= 1235.32/19.35

= 63.84 N/mm

We will use these values to create the spring in the next section.

CREATE SPRING

IMPORTANT – If your values differ, use them to calculate stiffness, spring free length, and preload values for the spring to work properly.

33. Minimize the Output Grapher > Select **Construction Mode**
34. Select **Insert Joint** from the **Dynamic Simulation** panel
35. Select the **Spring/Damper/Jack** joint from the list or joints table
36. Select the edge of the spring top to define **Point 1**

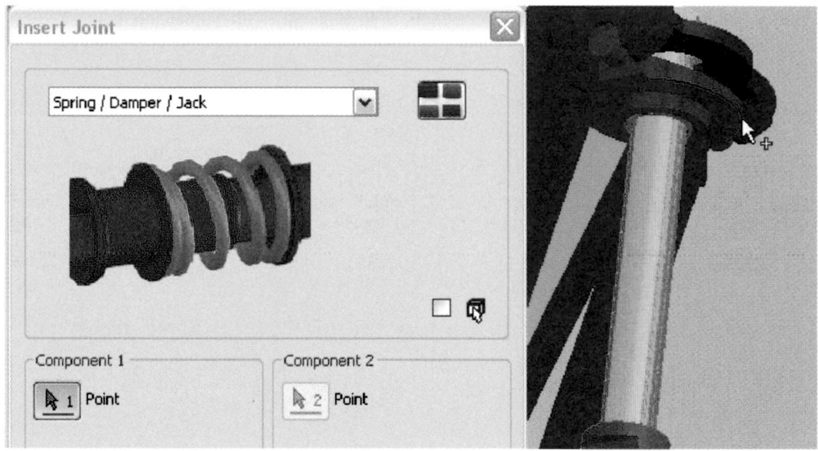

37. Select the edge of the spring to define **Point 2** > Click **OK**

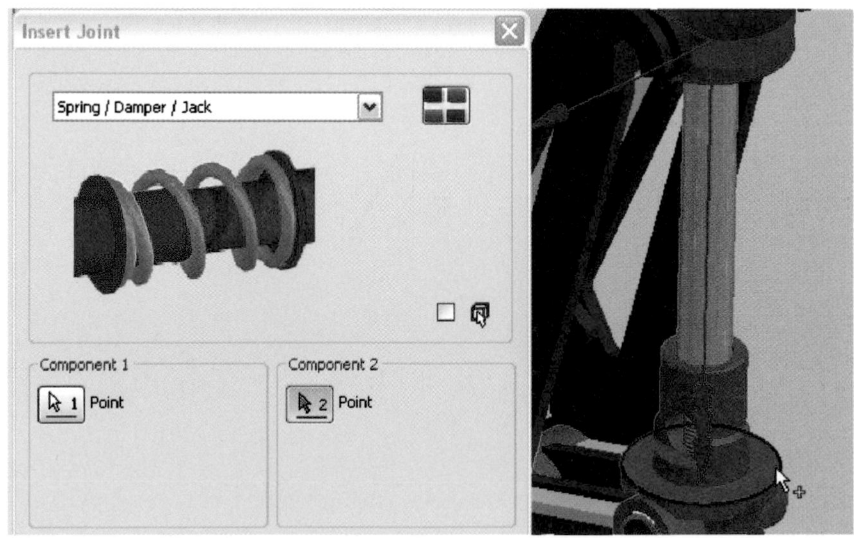

Stiffness = 63.84 N/mm

Free length = 407.6 mm

The spring will initially be compressed to give it strength and rigidity in order to react against the preload of 1500 N. This will be created by defining a new free length.

Free length = (6231.64/ 63.84) + 310 = **407.6 mm**

Damping = 6 Ns/mm

As a rule of thumb, apply damping in the ratio 1:10 to value of stiffness.

38. Double click on the spring joint

39. Specify the following values in the spring's dialog box

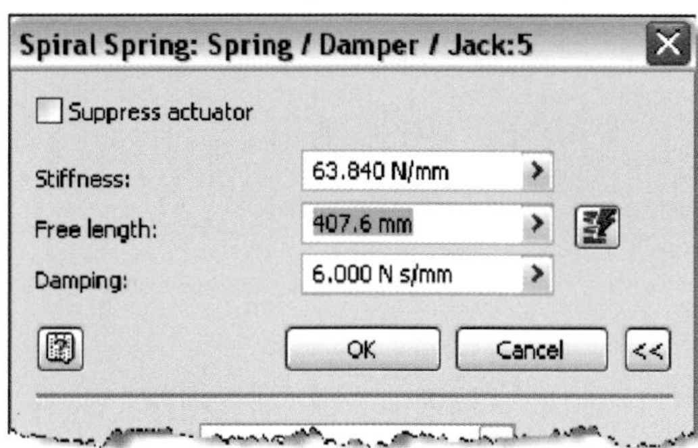

40. Specify the following visual properties of the spring:

Radius = 40 mm

Facets = 6

Turns = 8

Wire radius = 8 mm

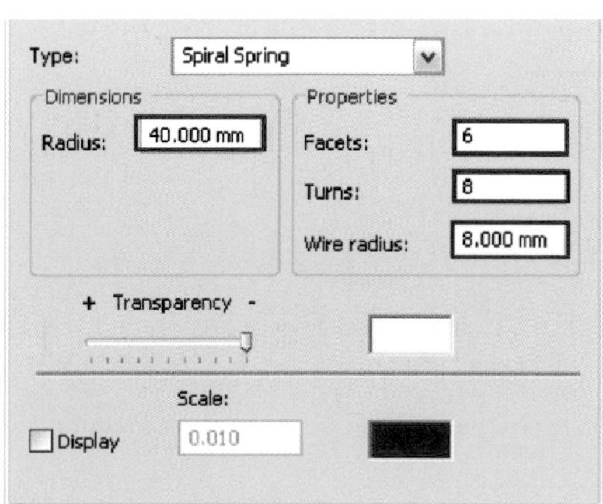

41. Click **OK**

42. Double click on **Force** > Change **Magnitude** to 1500 N > Click **OK**

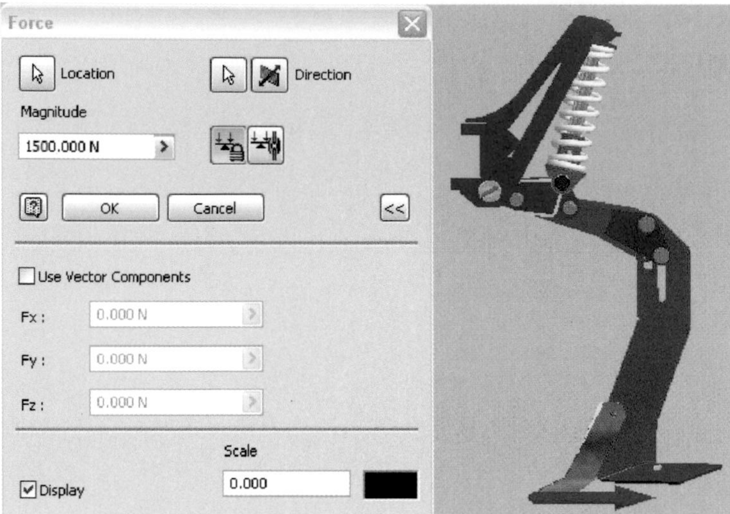

43. Play the simulation

The tine tip will not move as expected because of the preload in the spring. There may be slight movement.

44. Select **Construction Mode** > Double click on **Force** > Change **Magnitude** to 1750 N > Click **OK**

45. Play the simulation > Select the Output Grapher

46. Right click the Spring Length variable and unselect **Set as Reference** > Select P[Z] under Trace:1

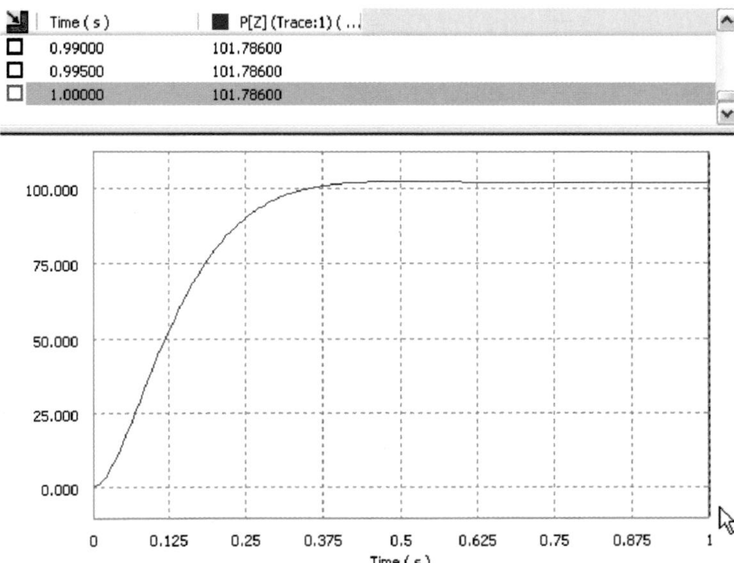

The tine tip displaces horizontally by 101 mm when subjected to a load of 1750 N, which satisfactorily indicates that the spring is working under normal operating conditions and also resembles the graph below.

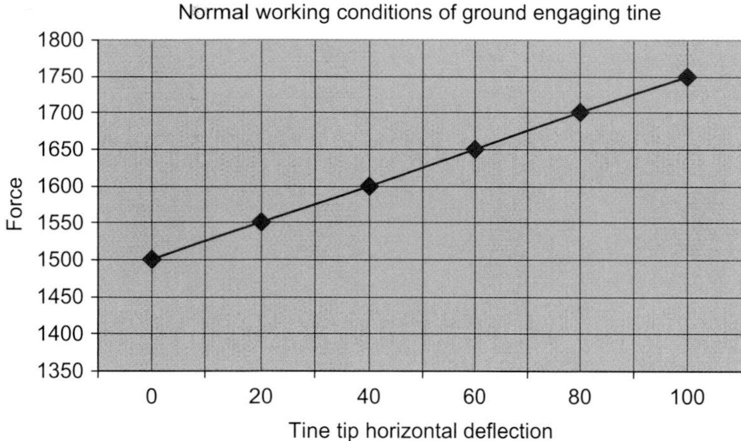

47. Close the Output Grapher > Select **Construction Mode**

Now, since we have defined and created the spring, we need to determine the maximum tip force and height of the mechanism.

DETERMINE MAXIMUM TIP FORCE AND TIP HEIGHT OF THE GROUND ENGAGING TINE

48. Right click Force:1 (P15958:1) > Select **Suppress Force**

49. Select **Unknown Force** from the **Dynamic Simulation** panel

50. In the **Unknown Force** dialog box, select **Force** > For the **Location** select the point as shown

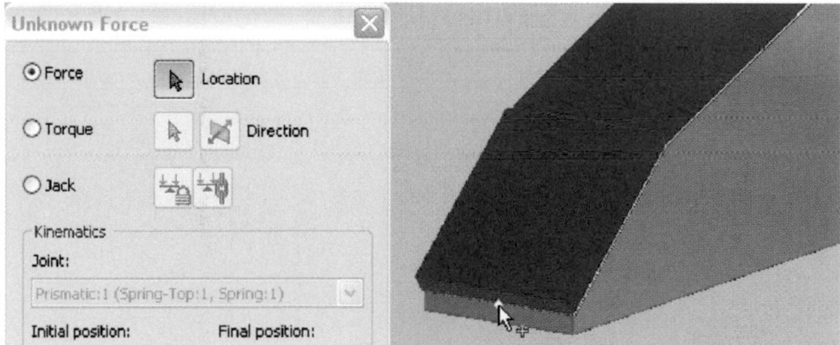

51. For **Direction**, select the edge of the body, as shown

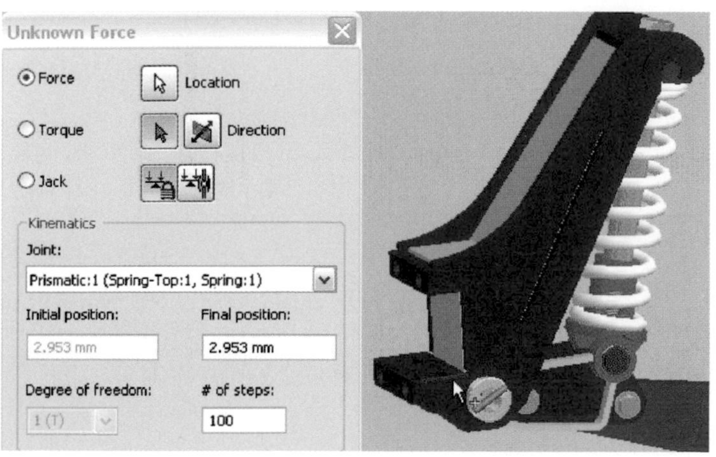

52. Change the **Final position** of the Prismatic:1 joint to 80 mm > Change # **of steps** to 20

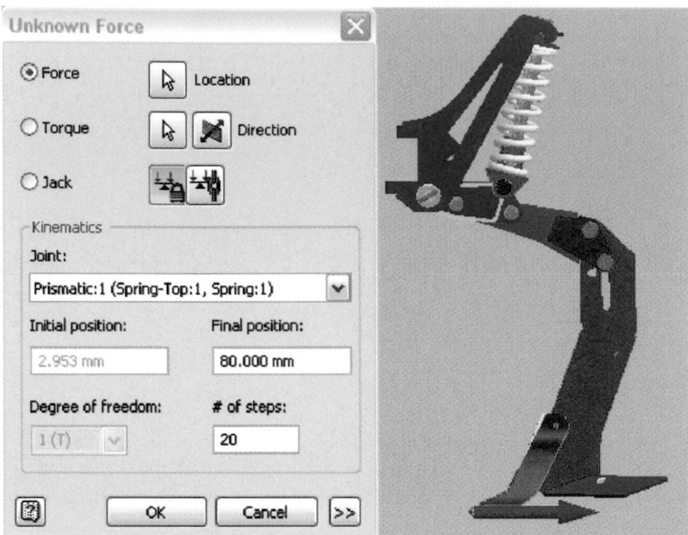

The value of 80 mm represents the maximum safe compression of the spring before it is over-stressed. This value has been determined using the compression spring component generator, within Design Accelerator.

53. Click **OK**

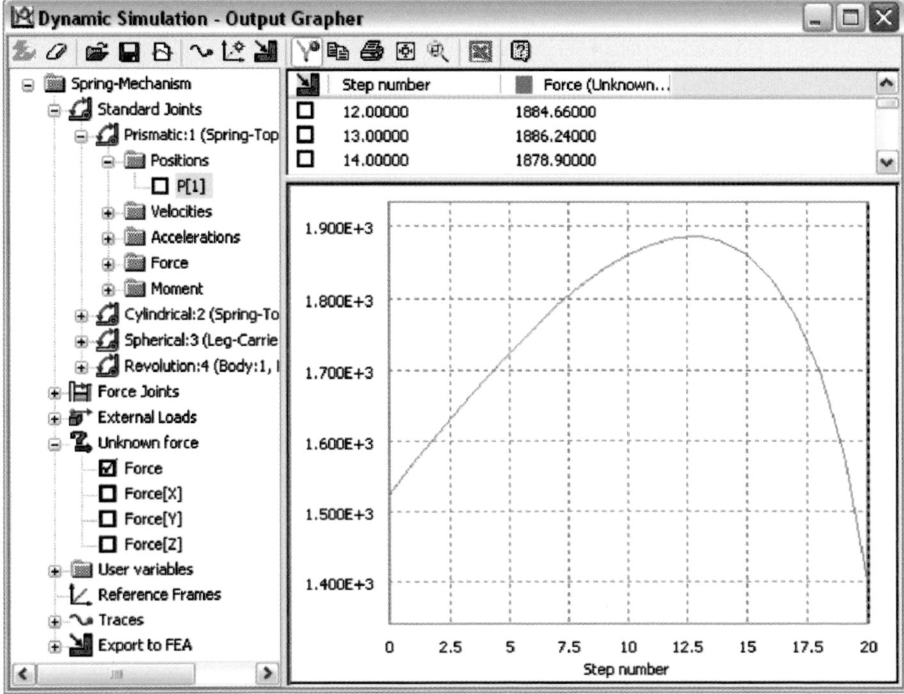

You may need to delete all other curves by right clicking and selecting **Delete**. You also may need to unselect **Spring Length** as **Set as Reference**.

In the Output Grapher, Step number is meaningless as we also need to determine the maximum tine height.

54. Right click P[Y] > Select **Set as Reference**

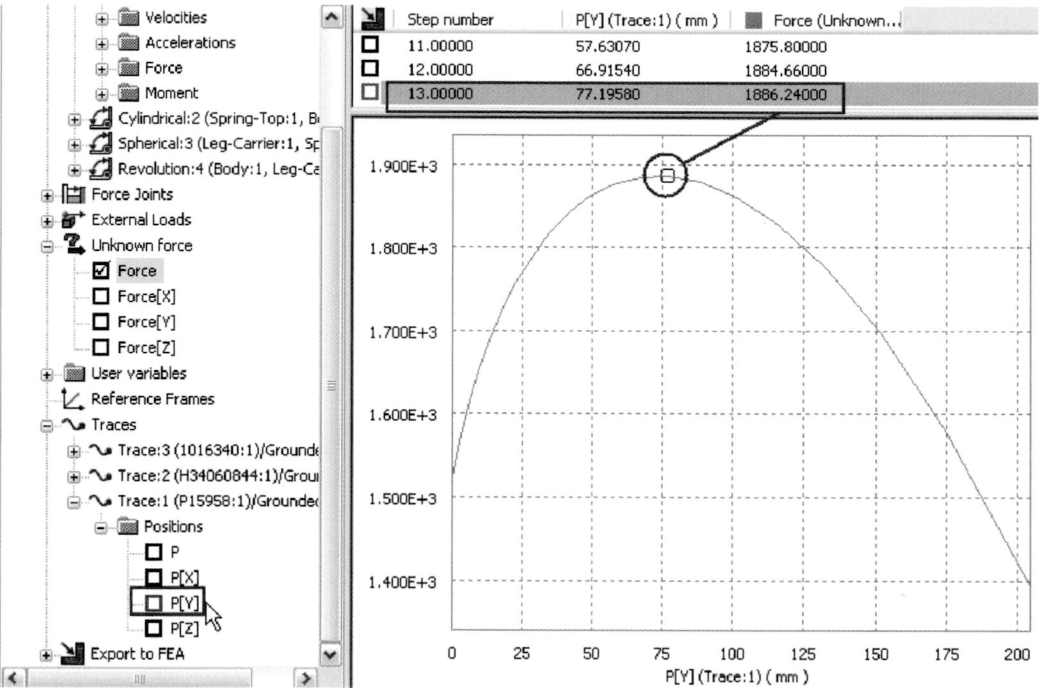

This graph now shows tip force in relation to tip height.

55. Right click P[Y] > Unselect **Set as Reference**

56. Right click P[Z] > Select **Set as Reference**

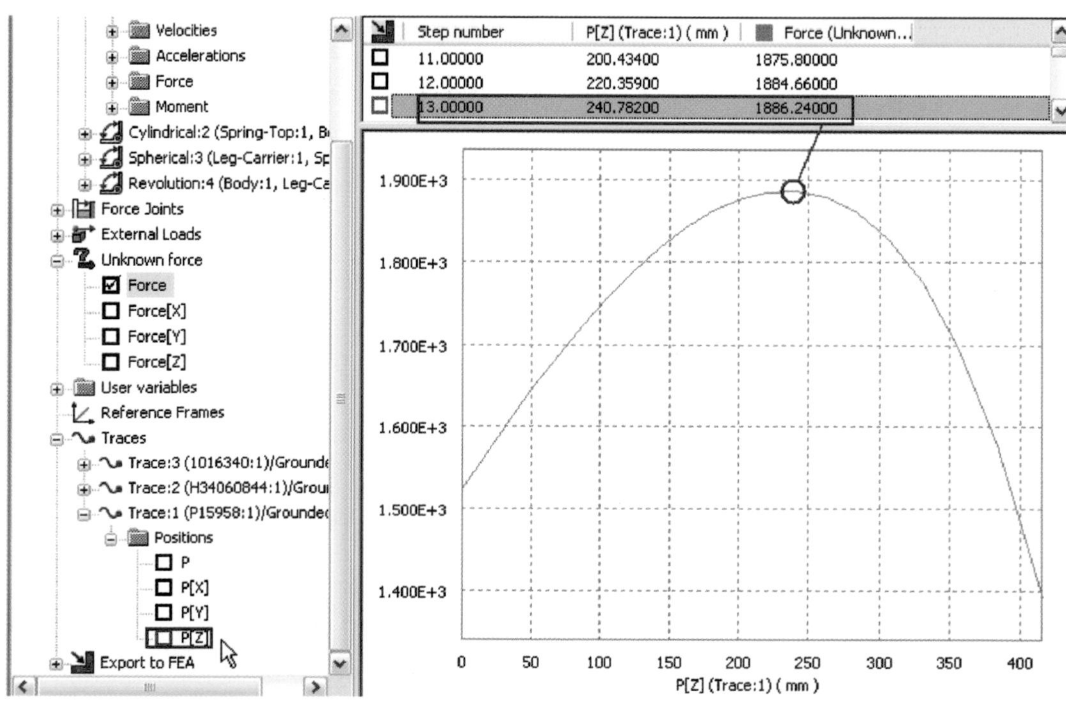

This graph now shows tip force in relation to the horizontal deflection of the tip.

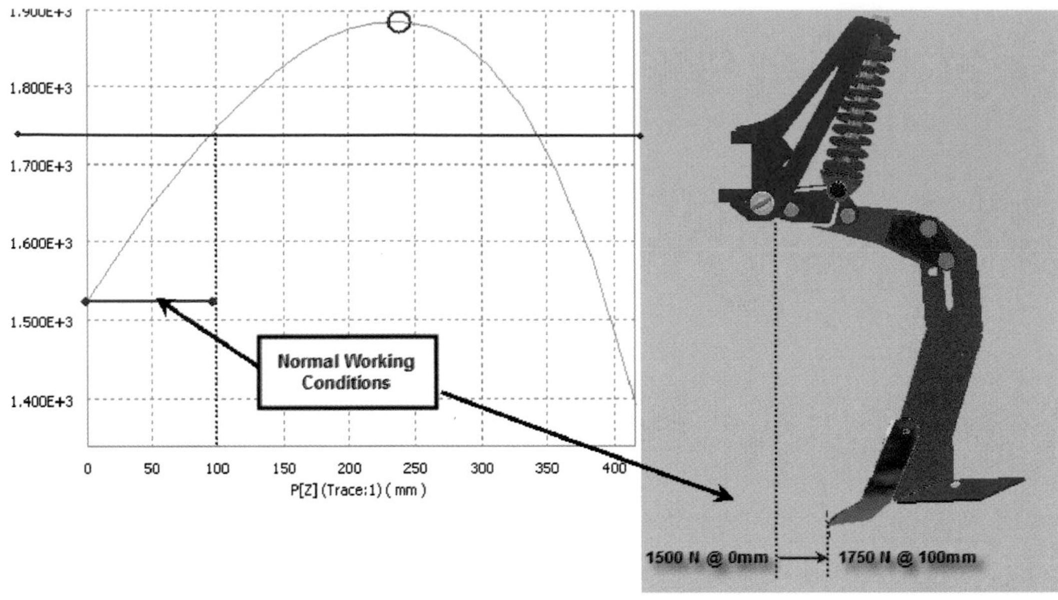

57. Close the Output Grapher > Click on **Construction Mode**

58. Close *Spring-Mechanism.iam*

Sizing of springs and determining maximum loads and trip heights for the resulting assembly can mean that such design exercises become iterative processes. This is often the case where springs are part of a design.

By simulating the action of the assembly through its working and tripping ranges, it is possible to consider alternatives to the spring specification and clamp geometry at an early stage of the design. In turn, this helps to shorten development time during prototyping, which can be valuable where costs and lead times for prototype components such as springs and fabrications are high.

CHAPTER 7

DP6 – Size a Spring

Rotary Compressor Design

(Design Problem courtesy of In-CAD Services Ltd)

JOINTS INTRODUCED/COVERED IN THIS DESIGN PROBLEM

	Joints	Joint creation process
1	Cylindrical	Automatically created
2	Revolution	Automatically created
3	Prismatic	Automatically created
4	Sliding point on curve	Manually created

KEY FEATURES AND WORKFLOWS INTRODUCED IN THIS DESIGN PROBLEM

	Key features/workflows
1	Repairing redundant joints by altering assembly constraints
2	Locking degrees of freedom
3	Imposed motion – constant velocity
4	**Export to Sketch**

INTRODUCTION

In-CAD Services, a specialist consultancy provider, came up with a novel rotary compressor design. One of the key benefits of this design is the small number of moving parts, thus reducing maintenance costs.

To enable the successful operation of the compressor, the production of the complex shape of the rotor is critical as slight imperfections in the shape can have a significant impact on the efficiency of the compressor. The compressor can be broadly split into two sections: compressed and uncompressed.

Up and Running with Autodesk Inventor Simulation 2011. ISBN: 978-0-12-382102-7
Copyright © 2010 Elsevier Inc. All rights of reproduction, in any form reserved.

Based on mathematical calculations, the compressed side can be accurately defined; however, the uncompressed side cannot be easily defined by mathematical calculations and can lead to errors in the design.

$$\text{Compression ratio} = \frac{\text{Uncompressed volume}}{\text{Compressed volume}}$$

where the volume is between the rotor and the compressor body on either side.

Generally, a higher compression ratio is highly desirable; however, this can generate higher centrifugal forces, resulting in increased wear of the rotor blades and body due to heat generation between these components.

Another major factor is the design of the springs. These springs allow for a small amount of movement in the blades to allow for slight imperfections in the complex compressor shape. These imperfections can be due to manufacturing tolerances and mathematical calculations. Overdesigning these springs can lead to excess heat generation. However, underdesigning these springs can lead to unsatisfactory operation of the compressor due to lack of contact between the blades and rotor. Hence, the design of these springs is critical.

Currently, the design is in the prototype stages and the spring is being modified as the original spring keeps failing at desired operating conditions of the rotor.

Hence, the main requirements for the compressor design are to:

- Simulate and create the complex shape of the uncompressed section of the rotor body, based on the predefined shape of the compressed section.
- Determine the maximum centrifugal forces acting on the rotor blades.
- Use the calculated force to determine the optimum size of the spring.

Assumptions/restrictions

The following will be taken into account when designing the compressor:

- The operating rotational velocity of the compressor is 2700 rpm.
- The compressed section of the rotor is assumed to be accurate.
- The simulation will be undertaken for one complete revolution of the rotor.

Therefore

$$2700\ \text{rev/min} = 2700/60\,\text{rev/s} = 45\,\text{rev/s}$$

So, for one complete revolution, the time taken will be

$$\frac{1}{45} = 0.0222\ \text{seconds}$$

Hence, the final simulation time to be used will be 0.0222 seconds.

WORKFLOW OF DESIGN PROBLEM 6

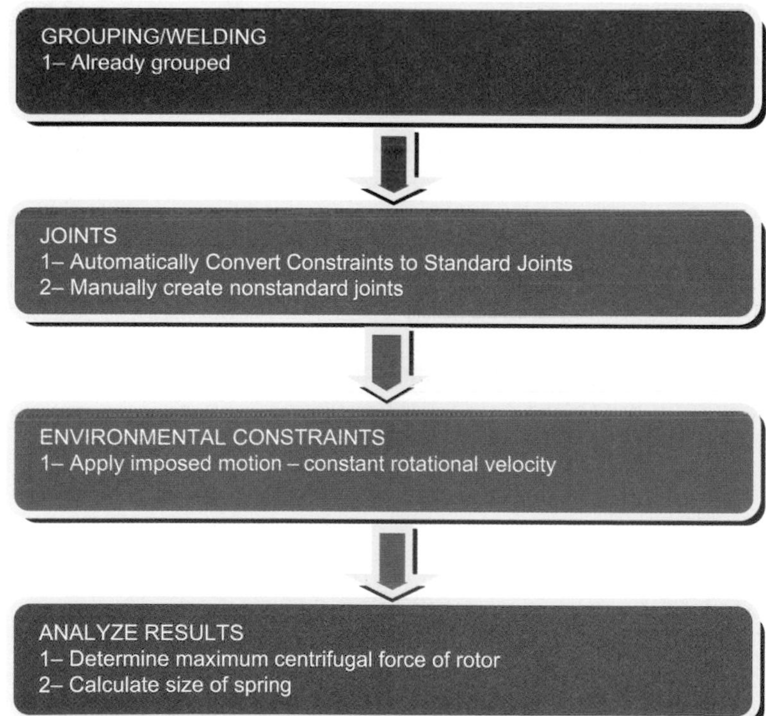

There are not many components to group together, and the components that make up the left and right sides of the rotor are already grouped together via the rotor subassembly.

Joints

Initially, we will create standard joints automatically, and then the remaining joints will be created manually.

AUTOMATICALLY CONVERT CONSTRAINTS TO STANDARD JOINTS

1. Open *Compressor.iam*

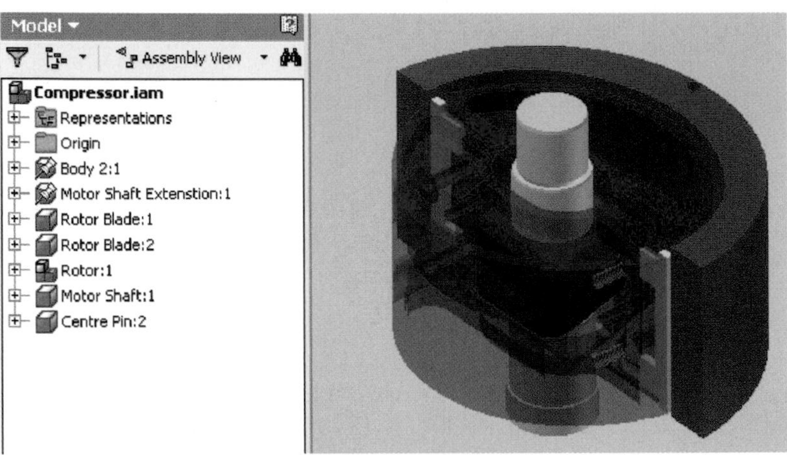

2. Select **Environments** tab > **Dynamic Simulation**

3. Select **Simulation Settings**

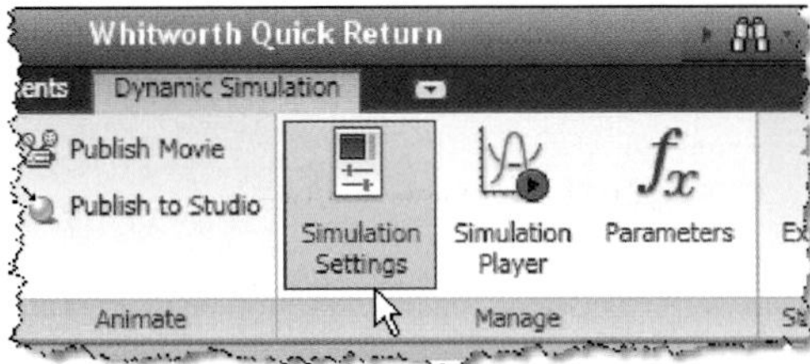

4. Select **Automatically Convert Constraints to Standard Joints** > Select **Input angular velocity in revolutions per minute (rpm)**

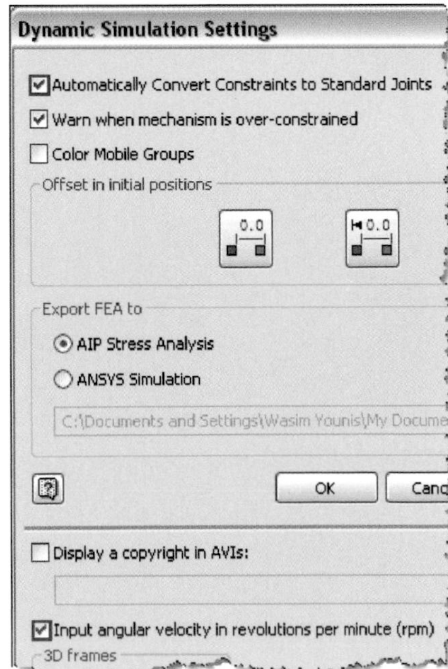

5. Click **OK** > Accept the warning

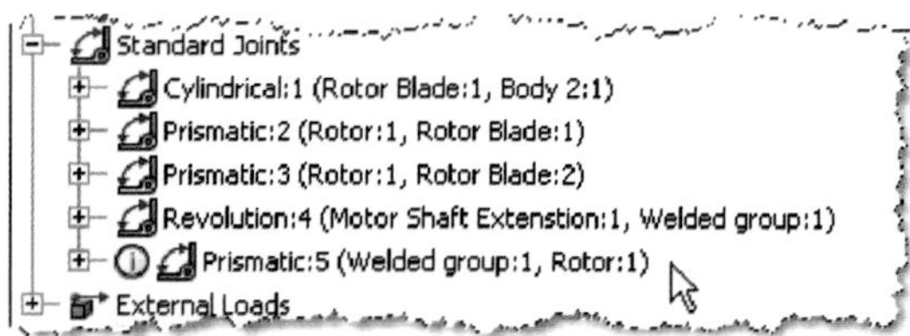

The quickest way to remove redundancy is to modify or suppress assembly constraints within the Simulation environment.

6. Right click the Mate:6 constraint > Select **Suppress** > Click **OK**

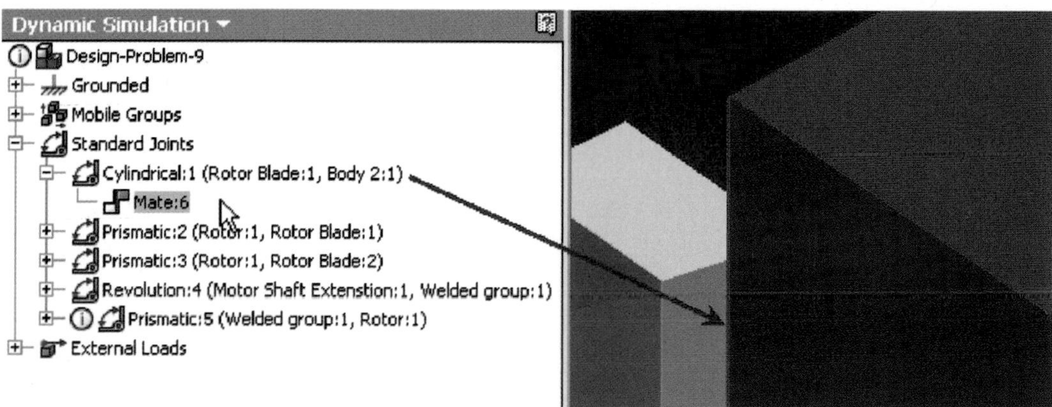

This was the constraint used to initialize the tip position of the blade with the rotor body. As this was an axis-axis constraint, Dynamic Simulation has created a cylindrical joint between the two components.

7. Select **Mechanism Status** to determine the redundancy and mobility of the assembly > Click **OK**

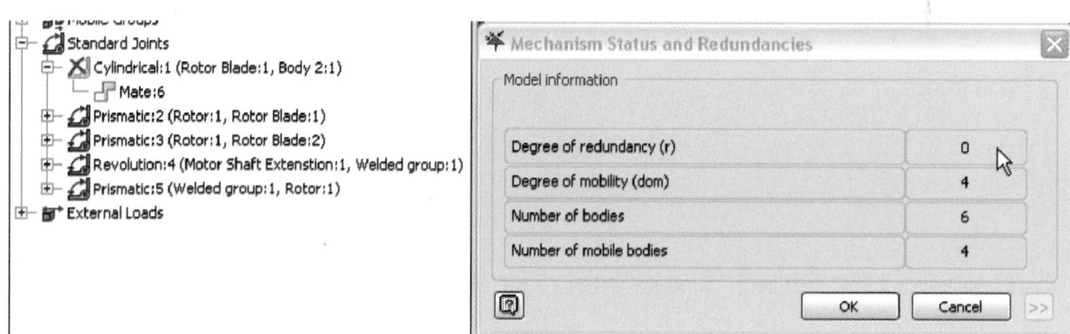

If we had initially suppressed the Mate:6 constraint within the Assembly environment, the cylindrical joint would have not been created automatically.

Suppressing the constraint within Dynamic Simulation has also suppressed the associated joint.

MANUALLY CREATE NONSTANDARD JOINTS

Now we need to create joints that enable contact between the blade and the rotor body.

8. Select **Insert Joint** > **Sliding: Point Curve**
9. For **Curve 1**, select the face and edge of Body 2

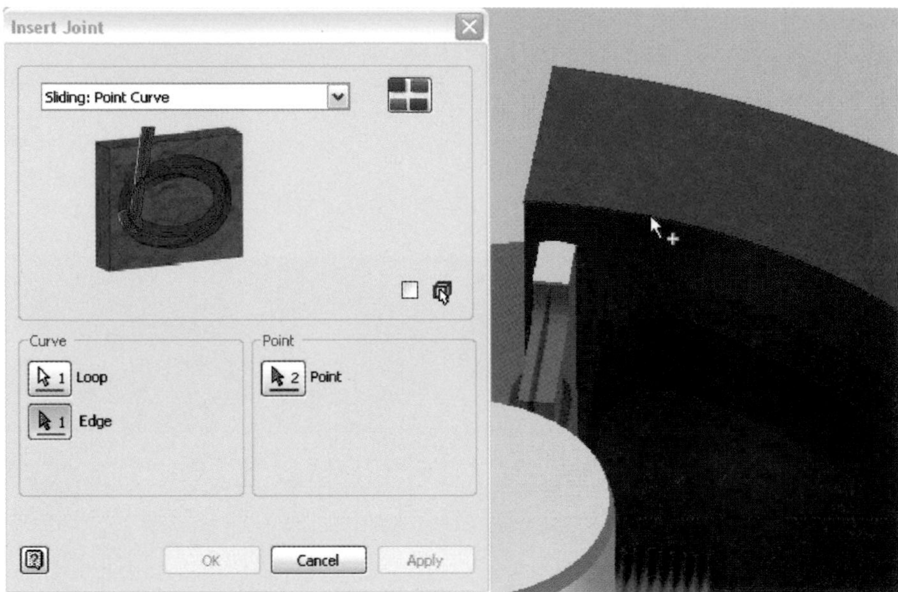

10. For **Point 2**, select the tip of Rotor Blade:2

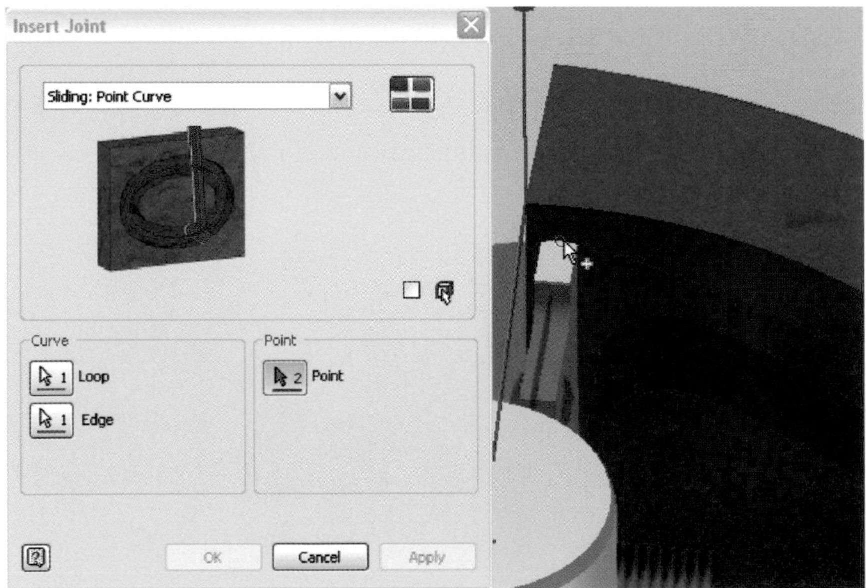

 It is not necessary to have components on the same plane. Once the joint has been created, Dynamic Simulation will treat them as if they were on the same plane.

 The chosen plane of the joint is dependent on the first component selection.

11. Click **OK**

Environmental constraints

In the following sections, we will define the rotational speed of the compressor and determine the complex shape of the uncompressed section of the rotor using the **Trace** tool.

12. Double click the Revolution:4 joint to edit the properties > Select **dof 1 (R)** > Select **Enable imposed motion** > Specify 2700 for **Velocity** > Click **OK**

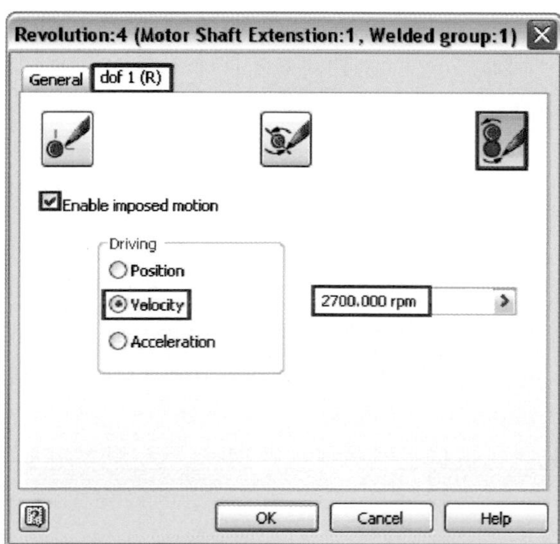

13. Set the simulation time to 0.022 seconds and **Images** to 2200 > Play the simulation > Accept the warning when it appears

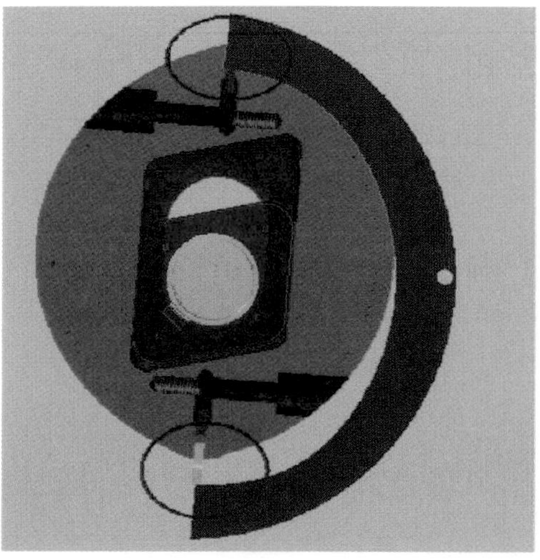

The simulation did not behave as expected, as the rotor is overlapping the compressor body. The reason for this is that we have not specified how far the blades can move in relation to the rotor. In reality, this motion is restricted by a spring that allows a small amount of lateral movement in the blades to overcome imperfections due to manufacturing tolerances. Since the main aim here is to determine the other side of the compressor, based on the compressed section, we will lock the initial positions of the joint.

14. Select **Construction Mode**

15. Select the Prismatic:2 and Prismatic:3 joints > Right click and select **Lock all dofs**

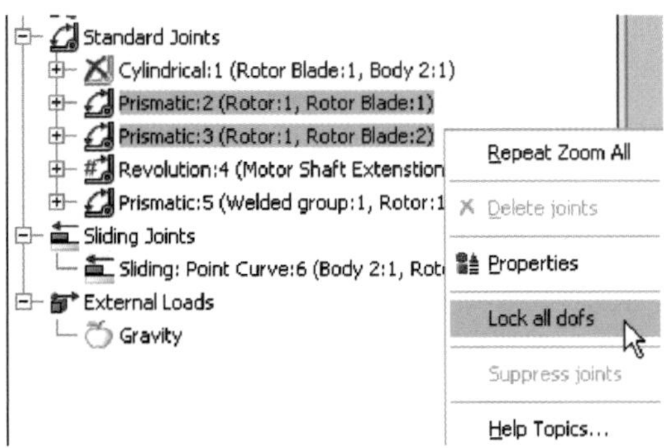

16. Play the simulation > Accept the Locked dof warning

After about 0.011 seconds, the following warning will appear:

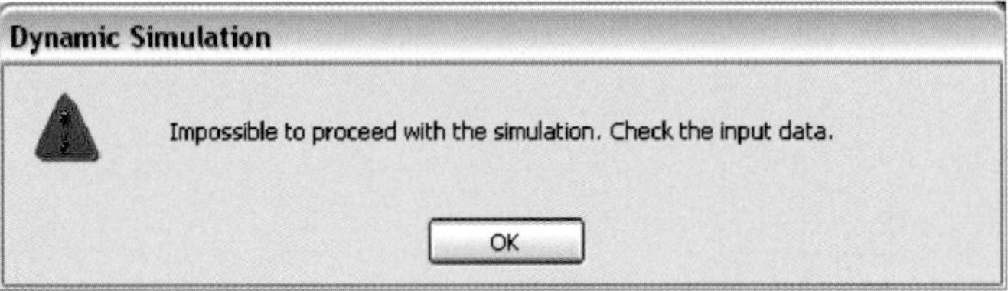

The reason that this warning appears is that the end of the curve has been reached.

17. Click **OK**

Analyze results

Using the Output Grapher, we will now determine the position of the other blade via a trace.

18. Select the Output Grapher > Select **Add Trace** > Select the tip of Rotor Blade:1

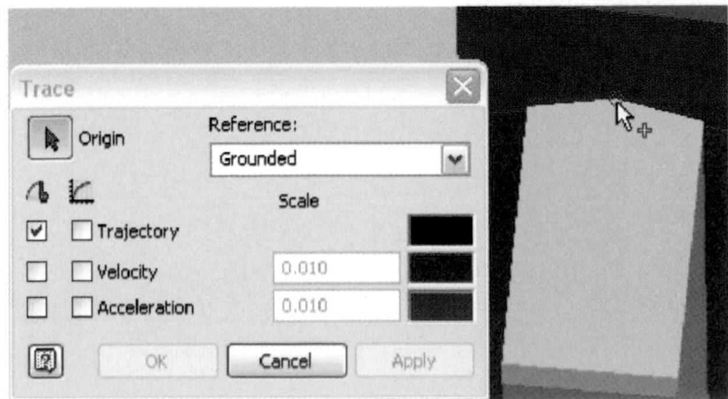

19. Click **OK**

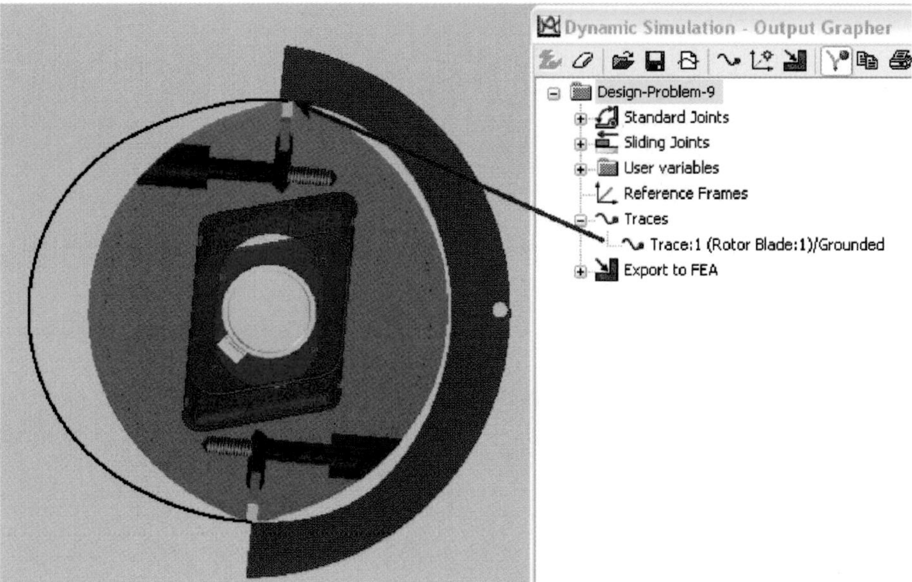

The trace of the other blade is displayed. We will now use this trace to complete the design of the compressor body.

20. Right click Trace:1 > Select **Export to Sketch** > Select the Compressor Body-Half (also referred to as Body 2)

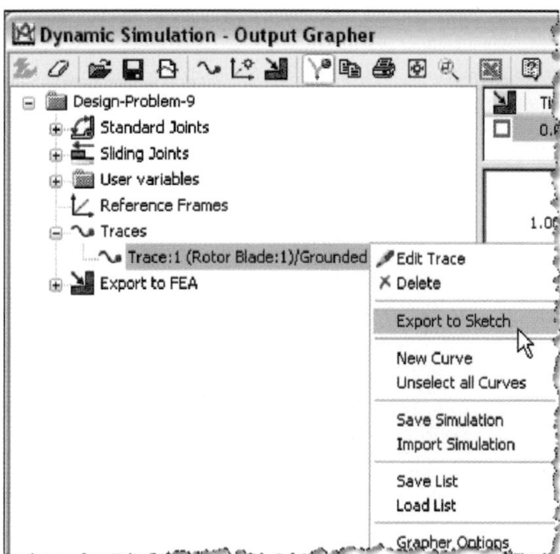

This creates a sketch, based on the trace, in the Compressor Body-Half component.

21. Close the Output Grapher > Select **Construction Mode**

22. Double click the Compressor Body-Half

23. Double click Sketch46

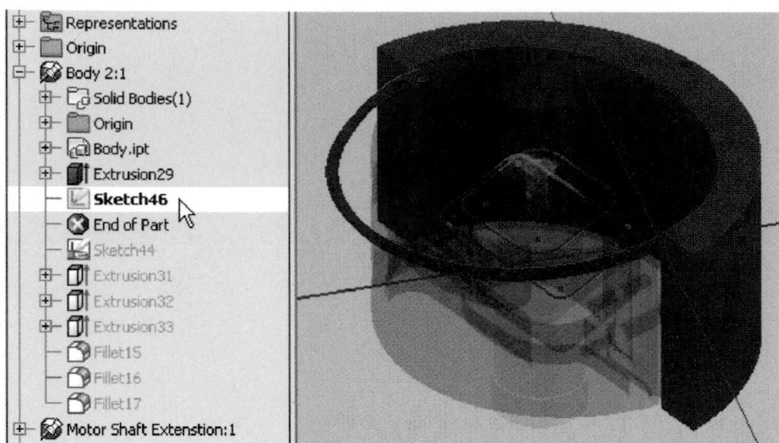

Each image set in the simulation player, 2200 in this case, generates a point along each image frame and, hence, a lot of points have been produced. A lower image setting will produce a reduced number of points. However, it is important to note that a more accurate curve will be produced with a higher image setting but it will take longer to calculate the sketch when exporting.

As we have already created the body from the exported trace, we can now open another assembly.

24. Close *Compressor.iam*

25. Open *Compressor-2.iam*

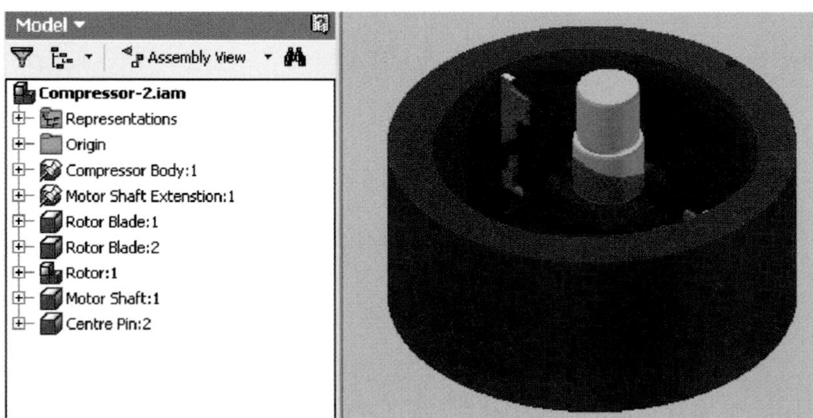

26. Select **Environments** tab > **Dynamic Simulation**

27. Set the simulation time to 0.022 seconds

28. Select **Display to Wireframe**

Viewing the assembly in wireframe will quickly help to identify any overlaps between the blades, rotor, and compressor bodies.

29. Play the simulation > Accept the Locked dof warning

Very quickly the following warning appears:

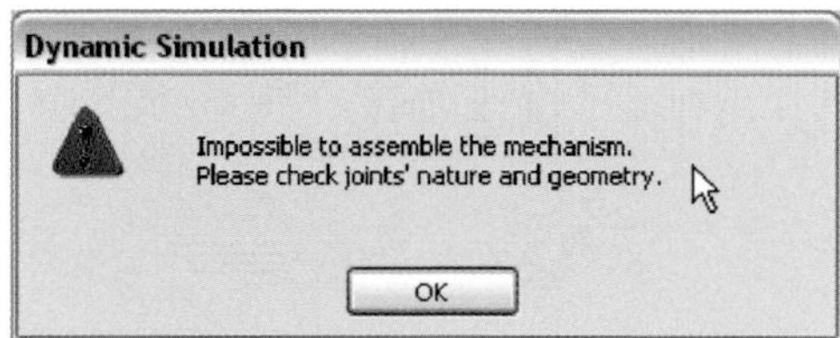

This warning appears because the rotor blade's lateral movement is locked and needs unlocking. Either the Prismatic:1 or :2 joint can be used. In reality, these blades will have some movement – restricted by a spring attached to the blade and rotor.

30. Select **Construction Mode**

31. Right click the Prismatic:1 joint > Unselect **Lock all dofs**

In this part of the example, Rotor Blade:2 is allowed to move so that it maintains its contact position. Any of the prismatic joints could have been chosen, as we are interested in the maximum lateral movement of the blades.

32. Play the simulation > Accept the warning (note: it may take a while to complete the simulation)

33. Select the Output Grapher > Select the **Position** values for both the Prismatic:1 and Prismatic:2 joints

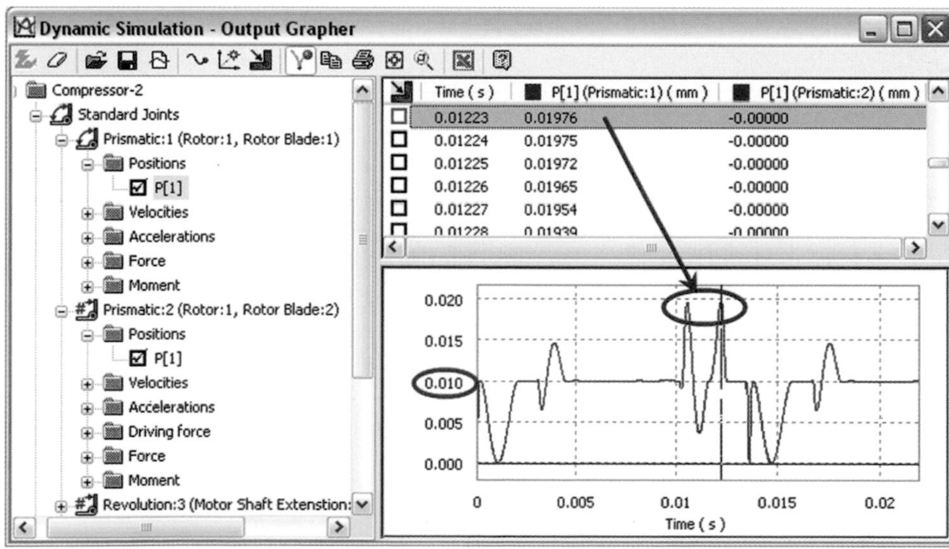

By analyzing the graph, we can see that the blade had to move laterally by at least 0.005 mm at the beginning to enable the rotor to rotate. By further analyzing the graph, we can see that the blade needs to move by at least 0.01 mm for most of the time during the complete revolution. The graph also illustrates that the curve needs to be further refined; for example, it needs smoothing, especially at the areas where the blade has to move 0.02 mm to maintain contact with the body. Therefore, in reality, these blades have springs attached to them to overcome these small manufacturing imperfections/tolerances, due to the complex shape. Further, the maximum distance of travel of the blades will be equally shared by both blades, as both blades will be allowed to travel.

We will continue with the design problem to determine the maximum centrifugal force. However, you can, at the end of the design problem, try to further refine the curve.

DETERMINE MAXIMUM CENTRIFUGAL FORCE OF ROTOR

34. Select **Add Trace** > Select **Origin** as shown > Select Rotor:1 for **Reference** > Select **Trajectory** to display data in Output Grapher > Click **OK**

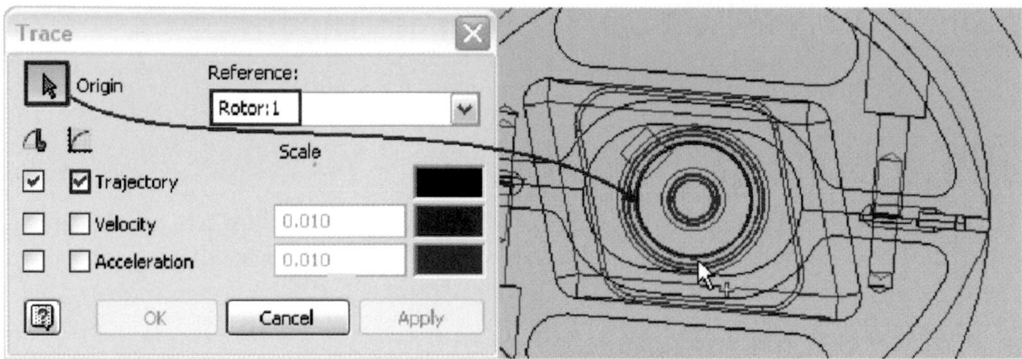

35. Select **Unselect all Curves**

36. Select P[Y] (positions) for Trace:1

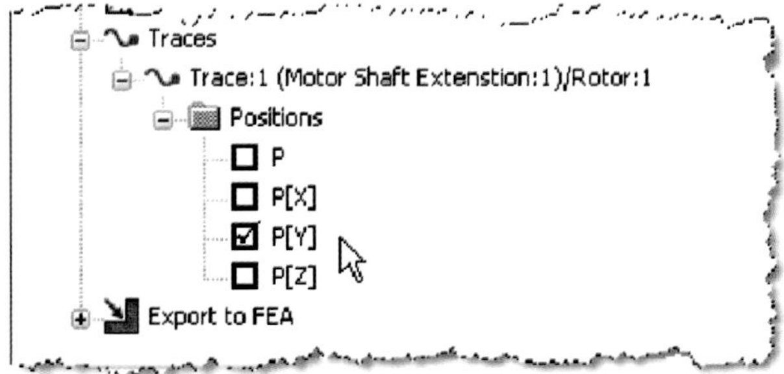

The Position graph displays two peaks (positive and negative). The maximum centrifugal forces will occur at these maximum peak positions (time at 0.00663 seconds and 0.02026 seconds). These are the maximum distances of the center pin from the center and thus they create a higher out-of-balance force. In one complete revolution, there will be two maximum forces acting on each blade, as illustrated below.

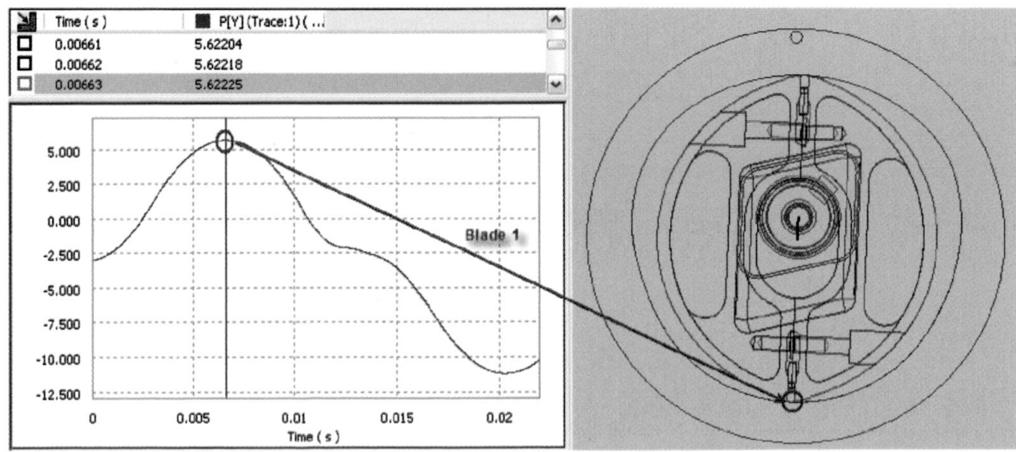

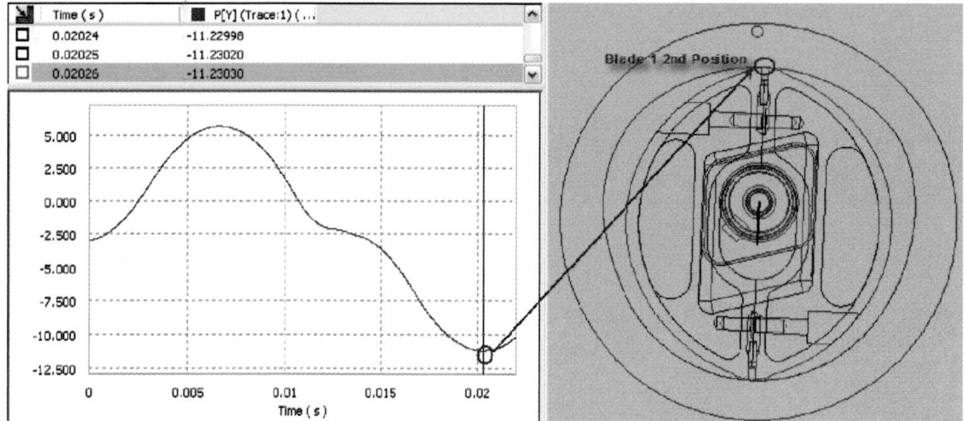

Now, we can determine the forces acting on the blades at these key positions. It is important to note that we are interested in the Rotor Blade:2 forces, because Rotor Blade:1 is free to move, by unlocking its degrees of freedom, meaning the forces will be minimal.

37. Select the Force values for the Sliding Point Curve:5 joint

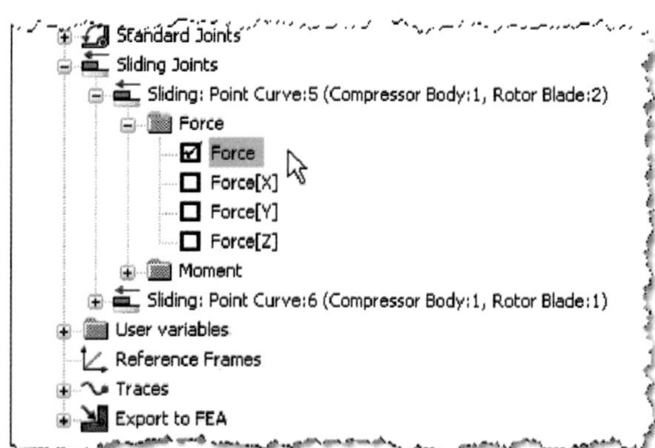

38. Select **Search Max.** in the P[Y] (Trace:1) column

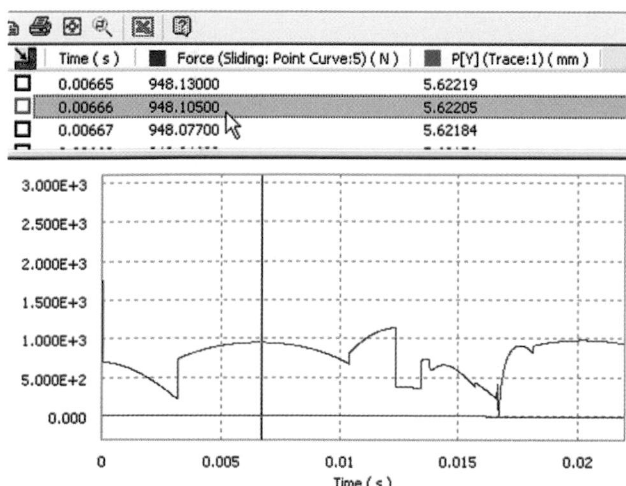

The maximum centrifugal force acting on the blade is 948 N.

39. Select **Search Min** in the P[Y] (Trace:1) column

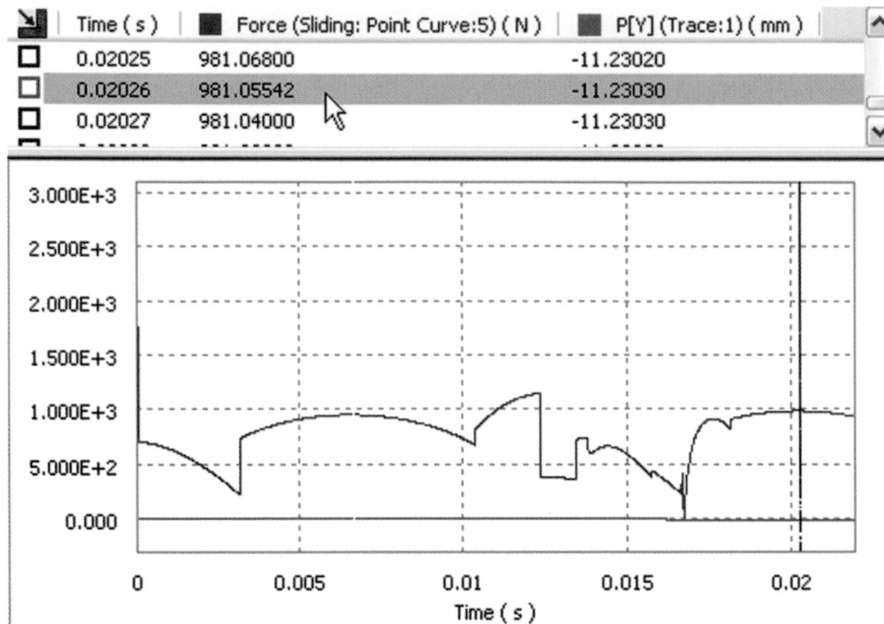

The maximum centrifugal force acting on the same blade on the other side is 981 N.

The sudden peaks in the forces are due to intangencies in the curve shape of the rotor. These peak forces can be removed by further smoothing and refining the curve.

40. Close Output Grapher

41. Close *Compressor-2.iam*

CALCULATE SIZE OF SPRING

For the purposes of calculation, we will use the maximum lateral movement of the blade, which was 0.02. Since in reality both blades will have lateral movement, a value of 0.01 will be used to calculate the spring. However, it is important to note that this is allowing for the maximum imperfections. Obviously, if the shape is refined then the lateral movement of the blades will reduce. Now, based on the calculated force, we can determine the size of the spring stiffness (K value):

$$\text{Force} = K(L - L_0)$$

where

Force = 981 N
K = Needs calculating
L = Distance the blade needs to move; that is, 0.01 mm target
L_0 = Start position of blade, initialized to zero

$$\text{Spring stiffness} = \frac{981}{0.0120} = 98100 \frac{\text{N}}{\text{mm}^2}$$

The smaller the distance the blade tip needs to travel, the smaller the spring required, resulting in a longer working life of the compressor.

The greater the distance blade tip needs to travel, the bigger the spring required, resulting in more wear and tear.

CHAPTER 8

DP7 – Simulate a Sprocket Chain

Design of a High-Speed Bottle Transfer Unit

(Design Problem courtesy of Sheppee International)

JOINTS INTRODUCED/COVERED IN THIS DESIGN PROBLEM

	Joints	Joints creation process
1	Revolution	Automatically created
2	Planar	Automatically created
3	Cylindrical	Automatically created
4	Point-line	Automatically created
5	Sliding cylinder on curve	Manually created
6	Sliding point on curve	Manually created
7	Rolling cylinder on curve	Manually created

KEY FEATURES AND WORKFLOWS INTRODUCED IN THIS DESIGN PROBLEM

	Key features/workflows
1	Imposed motion – constant velocity
2	Simulating a sprocket chain mechanism

INTRODUCTION

Sheppee International Ltd is a world leader in hot glassware handling for both the container and the tableware industries, with over 50 years of experience providing solutions for

Up and Running with Autodesk Inventor Simulation 2011. ISBN: 978-0-12-382102-7
Copyright © 2010 Elsevier Inc. All rights of reproduction, in any form reserved.

difficult ware handling in applications of up to 1000 containers per minute. The high speed bottle transfer unit illustrated below is a typical product from Sheppee International.

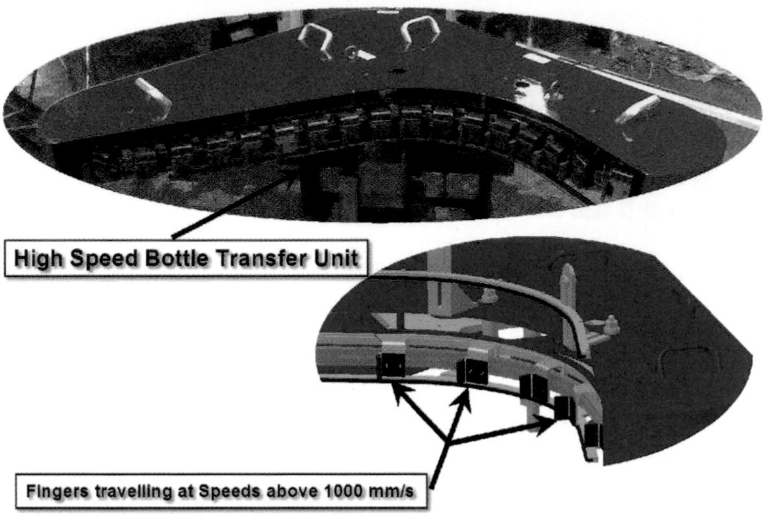

This high speed bottle transfer unit transfers randomly spaced bottles on the first conveyor to equally spaced bottles on the second conveyor. It can transfer 700 bottles per minute and these high speeds are achieved by a sprocket and chain mechanism.

The requirement of this design problem is to simulate this high speed sprocket chain mechanism transfering 700 bottles per minute.

In addition to the main requirement, the following criteria will be taken into account:

Dynamic Simulation does not have any joints that directly simulate a sprocket and chain mechanism. For this reason, the first task is to devise a methodology and process to simulate this sprocket and chain mechanism. Once this process has been achieved, the next step is to calculate the velocity of the mechanism based on the following formula:

$$\text{Chain speed} = \frac{\text{Bottles}}{\text{min} \times \text{Article spacing}}$$

$$\text{Link size} = \frac{3}{4}\ \text{in.} \quad \text{or} \quad 19.05\ \text{mm}$$

So, for transferring 700 bottles per minute, we need a chain speed of

$$\text{Chain speed} = 700 \times 3\frac{3}{4}\ \text{in.} = 700 \times 95.25\ \text{mm} = 66.675\ \text{mm/min}$$

$$\text{Chain speed} = 1111.25\ \text{mm/s}$$

WORKFLOW OF DESIGN PROBLEM 7 – STAGE 2

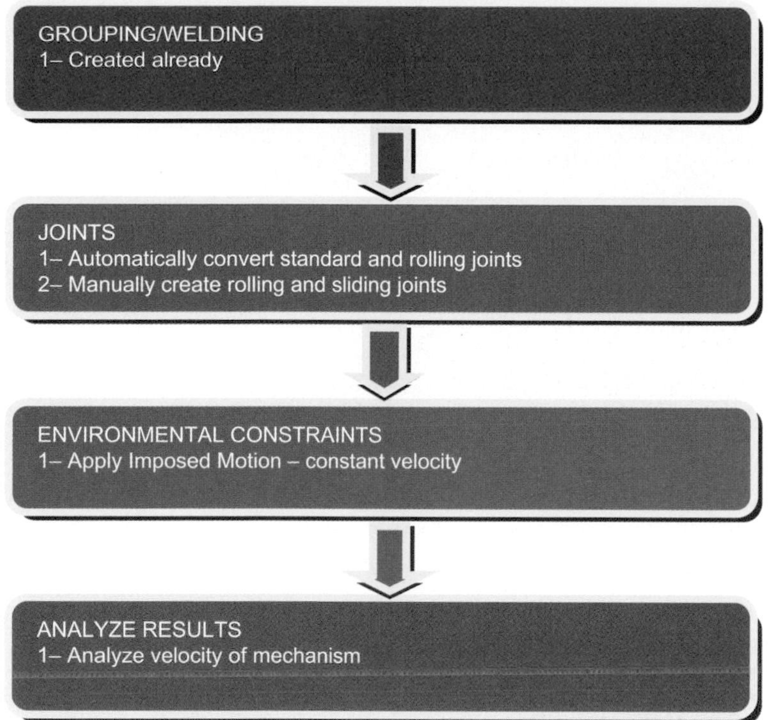

Stage 1 – Devising a process for simulating a sprocket chain mechanism

In this stage, we need to devise a method that allows us to maintain a constant velocity of a component along a curve or path. The following simple chain link example will be used to determine a process for simulating a sprocket and chain mechanism.

1. Open *Simulating Sprocket Chain1.iam*

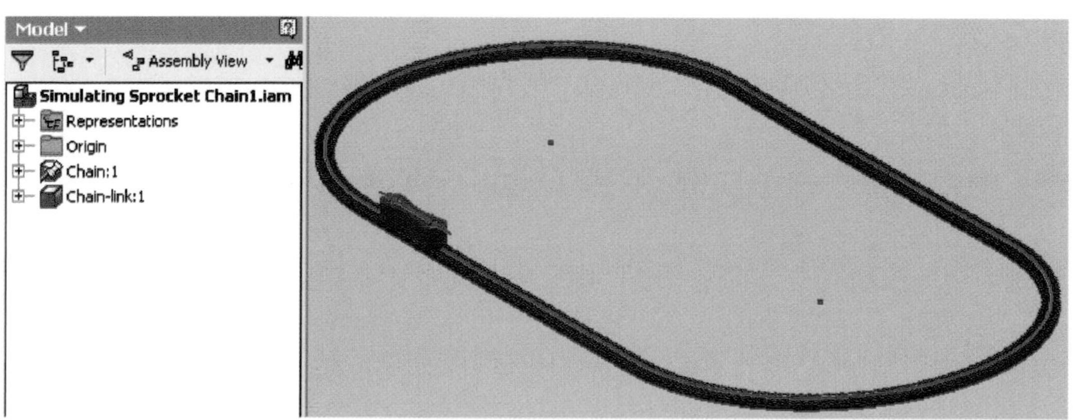

2. Select **Environments** tab > **Dynamic Simulation**

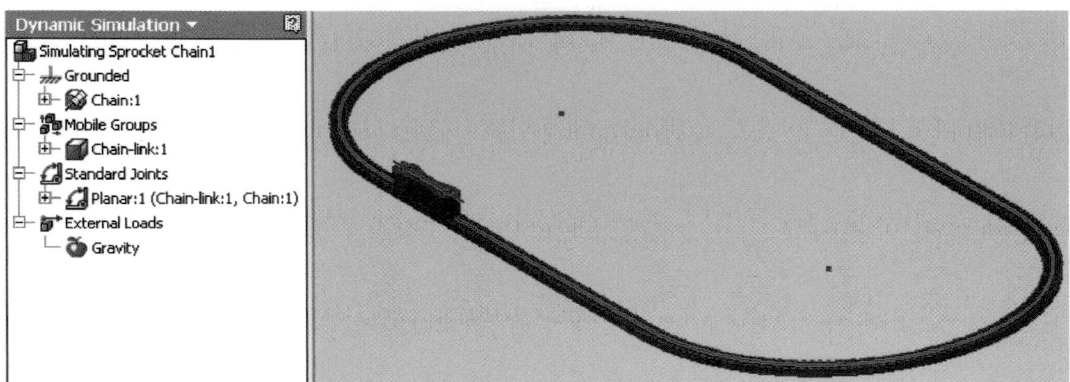

Note that only one planar joint is created between both components.

As the link needs to follow the chain, we need to constrain the link to the chain.

3. Select **Insert Joint** > **Sliding: Point Curve**

4. For **Curve 1**, select the projected Loop on chain component

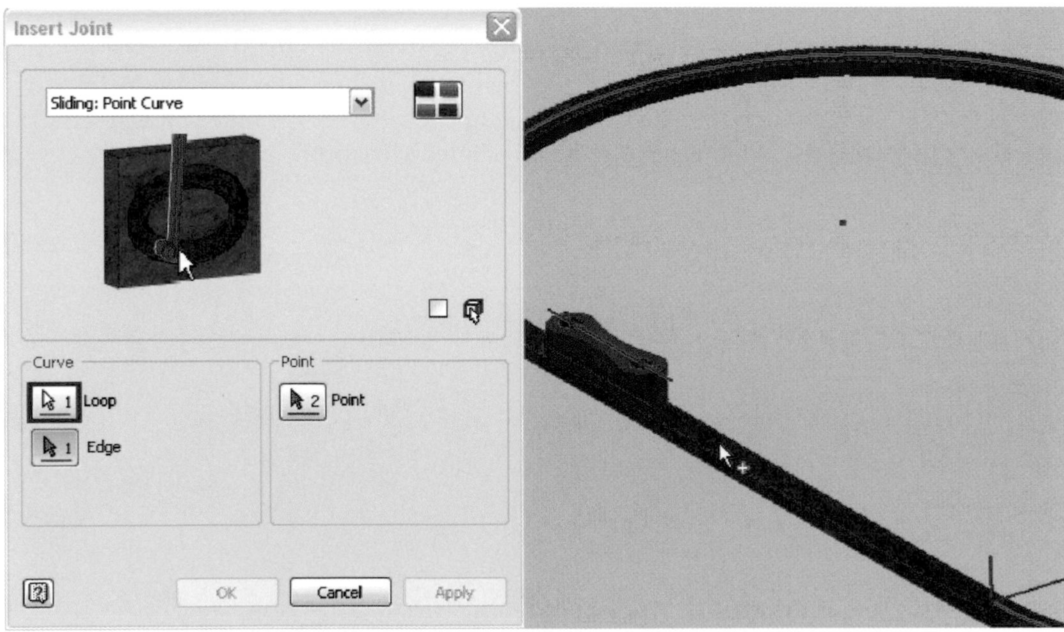

5. For **Point** 2, select the edge of the chain-link component

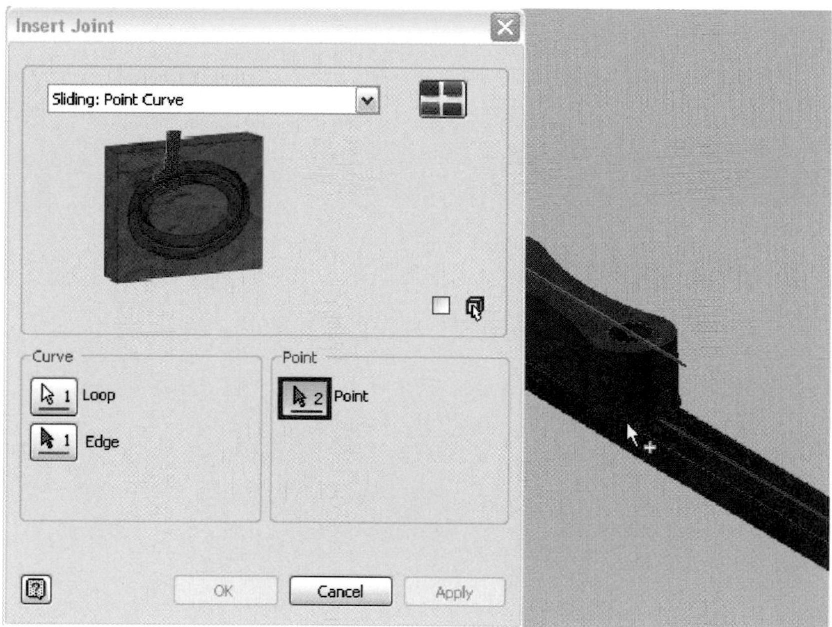

6. Click **Apply**

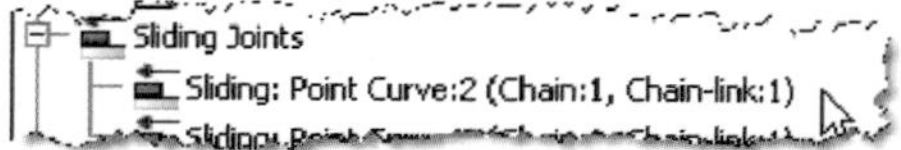

Now we need to constrain the other end of the link to the curve.

7. Repeat Steps 4–5

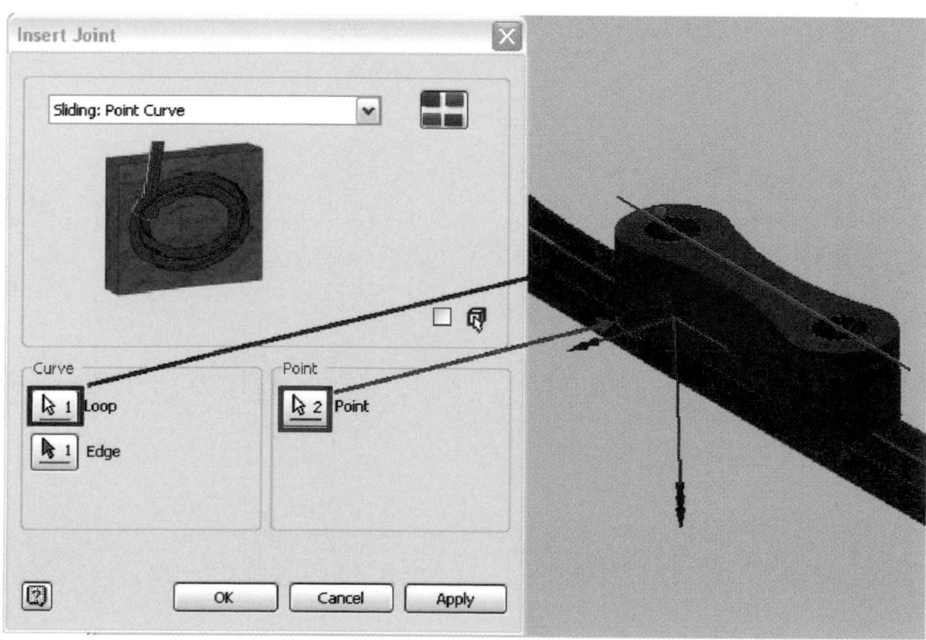

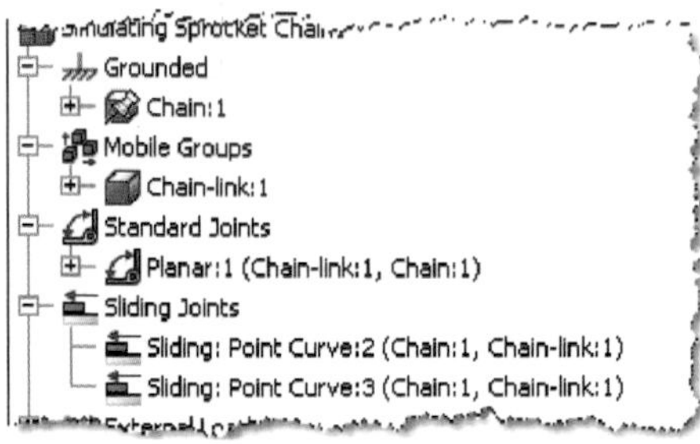

8. Click **OK**

We need to impose motion on the link at a constant velocity of 10 mm/s.

9. Double click the Planar:1 joint > Select the **dof 3 (T)** tab > Select **Enable imposed motion** > Apply a **Velocity** of 10 mm/s > Click **OK**

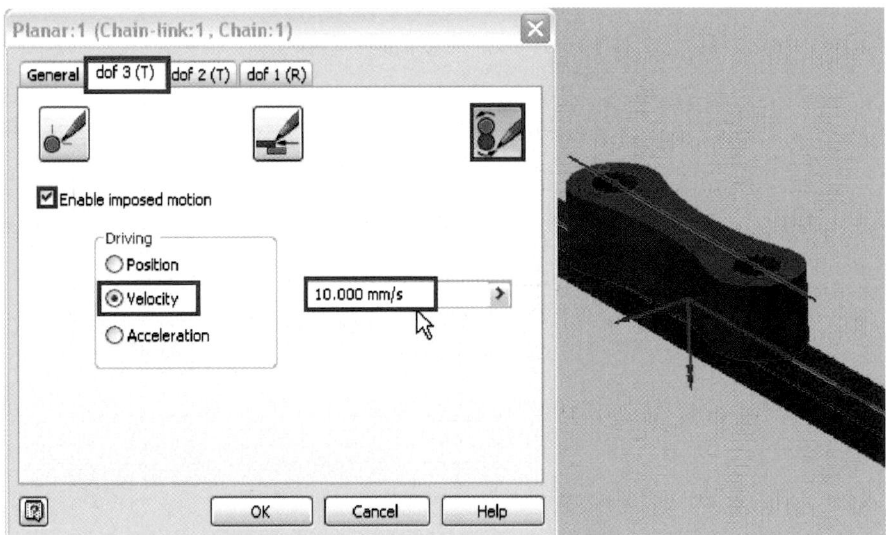

10. Play the simulation

After approximately eight seconds, you will get the following warning:

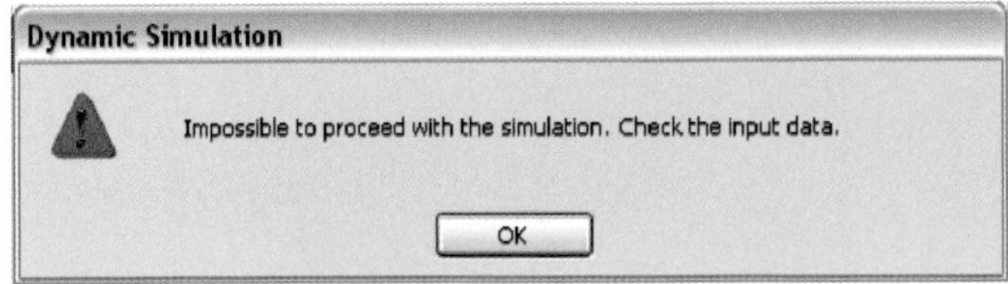

The warning appears because the direction of the imposed motion has changed. Currently, we cannot apply motion along the curve in which direction can also change.

The next steps will take you through an alternative process of simulating imposed motion along a curve, as this motion is required to simulate the sprocket and chain mechanism.

11. Close *Simulating Sprocket Chain1.iam*
12. Open *Simulating Sprocket Chain2.iam*

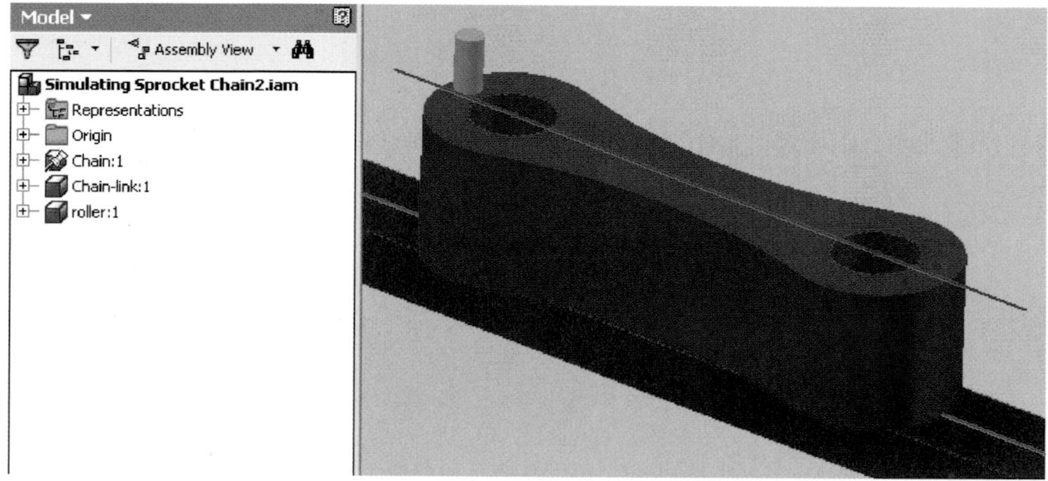

13. Select **Environments** tab > **Dynamic Simulation**
14. Create a revolution joint between roller:1 and Chain-link:1 > Click **OK**

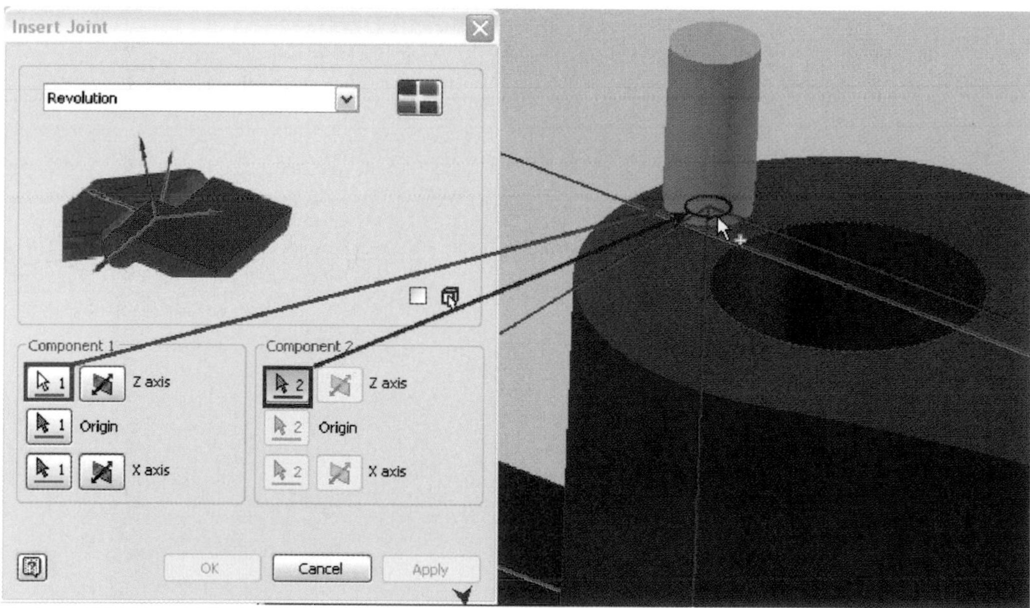

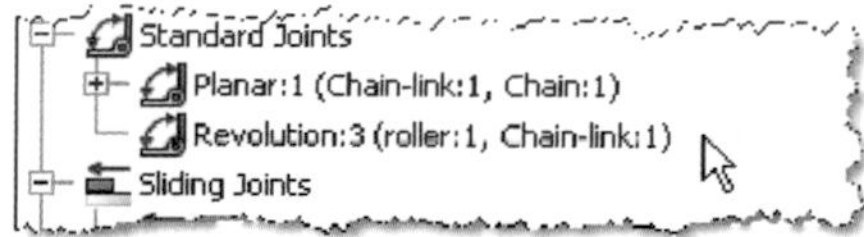

We are now going to constrain the back end of the Chain-link:1 component with a rolling cylinder curve rather than a sliding cylinder curve, as previously done.

15. Select **Insert Joint** > Select the Rolling: Cylinder Curve joint > For **Curve 1**, select the projected loop on Chain:1

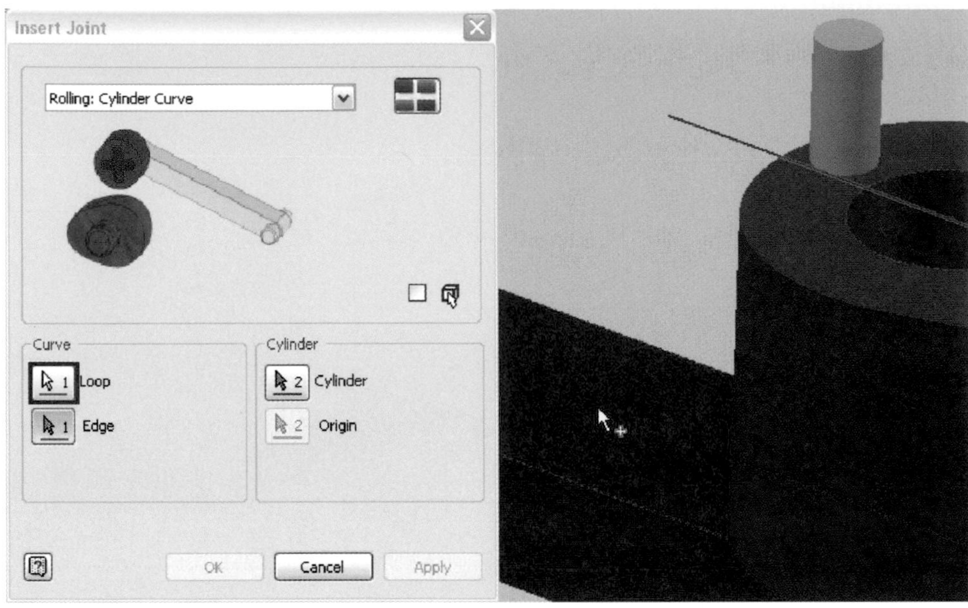

16. For Cylinder 2, select the roller:1 component

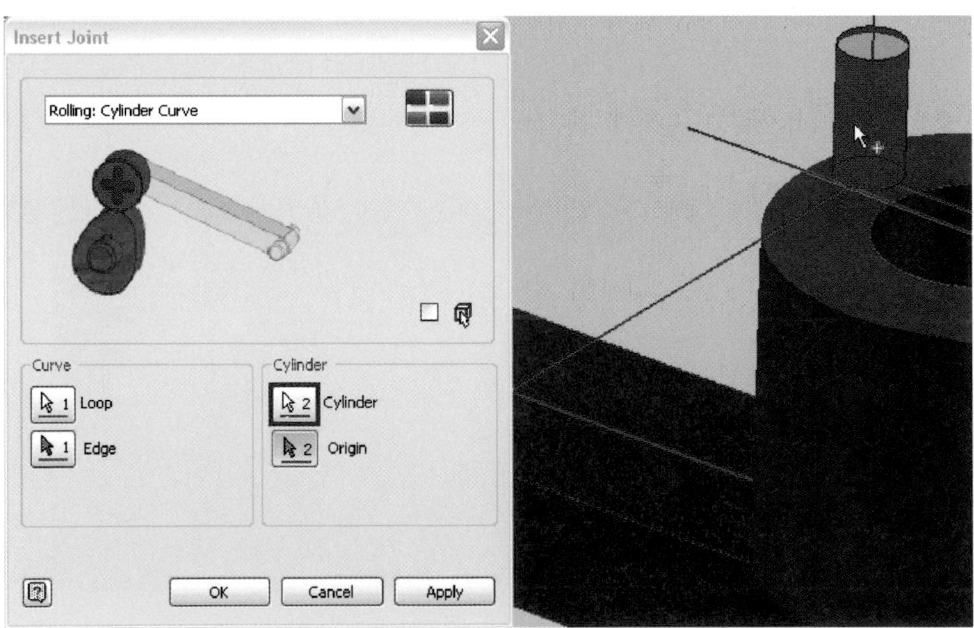

17. Click **OK**

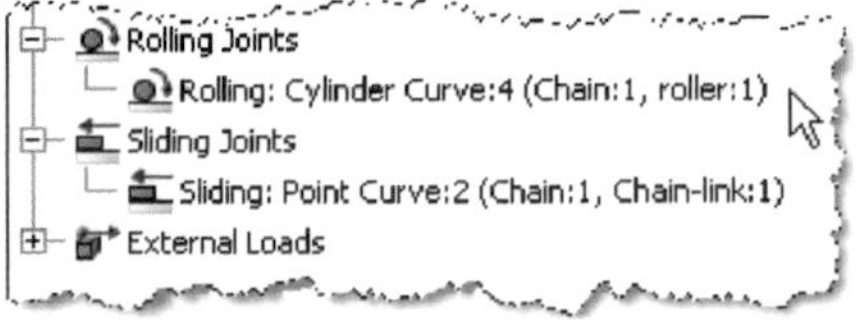

Now we are going to impose a rotational velocity on the roller joint.

18. Double click the Revolution:3 joint > Select the **dof 1 (R)** tab > Select **Enable imposed motion** > Apply a **Velocity** of 20 rad/s > Click **OK**

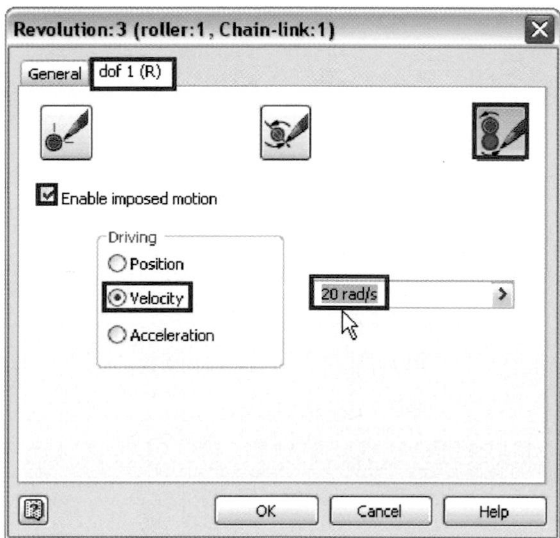

We have specified rad/s instead of deg/s because we need to make sure the chain-link to which the roller is attached travels at a velocity of 10 mm/s. This is mathematically explained below:

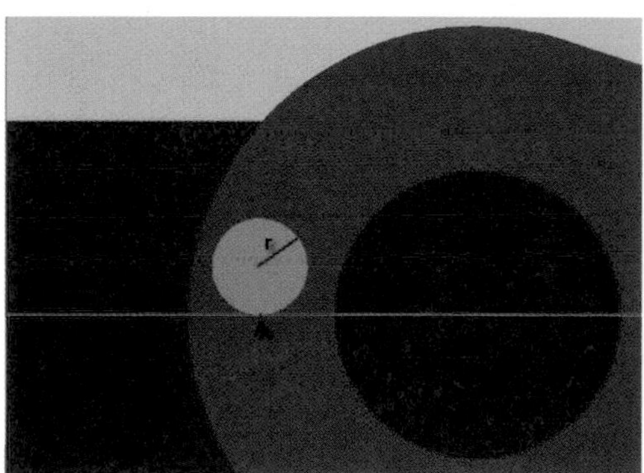

$$w = \frac{V}{r}$$

where

$V = \text{mm/s}$
$w = \text{rad/s}$
$r = \text{mm}$

So, to obtain a constant velocity of 10 mm/s, we need to specify w as

$$w = \frac{10}{0.5} = 20 \text{ rad/s}$$

19. Set the final time to 40 seconds > Play the simulation

20. Select Output Grapher

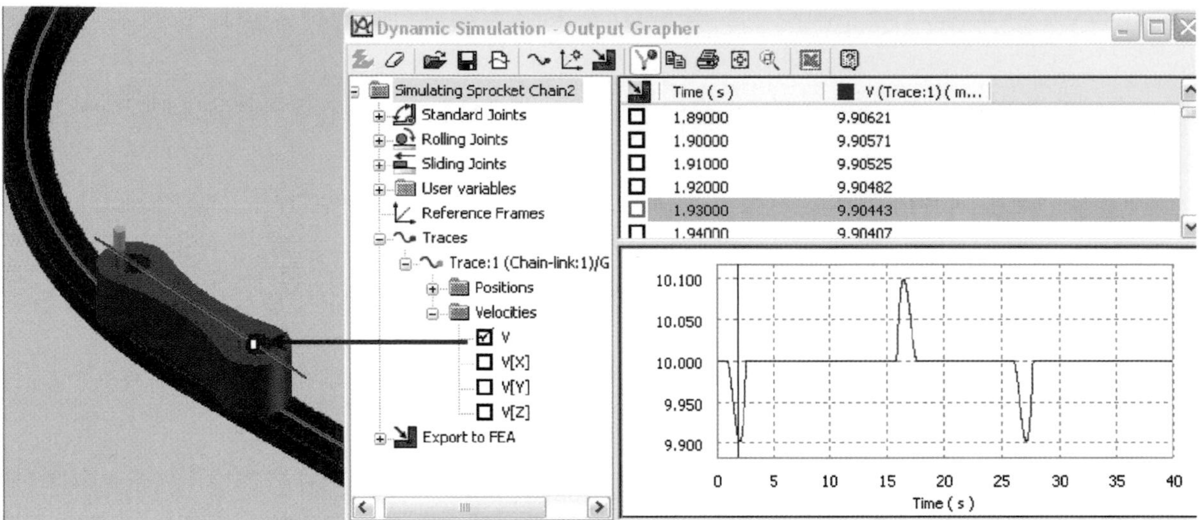

The Chain-link:1 component now completely follows the curve and maintains the velocity of 10 mm/s. However, there is a slight deviation of 0.1 mm/s (1%) when the link changes direction, for example going from a straight to a curve. This is a result of the geometrical radial offset of the roller from the curve. This error can be further reduced by reducing the radius of the roller.

Since we now have a method to simulate a component along a curve, we will use this method for the next stage of the design problem.

Stage 2 – Simulate the sprocket chain mechanism

21. Open *Transfer-Unit.iam*

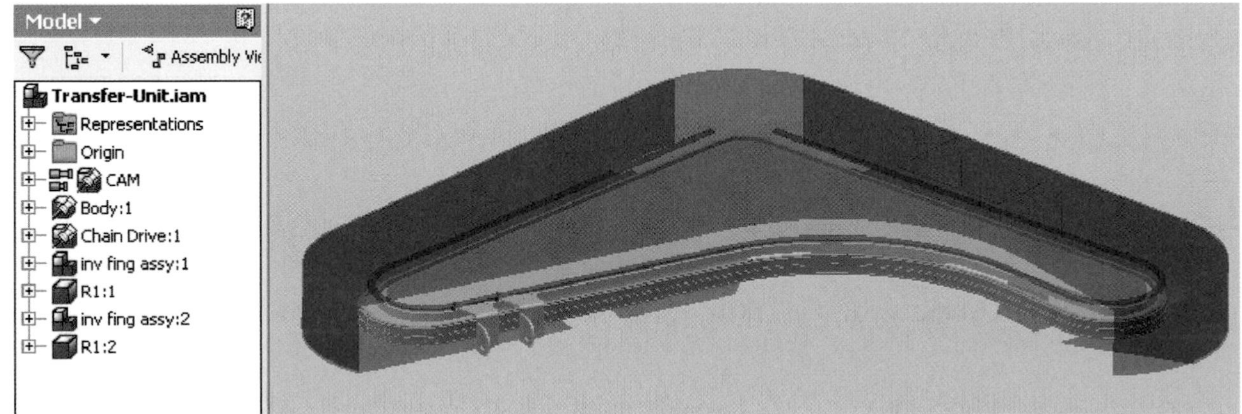

Joints

22. Select **Environments** tab > **Dynamic Simulation**

23. Select **Simulation Settings**

24. Select **Automatically Convert Constraints to Standard Joints** > Click **OK**

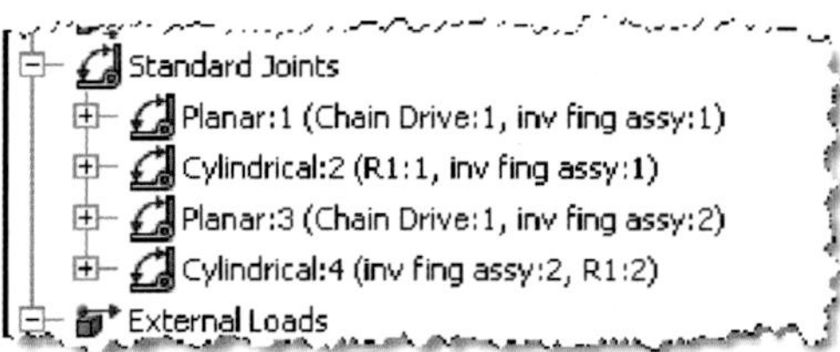

Since the standard joints have been automatically created, we next need to create nonstandard joints.

25. Switch off the visibility of the body component

This will aid in creating joints with ease.

26. Select **Insert Joint** > **Sliding: Cylinder Curve**

27. For **Curve,** select the face then select the edge

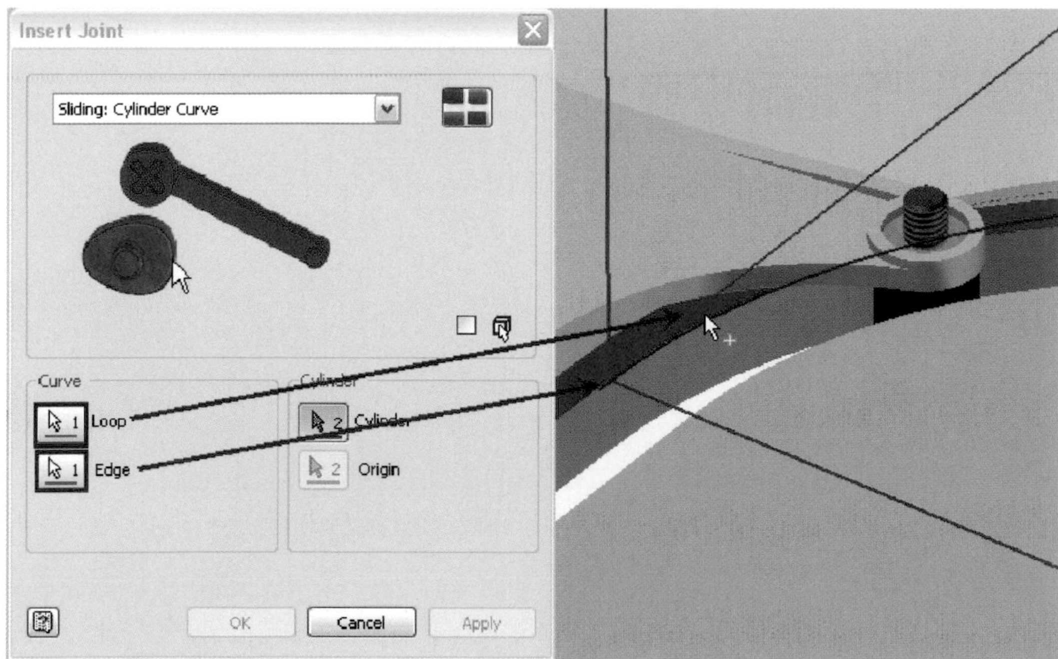

28. For **Cylinder,** select the component of the first finger subassembly > Click **Apply**

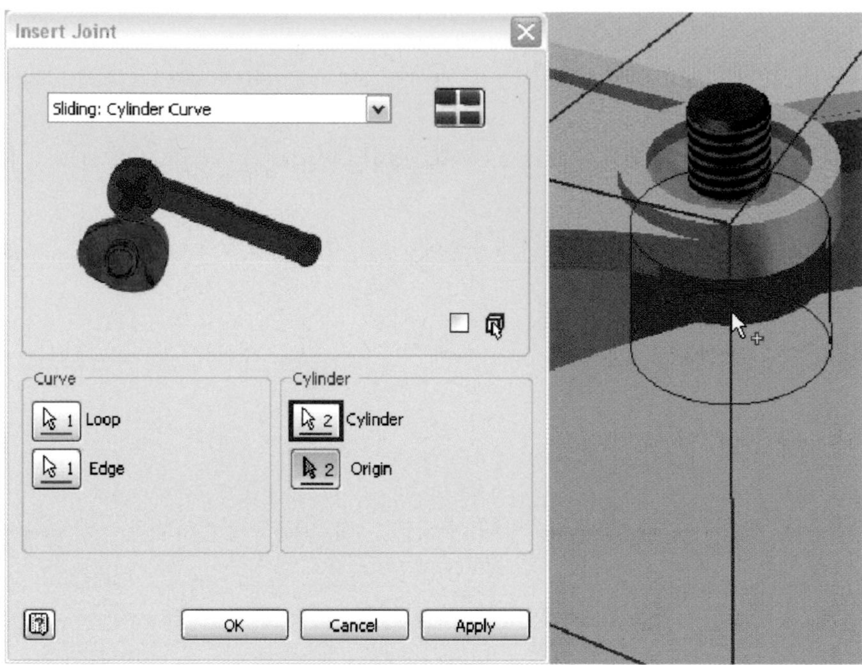

29. Repeat Steps 27–29 for the other finger

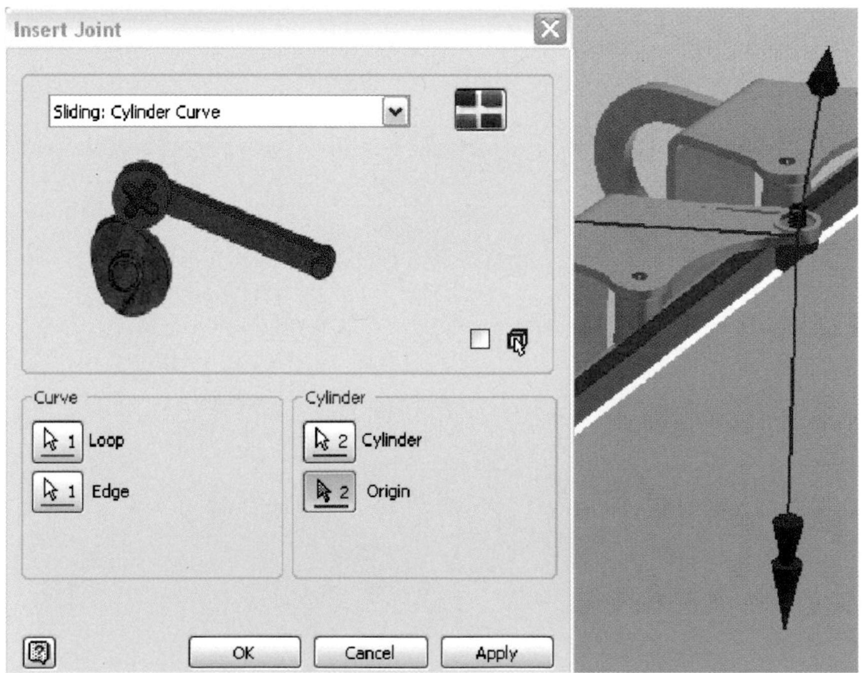

30. Click **OK**

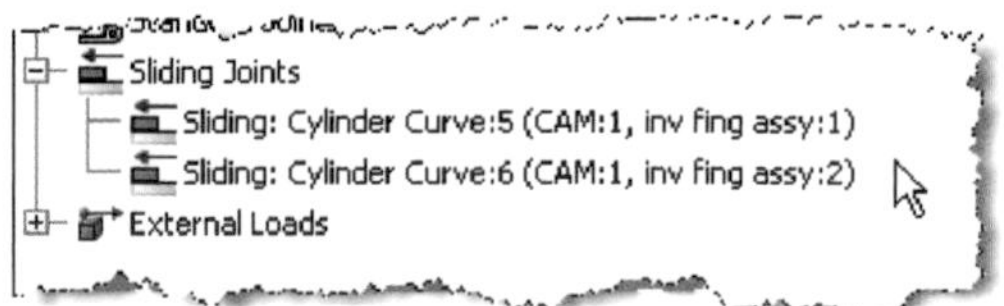

Now we need to create a Rolling: Cylinder Curve. This is the joint that drives the link at constant velocity.

31. Select **Insert Joint** > **Rolling: Cylinder Curve**

32. For **Curve 1**, select the projected loop

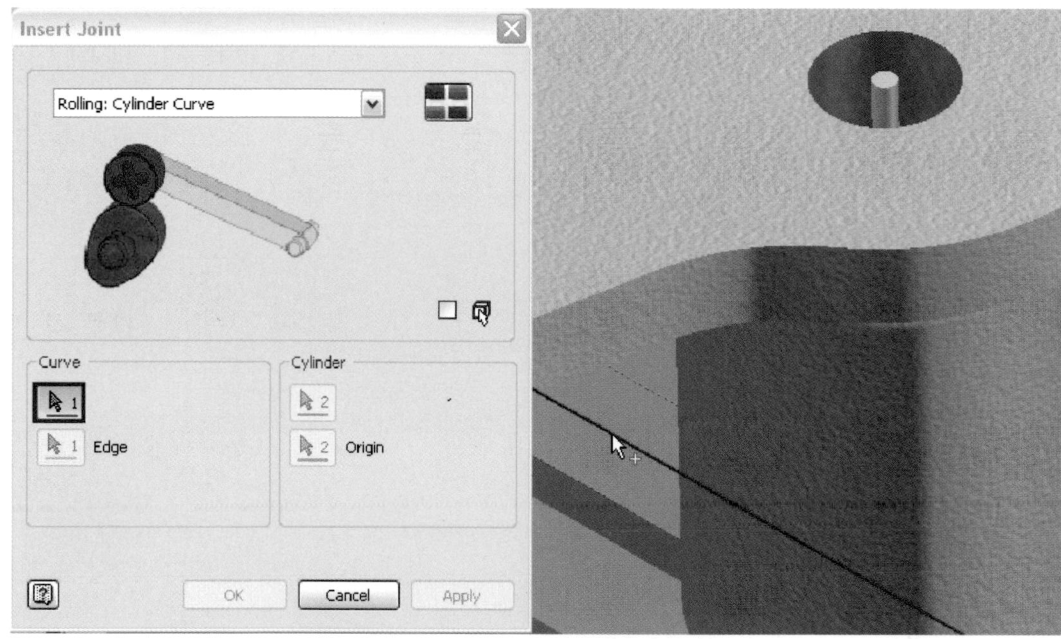

33. For **Cylinder**, select the small cylinder inside finger 1

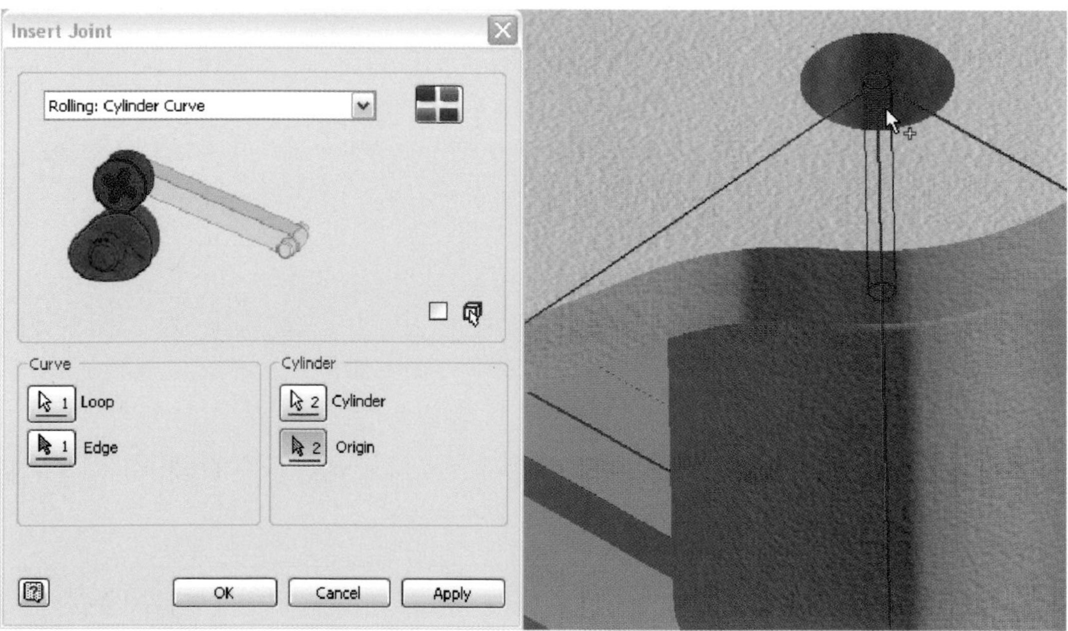

34. Click **OK**

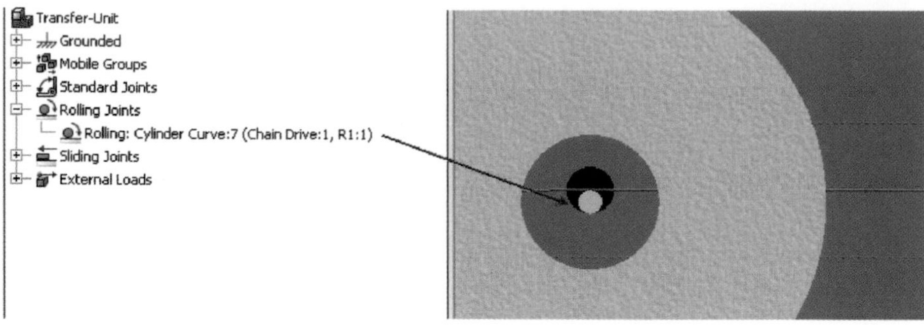

35. Repeat Steps 31–34 for the other finger

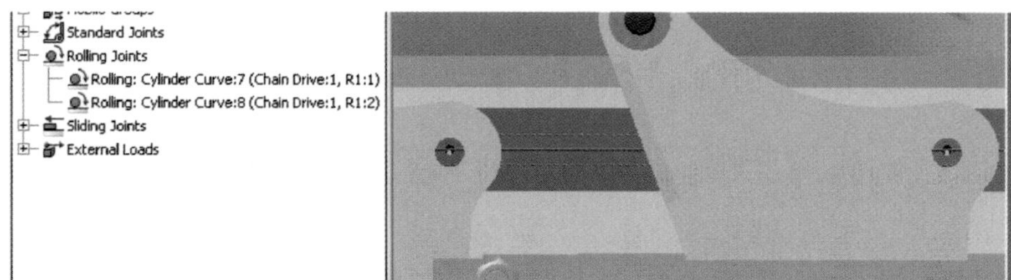

Environmental constraints

36. Select both cylindrical joints

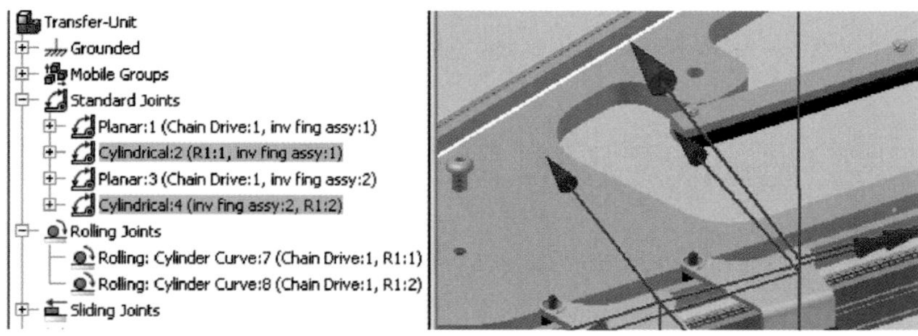

37. Right click > Select **Properties**

38. Select the **dof 1 (R)** tab > Select **Enable imposed motion** > Specify 2222.5 rad/s for constant **Velocity** > Click **OK**

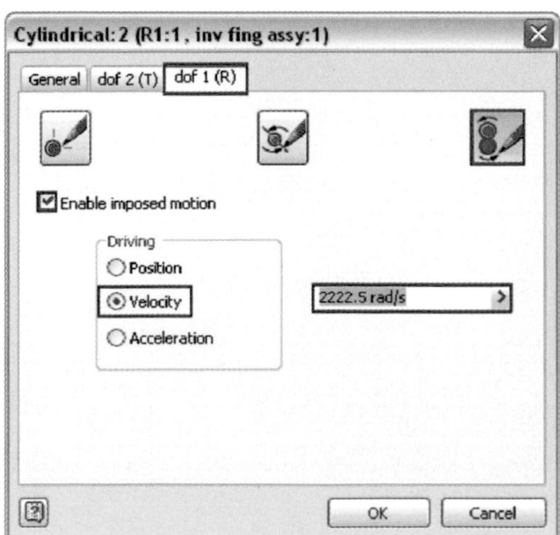

This specifies the rotational velocity of both joints.

39. Play the simulation

The finger moves in the opposite direction and we need to modify the Z axis orientation of one of the fingers so that they both move in the same direction.

40. Select **Construction Mode**

41. Right click the Cylindrical:4 joint > Select **Edit** > Select the Z axis to switch direction > Click **OK**

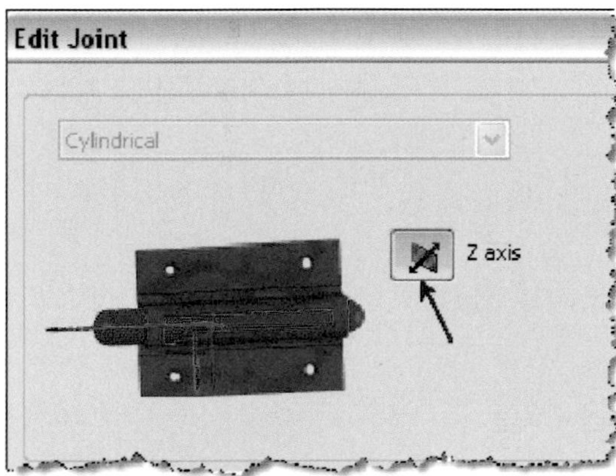

Analyze results

42. Play the simulation for 30 seconds

Both fingers now move in the same direction and at the same speed.

43. Select the Output Grapher > Select V[1]

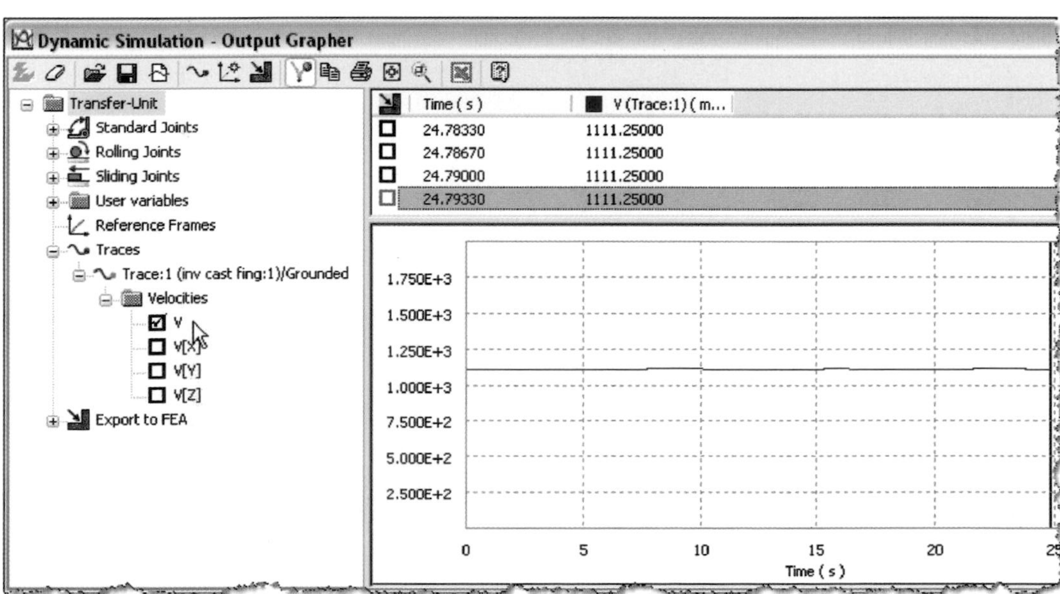

The velocity seems to be constant throughout the simulation because the range of the scale on the Y axis is large, with values between 250 and 1750.

44. Double click in the Output Grapher. Now the scale on the Y axis is very small, with values between 1110 and 1120, enabling you to analyze any small deviations in the results

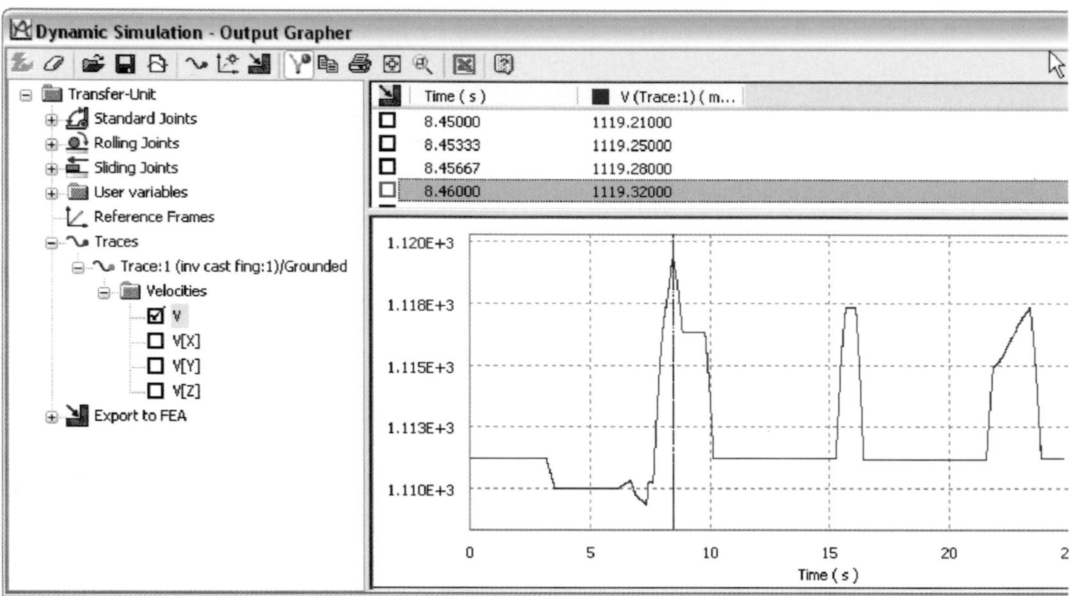

The velocity has a small deviation of about 0.6%.

This deviation is small and is acceptable for the purposes of simulating the sprocket and chain mechanism.

At this stage, you may continue to insert futher fingers into your simulation. Each finger will need three constraints, of which one is with the roller, as shown below.

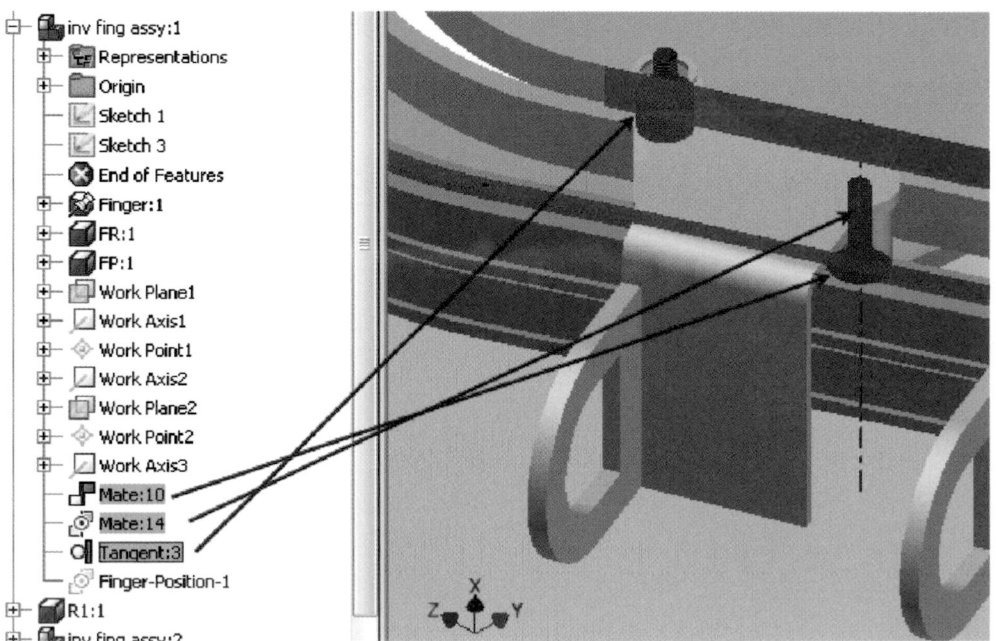

The fingers can be positioned using the black boss features indicating the start of each finger.

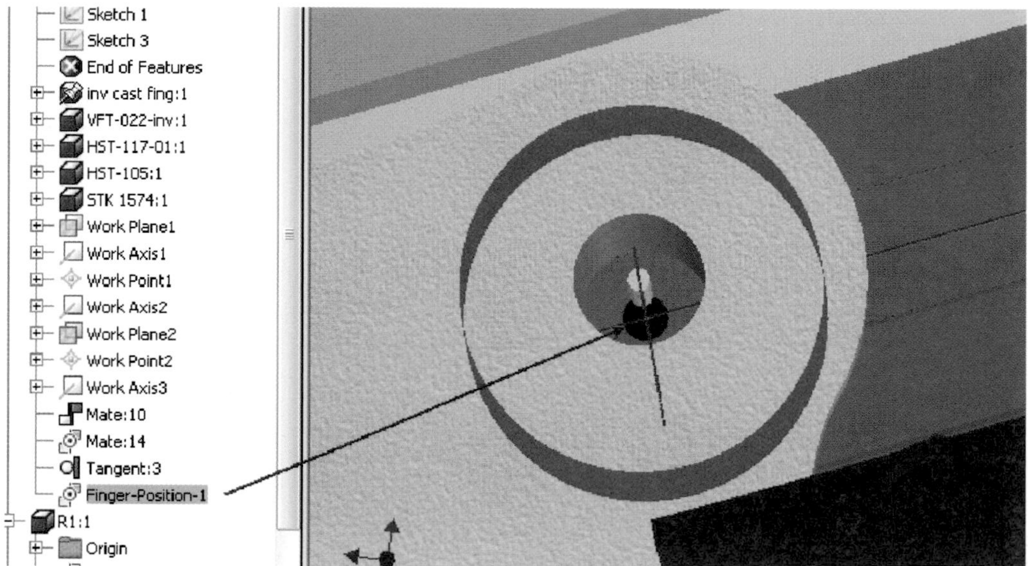

You will need to suppress this constraint created to position the fingers; otherwise, no joints will be automatically created in Dynamic Simulation.

These boss features are equally spaced out 42 times using a rectangular pattern along the projected curve and using the equation below:

$$\text{Number of fingers} = \frac{\text{Length of chain}}{\text{Article spacing}} = \frac{4039.718}{95.25} = 42.4 \sim 42$$

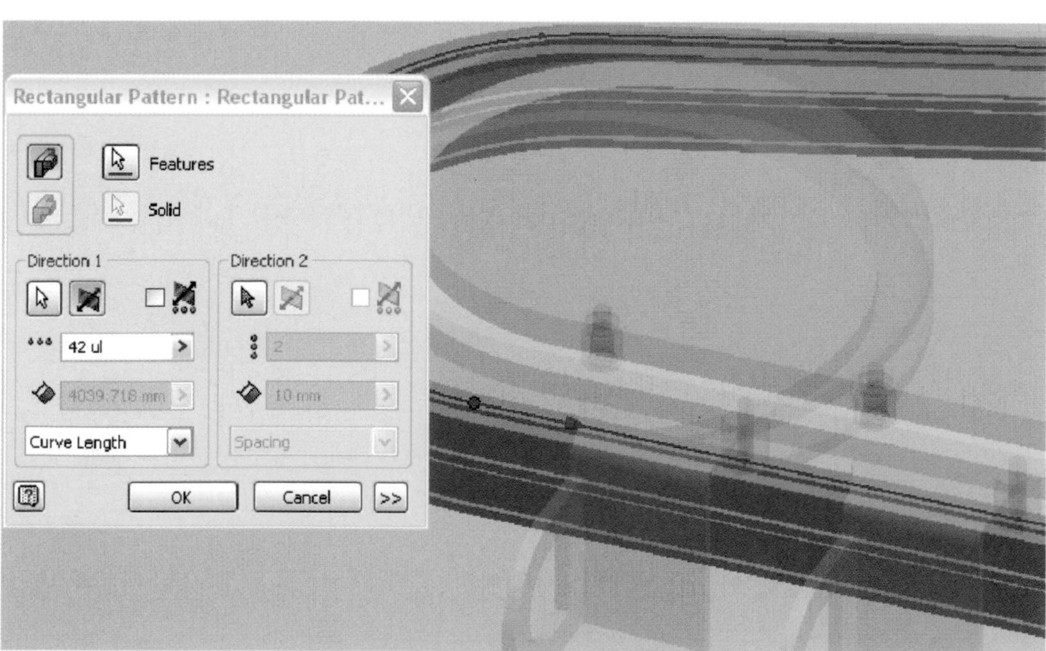

This means that 42 fingers are required in this example to transfer 700 bottles per minute.

45. Close the file

The next stage attempts to simulate 20 fingers of the sprocket and chain mechanism equally spaced out.

Stage 3 – Simulate the complete sprocket and chain mechanism

46. Open *Transfer-Unit-Complete.iam*

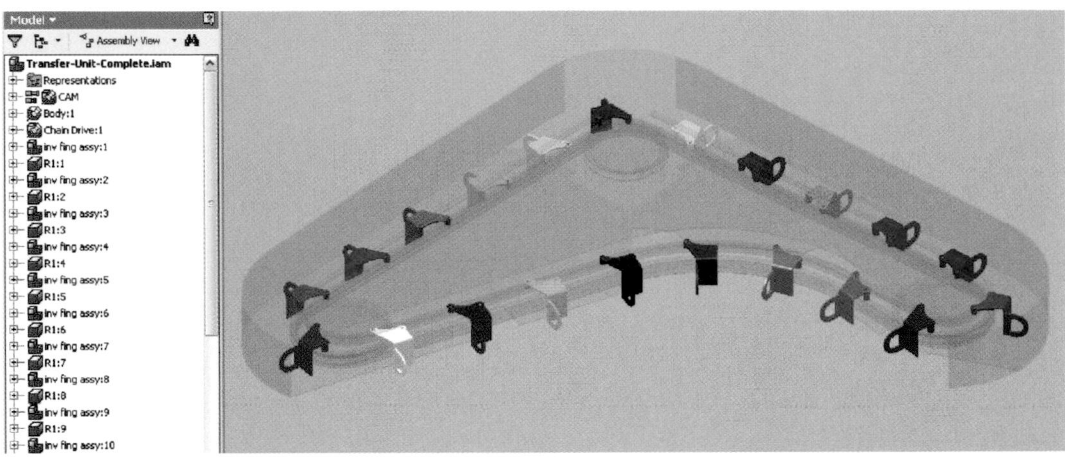

47. Select **Environments** tab > **Dynamic Simulation**

48. Set the simulation time to 3.635 seconds

$$\text{Time taken for finger to return to original position} = \frac{\text{Length of chain}}{\text{Chain speed}}$$

$$= \frac{4039.718}{1111.25}$$

$$= 3.635 \text{ seconds}$$

49. Play the simulation

50. Change the **Filter** value from 1 to 50 > Select **Continuous Play**

51. Close the Output Grapher

CHAPTER 9

The Stress Analysis Environment

THE FINITE ELEMENT METHOD (FEM) – AN OVERVIEW

The finite element method (FEM) is a mathematical/computer-based numerical technique for calculating the strength and behavior of engineering structures. Autodesk Inventor – and much other analysis software – is based on the FEM, where, simply, a component is broken down into many small elements, as shown below.

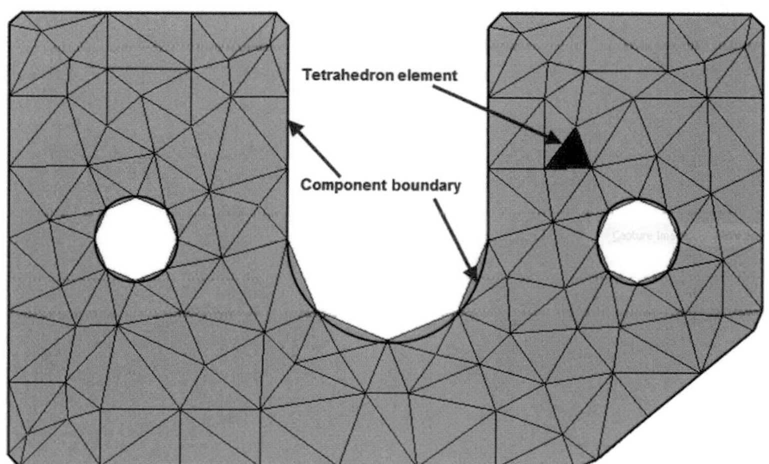

Discretization of a component into a number of tetrahedron finite elements

Let's assume that we need to determine the displacement of the component. This displacement (unknown quantity) acts over each element in a predefined manner – with the number and type of elements chosen so that overall distribution through the component is sufficiently approximated. This distribution across each element is commonly presented by a polynomial – whether it is linear, quadratic, or even cubic. It is important to note that FEM is always an approximation of the actual component and by its very nature will have errors due to discretization – particularly around curved boundaries (as shown above) or geometrically complex components.

These errors due to discretization can be reduced by either specifying more elements or using higher order polynomials to approximate the distribution of the unknown quantity over the elements – also referred to as polynomial interpolation function. Most finite element software uses the former method, specifically known as the H-refinement process, in

Up and Running with Autodesk Inventor Simulation 2011. ISBN: 978-0-12-382102-7
Copyright © 2010 Elsevier Inc. All rights of reproduction, in any form reserved.

which the software goes through an iterative process of reducing the number of elements at each iteration until the results have converged. The latter method, of using higher order polynomials, is called the P-refinement process, in which the software increases the order of the polynomial at each iteration starting from 1(linear) to 2(quadratic), 3(cubic), and so on.

Another approach to reducing errors due to discretization is to use higher order elements; this is discussed in the next section in more detail.

TYPES OF FINITE ELEMENT METHOD (FEM) ELEMENTS

Autodesk uses first and second order tetrahedron elements, as shown below.

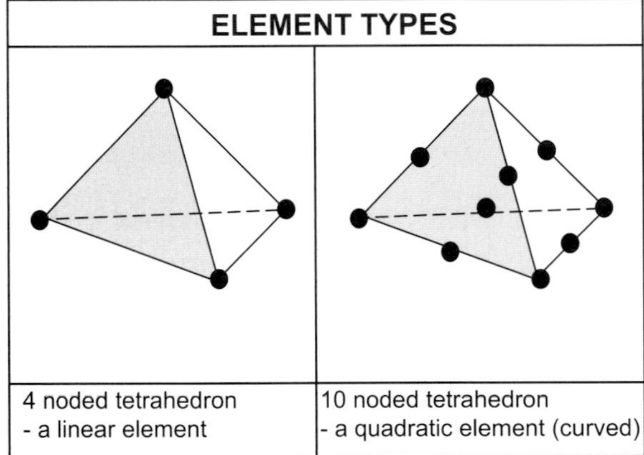

The following diagram illustrates that one quadratic element around a 90° circular object/component is better than two linear elements, as the quadratic element tries to match the 90° arc more closely and also can affect the accuracy of the result.

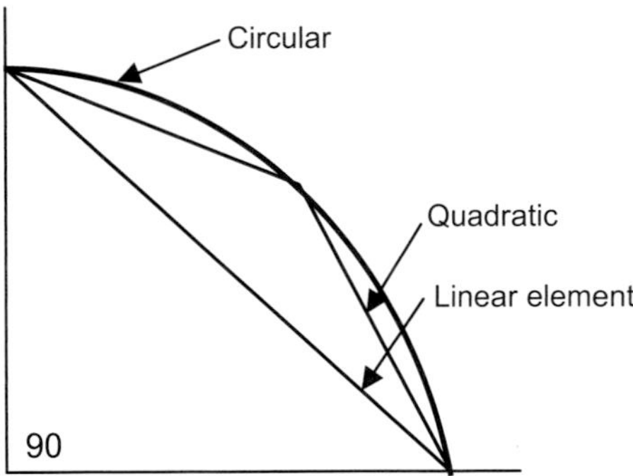

Also it is worth noting that the curved element almost matches the true profile of a 45° curved object (<1% geometrical error). Therefore, it is advisable to have at least two quadratic elements around a 90° arc, whereas there should be at least three linear elements, preferably four, around a 90° circular object.

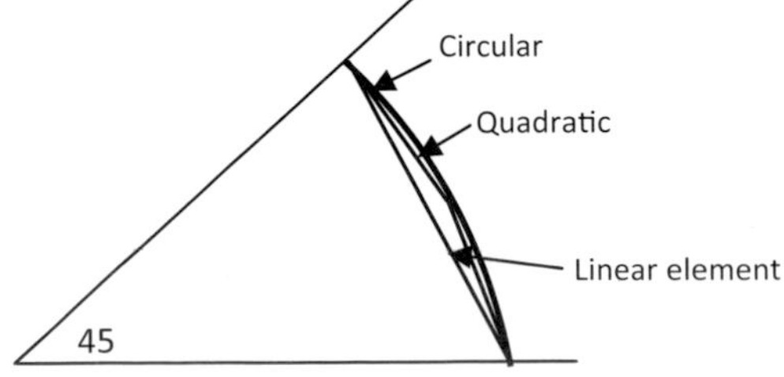

METHODS TO ENHANCE FINITE ELEMENT METHOD (FEM) RESULTS

In summary, there are three methods within Autodesk Inventor Simulation that can be used to enhance the accuracy of the results:

1. P-refinement
2. H-refinement
3. Higher order elements

There are pros and cons of using both P- and H-refinement:

	H-refinement	P-refinement
Results convergence	Slower – polynomial rate of convergence	Faster – exponential rate of convergence
Analysis time	Faster – in comparison to P-refinement	Slower – especially as P-order increases
Stress singularities	Can converge – with careful consideration given to settings	Never converges

P-refinement is automatically controlled by the software.

It is worth noting that curved elements can take more time to produce results when compared to linear elements, especially for a large or complex model.

For complex shapes it is always advisable to use quadratic elements.

When using quadratic elements it can take twice as long to analyze the results, as compared to linear elements.

Autodesk Inventor Simulation 2011 overcomes the pros and cons of both methods by using an H–P refinement approach, with some benefits being:

1. Exponential convergence in practical calculations (in cases with stress concentrations and stress singularities).
2. Potential of exponential convergence and maximal sparseness of the stiffness matrix.

Autodesk Inventor Simulation takes this H–P refinement approach one step further by making the H–P approach adaptive. This means that the software will only refine the elements around the high stress areas – rather than the whole model – meaning that the results convergence process will be further enhanced. This process is explained in the next section.

H–P convergence

Within Inventor Simulation 2011, the user can only control the H-refinement part of the H–P refinement convergence process. The software automatically increases P-order from one to three for every part analysis and from one to two for assembly analysis. The assembly analysis does not use a P-order of three because, as P-order gets higher than two, the analysis time can become exponentially longer – especially when there are a lot of parts to analyze.

If the user has specified two iterations for H-refinement in the **Convergence** dialog box, the software will perform the following H–P refinement:

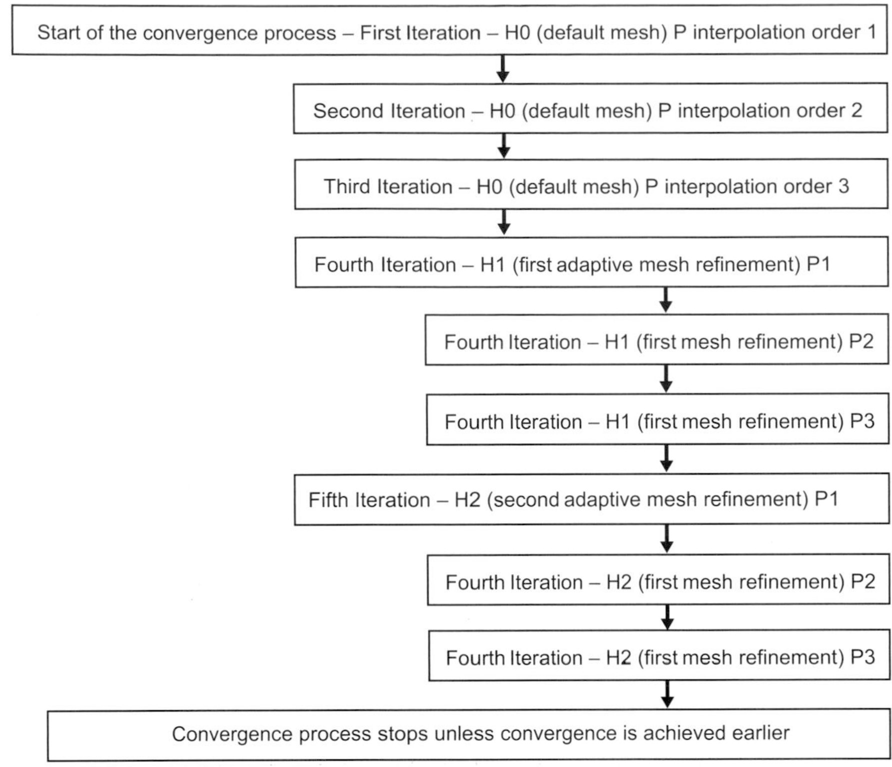

This H–P convergence process is very efficient, except when the model does not have stress singularities present. Stress singularities and methods to overcome them are explained later.

LINEAR AND NONLINEAR ANALYSIS

Autodesk Inventor Simulation is only capable of performing linear analysis where components have small deformations, under operational loading conditions. On the other hand, nonlinear analysis is typically involved when components are experiencing large deformations and thus component material can deform beyond the elastic limit.

Linear analysis

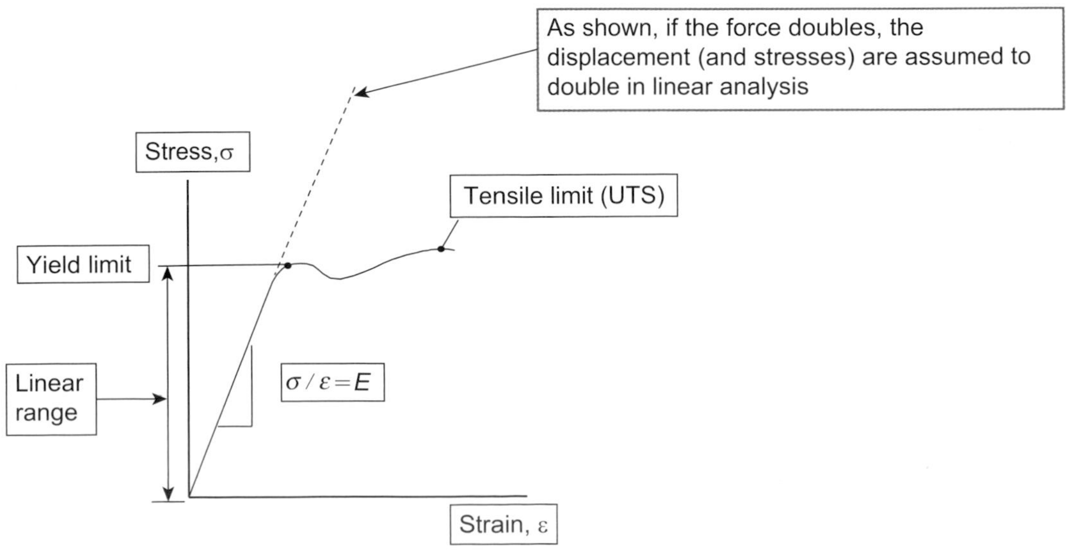

Young's modulus provides the stiffness of the material; for example a higher Young's modulus will produce a stronger material and a lower Young's modulus will produce a weaker material.

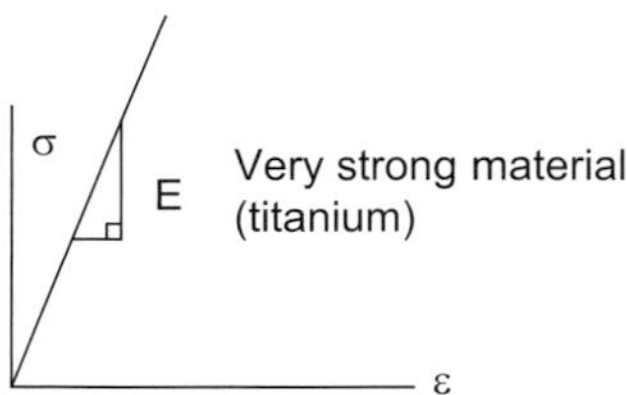

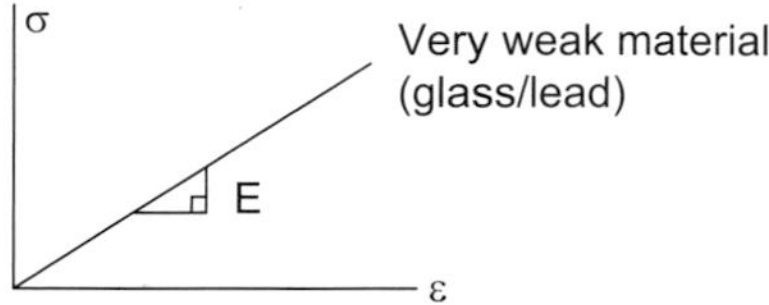

$$\text{Strain} = \frac{\text{Change in length}}{\text{Original length}} \quad \text{and} = \frac{\text{Force}}{\text{Area}}$$

(Note: for linear analysis it is assumed that the change in length is very small compared to the original length.)

Assumptions normally made when conducting a linear analysis	
1	The material properties of the component remaln linear after the yield limit. Hence, results beyond this limit are not valid using Autodesk Inventor Simulation Suite
2	The deflections of components are small compared to the overall component size
3	The components are rigid and ductile like metal (not rubber)
4	The components deform equally in all three directions; that is, the material properties are isotropic

Nonlinear analysis

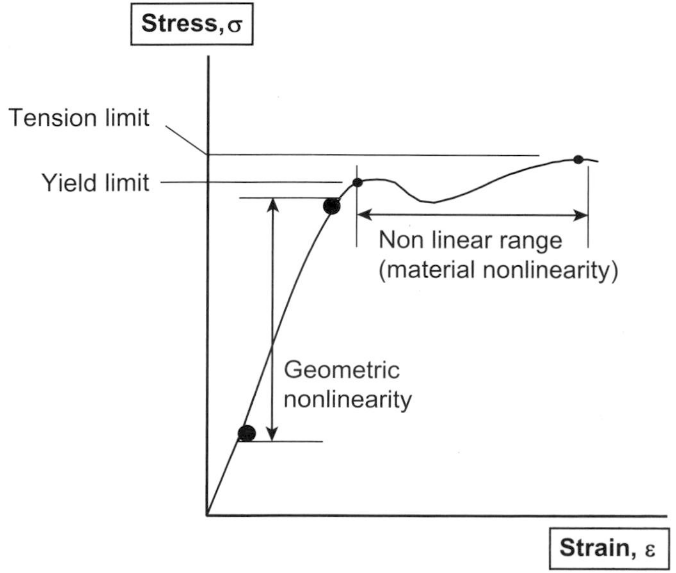

Nonlinear analysis falls into the following three categories:

Geometric nonlinearity – Where a component experiences large deformations and as a result can cause the component to experience nonlinear behavior. A typical example is a fishing rod.

Material nonlinearity – When the component goes beyond the yield limit, the stress/strain relationship becomes nonlinear as the material starts to deform permanently.

Contact – Includes the effect of two components coming into contact; that is, they can experience an abrupt change in stiffness resulting in localized material deformation at the region of contact.

Currently, Autodesk Inventor Simulation 2011 allows performance of linear static and modal analysis; both are discussed in the next sections.

STATIC ANALYSIS – AN OVERVIEW

Static analysis is an engineering discipline that determines the stress in materials and structures subjected to static or dynamic forces or loads. The aim of the analysis is usually to determine whether the element or collection of elements, usually referred to as a structure or component, can safely withstand the specified forces and loads. This is achieved when the determined stress from the applied force(s) is less than the yield strength the material is known to be able to withstand. This stress relationship is commonly referred to as factor of safety (FOS) and is used in many analyses as an indicator of success or failure in analysis.

$$\text{MFOS} = \frac{\text{Yield stress}}{\text{Calculated stress}} = \frac{\text{Ultimate stress}}{\text{Calculated stress}}$$

Factor of safety can be based on either the yield or ultimate stress limit of the material. The FOS on yield strength aims to prevent detrimental deformations and the FOS on ultimate strength aims to prevent collapse, and can only be conducted by nonlinear analysis software. Autodesk Inventor 2011 can only perform linear analysis and hence FOS will more commonly be based on yield limit.

Below are some examples where static analysis can be useful.

The canal bridge is a typical example of static analysis. Here, one will be interested to know whether the bridge will withstand the load of a vehicle when it crosses the bridge. This will also help us identify weak spots of the structure, ultimately allowing us to design a bridge to carry the maximum physically possible load.

Canal Bridge

For Halifax fans we need to be able, for example, to determine the maximum deflection of the blade, which can have an impact on the efficiency of the fan. With the help of static analysis, the blade can be studied and analyzed to reduce deformation, for example by using different materials, increasing the thickness, or adding structural stiffeners.

One of the major obstacles when conducting static analyses is stress singularities, which can significantly distort results and may reduce confidence in the results, as illustrated and discussed in the next section.

Stress singularities

Stress singularities are a major concern when analyzing results as they can considerably distort results. They are also a main cause for nonconvergence of results. So, the first question is: what is stress singularity? This can be best explained by the following example.

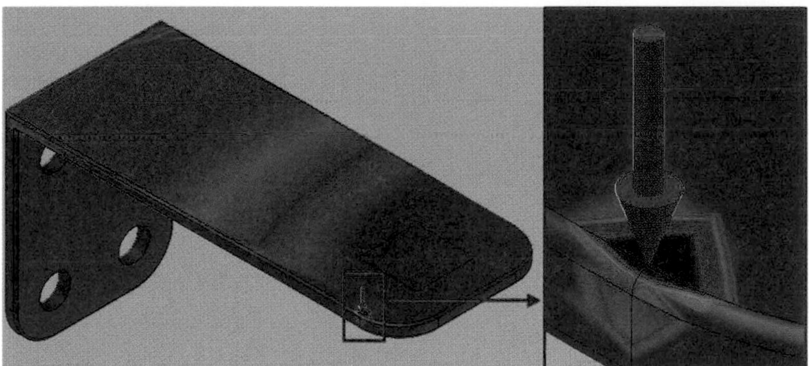

The above bracket has a localized high stress around the force applied on a point. This stress can be considerably higher than the operational stress and applying a more dense mesh around this simply leads to a much higher stress. This phenomenon is known as stress singularity where the stress becomes infinite, as illustrated by the following formula:

$$\text{Stress (infinite)} = \frac{\text{Force}}{\text{Area of point (almost} = 0)}$$

Therefore, to avoid stress singularities when applying loads, it is recommended *not to apply loads at points and small edges.*

Stress singularities can also occur *by applying constraints on points and small edges* – even faces with sharp corners, as illustrated below.

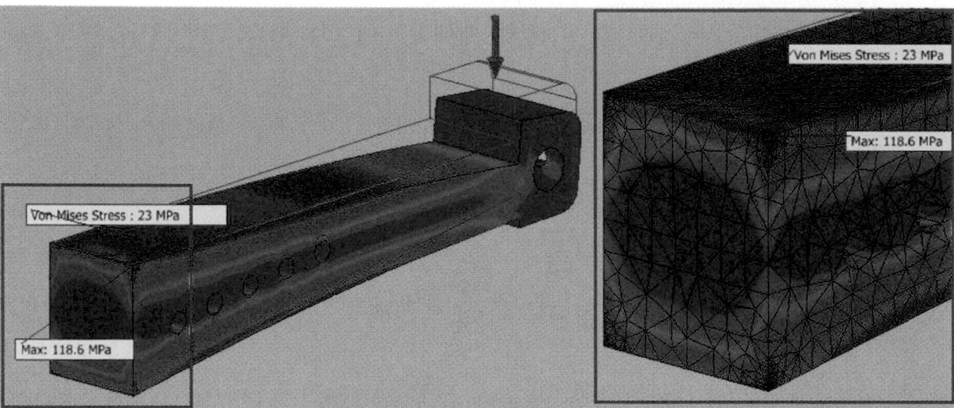

In the above example, stress singularities resulted from using automatic convergence, whereas in the image below the same model is showing the same stress in the area of interest by using the default mesh and no automatic convergence. Therefore, interpret results with care.

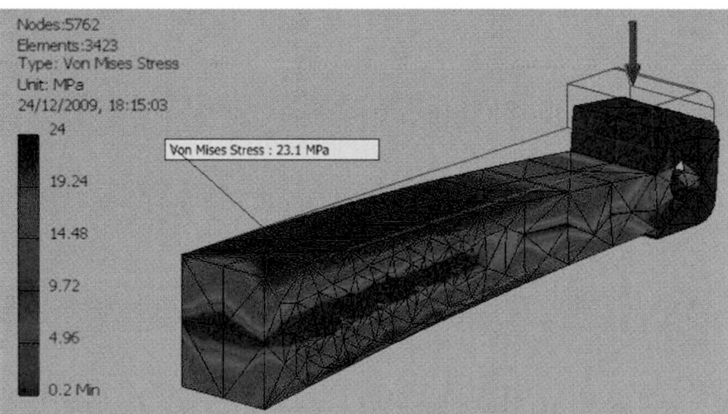

Gain further confidence in your results by using manual convergence, mentioned later in this chapter, when models have stress singularities present.

Finally, another cause of stress singularity *is over-simplification of components.* Let's look at the following example:

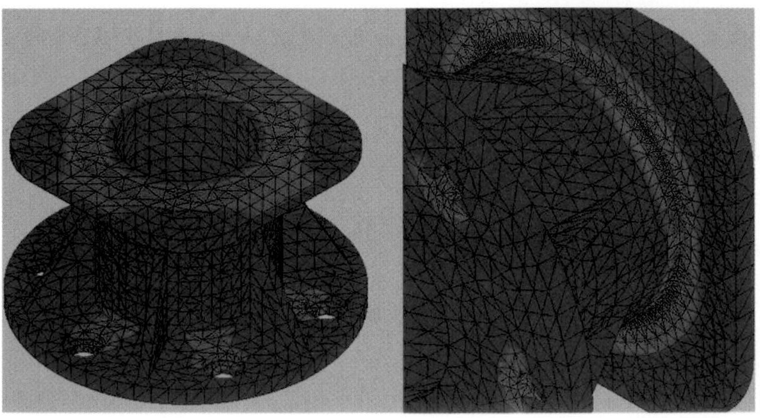

In this example, the fillets have been removed to simplify the analysis; however, when using automatic convergence, the maximum stress value does not converge as all the stress is concentrated around the edge, as shown. In this scenario it would be advisable to unsuppress the fillets (or, in cases when fillets are not modeled, use fillets to distribute loads).

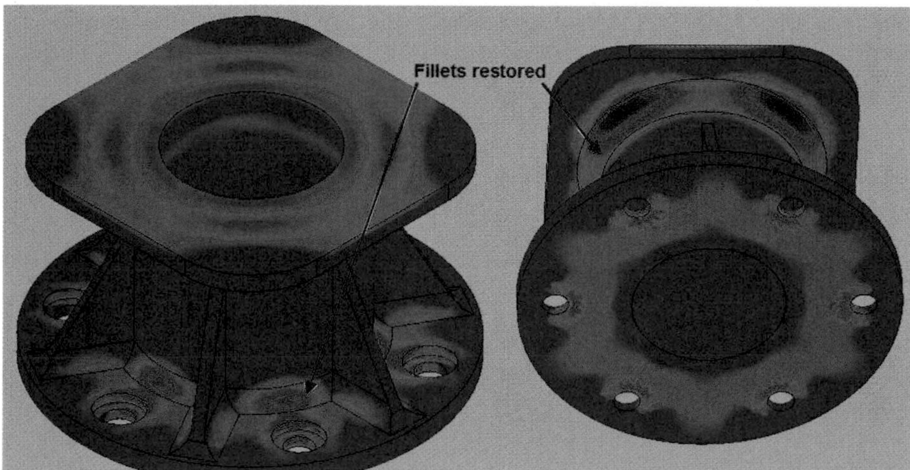

So, in brief to avoid stress singularities within models:

1. Avoid applying loads on points and small edges.
2. Avoid restraining faces with sharp corners, including points and small edges.
3. Apply fillets and chamfers to evenly distribute loads.

Use linear elements when a model has stress singularities present, as they can capture stress singularities much better than the curve elements.

In some cases it is impossible to remove stress singularities, in which case careful interpretation of results is very important. One approach to this is detailed in Chapter 14.

MODAL ANALYSIS – AN OVERVIEW

Modal analysis determines modes to better understand the behavior of components and structures under free vibration. Geometry, mass, and constraints are the only factors that can affect modal analysis. Modes are inherent properties of a structure, and are determined by the material properties and boundary conditions of the structure. Each mode is defined by a natural frequency and a mode shape. Frequency is defined as cycles/s; for example, 10 cycles/s is equivalent to 10 Hz. It is these frequencies that cause vibrations in components/ structures. Most, if not all, engineered products cause vibration in today's world for example the vibration felt through the steering wheel of a tire caused by unbalanced tires; the vibration felt through the floor when a passenger train goes past; and noise in airplanes especially at take-off, caused by revving of the aero-engines. By analyzing these modal shapes and frequencies, one can try to minimize these vibrations as they can cause failure in products by weakening the components and structures due to fatigue. Another cause of failure due to vibrations is resonance – this is where two components have the same natural frequencies, resulting in excessive vibration and ultimately leading to destruction. Following are some examples of structures that have been affected by resonance components that in some cases, leading to destruction or excessive vibration.

The Tacoma Bridge in Washington, USA, is a famous example of bridge failure due to resonance induced by wind.

The bridge was completely reconstructed to better withstand variations in the wind speed etc. and with better damping to minimize and isolate vibrations in the bridge.

The Millennium Bridge in London, UK, was another example, in which lateral vibration was caused in the bridge as pedestrians walked over it. The greater the number of people walking on the bridge, the greater was the lateral movement.

The bridge was closed soon after it was opened and remained closed for two further years. The problem was rectified by using a damping solution to absorb the movements, as stiffening the structure would have meant considerably altering the bridge.

Washing machines, which are used in many households, today can lead to excessive vibration of the drum induced by the full cycle spinning speed, in some cases in combination with the load weight of the wash. This in some extreme cases can lead the door to open, or even the machine to move from its original position, particularly in older machines.

Helicopter design is another field where vibration and resonance are critical issues. For example, if any of the components of a helicopter have frequencies that are close to the rotational speed of the rotors, then resonance of a component could occur, leading, for example, to a possible fatigue failure.

Thus, modal analysis is instrumental in helping us to better understand the structural flexibility and potential vibratory issues related to noise, fatigue, and resonance failures.

Natural frequencies – Basic theory

Theory for vibrations of continuous beams can be found in standard engineering textbooks. The natural frequencies of a simple cantilever can be determined theoretically using the following equation:

$$M \frac{K^2}{2\pi}\sqrt{\frac{EI}{\rho A L^4}}$$

where the K values for the first four modes are
1 – 1.8751
2 – 4.6941
3 – 7.8547
4 – 10.9955

and

E = Young's modulus
I = Area moment of inertia
p = Density
A = Area
L = Length

For a simple plate, 30 mm × 10 mm × 300 mm, made out of nylon 66, the first two calculated natural frequencies are 5.75 Hz and 36.04 Hz.

The following is a summary of the results:

	Theoretical (Hz)	Modal analysis (mesh size 0.1) (Hz)	Modal analysis (mesh size 0.05) (Hz)	Modal analysis (mesh size 0.025) (Hz)
Mode 1	5.75	6.33	5.9	5.85
Mode 2	36.04	43.87	37.09	36.67

Note: Mesh size refers to average element size.

The followings settings were also used:

1. Enhanced accuracy
2. Curved elements

For modal analysis, the mesh size can have an impact on the accuracy of the results. An average element mesh size of 0.025 produces results within 2% when compared with theoretical results.

Preloaded modes

In some situations, however, the loads will affect the natural frequencies. An example would be a guitar string: as tension is applied, the frequency changes. Loads that produce membrane stresses will affect the natural frequency of the object. Tensile member stresses will increase the natural frequencies and compressive membrane stresses will lower them, whereas pure bending stress will not affect natural frequency.

Suspension bridge designs are classic examples of where extensive use is made of tensile members (cables) suspended via towers to hold up the road deck. The weight is held by the cables via the towers, which in turn transfer the weight to the ground. Tension within cables also provides rigidity to the structural integrity of the bridge.

Let's look at a simple tie rod example in which the tie rod is not prestressed; the first mode and shape of the rod are shown below, giving a natural frequency of 32.63 Hz.

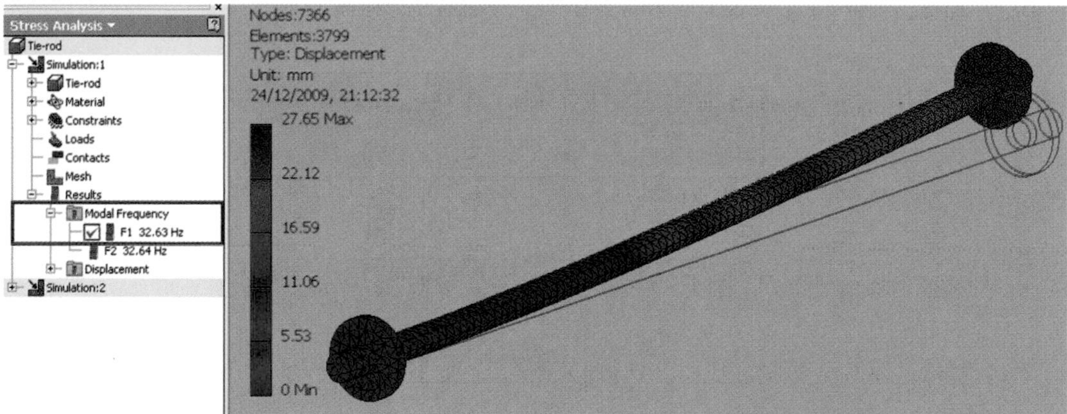

On the other hand, if a tensile load of 1000 N is applied to prestress the tie rod, the natural frequency of the first mode almost doubles to 60.24 Hz.

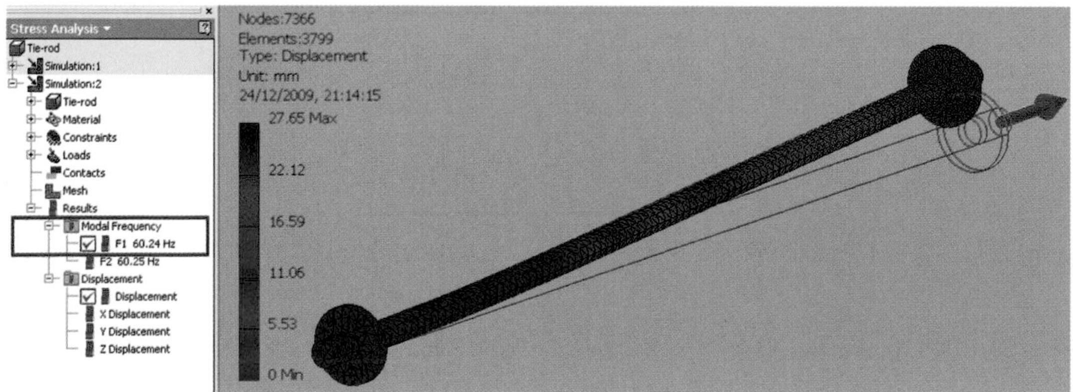

As we have now covered the basics of stress and modal theory, we will now go over the user interface of Autodesk Inventor Simulation 2011.

STRESS ANALYSIS WORKFLOW

The process of creating a dynamic simulation study involves four core steps:

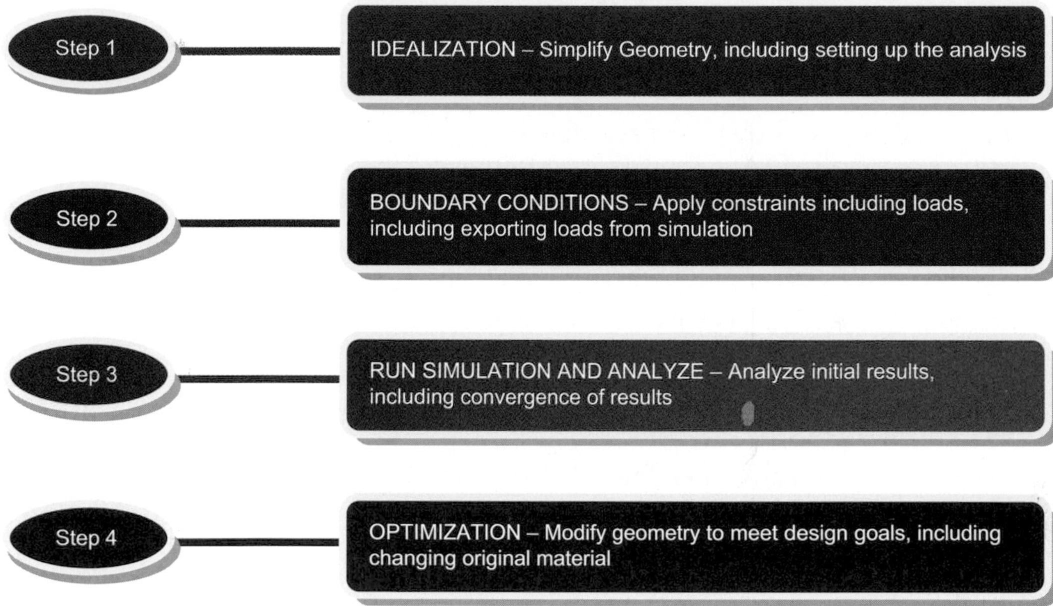

STRESS ANALYSIS USER INTERFACE

Stress Analysis can be accessed from both the Part and the Assembly environments via the **Analysis** tab.

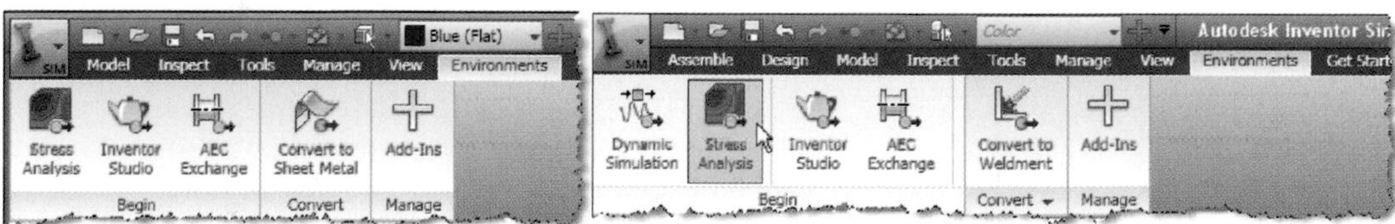

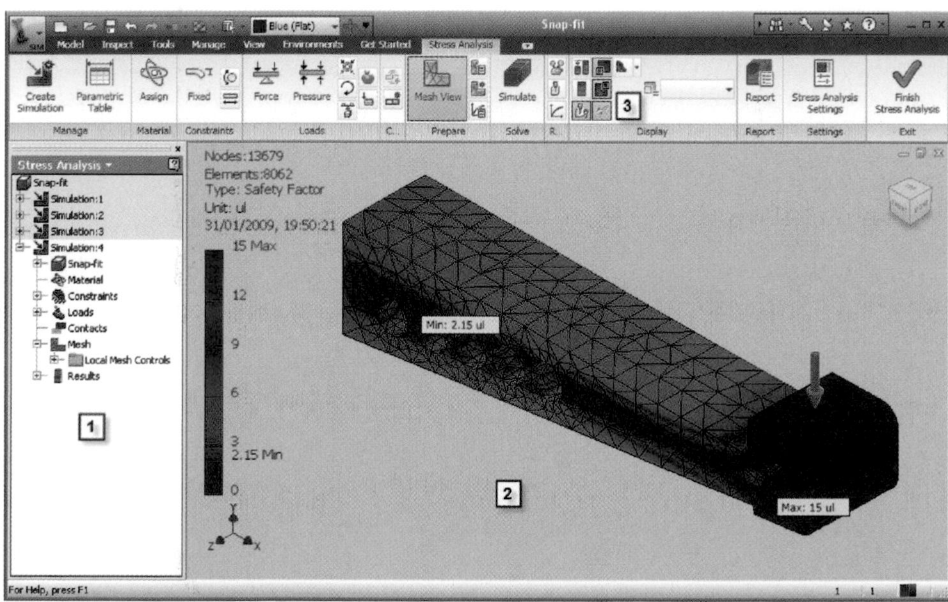

1. **Stress Analysis** browser
2. **Stress Analysis** graphic window
3. **Stress Analysis** panel

Stress analysis browser

Displays the simulations with part or assembly and simulation parameters in a hierarchical view with nested levels of feature and attribute information. You can:

- Copy whole simulations or simulation objects between simulations.
- Right click on a node for context menu options.
- Expand the folders, select the nodes, and see the selection cross-highlight in the graphic region.

Stress analysis graphic window

Displays the model geometry and simulation results. Updates to show the current status of the simulation, including applying boundary conditions and loads with the help of the view manipulation tools.

Stress analysis panel

Stress Analysis tab	Workflow stage	Description
Create Simulation, Parametric Table (Manage)	Step 1 Step 4	**Create Simulation** – Here you decide whether you need to create a stress, modal, or parametric analysis. **Parametric Table** – Define design constraints including mass, stress, deformation, etc.
Assign (Material)		**Material** – Create and apply material for the components if not already defined in the Part environment.
Fixed, Pin, Frictionless (Constraints)		**Constraints** – Represent how a part is fixed or attached to other parts in reality, and thus restrict their motion.
Force, Pressure (Loads)	Step 2	**Loads** – Represent the external forces that are exerted on a component. During normal use, the component is expected to withstand these loads and continue to perform as intended.
Automatic, Manual (Contacts)		**Contacts** – Create contacts between components automatically or manually. There are seven types of contact, including bonded.

Stress Analysis tab	Workflow Stage	Description
Mesh View Prepare		**Mesh** – Preview and create mesh, including global and local mesh refinement.
Simulate Solve	Step 3	**Solve** – Run the simulation to analyze the results as a consequence of defining materials, constraints, and loads.
Animate Probe Convergence Plot Result		**Results** – View the stress and deformation results to provide an informed decision on whether the component will function under the defined loads and constraints.
Actual Display	–	**Display** – Modify color plots, including displaying maximum and minimum values.
Report Report	–	**Report** – Generate an html report of the results to share.
Guide Guide	–	**Guide** – Provides guidance, when activated, on how to best set up and run an analysis.
Stress Analysis Settings Settings	–	**Settings** – Can predefine initial settings, including contact tolerance and mesh settings.

MANAGE TAB

This is the first step in creating a stress analysis study.

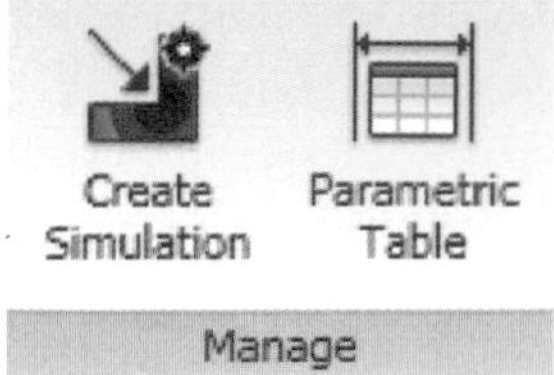

Create simulation

Here you can define whether you want to carry out single static analysis, modal analysis, or a parametric study, including the option of selecting different levels of detail.

Design Objective – Here you define whether you want to carry out a single or parametric optimization; this is discussed in the next section.

Simulation Type – Here you define whether a stress or modal analysis is to be carried out.

Model State – For an assembly, you can choose any **Design View** and **Level of Detail** on which to perform analysis.

Use **Level of Detail** (with all parts suppressed, except one) when there is a need to analyze a component, which has loads exported from Simulation.

STATIC ANALYSIS

When performing stress analysis there are three settings that can be defined.

Detect and Eliminate Rigid Body Modes – It is possible that a model may not have enough structural constraints to fix it completely in space. For example, imagine a cube the top face of which is loaded with normal pressure, and the bottom face of which is constrained by a frictionless constraint. One frictionless constraint is not enough to uniquely define the position of the cube; it can slide sideways as a whole, and we call such movement a rigid body mode. For such cases of incomplete constraints we have a special algorithm that eliminates rigid body movements from the displacements if **Detect and Eliminate Rigid Body Modes** is selected. The cube as illustrated on the next page will compress and expand sideways, but its center of mass will stay in place.

Select **Detect and Eliminate Rigid Body Modes** if you intend to use frictionless constraints only.

If rigid body motion is detected in an assembly analysis, this option will automatically be switched on if it was not initially selected.

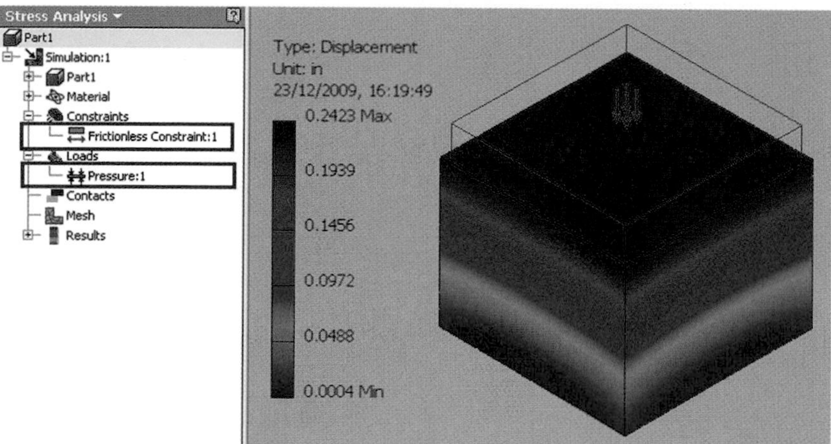

Separate Stresses Across Contact Surfaces – If two bodies have the same material and are connected by the bonded contact, theoretically both displacements and stresses should be continuous across the boundary. In FEA solution, because the meshes on the bodies do not exactly match, we may end up with different stresses on different sides of the boundary. By default, we compute the average of the two sides and show it as the stress at both sides of the boundary. However, when elements on one side are substantially smaller than on the other, and the distribution of the stress on the contact is important, the user can turn the **Separate Stresses Across Contact Surfaces** option on, and have each side's stress computed, resulting in differing stress plots on adjacent contact faces.

This option only applies to bonded contacts and same materials.

Motion Loads Analysis – This option will only be available if the part to be analyzed has its loads transferred from the Dynamic Simulation study. If multiple time steps have been transferred then the user can select the specific time to be used for the stress analysis.

You can copy and edit the first simulation and select another time step in order to compare the results with the first. When copying, all the boundary conditions including the mesh and loads will also be copied.

MODAL ANALYSIS

When performing modal analysis there are four settings that can be defined.

Number of Modes – Here you define how many modes you want the software to calculate. You can specify any value between 1 and 200, with 8 being the default value. The following shows one mode, as one mode was chosen.

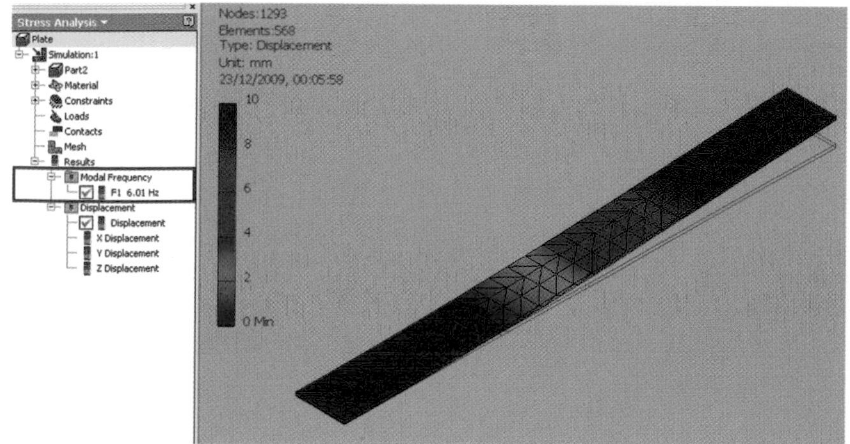

Frequency Range – Here you can specify the natural frequency range you want the software to calculate. If you have not constrained your model then you can specify a higher value than zero for your initial value as this will not calculate the first six modes, which will be zero due to rigid body motion reflecting the six degrees of freedom, with no distortion of the body shape.

Compute Preloaded Modes – Select to compute stress on the model and then compute modes for the prestressed condition. The following example illustrates that natural frequency increases from 6.01 to 105.52 Hz as a result of applying a tensile force of 1000 N.

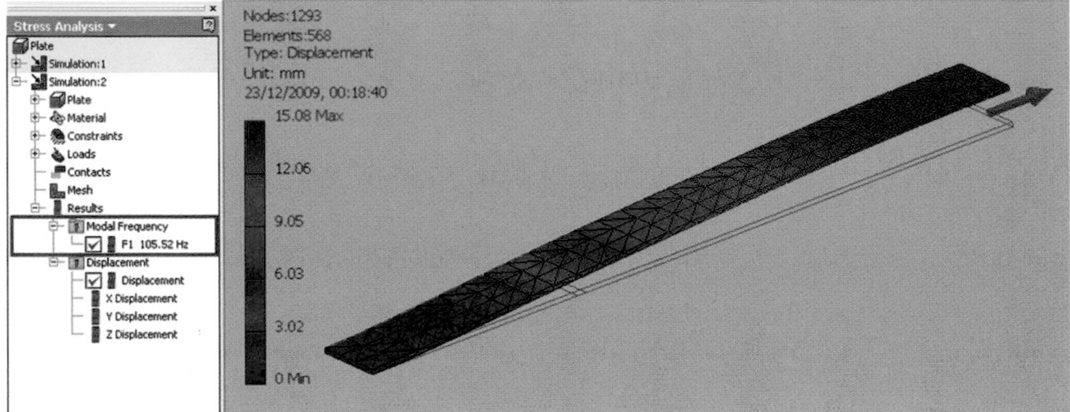

You cannot run a preload modal analysis if you apply a compressive and bending load within Autodesk Inventor Simulation 2011.

Enhanced Accuracy – This option, if selected, increases the accuracy of the calculated frequency values by an order of magnitude (10). The following example illustrates that the frequency is very similar at 104.7 Hz (less than 1% difference); the result can be assumed to be converged.

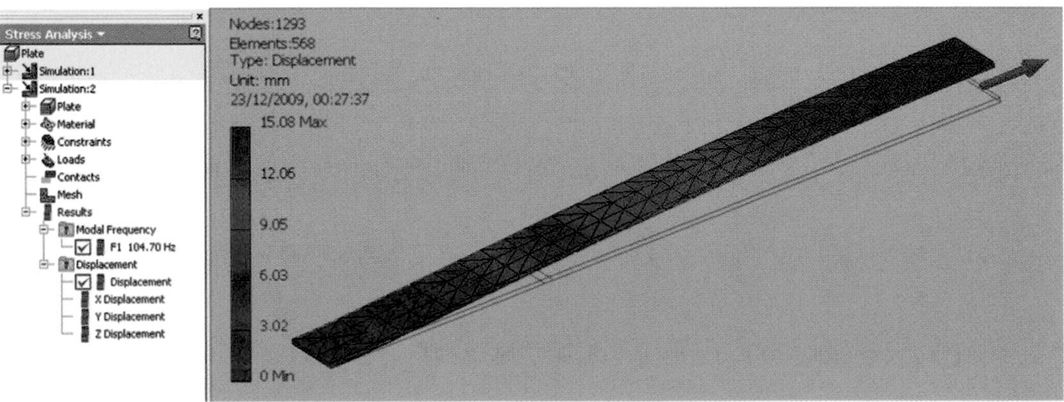

CONTACTS

If an assembly is being analyzed then you can also define a contact tolerance setting and type of contact to be automatically created. A contact tolerance of 0.1 mm will create bonded contacts between all components that have gaps of less than or equal to 0.1 mm.

Parametric table

One of the unique and powerful features of Inventor Simulation is the ability to perform parametric optimization studies.

Design constraints including mass and others can be accessed and selected by right clicking in the **Design Constraints** row.

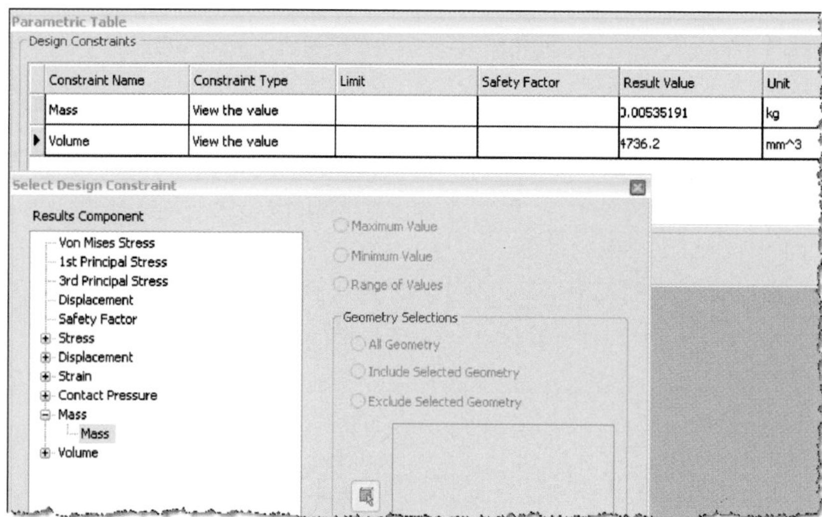

The **Constraint Type** values can be set to any of the following:

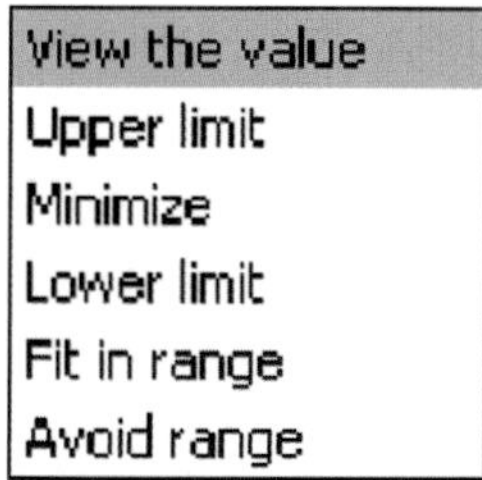

For example, if the criteria is to minimize the mass, we would select **Minimize** and then the optimum design configuration would be selected automatically.

By right clicking on any component within the browser, we can select **Show Parameters** and then choose any parameters we need to optimize.

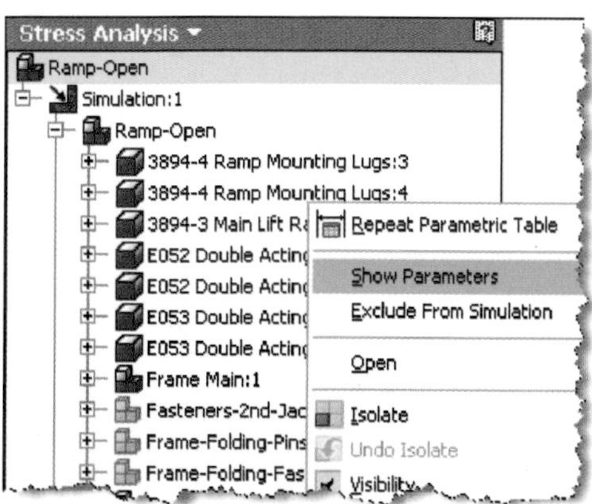

Once the parameters have been selected, the parameter range can be produced by either of the following two methods:

1. If specific values are required, specify the values separated by commas, as illustrated below:

> 1, 4, 6, 13 will produce the specified individual values

2. If you are generally interested in seeing the effect of a parameter, the parameter range can be produced as illustrated below:

> 1–9 : 5 will produce three more values equally spaced between 1 and 9; that is, 3, 5, 7

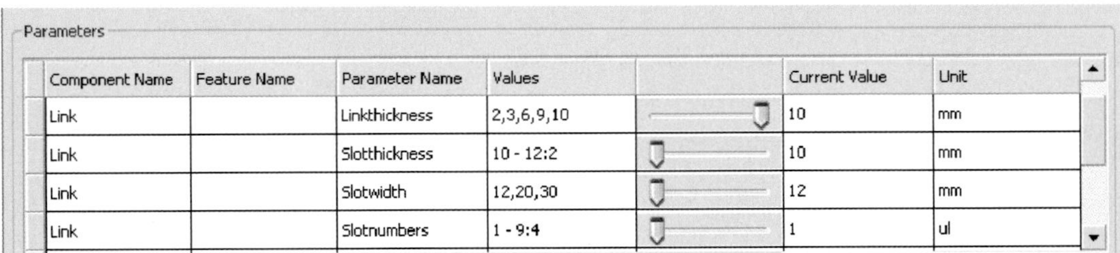

Parameters

Component Name	Feature Name	Parameter Name	Values		Current Value	Unit
Link		Linkthickness	2,3,6,9,10		10	mm
Link		Slotthickness	10 - 12:2		10	mm
Link		Slotwidth	12,20,30		12	mm
Link		Slotnumbers	1 - 9:4		1	ul

Once the design constraints and parameters have been defined, the parameter configurations can be produced by right clicking anywhere in the parameter rows and selecting any of the following:

- Promote configuration to model
- Remove Parameter
- Show Base Configuration
- Generate Single Configuration
- Generate Range Configurations
- Generate All Configurations
- Simulate this configuration

Promote configuration to model – Promotes the value to the part parameter table, overriding the original value.

Remove Parameter – Removes the parameter from the parametric table and updates the geometry with the parameter base value.

Show Base Configuration – Displays the base configuration of the model in the graphics region.

Generate Single Configuration – Generates and displays the current value, if not already selected.

Generate Range Configurations – Generates a configuration for each value in the specified range for that parameter row.

Select **Generate Range Configurations** for each range individually rather than generate all.

Generate All Configurations – Creates configurations for all the values in the parametric table.

Selecting **Generate All Configurations** can take a very long time, especially if there is a large number of parameters.

Simulate this configuration – Simulates the selected configuration only.

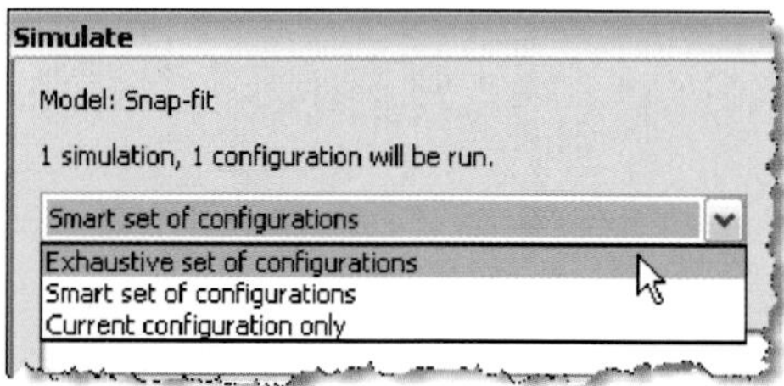

Exhaustive set of configurations – Simulates all the configurations and can take a very long time.

Smart set of configurations – The software will determine and simulate the optimum number of configurations, not necessarily all.

MATERIALS TAB

Normally, most components will have their materials assigned within the Part environment, thus removing the need to assign materials, as they will come across directly from the Part environment.

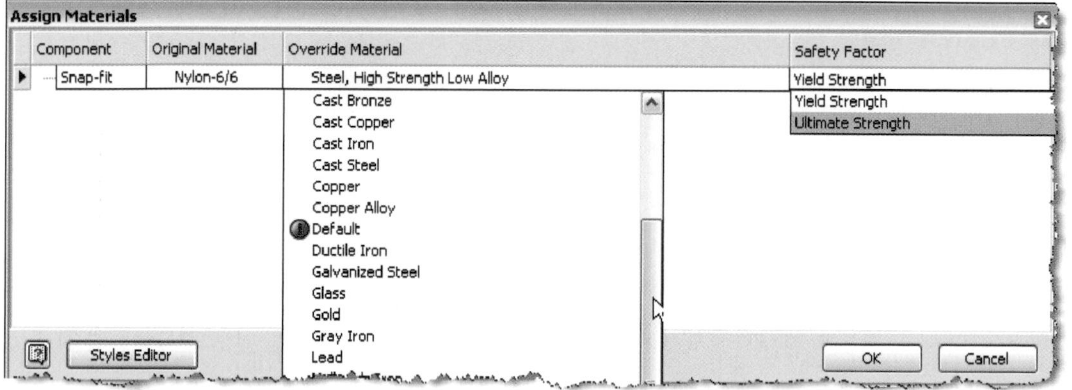

However, the materials can be overridden by selecting other materials from the styles library. New materials can also be created via the **Styles Editor** button.

Further, the safety factor can be calculated from either the **Yield Strength** or **Ultimate Strength** values.

Factor of safety is normally calculated based on yield strength.

Safety factor values below zero will not produce valid stress results.

CONSTRAINTS TAB

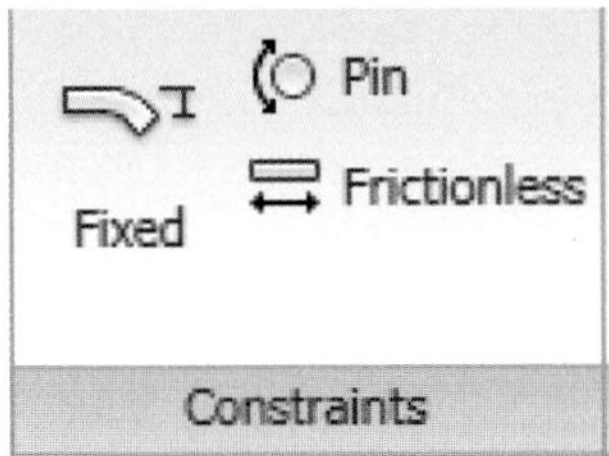

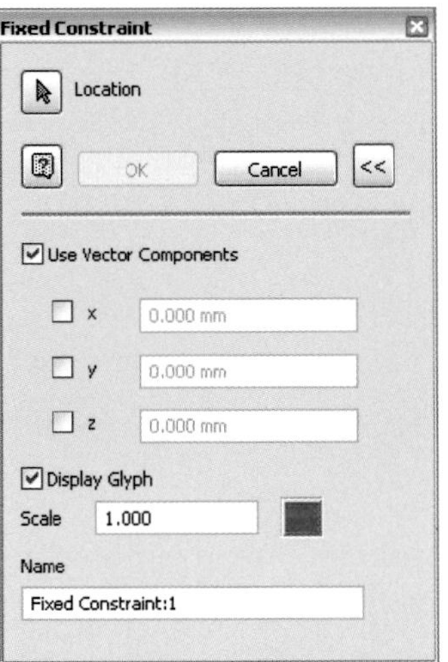

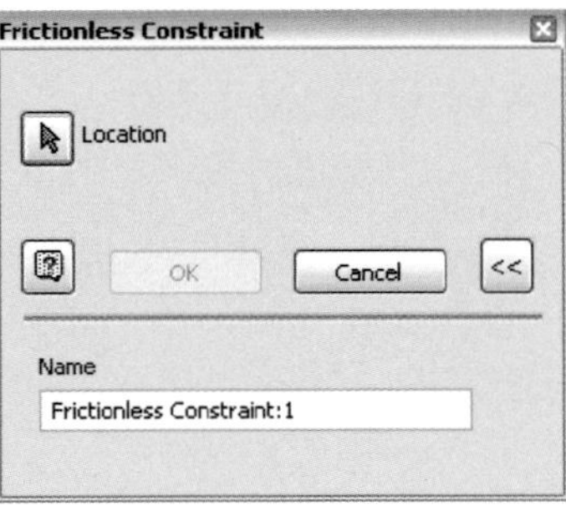

Fixed constraint

The location can be defined by specifying either a point, an edge, or a face. A fixed constraint allows you to restrict the translational direction of the component in the x, y, z direction. For example, if a component is fixed or bonded, you will normally fix all three directions.

Pin constraint

The location can only be defined by a cylindrical face and this constraint is typically used where holes are supported by bearings or pins. Typically, for a bearing or pin, you free the tangential direction to enable the surface to rotate freely.

A pin constraint is the same as a fixed constraint if the tangential direction is also fixed.

Frictionless constraint

The location can only be defined by a planar face and enables a component to freely slide along a plane and prevent motion normal to the sliding plane or surface.

Frictional constraints can also be used to model symmetry boundary conditions, for example a quarter or half model.

LOADS TAB

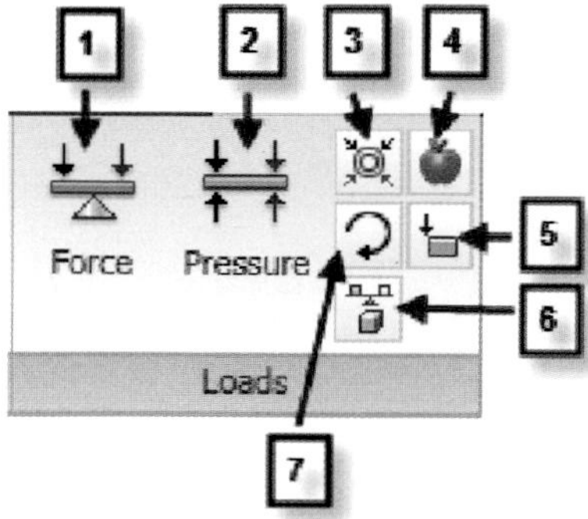

1. Force
2. Pressure
3. Bearing Loads
4. Gravity
5. Remote Loads
6. Body Load
7. Moment

These loads can be generally categorized into

- General loads
- Face loads
- Body loads

General loads

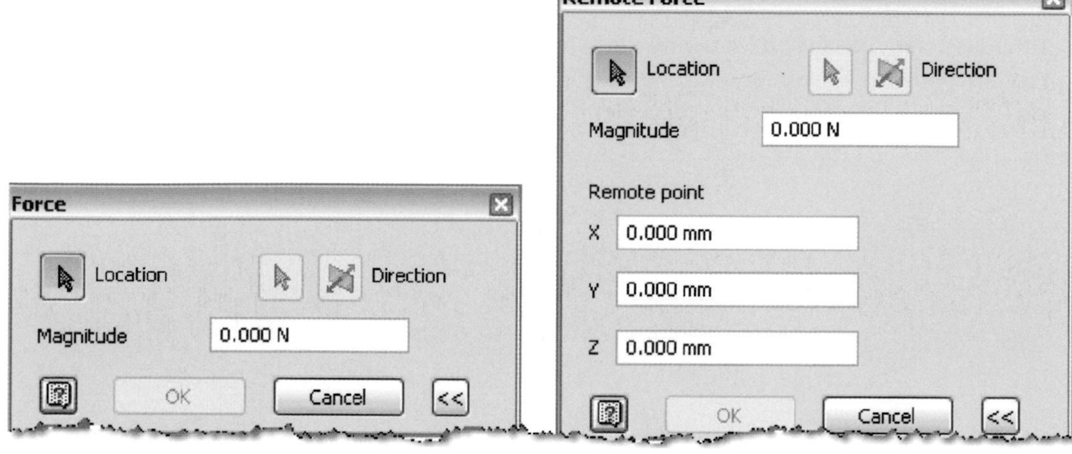

To fully define general loads, a location, direction, and magnitude are all required. Location can be defined by a planar face and, in the case of force, can also be defined by an edge or point. Direction can be defined by either a planar face, work plane, edge, or work axis.

Do not apply force on holes as this will not simulate reality. This is because the force will be applied on the complete hole whereas, in reality, the force is only applied on a portion of the hole via, for example, a pin.

Face loads

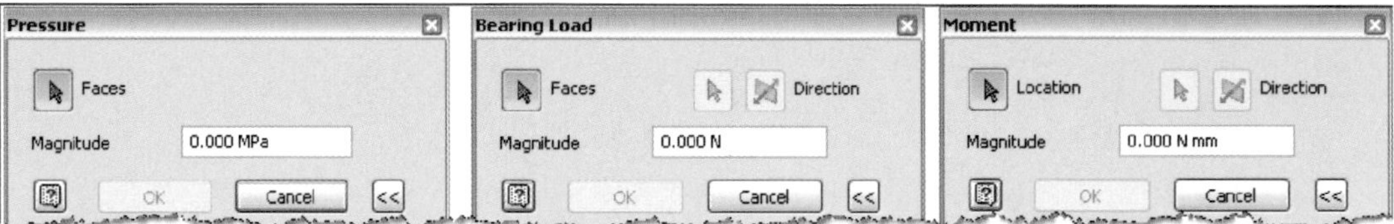

With the exception of pressure, to fully define other face loads, a location, direction, and magnitude are all required. Location can only be defined by a planar face for pressure and moment and, in the case of bearing load, the face needs to be cylindrical. With the exception of pressure, direction can be defined by a planar face, work plane, an edge or work axis. The direction of the pressure is always normal to the face.

Always use bearing loads to specify force in holes.

Pressure is related to area, so, if a component is being parametrically optimized, take care as pressure can also change if the area changes.

Body loads

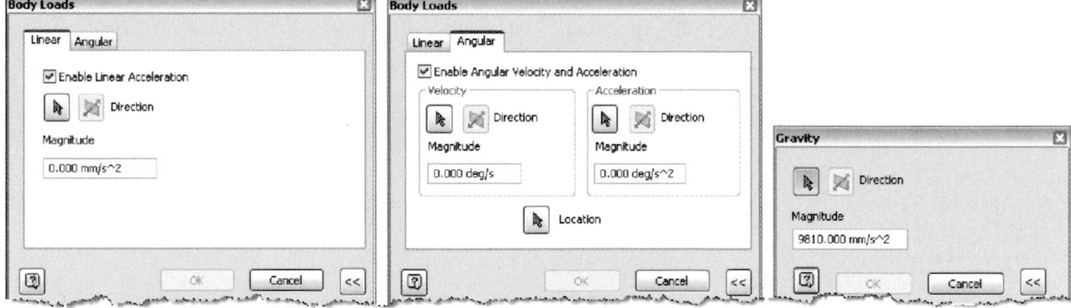

To fully define body loads, a direction and a magnitude are required. Direction can be defined by either a planar face, work plane, edge, or work axis.

For all loads, magnitude can be specified by entering an absolute value or a mathematical expression. An example of a mathematical expression could be 100 × sin (45 degrees).

With the exception of pressure, the direction and magnitude can alternatively be specified by using vector components.

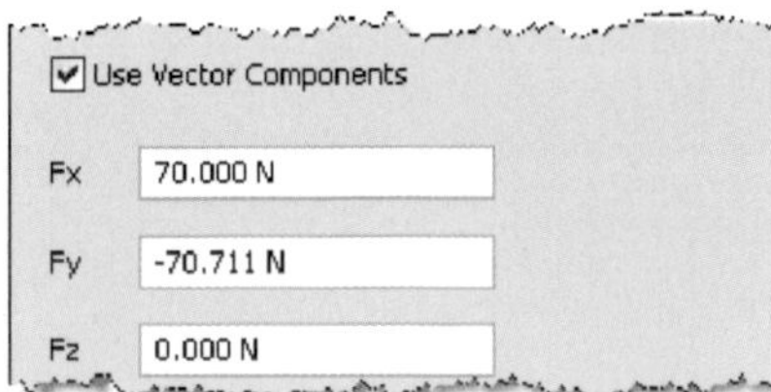

With the exception of body loads and gravity, the display glyph color and scale can be altered in addition to the name.

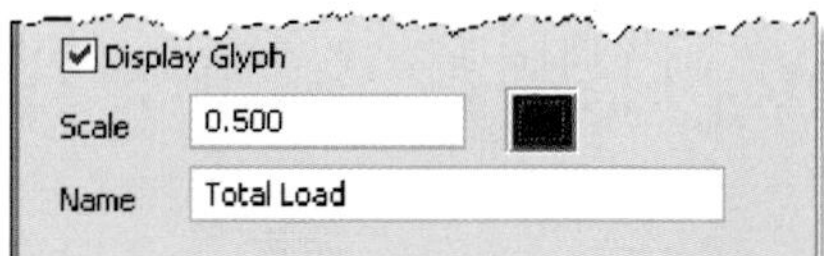

When applying loads, it is advisable not to apply loads at points or small edges as this can produce very high localized stresses.

CONTACTS TAB

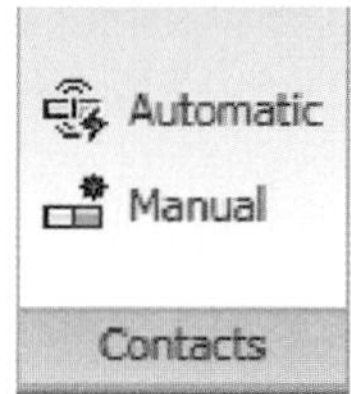

There are seven types of contacts in the Inventor Simulation Suite.

Types of contacts

1. **Bonded** – Bonds contact faces to each other; for example, in fabricated structures.

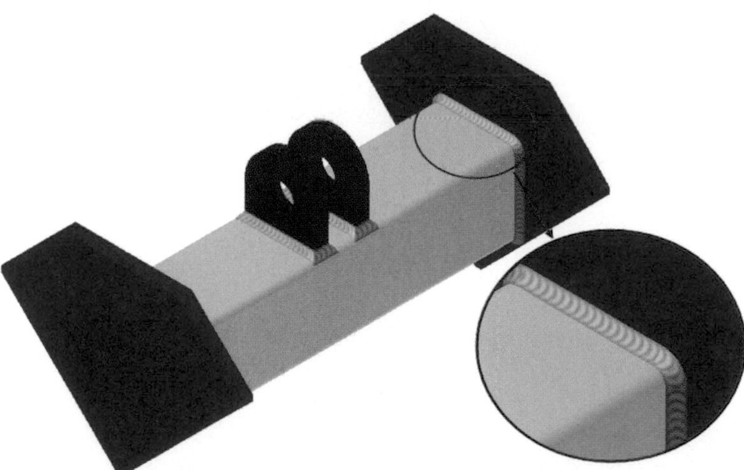

2. **Separation** – Allows adjacent contact faces to separate and slide under deformation; for example, loose bolt hole connections.
3. **Sliding/No Separation** – Maintains contact between adjacent faces and allows sliding when under deformation; for example, tight bolt hole connections.

4. **Separation/No Sliding** – Separates contact faces partially or fully without sliding.
5. **Shrink Fit/Sliding** – Similar to **Separation** contact, with the addition of allowing for initial overlaps between components, creating prestress conditions.
6. **Shrink Fit/No Sliding** – Similar to **Separation/No Sliding** contact, with the addition of allowing for initial overlaps between components, creating prestress conditions; for example, in seal and pipe/clamp connections.

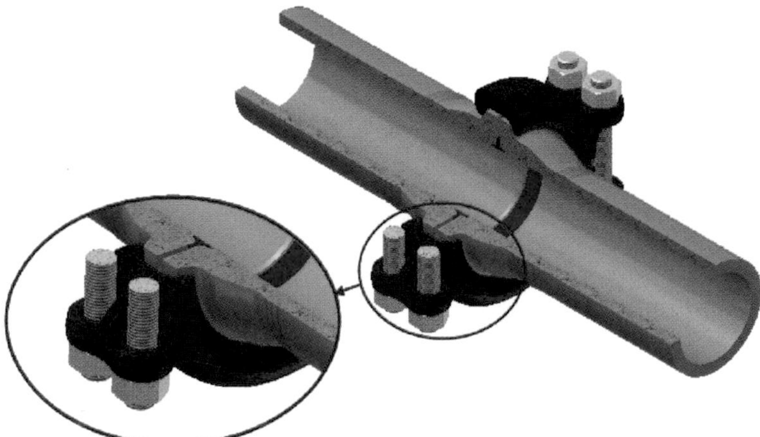

7. **Spring** – Creates spring conditions between two components by applying stiffness properties.

The process of creating contacts

There are two ways to create contacts: automatically and manually. The automatic method is by far the quickest and creates contacts between adjacent faces within the predefined settings, as below.

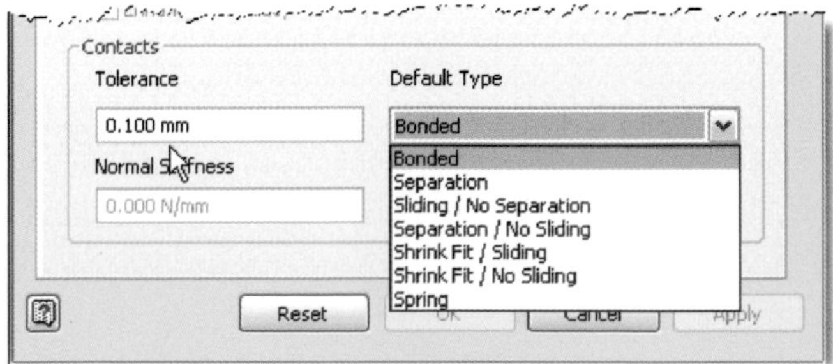

In some cases, the automatic method of creating contacts will not detect adjacent faces that have a higher gap than the predefined contact tolerance settings. In this scenario, you can use the manual method of creating contacts to create a contact.

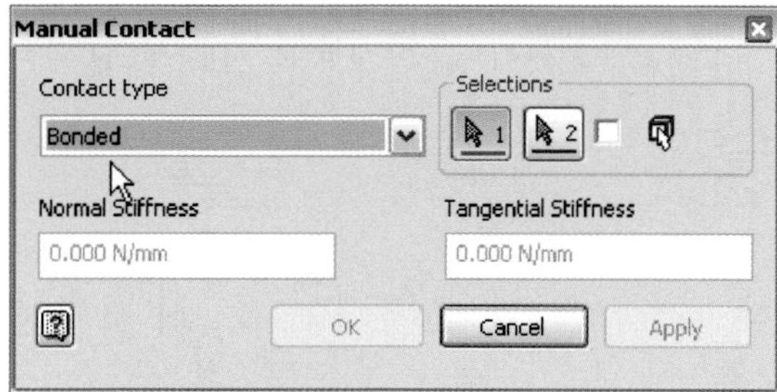

PREPARE TAB

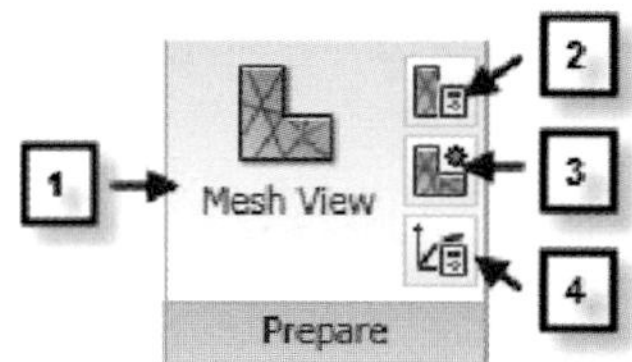

1. Generate and Preview Mesh
2. Mesh Settings
3. Local Mesh Control
4. Convergence Settings

The above tools can be further categorized into the following:

- Manual mesh refinement
- Automatic mesh refinement (or automatic convergence)

Manual mesh refinement

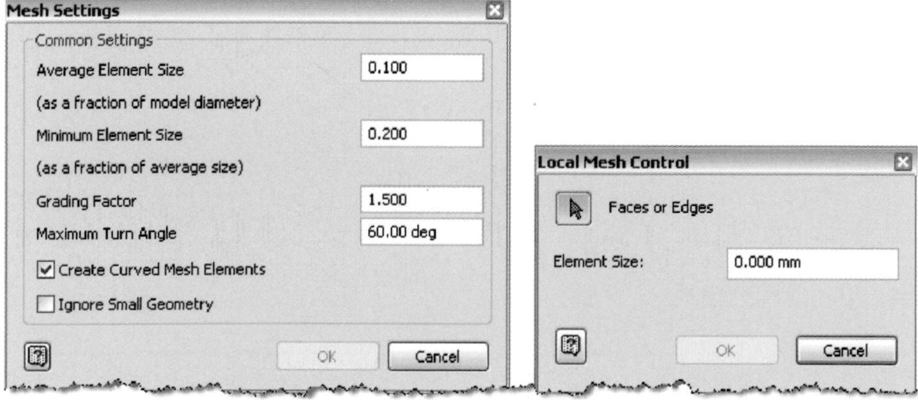

Here, an example will be used to explain the manual mesh refinement tools.

EXAMPLE 1 – MESH SETTINGS

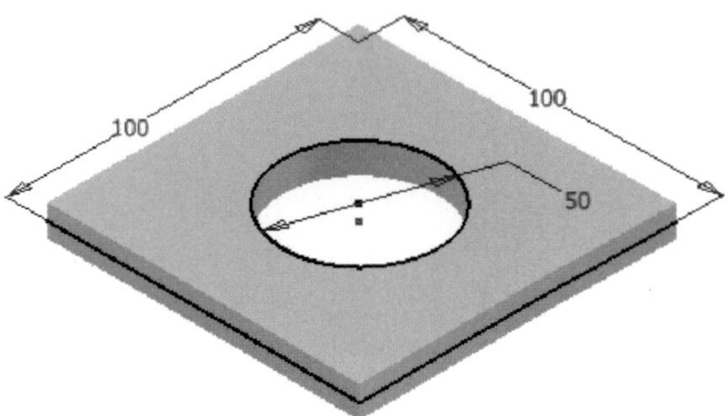

where the thickness of the component is 10 mm.

Average Element Size

Initially, we will check the effect of altering the **Average Element Size**.

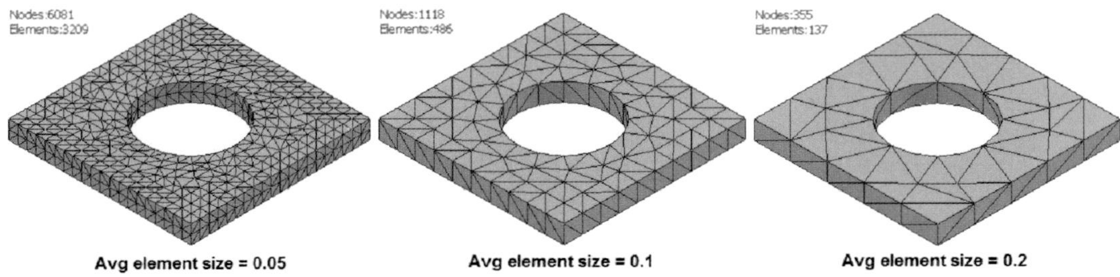

Using a smaller number will produce a denser mesh, as illustrated above.

As a guide, to determine the size of an element, the following can be used:

Size of mesh element = Longest parameter of object × Average element size

So, for an average element size of 0.2, the mesh size, for example, would be approximately

Size of mesh element = 100 × 0.2 = 20

The maximum **Average Element Size** that can be specified is 1.

A denser mesh will take longer to analyze.

Minimum Element Size

Minimum Element Size is a highly sensitive parameter and, as a rule of thumb, can remain unaltered at a value of 0.02. If the value needs changing, use any number in the following range:

$$0.01 \geq \text{minimum element size} \geq 0.02$$

Grading Factor

Grading Factor specifies the maximum ratio of adjacent mesh edges for transitioning between coarse and fine regions. A smaller grading factor produces a more uniform mesh.

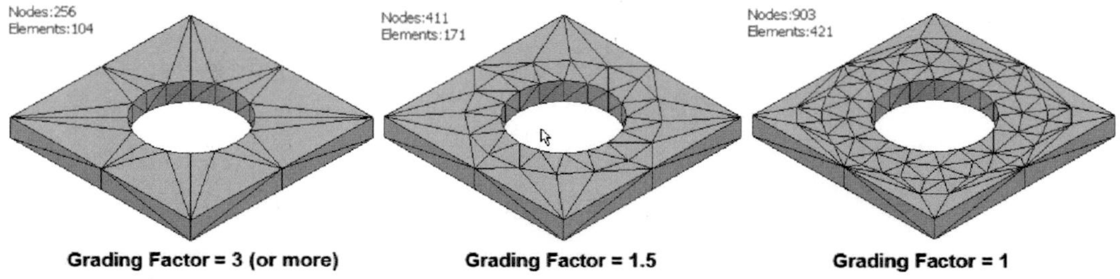

Grading Factor = 3 (or more) **Grading Factor = 1.5** **Grading Factor = 1**

Using a smaller number will produce a denser mesh, as illustrated above.

The value for the grading factor can be specified between 1 and 10. The recommended range is:

> 1.5 ≥ Grading factor ≤ 3

Maximum Turn Angle

The **Maximum Turn Angle** allows you to control the number of elements along a 90 degree arc. Specifying 60 degrees will at least create two or more elements to fill a 90 degree arc, whereas a maximum turn angle of 30 degrees will create at least three or more elements to fill a 90 degree arc.

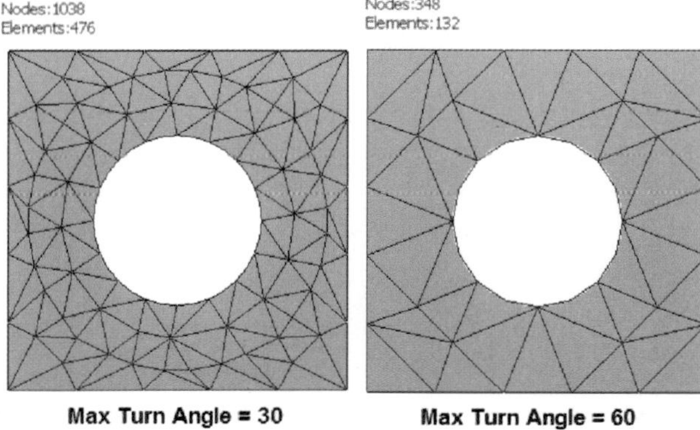

Max Turn Angle = 30 **Max Turn Angle = 60**

A small angle value of 15, for example, can produce a very dense mesh, especially when the model contains holes and radii. The recommended range is:

> 30 ≥ Maximum Turn Angle ≤ 60

Curved mesh elements

Curved mesh elements represent models with circular features more accurately than straight elements.

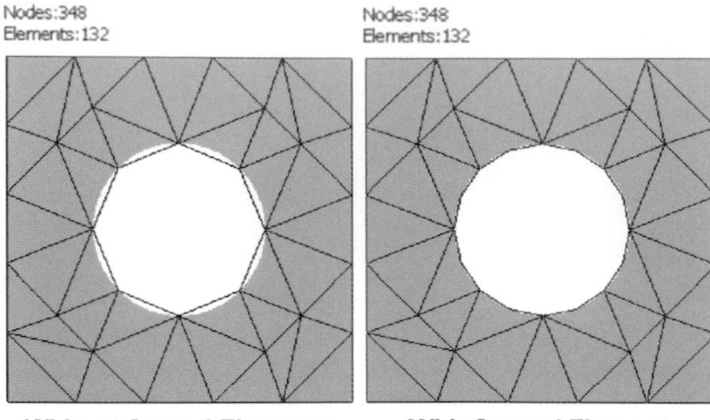

Without Curved Elements With Curved Elements

Curved elements may help to produce more accurate results.

Ignore Small Geometry

If selected, removes elements in the model that are close to the model tolerance.

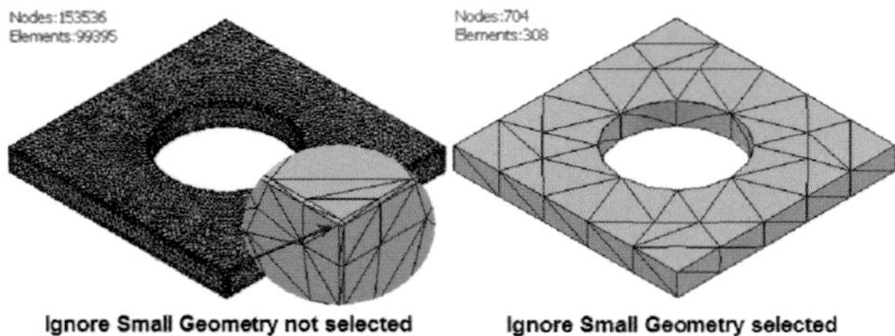

Ignore Small Geometry not selected Ignore Small Geometry selected

It is advisable to suppress small features (e.g. fillets, to avoid creating significantly more elements). The other option would be to select **Ignore Small Geometry** as this will remove small features and ignore small kinks as a result of poor modeling.

Local Mesh Control

Local Mesh Control is used to further refine the model by specifying an absolute value on faces or edges.

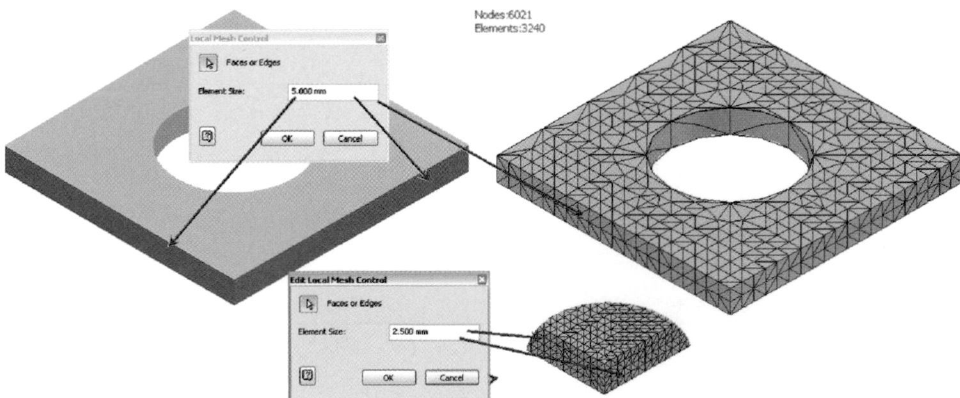

Specifying a value of 5 mm will create two elements vertically on the side faces, as the height of the base is 10 mm. A local mesh size of 2.5 mm will create four elements vertically on the selected side faces.

Try it out yourself: Open *plate.ipt*

Automatic mesh refinement (or automatic convergence)

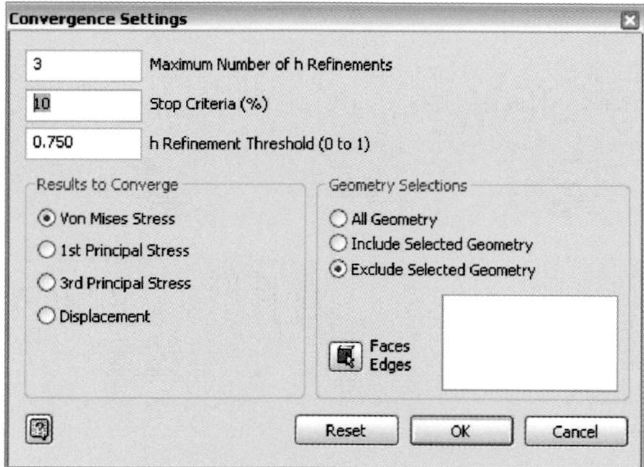

Maximum number of h refinements

Here, you specify the maximum number of mesh refinements based around maximum stresses. Values higher than 5 may result in stress singularities and take a long time to analyze.

Stop Criteria (%)

Stop Criteria (%) is used for convergence between two consecutive refinements. If the difference between the two refinements is less than 10%, the convergence process will stop.

h Refinement Threshold (0 to 1)

A value of 0 will include all elements in the model as candidates for refinement, whereas a value of 1 will exclude all elements from the H-refinement process. The default value is 0.75, which means that the top 25% of elements around the high stress area will likely be candidates for refinement.

Use **Exclude Selected Geometry** where models have stress singularities.

Use a lower value if the model has multiple stress singularity areas.

Automatic convergence may not necessarily result in convergence of results, especially where models have sharp and small edges, including pointed corners. The solution goes through H–P adaptive refinements.

Here, again another example will be used to explain the convergence settings required to automatically refine the mesh and convergence of results.

EXAMPLE 2 – CONVERGENCE SETTINGS

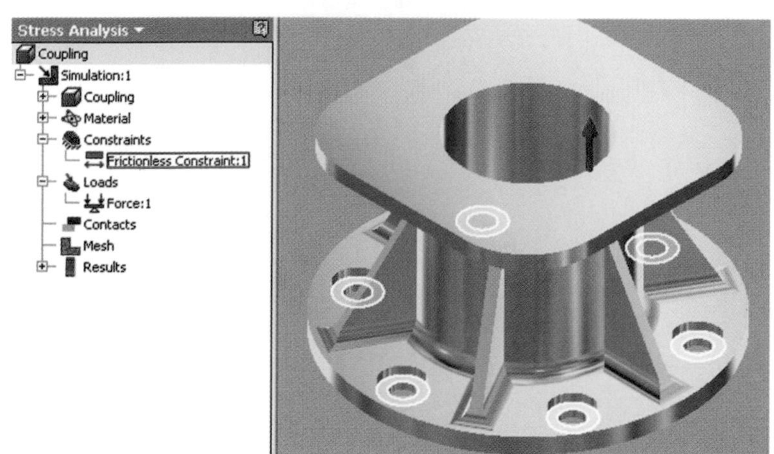

In this example we need to determine whether the component can withstand a load of 1000 N, which is fixed at each of the bolt holes. Secondly, we need to determine the maximum stress, which is required, for example, to determine fatigue life.

Using a mesh setting of **Average Element Size** of 0.05, the example is analyzed with peak stresses around all the bolt holes – fixed using frictional constraints. The convergence plot shows that the results have not converged with the initial P-refinement (with H-refinement set to 0).

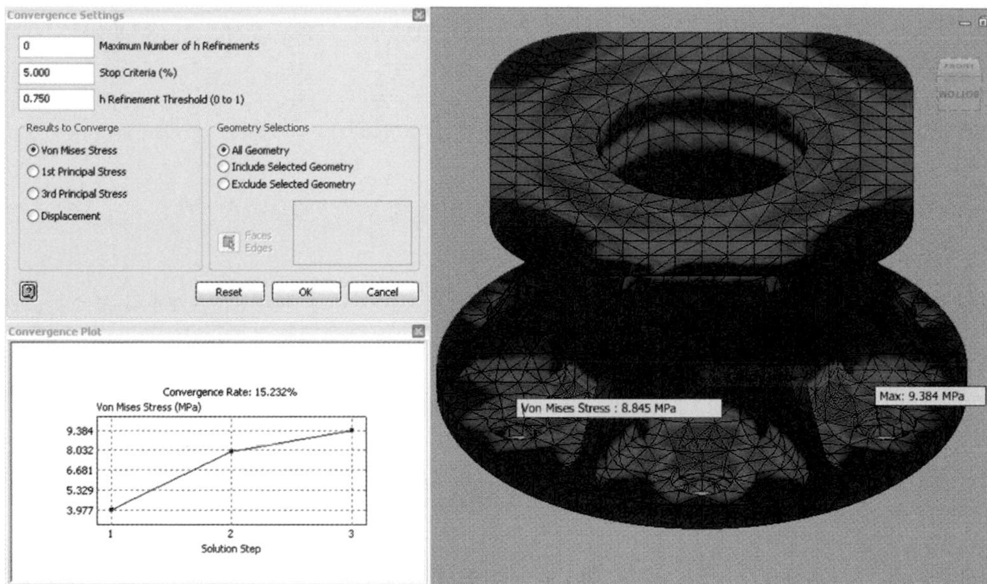

To obtain convergence we will rerun the analysis, this time with H-refinement set to 2 and the **Stop Criteria (%)** set to 4. The H-refinement threshold will be reduced to 0.5, as we have multiple areas of high stress. This value will refine at least 50% of the model mesh around peak stress regions.

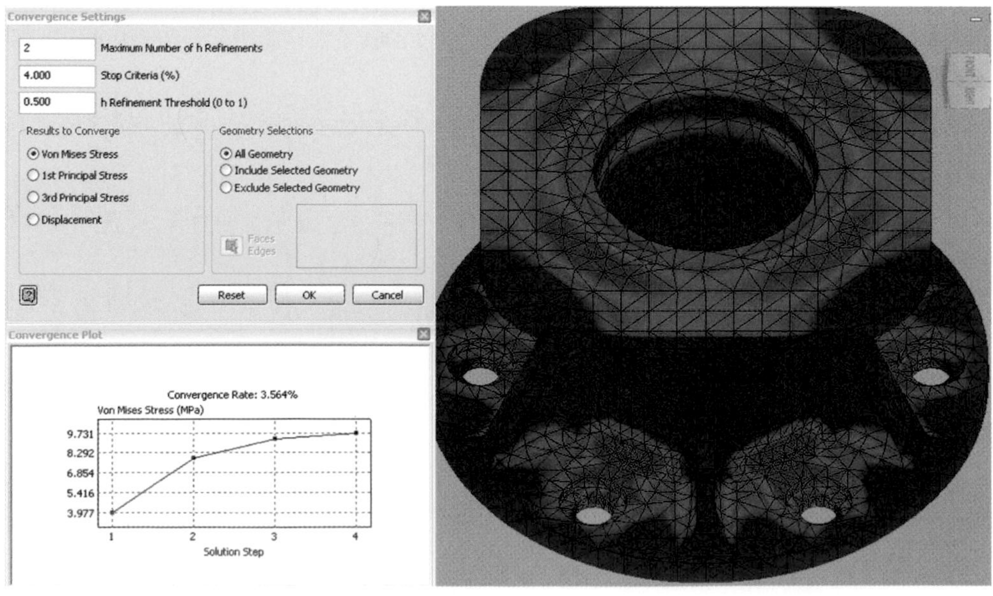

From the results we can see that the stress has converged at the first iteration of the H-P Refinement Process (point 4) and therefore does not need to go to the second iteration of the H-P Refinement Process (point 5). Further, it is important to note that the mesh has been refined around the bolt holes, and other areas of the model, where there was high stress.

In cases when the model has stress singularities, you can still use automatic convergence with the **Excluded Selected Geometry** option selected to obtain automatic convergence of results in key areas of interest, as illustrated below.

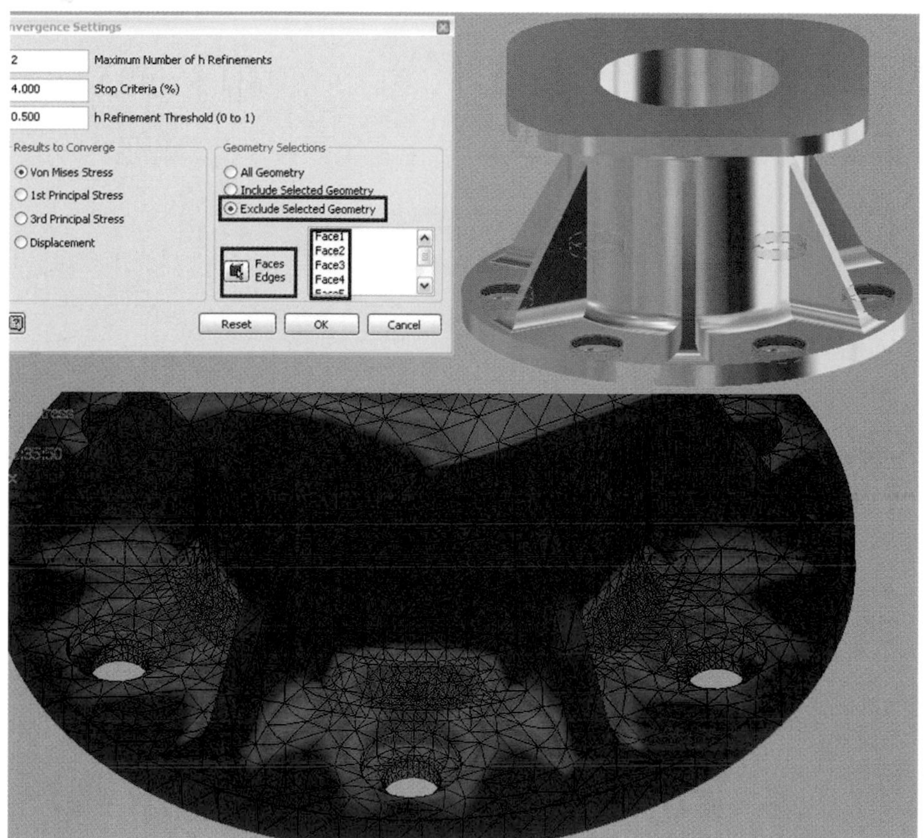

The mesh is not refined around the areas of excluded geometry (the top faces of the bolt hole).

Try it out yourself: Open *Coupling.ipt*

An alternative process to using the automatic convergence where models have stress singularities is to use manual convergence.

Manual convergence

1. Run analysis with **Average Element Size** of 0.1
2. Rerun analysis with **Average Element Size** of 0.05
3. Rerun analysis with **Average Element Size** 0.025

If the difference between the first and last analyses is within 10%, you can assume that your results have converged. Use the color bar to modify legend values to help visualize results better by isolating the stress singularity results.

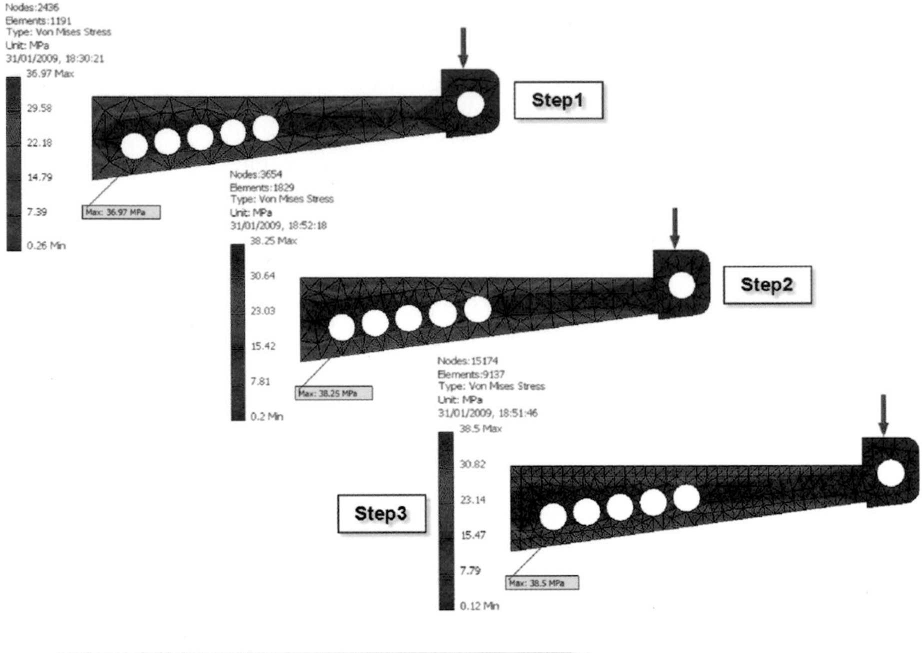

Try it out yourself: Open *Snap-fit.ipt*

RESULT TAB

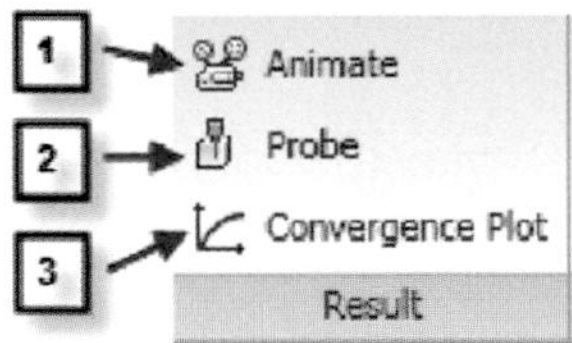

1. Animate
2. Probe
3. Convergence Plot

Inventor Simulation now offers many more result displays, including planar (XX, YY, ZZ) and shear stresses (XY, XZ, YZ).

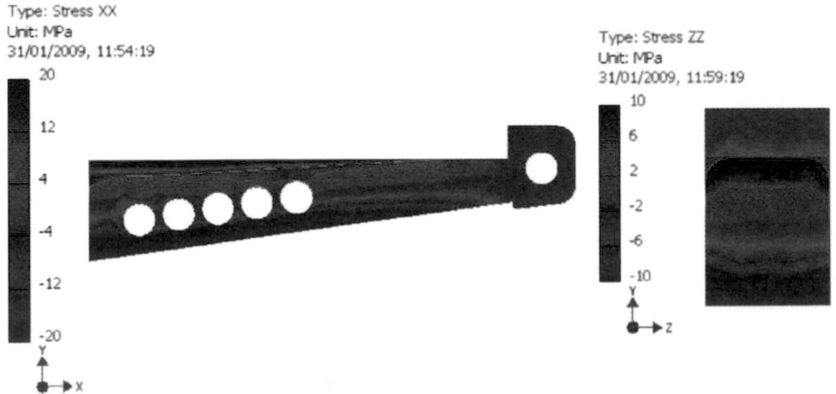

The complete list of result displays available is shown below.

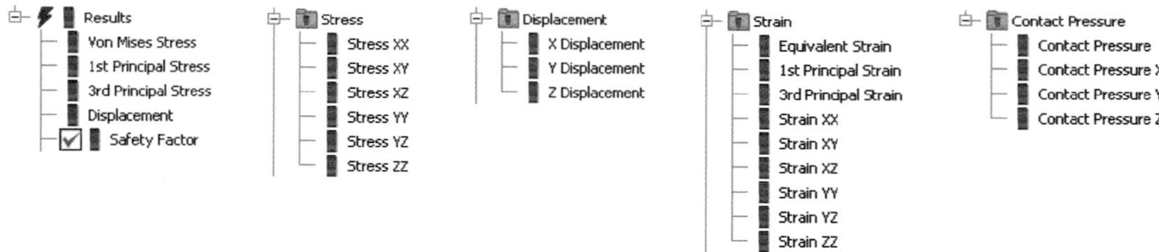

Animate

Creates a video file of the animation.

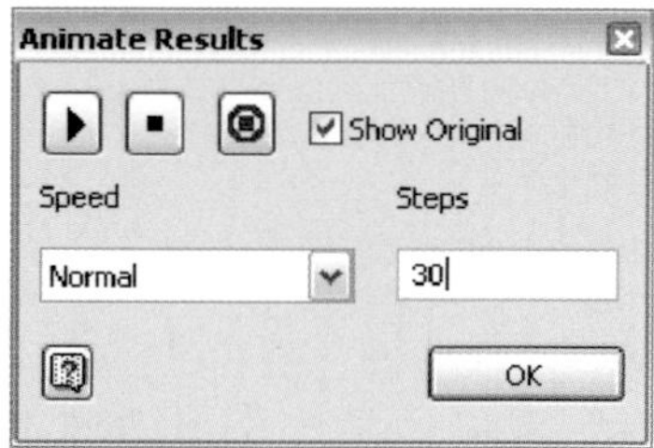

For a smoother display, increase the number of steps.

The valid range of steps is $3 \leq$ **Steps** ≤ 30.

Probe

Helps to pinpoint the key areas of interest in the model, especially when the model has maximum results distorted due to stress singularities.

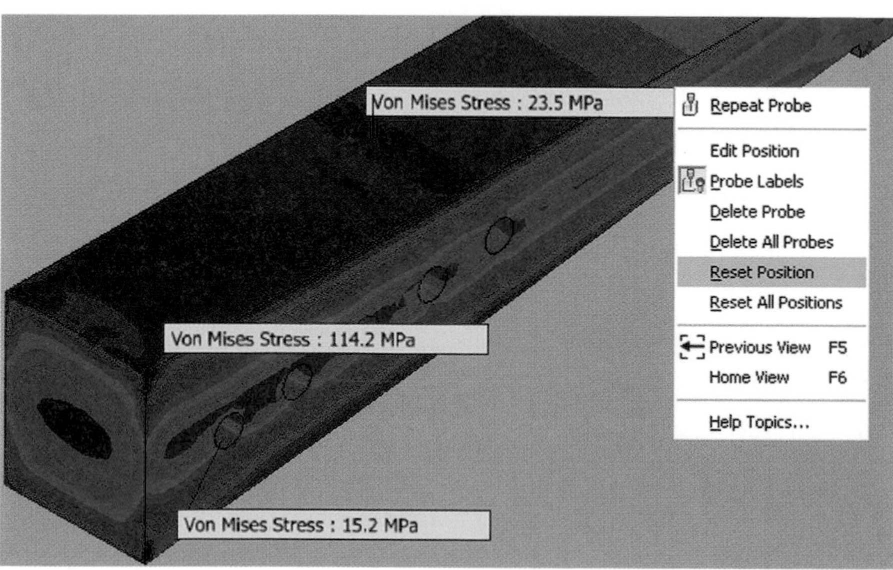

Convergence plot

Helps us to gain confidence by illustrating that the results in the area of interest have converged, as illustrated below.

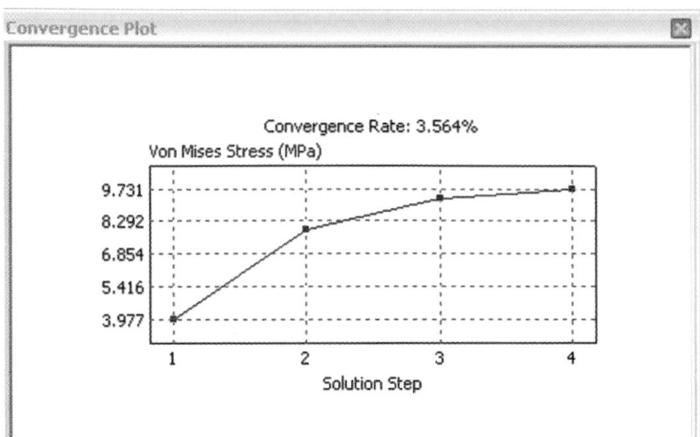

Every analysis goes through an automatic P-refinement. If the results have not converged, then H-refinement can be activated. The above example shows that the results have converged within four iterations – the first three being of P-refinement and the fourth being of H-refinement (for example, H-refinement set to 1). If the results do not converge then the H-refinement value can be further increased to 2, 3, or 4.

For parts, the first three convergence plot points are related to P-refinement.

For assemblies, the first two convergence plot points are related to P-refinement.

DISPLAY TAB

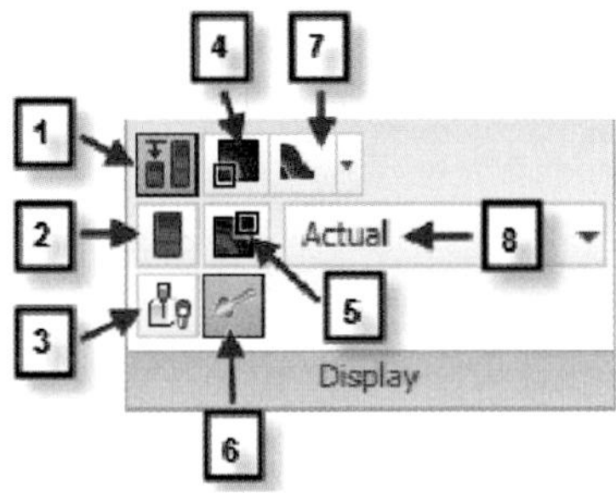

1. Apply Uniform Scale
2. Color Bar
3. Show Probe Labels
4. Show Maximum Value
5. Show Minimum Value
6. Show Boundary Conditions
7. Display Results
8. Adjust Displacement Scale

Apply uniform scale

This is switched off by default and can be useful when carrying out a parametric optimization study. When activated, the color bar scale remains the same when viewing different configurations and thus allows you to compare results visually.

The color bar is scaled based on the maximum and minimum values within the parametric configuration results.

Use **Apply Uniform Scale** when viewing a component when the rest of the assembly is excluded from the results.

Color Bar

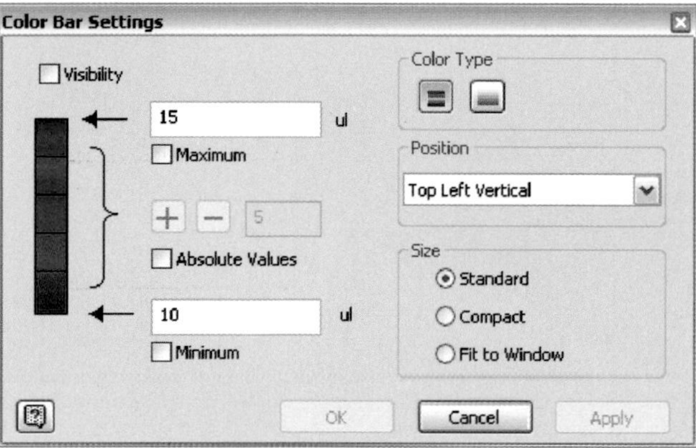

The color bar is probably the most important tool within the Display panel and, when effectively used, can help you to understand the results with ease. It can be displayed in various locations in the graphic window using the **Position** setting. The **Maximum** and **Minimum** threshold values can be altered by unchecking the **Maximum** and **Minimum** values.

The number of color legends can only be changed when **Contour Shading** is selected. **Smooth Shading** by default will use the maximum number of color legends.

Alter the **Maximum** and **Minimum** values to help isolate stress singularities.

Show probe labels

Displays all the probe labels created by the user.

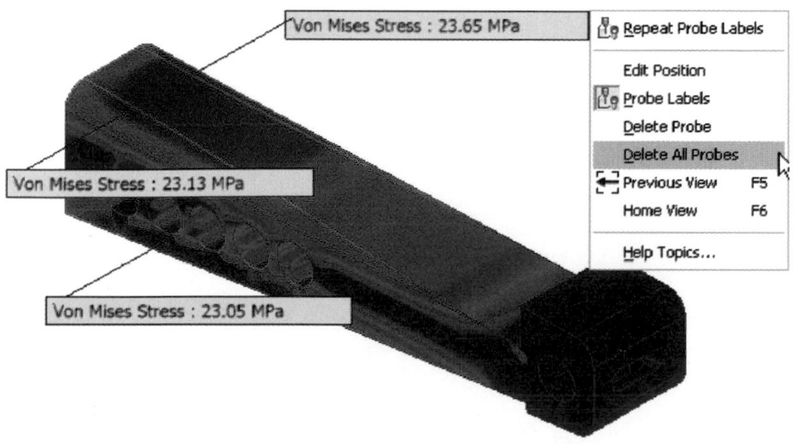

The position of a probe can be altered by right clicking its label and selecting **Edit Position**. This will help to identify whether the value has increased or decreased around the original selection area.

Individual probes can be deleted by right clicking the probe and selecting **Delete Probe** or **Delete All Probes**.

Show maximum and minimum values

Displays the maximum and minimum values and their locations on the model, as illustrated below.

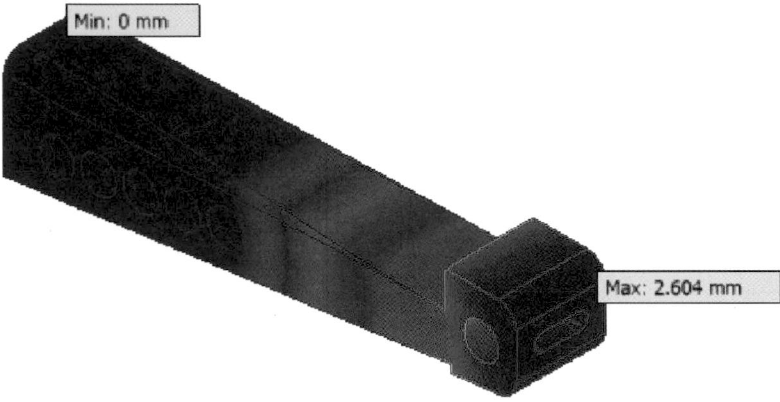

Show boundary conditions

Displays all the boundary conditions, including the loads applied on the model.

Display results

Here, you can decide whether you want **Smooth**, **Contour**, or **No Shading** display.

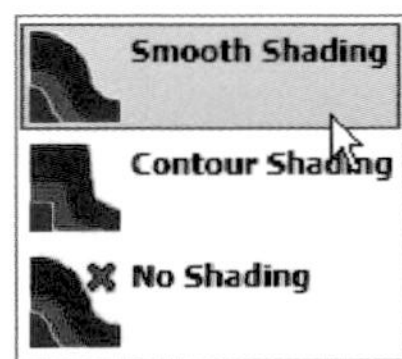

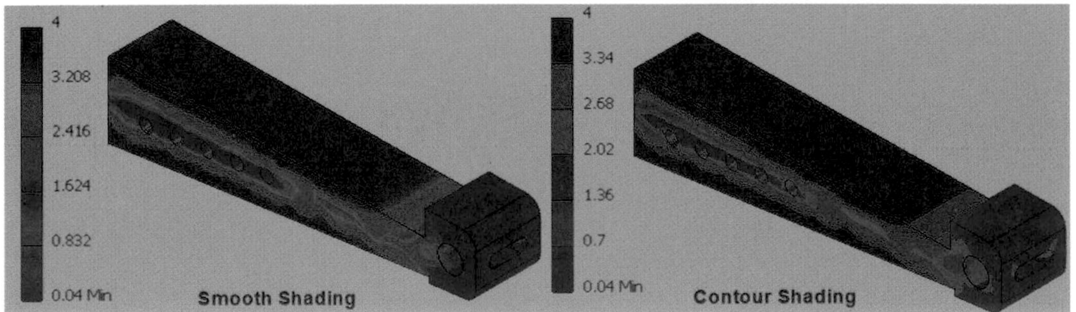

Adjust displacement display

You can adjust the scale of the results to obtain a better indication of whether the boundary conditions applied are correct.

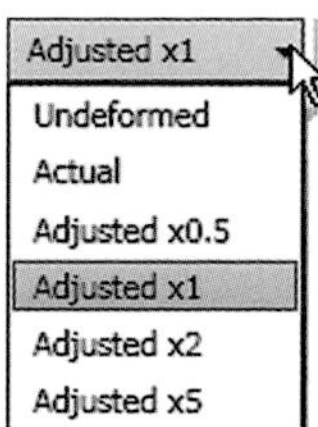

Adjust the scale so that the deformation is visible before selecting **Animate results**, as animations without visible deformation are not very visual.

REPORT TAB

Report

Report

Autodesk Inventor 2011 – in addition to standard html format – now lets you create reports in mhtml (single web page) and rich text formats (Microsoft Word documents), making it very easy to customize the reports to specific requirements.

Microsoft Word is required to generate the RTF file.

In addition to the ability to customize settings from the **General**, **Properties**, and **Simulation** tabs from with the **Report Generator** dialog box, there are now additional settings within the format tab:

Use Dynamic Content – Select to include size buttons for image width and buttons that you can click to collapse or expand the associated sections.

Not available for RTF format.

Create OLE Link – Select to create an OLE link from the model browser to the report. The report icon displays under the Third Party folder in the model browser. To edit the report, double click the icon or right click and select **Edit**.

Not available for HTML format.

Stress Analysis Report

Autodesk®

Analyzed File:	Snap-fit-feA.ipt
Autodesk Inventor Version:	2011 Beta2 (Build 150194000, 194) Debug 0
Creation Date:	25/12/2009, 14:16
Simulation Author:	Wasim
Summary:	

Project Info (iProperties)

Summary

Author	Wasim Younis

Project

Part Number	Snap-fit-feA
Designer	WY
Cost	£0.00

Status

Design Status	WorkInProgress

Physical

Material	Nylon-6/6
Density	1.13 g/cm^3
Mass	0.00532516 kg
Area	2804.8 mm^2

Results

Result Summary

Name	Minimum	Maximum
Volume	4712.43 mm^3	
Mass	0.00532504 kg	
Von Mises Stress	0.0396384 MPa	4.84543 MPa
Displacement	0 mm	0.565338 mm
Safety Factor	15 ul	15 ul

Figures

Von Mises Stress

GUIDE TAB

The Guide is a useful tool for novice and intermediate users who want advice on certain aspects of simulation. The Guide tool is accessible from the **Analysis** panel and by right clicking Loads, Constraints, Contacts and Results Guide.

Below is an example: the Constraints Guide.

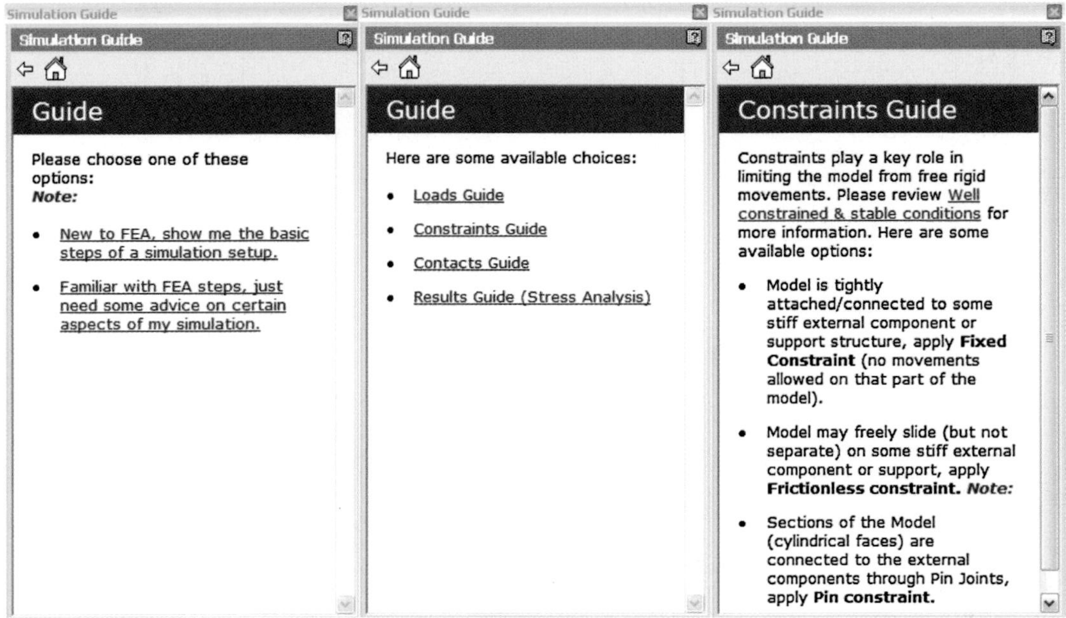

STRESS ANALYSIS SETTINGS TAB

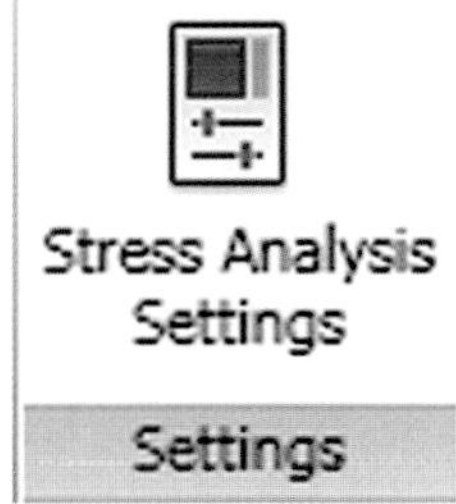

Allows you to predefine settings for current and preceding analyses.

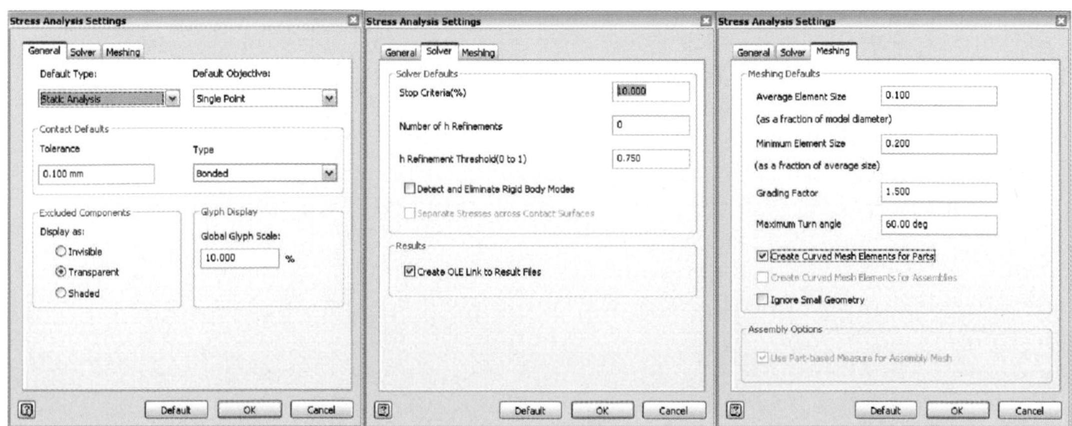

Refer to the specific sections for a detailed explanation of the individual settings.

CHAPTER 10

DP8 – Motion Load Transfer Analysis

Structural Validation of Mounting Lugs

(Design Problem courtesy of In-CAD Services Ltd)

KEY FEATURES INTRODUCED IN THIS DESIGN PROBLEM

	Key features
1	Analysis of different assembly **Level of Detail** options
2	Motion loads

INTRODUCTION

This design problem is a follow-on from Design Problem 3 and will look at the effective use of Dynamic Simulation to validate the structural integrity of the mounting lugs. The force will be exported from the simulation study and will be directly used in the Stress Analysis environment, removing the need to apply loads and restraint.

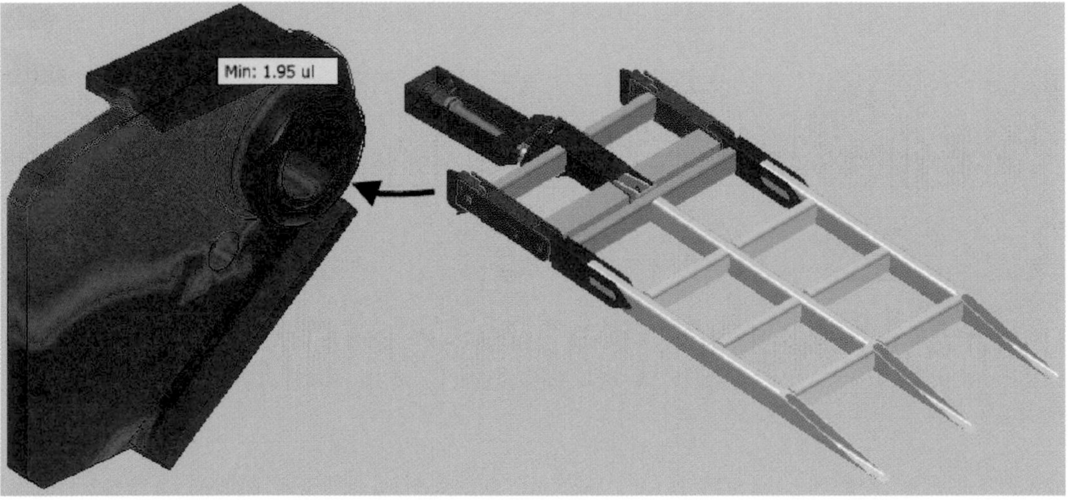

Up and Running with Autodesk Inventor Simulation 2011. ISBN: 978-0-12-382102-7
Copyright © 2010 Elsevier Inc. All rights of reproduction, in any form reserved.

The main requirements of this design problem are to determine

- Maximum stress in the mounting lugs when the ramp is fully loaded.
- Factor of safety – for example, the fatigue life could be predicted.

In addition to the above requirements, the design criteria to be used for this design problem are:

- The material to be used is mild steel.
- The factor of safety required is 2.
- Impact loading will not be taken into account.

WORKFLOW OF DESIGN PROBLEM 8

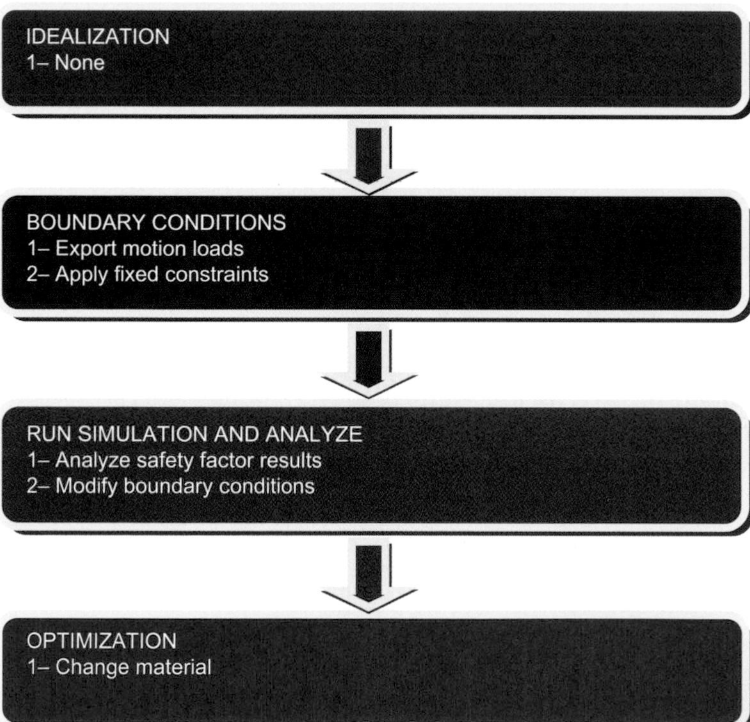

Idealization

The geometry of the model to be analyzed is simple and therefore there is no need to further simplify. However, the loads in this example will be transferred from the Dynamic Simulation environment.

1. Open *Ramp-Open.iam*

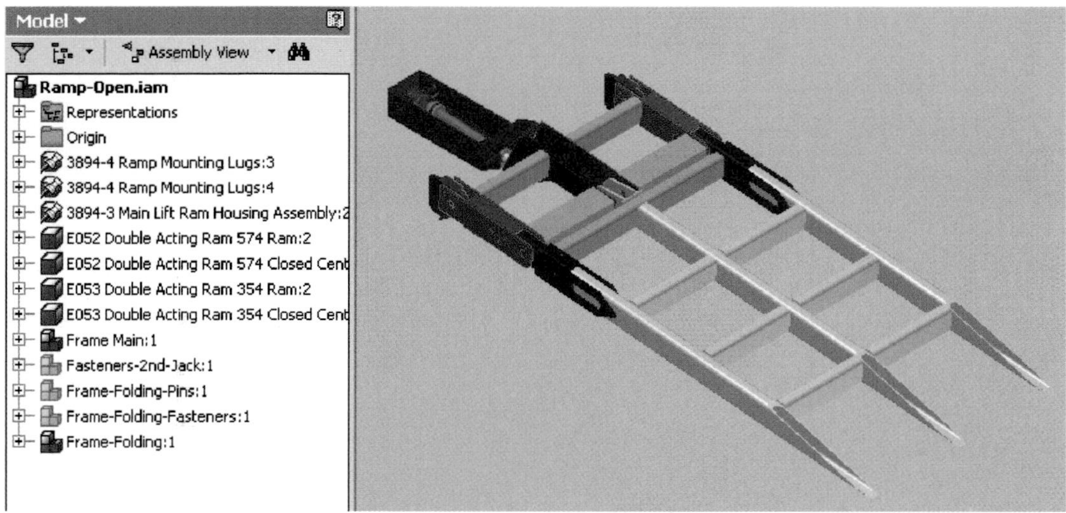

2. Select **Environments** tab > **Dynamic Simulation**

Boundary conditions

3. Play the simulation > Select **Output Grapher**
4. Right click the Force (Spherical:3) column > Select **Search Max.**

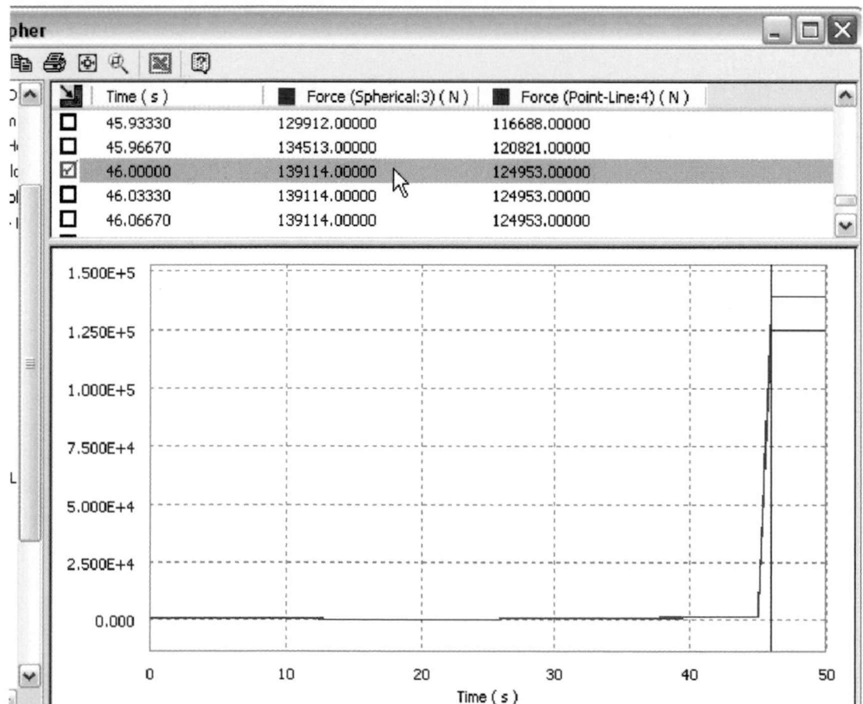

The maximum force for the Force (Spherical:3) joint is 139 114 N and it occurs after a time of 46 seconds. We will export this force to the Stress Analysis environment.

5. Tick in the **Export to FEA** column at 46 seconds to export loads at this time frame

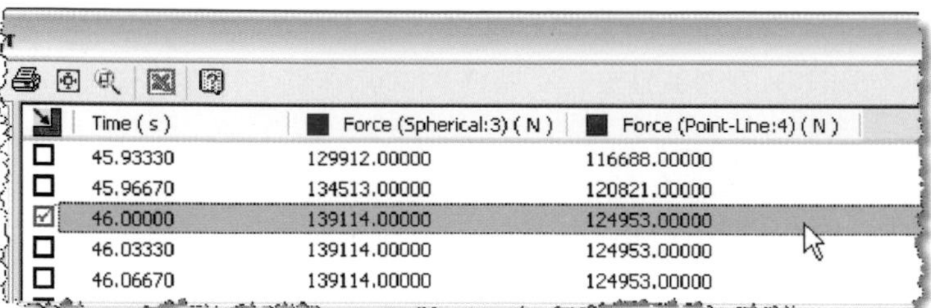

	Time (s)	Force (Spherical:3) (N)	Force (Point-Line:4) (N)
☐	45.93330	129912.00000	116688.00000
☐	45.96670	134513.00000	120821.00000
☑	46.00000	139114.00000	124953.00000
☐	46.03330	139114.00000	124953.00000
☐	46.06670	139114.00000	124953.00000

This time step is now added. Note that any time step between 46 and 50 can be used.

6. Select **Export to FEA** in the Output Grapher

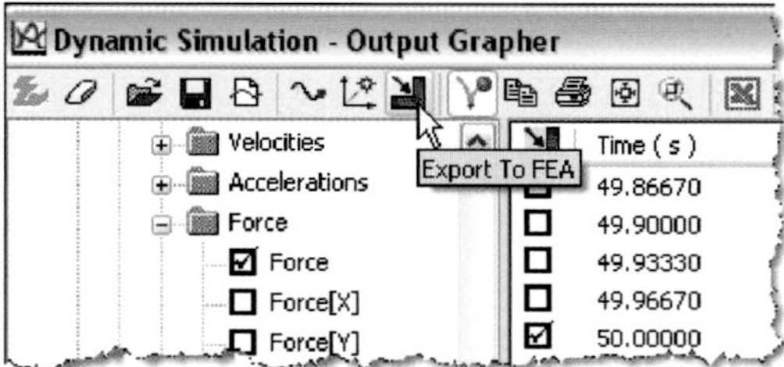

7. Select Mounting Lugs:3 > Click **OK**

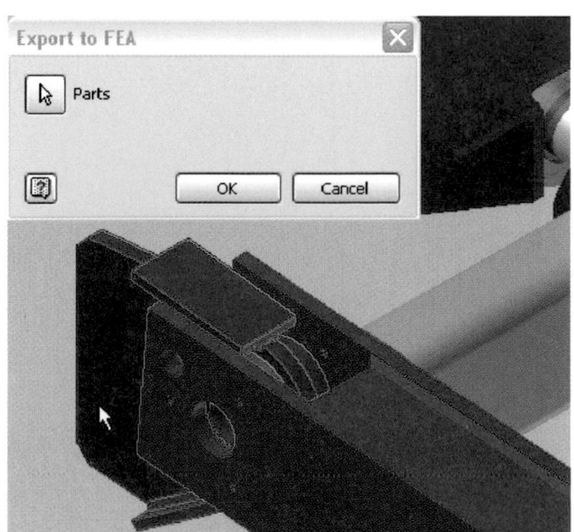

This component has now been added to the Stress Analysis environment.

There is no need to specify bearing load faces as they are already preselected. To see which face has been selected, right click the component (as indicated by the cursor above) and select **Edit Load Bearing Faces**.

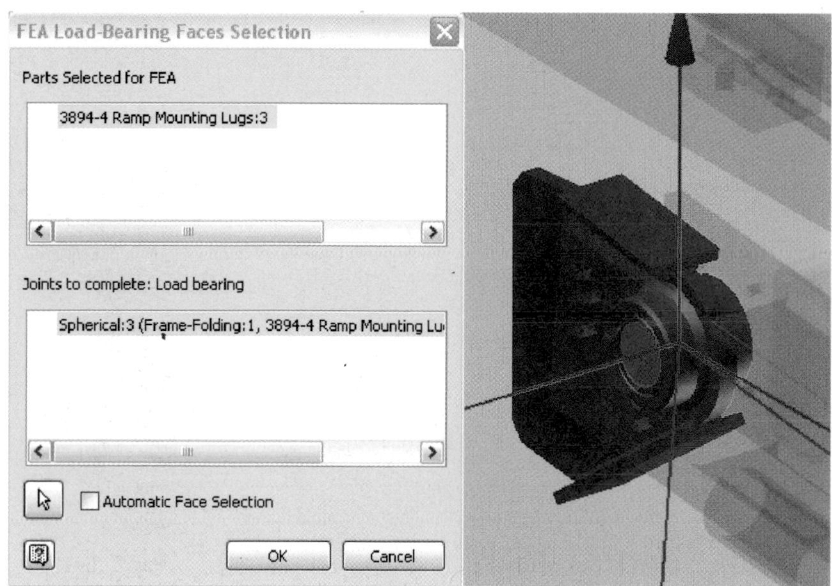

8. Close the Output Grapher > Select **Construction Mode**

9. Select **Finish Dynamic Simulation** > Select the **Environments** tab > **Stress Analysis**

10. Select **Create Simulation** > **Motion Loads Analysis**

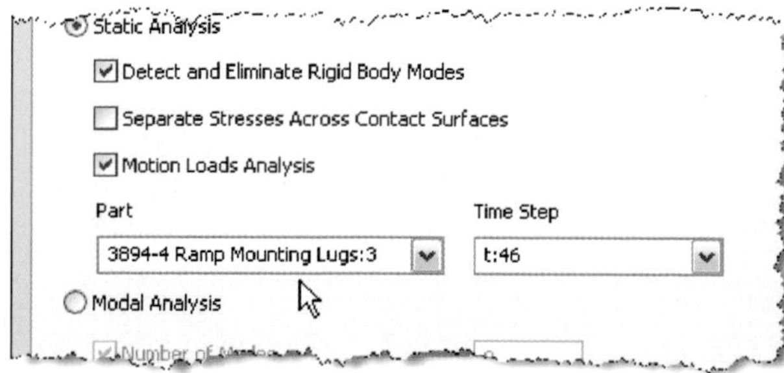

11. Select the **Model State** tab > Select **Mount-Lug** for **Level of Detail** > Click **OK**

At present, the software will try to mesh all components excluded from the simulation. This means that for a large assembly this can take a long time. In these scenarios, use **Level of Detail** representations.

12. Click **OK** to the **Grounded Part Warning**

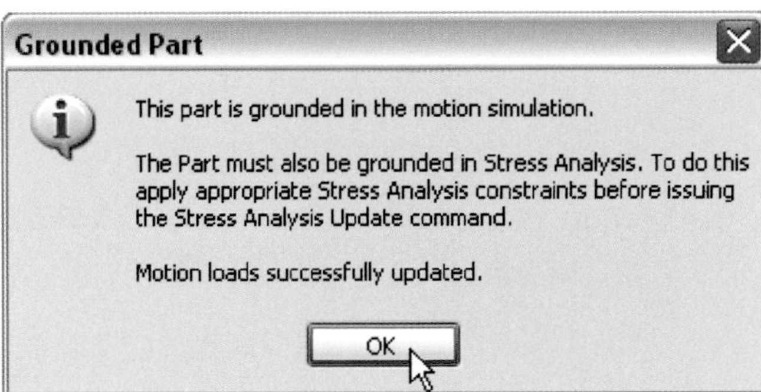

This message means that, since the part is grounded in Dynamic Simulation, we will also need to restrain it within the Stress Analysis environment.

13. Select **Fixed Constraint** from the **Stress Analysis** panel
14. Select the three faces of Component 3894-4 Ramp Mounting Lugs:3, as shown, to constrain the component

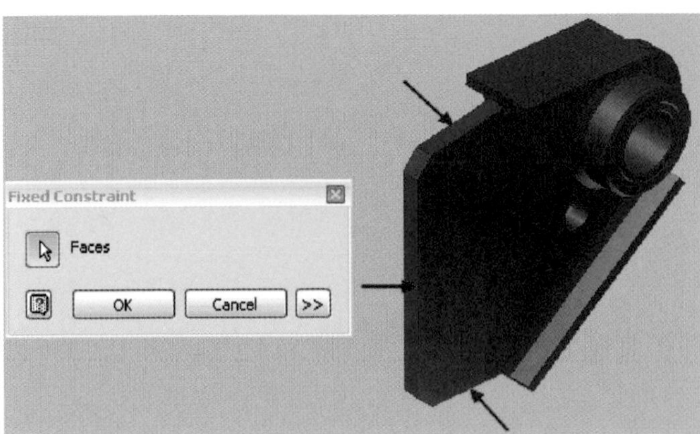

15. Select **Automatic Contacts** > **Mesh View**

Select **Automatic Contacts** if you are running the analysis for the first time, even if it is only a single part being analyzed. This may help to speed up the analysis.

Run simulation and analyze

16. Select **Simulate** > **Run**

The following stress results appear.

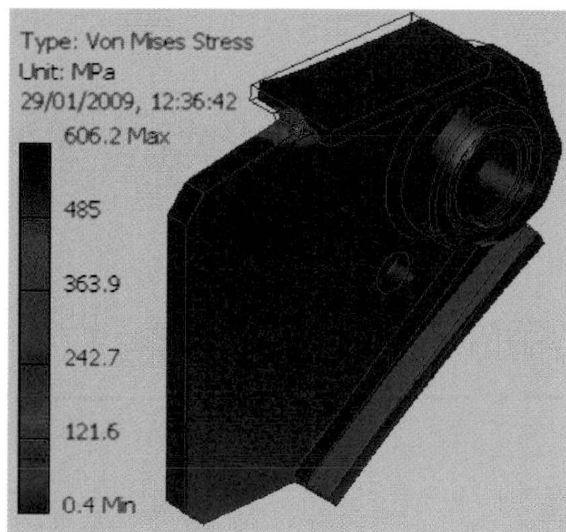

This stress display shows peak stresses around the top face, which was constrained. In reality, this lug is welded to a PFC channel at the fixed faces. Hence, the stresses that we will obtain will be slightly higher, as these faces will have no movement, whereas in reality these faces will transfer some movement into the PFC channel. To remove the peak stress and to allow extra movement in the lug, we will apply frictional constraints at the top and bottom face, instead of fixed constraints.

17. Double click Fixed Constraints:1 > Deselect the top and bottom faces > Click **OK**

18. Select **Frictionless Constraint** > Select the top and bottom faces > Click **OK**

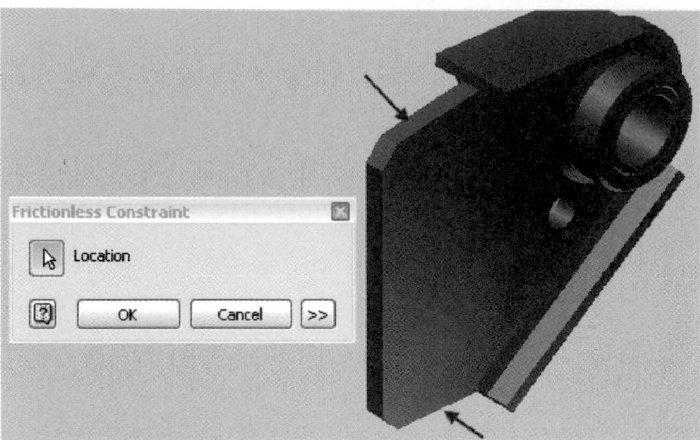

19. Select **Simulate** > Rerun the analysis

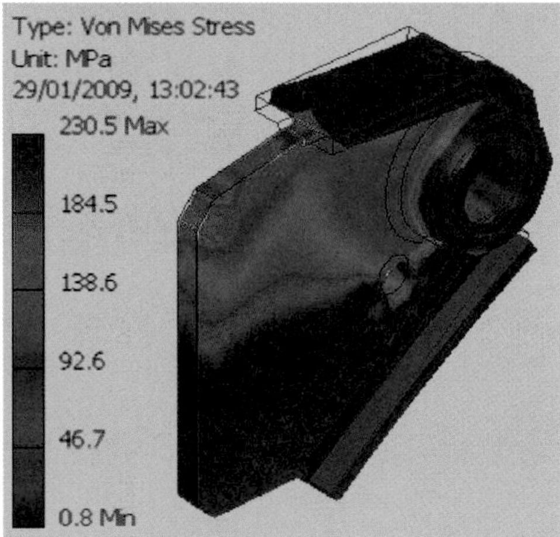

By applying frictionless constraints, we have removed the peak-localized stress and introduced some extra movement in the lug. However, care should be taken when manipulating boundary conditions as well as when interpreting results.

20. Double click **Safety Factor**

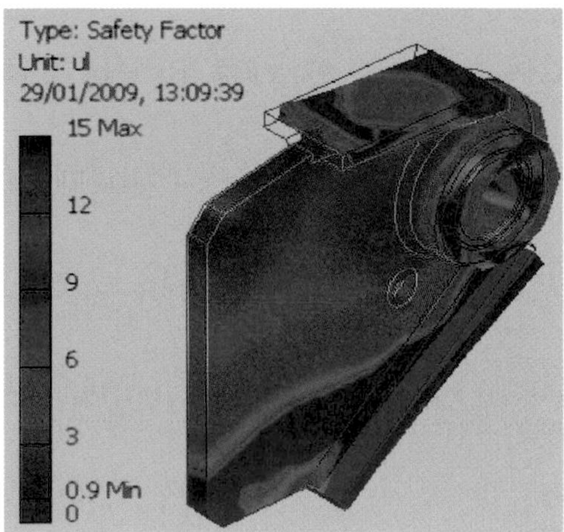

The minimum value of 0.9 suggests that the component has failed. This does not represent reality, as the load is twice as big, as explained in Chapter 4. So, we will alter the remote force by half. We do not need to alter the moment, as it is zero, and body load (gravity), as it is the same.

21. Right click Remote Force:1 > Select **Edit Remote Force**

22. Select **Use Vector Components**, specify half the value of the Fx and Fy vector components > Click **OK**

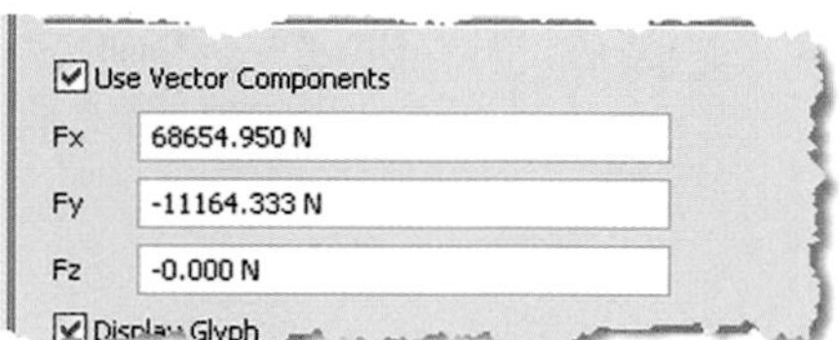

There is no need to edit remote point values.

23. Select **Simulate** > Rerun the analysis

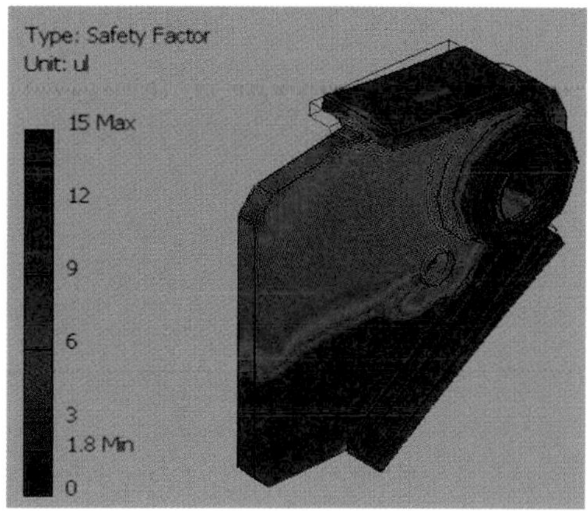

The safety factor has increased to 1.8. The safety factor is still lower than the limit and needs to be increased. One option is to use a higher strength material.

Optimization

24. Select **Assign Materials** > Change the material to **Steel, High Strength Low Alloy** in the **Override Material** column > Click **OK**

Assign Materials

Component	Original Material	Override Material	Safety Factor
Ramp-Open (Mount-Lug)			
3894-4 Ramp Mounting Lugs:3	Steel, Mild	Steel, High Strength Low Alloy	Yield Strength

25. Select **Simulate** > Rerun the analysis

26. Select **Safety Factor**

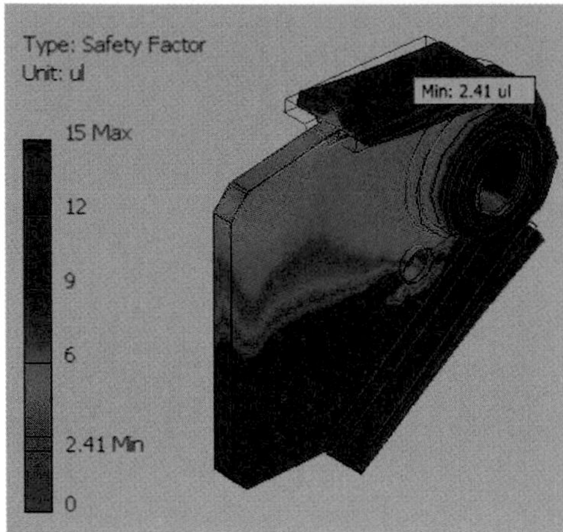

The safety factor has increased to 2.41, which means that the component is strong enough to withstand the full load.

27. Select **Finish Stress Analysis**

28. Close the file

CHAPTER 11

DP9 – Multiple Motion Load Transfer

Structural Validation of Connecting Rod

(Design Problem courtesy of Triple Eight Engineering Ltd)

KEY FEATURES INTRODUCED IN THIS DESIGN PROBLEM

	Key features
1	Motion loads – multiple time steps
2	Modify joint position
3	Automatic convergence of results

INTRODUCTION

This design problem is a follow-on from Design Problem 4 and will look at the effective use of Dynamic Simulation to validate the structural integrity of the connecting rod. The force will be exported from the simulation study and directly used in the Stress Analysis environment, removing the need to apply loads and restraints.

Up and Running with Autodesk Inventor Simulation 2011. ISBN: 978-0-12-382102-7
Copyright © 2010 Elsevier Inc. All rights of reproduction, in any form reserved.

The main requirements of this design problem are to determine:

- Maximum stress in the connecting rod while in operation.
- Maximum deflection in the connecting rod.
- Factor of safety – for example, the fatigue life could be predicted.

In addition to the above requirements, the design criteria to be used for this design problem are the following:

- The material to be used is mild steel[1].
- The factor of safety required is 5.
- Piston: minimum weight is 350 g.
- Conrod: minimum weight is 500 g (including all bolts and bearings).
- Crank-shaft: minimum weight is 11.0 kg.

WORKFLOW OF DESIGN PROBLEM 9

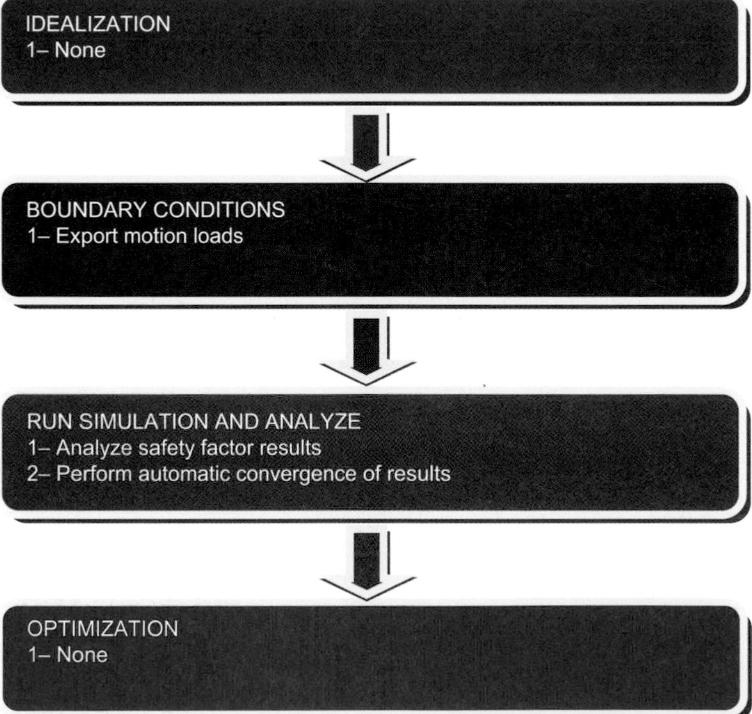

Idealization

The connecting rod has already been derived into a single part and hence no further simplification is required. In this design problem, the loads will be transferred from the Dynamic Simulation environment.

[1]Triple Eight Engineering will actually use a higher strength grade of steel.

1. Open *completed.iam*

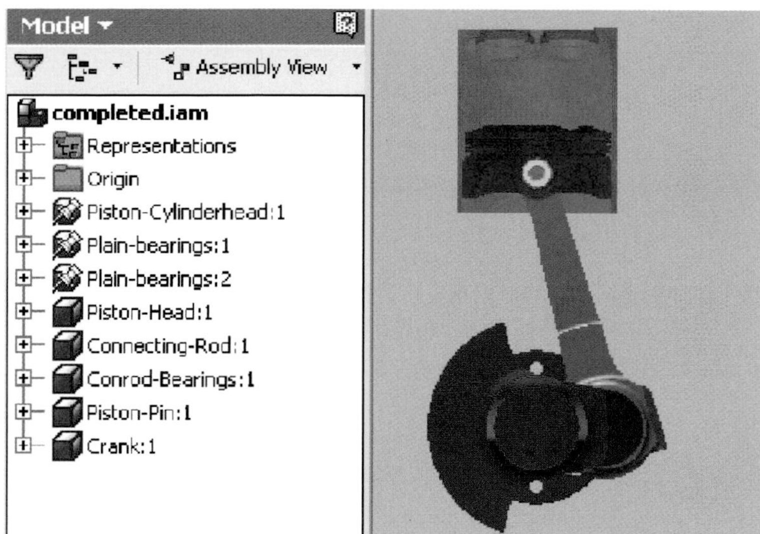

2. Select **Environments** tab > **Dynamic Simulation**

Boundary conditions

3. Play the simulation > Select the Output Grapher
4. Right click the Time data column > Select **Unselect all Curves**
5. Select Force for the Revolution:3 and Point-Line:4 joints

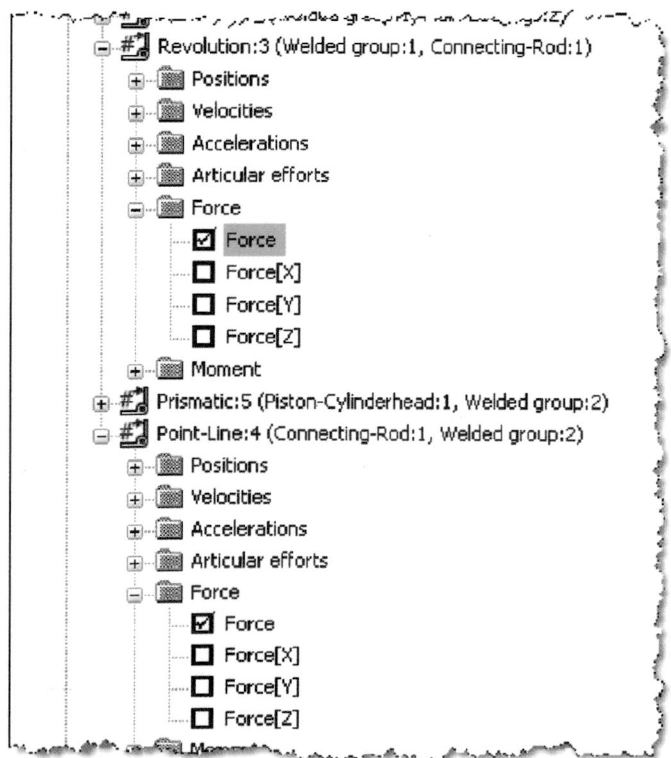

6. Right click the Force (Revolution:3) column > Select **Search Max.**

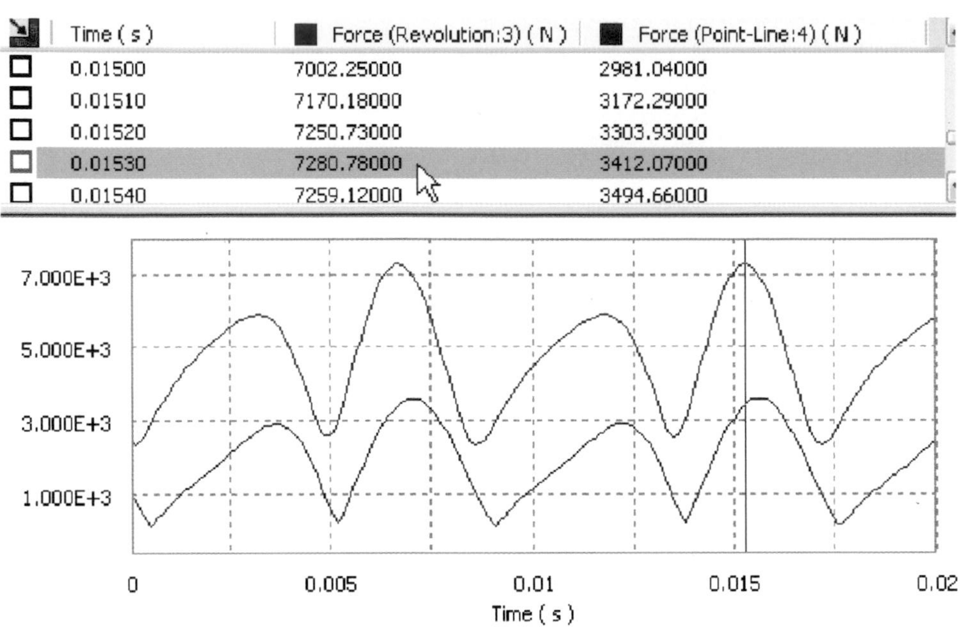

The maximum force for the Force (Revolution:3) joint is 7280 N and occurs at time 0.0153 seconds. We will export this force to the Stress Analysis environment. The value may differ.

7. Tick in the **Export to FEA** column at 0.0153 seconds to export loads at this time frame

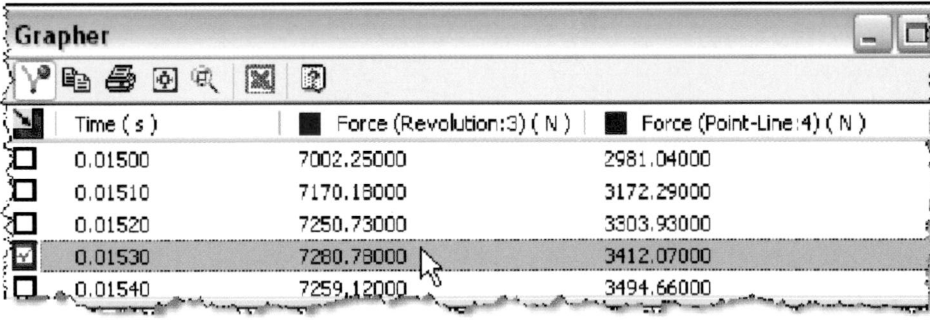

This time step is now added.

8. Right click the Force (Point-Line:4) column > Select **Search Max.**

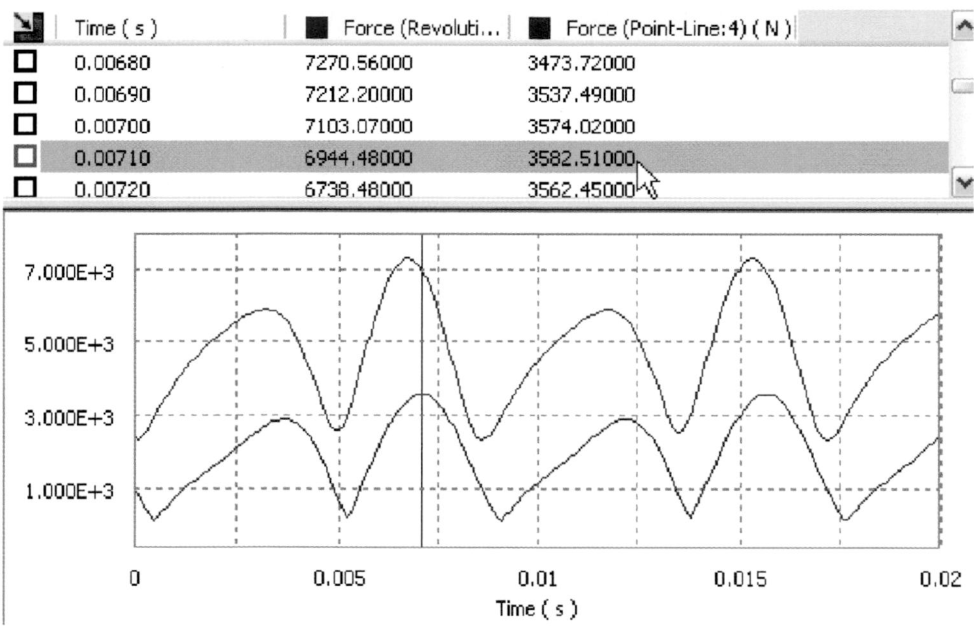

Time (s)	Force (Revoluti...	Force (Point-Line:4) (N)
0.00680	7270.56000	3473.72000
0.00690	7212.20000	3537.49000
0.00700	7103.07000	3574.02000
0.00710	6944.48000	3582.51000
0.00720	6738.48000	3562.45000

The maximum force for the Force (Point-Line:4) joint is 3582 N and occurs at time 0.0071 seconds. As this force occurs at a different time, we will also export this force to the Stress Analysis environment.

9. Tick in the **Export to FEA** column at 0.00710 seconds to export loads at this time frame

Grapher

Export	Time (s)	Force (Revoluti...	Force (Point-Line:4) (N)
☐	0.00680	7270.56000	3473.72000
☐	0.00690	7212.20000	3537.49000
☐	0.00700	7103.07000	3574.02000
☑	0.00710	6944.48000	3582.51000
☐	0.00720	6738.48000	3562.45000

This time step is also added.

10. Close the Output Grapher

11. Select **Export to FEA**

12. Select the Connecting Rod > Click **OK**

13. Select the face, as shown, to transfer the reaction loads of Joint 3 to this face

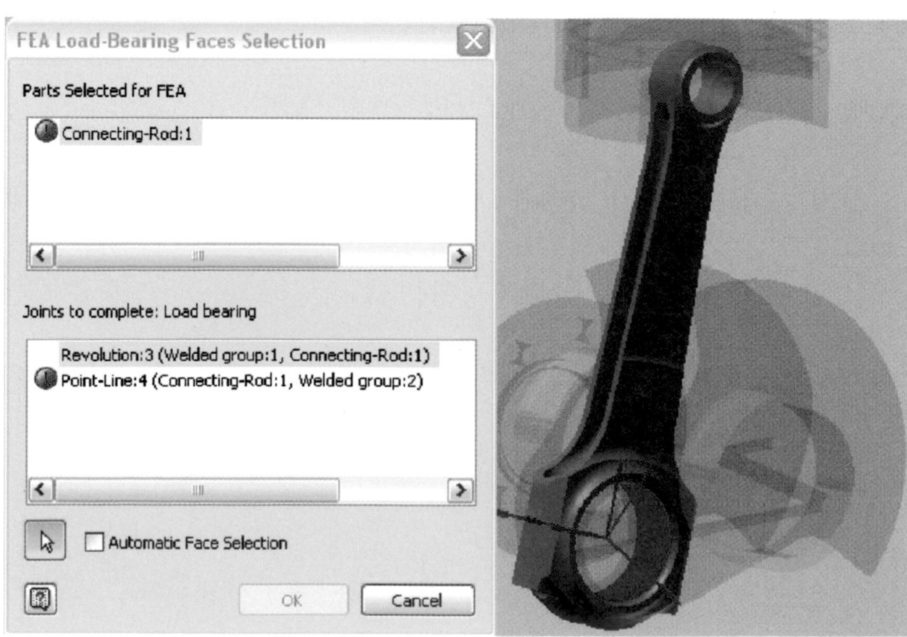

14. Select Joint 4 in the dialog box > Select the face on the other side of the connecting rod

15. Click **OK** > Select **Finish Dynamic Simulation**
16. Select **Environments** tab > **Stress Analysis**
17. Select **Create Simulation** > Specify 'Conrod-Analysis' for **Name** > Select **Motion Loads Analysis** > Click **OK**

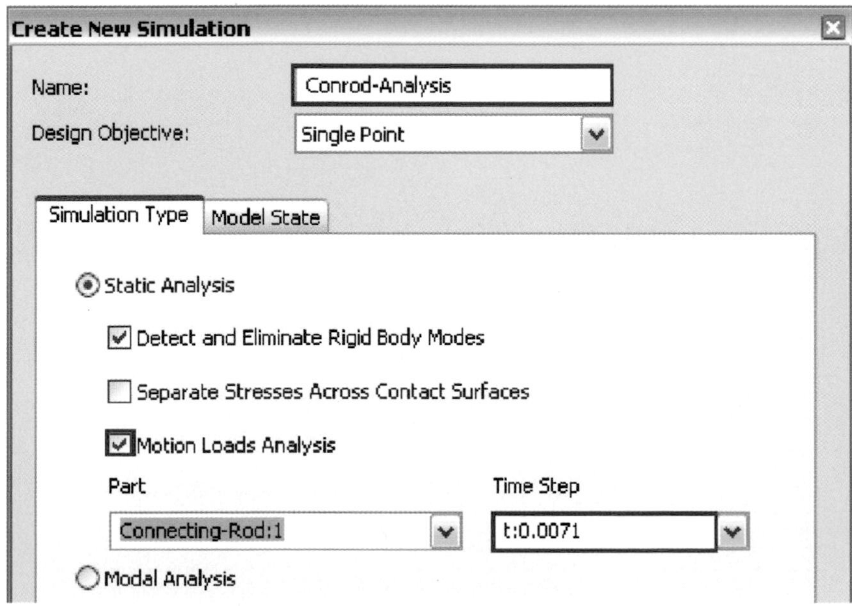

The following loads will be created:

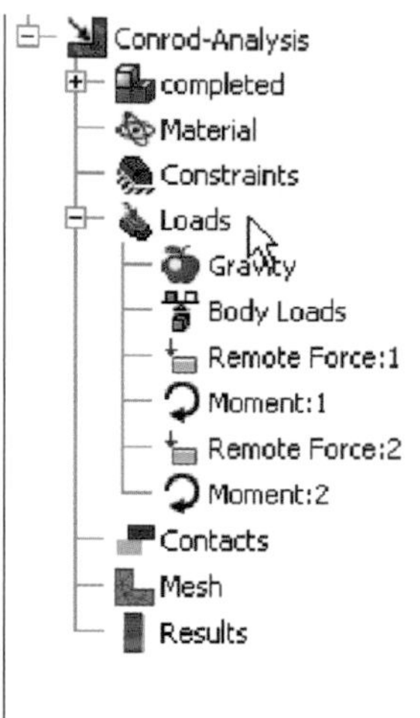

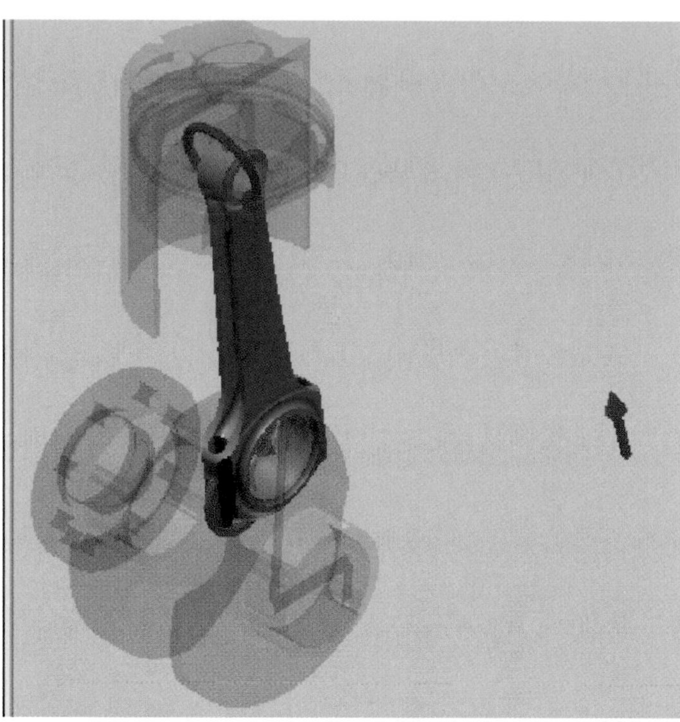

18. Select **Mesh View**

This may a take a little while. It is quicker if **Automatic Contacts** is selected first

Run simulation and analyze

19. Select **Simulate** > Run the simulation

20. Select **Actual** for **Displacement Display**. The value may slightly differ

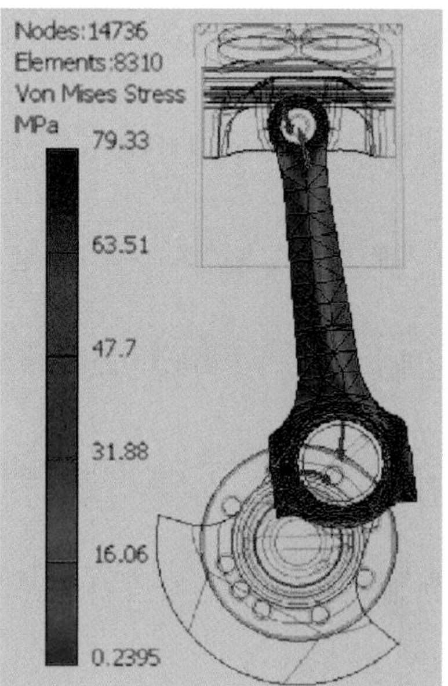

21. Double click **Displacement**. The value may slightly differ

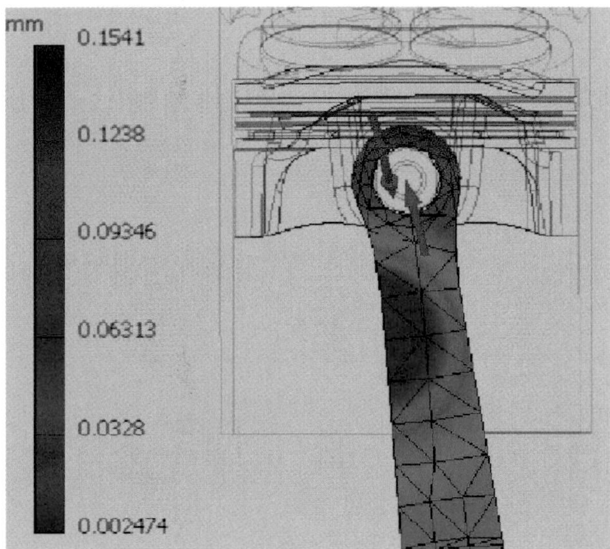

Although we have performed a stress analysis, the motion loads transferred from the top of the connecting rod are not correct as they have induced a moment and not acting directly through the center of the conrod.

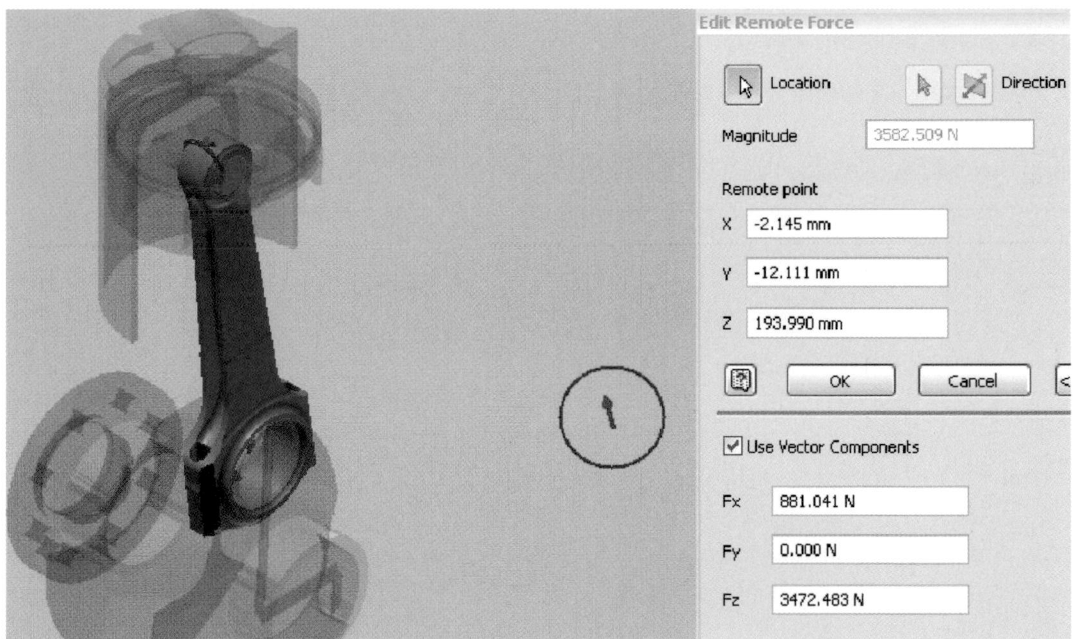

So, we need to alter this to see whether the results change.

22. Select **Finish Stress Analysis**

23. Select **Environments** tab > **Dynamic Simulation**

24. Select **Construction Mode** > Right click Point-Line:4 > Select **Edit**

25. For the Component 1 origin > Reselect the point on the conrod as shown > Click **OK**

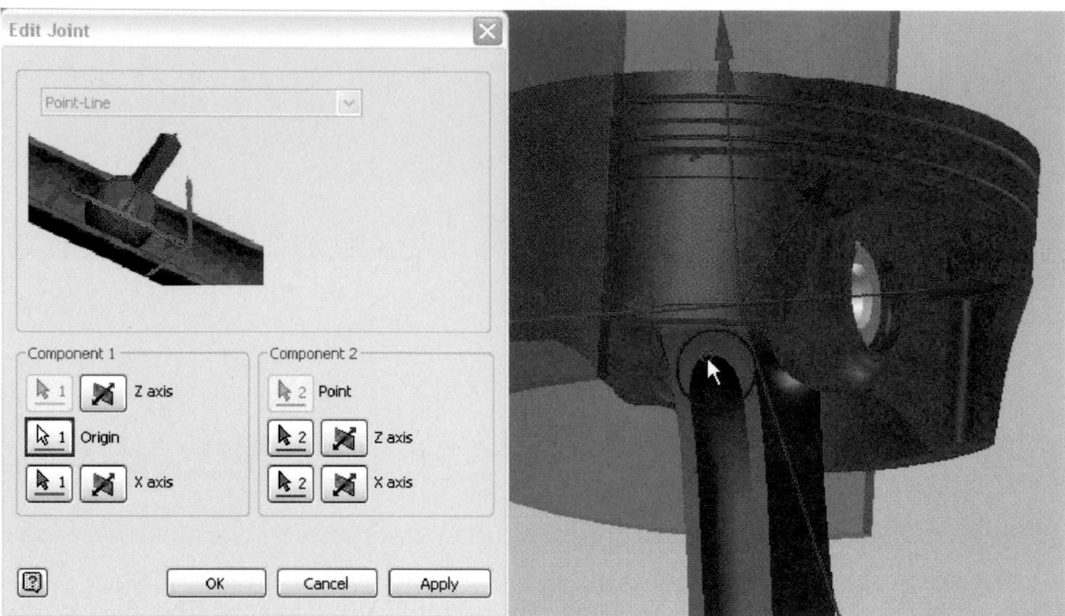

26. Play the simulation

27. Select **Finish Dynamic Simulation**

28. Select **Environments** tab > **Stress Analysis**

29. Right click **Loads** > Select **Update**

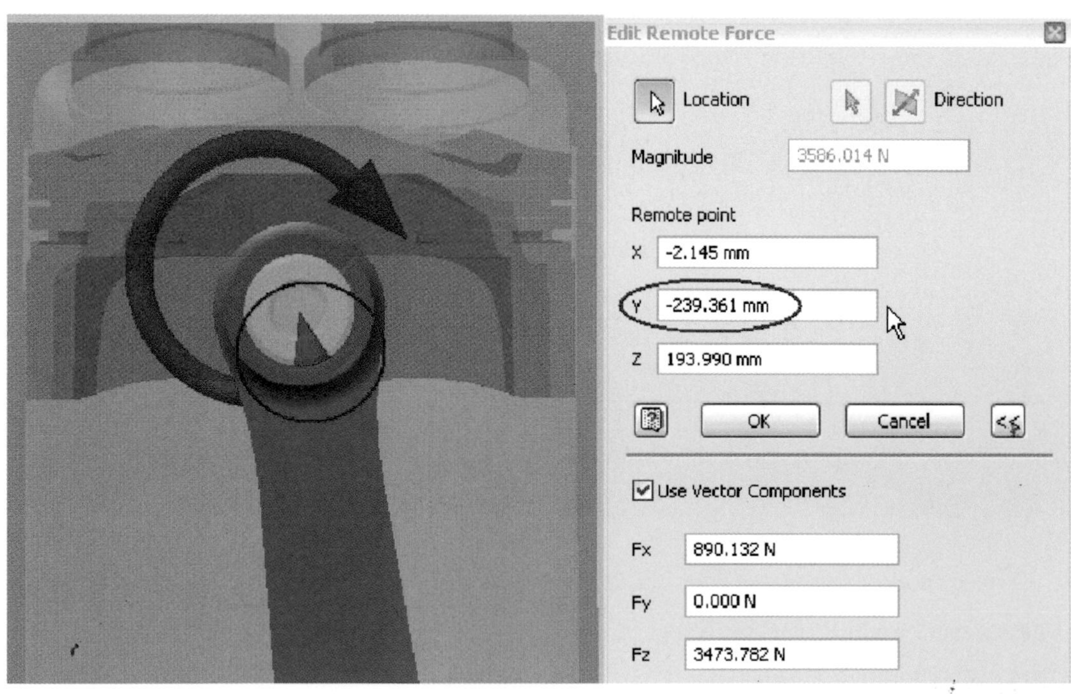

The force has now been moved in-line with the connecting rod instead of being offset.

30. Select **Simulate** > Run the simulation

31. Select **Actual** for **Displacement Display**. The value may slightly differ

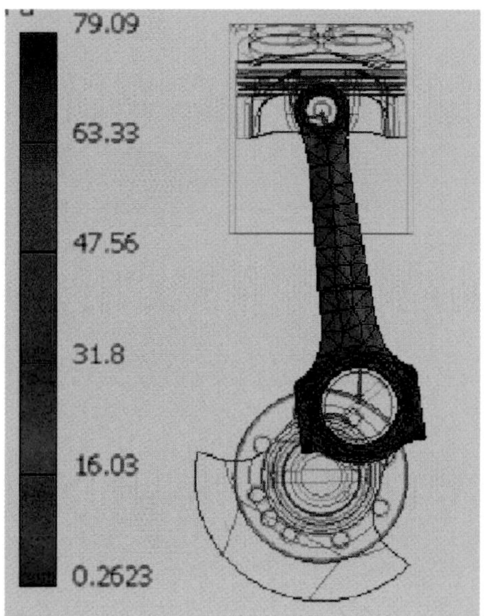

Even though the results did not change, you should be aware of the position of the joints and their potential impact on the results.

Next, we need to determine whether the results have converged.

32. Select **Convergence Settings**

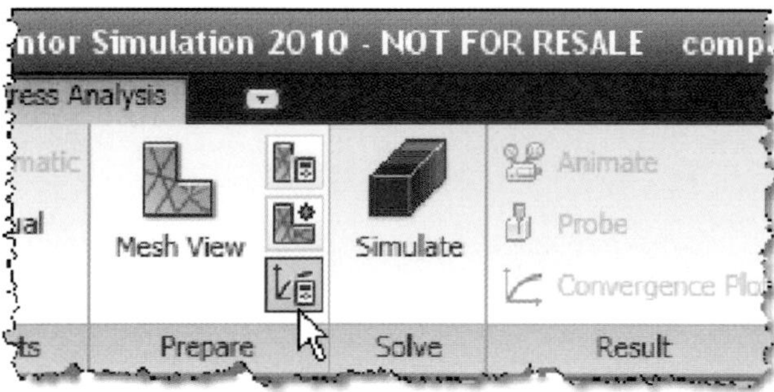

33. Specify 3 for **Maximum number of h refinements** > Click **OK**

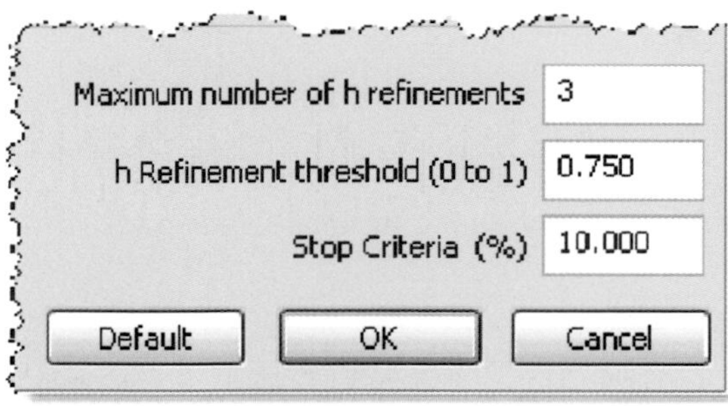

34. Select **Simulate** > Rerun the simulation

35. Select **Convergence Plot**

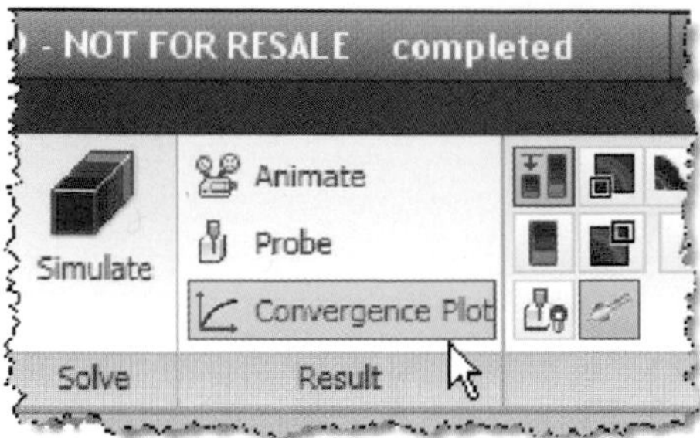

The graph below shows that the maximum Von Mises Stress results have converged. The value may slightly differ.

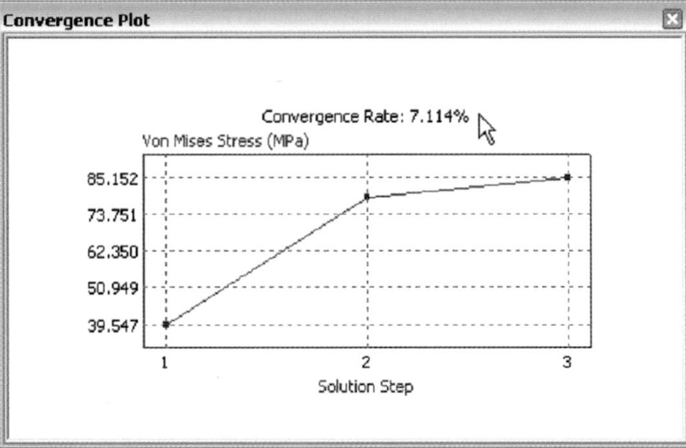

The convergence plot shows that the convergence has been achieved within the automatic P-refinement stage.

36. Close the convergence plot > Double click **Safety Factor**

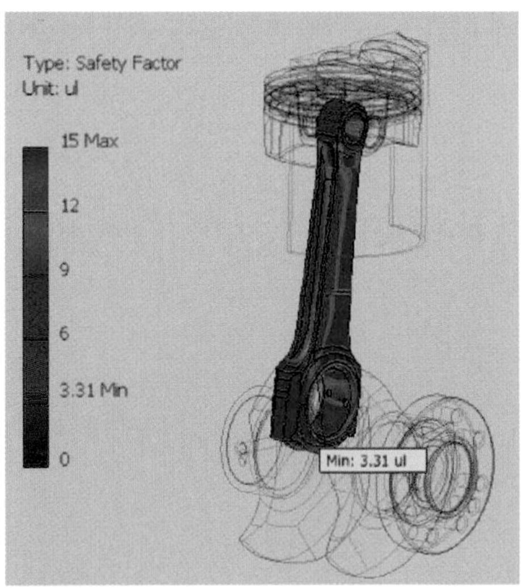

Now, repeat the above steps for the motion load transferred at time step 0.0153 seconds and see whether the safety factor changes.

Copy the simulation and change the time step from 0.00710 seconds to 0.0153 seconds. This will speed up the analysis.

37. Close the file

CHAPTER 12

DP10 – Cyclic Symmetry Analysis

Design of Industrial Fan Blades

(Design Problem courtesy of Halifax Fan Ltd)

KEY FEATURES INTRODUCED IN THIS DESIGN PROBLEM

	Key features
1	Cyclic Symmetry
2	Manual Convergence of Results

INTRODUCTION

Halifax Fan Ltd is one of the world's foremost manufacturers of industrial fans. They design and manufacture a full range of centrifugal fans from a wide range of materials, including mild and stainless steel, from their manufacturing operations in the UK and China. They supply a wide range of industrial customers, including power, pharmaceutical, chemical, nuclear, and marine markets all over the world.

Up and Running with Autodesk Inventor Simulation 2011. ISBN: 978-0-12-382102-7
Copyright © 2010 Elsevier Inc. All rights of reproduction, in any form reserved.

Halifax Fan is fully BSI certified to BS EN ISO9001 – 2000 and manufactures fans to many industrial standards including API 673, API 560, Shell DEP, and ATEX. Many of these designs are engineered to meet the customer's exact requirements and, thus, the company offers a wide range of services on- and off-site, including stress relief laser shaft alignment, site performance testing, vibration analysis, consultation, problem solving, repairs, and energy testing. As a consequence of offering special bespoke solutions, Halifax is regularly asked by its customers to validate their designs prior to delivery.

Some of the typical requirements include determining the following:

- The maximum stress and deflection of the fan blade.
- The factor of safety of the new design.

In addition to the above requirements, the design criteria to be used for this design problem are as follows:

- Material to be used is either mild steel or high strength low alloy steel[1].
- Factor of safety required is 1.75.
- Maximum deflection is 0.5 mm.
- Maximum blade thickness is 5 mm.

WORKFLOW OF DESIGN PROBLEM 10

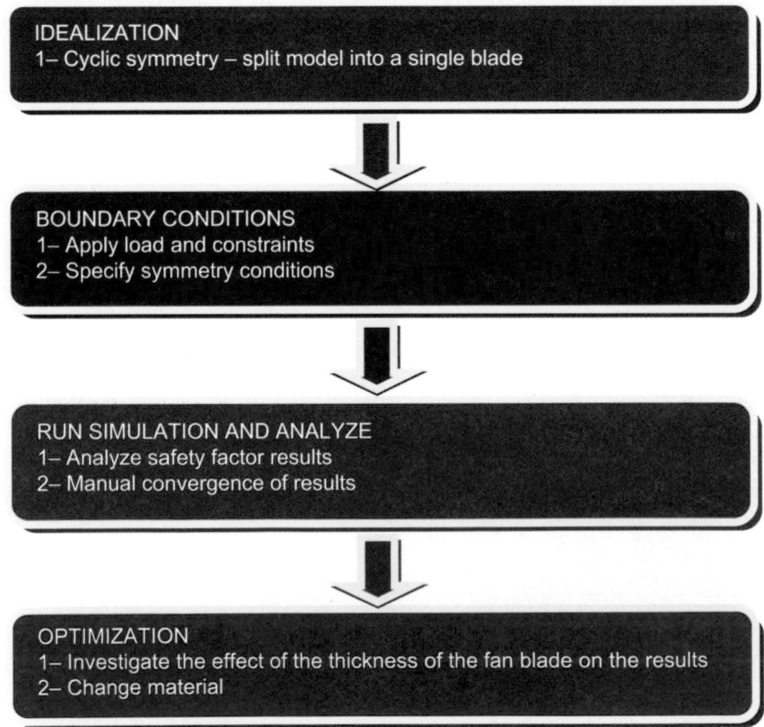

[1]Halifax Fan actually uses carbon steel to BS EN 10025 grade S275JR for its fans.

Idealization

Halifax Fans range from simple small fans to large detailed fans. In the case of large detailed fans, the size of the mesh can become very large and the time taken to analyze the results can become very lengthy.

Most fans comprise a number of similar blades and, when in operation, the deflection and stress induced in the blades are identical and for this reason it is only necessary to analyze one blade of the fan. This simplification approach is also referred to as cyclic symmetry and it significantly reduces the model size, giving more scope to refine and analyze the results efficiently. Therefore, in the following steps, the fan model is split such that only one blade remains.

1. Open *Fan-complete.ipt*

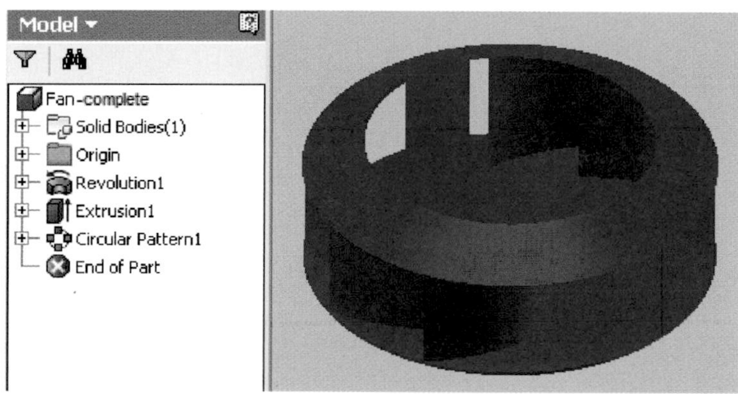

2. Create a new sketch on the YZ plane to the following dimensions:.

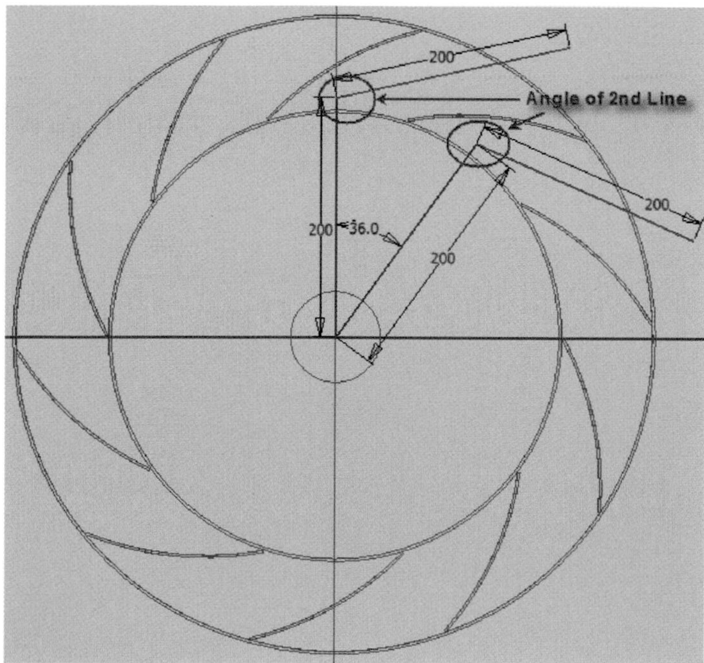

It will help to change the model display to wireframe or transparent color when creating the sketch, as this will allow you to see the top plate.

As there are 10 blades, we need to split the model by 36 degree angles.

$$\text{Angle of split to create single blade} = \frac{360}{\text{Number of blades}} = \frac{360}{10} = 36 \text{ degrees}$$

The angle of the second line is not critical as long as the line is more or less positioned in the middle of two blades.

3. Select **Finish Sketch** > Using the **Split** feature, split the part using the sketch created

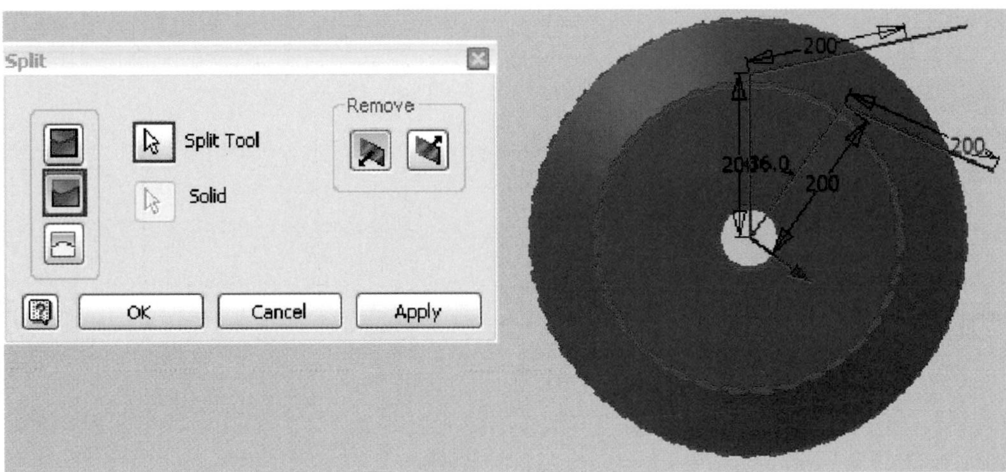

4. Click **OK**

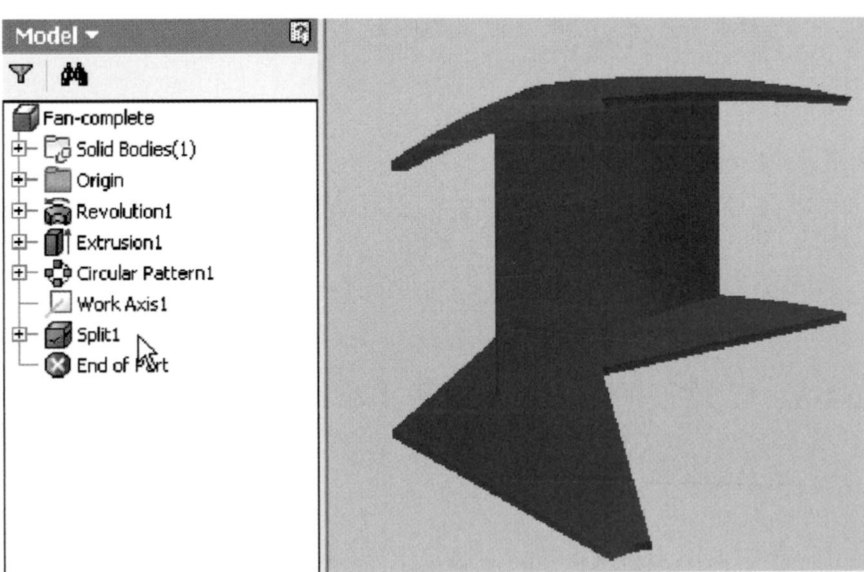

Now, in the next section, the boundary conditions will be applied to the single fan blade.

Boundary conditions

5. Select **Environments** tab > **Stress Analysis**

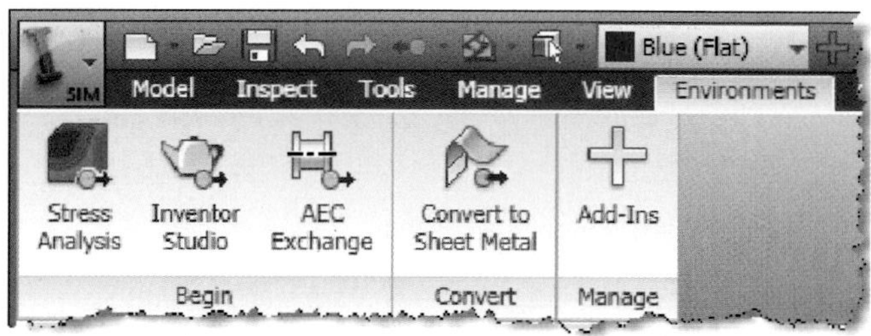

6. Select **Create Simulation** > Specify 'One-Blade' for **Name** > Click **OK**

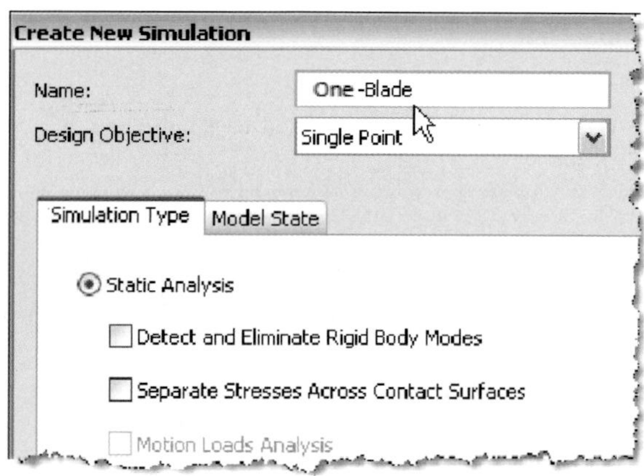

7. Select **Fixed Constraint** > Select the face as shown > Click **OK**

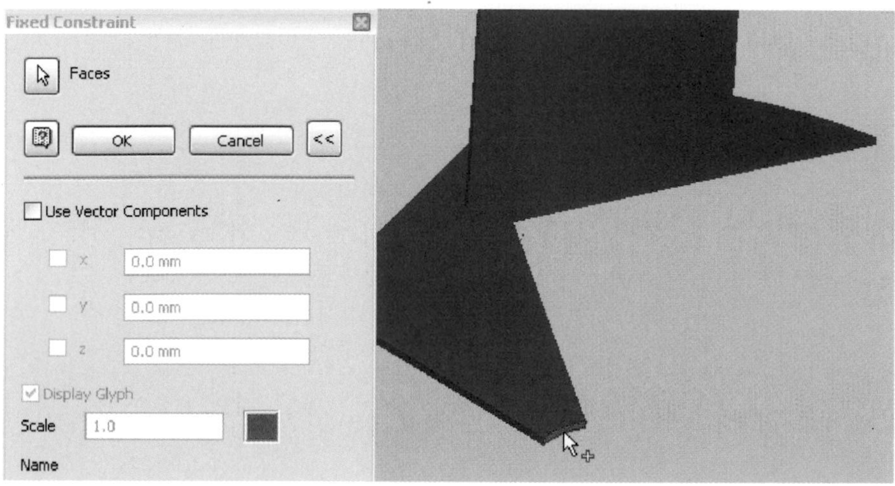

8. Select **Body Loads** > **Angular** tab > **Enable Angular Velocity and Acceleration** > Select the face to specify the direction of the fan speed > Specify 2000 rpm > Click **OK**

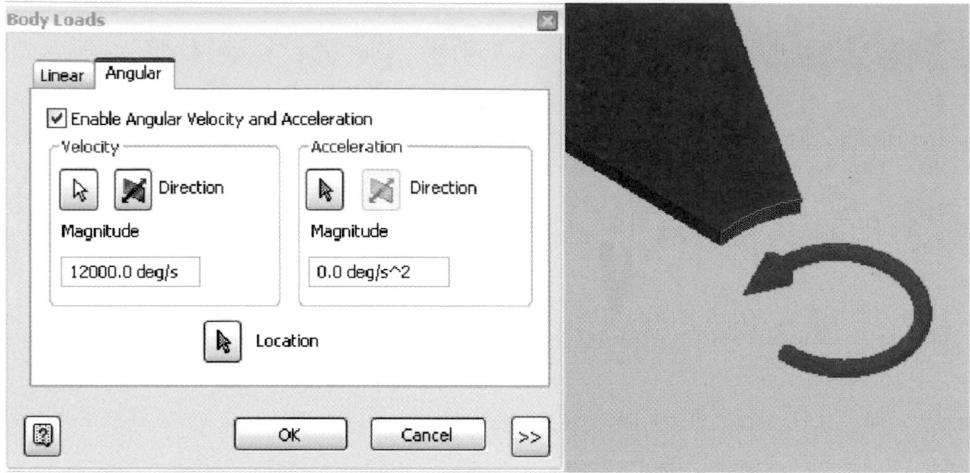

Specifying revolutions per minute after the value will convert the value to the default degrees/seconds.

If the complete fan were analyzed, the boundary conditions specified in Steps 7 and 8 would suffice. However, as we are only modeling a single blade, we need to specify extra boundary conditions to enable it to behave like a complete model. This can be achieved by applying frictional constraints on all faces that are created as a result of the **Split** feature.

9. Select **Frictionless Constraint** > Select all eight faces on the split planes > Specify 'Cyclic Symmetry' for **Name** > Click **OK**

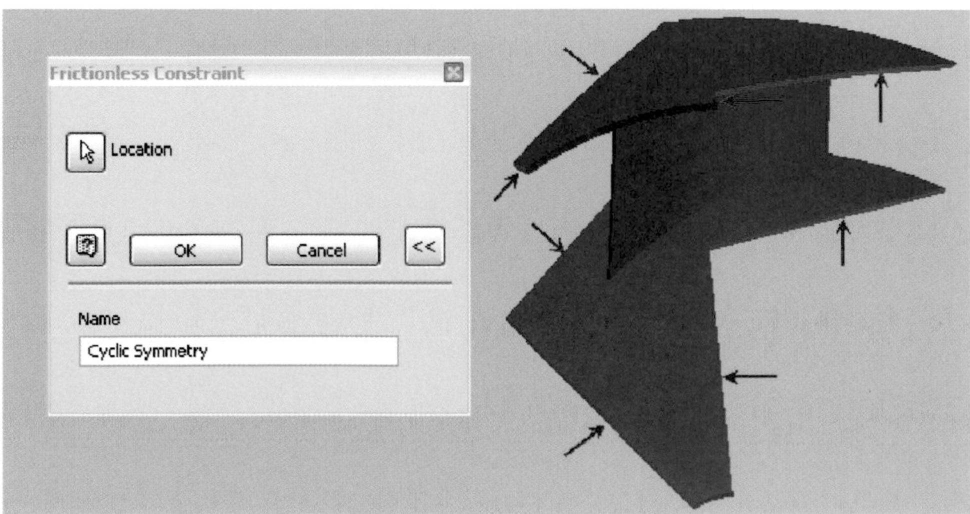

10. Select **Mesh View**

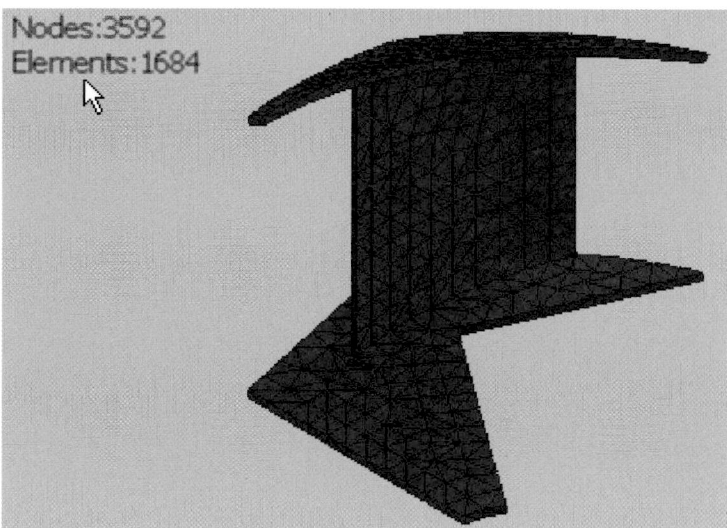

A complete model would create many more elements, as illustrated below:

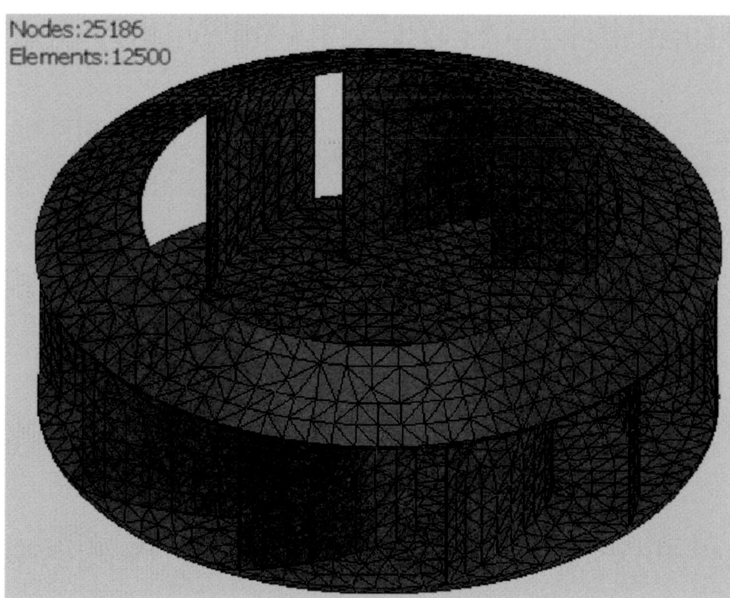

Run simulation and analyze

11. Select **Simulate** > Run the analysis
12. Select **Actual** for **Displacement Display** > Deselect **Mesh View**

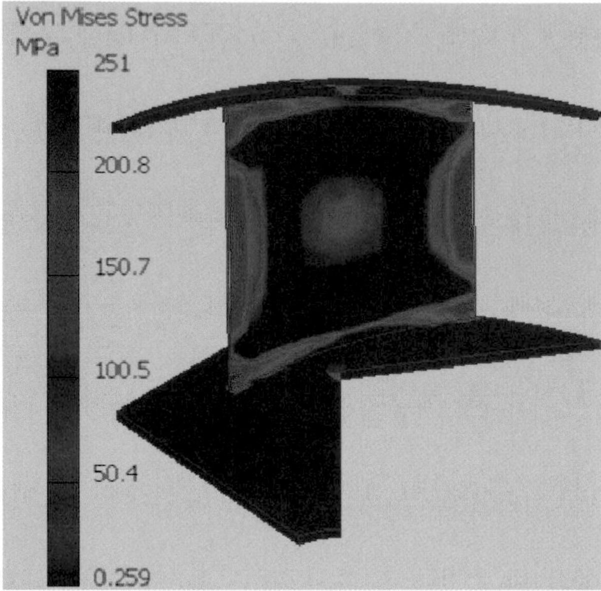

Stress singularities will appear at the blade and plate interface due to sudden geometrical discontinuities and will be ignored as the area of interest is in the middle of the blades.

Stress singularities may also occur in the area of the split faces and can be ignored as they would not have appeared if the complete fan had been analyzed.

13. Select **Color Bar** > Unselect **Maximum** > Specify 200 MPa > Click **OK**

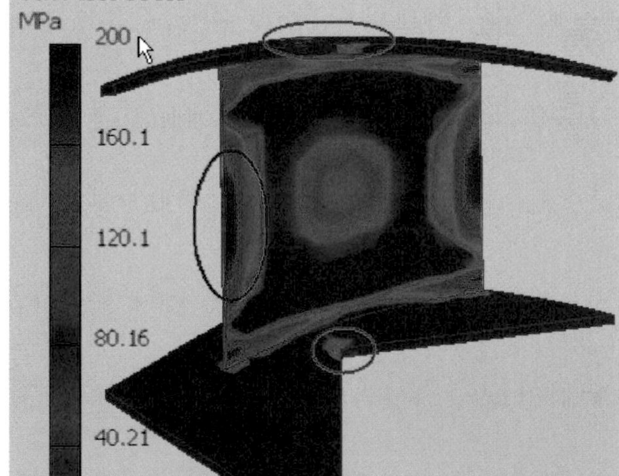

Use the color bar to pinpoint the stress display in the area of interest and to enhance the stress display.

As we are interested in the middle of the blade, we can use **Probe** to display stresses in the area of interest to us.

14. Select **Probe** > Click in the middle of the blade at the front and rear

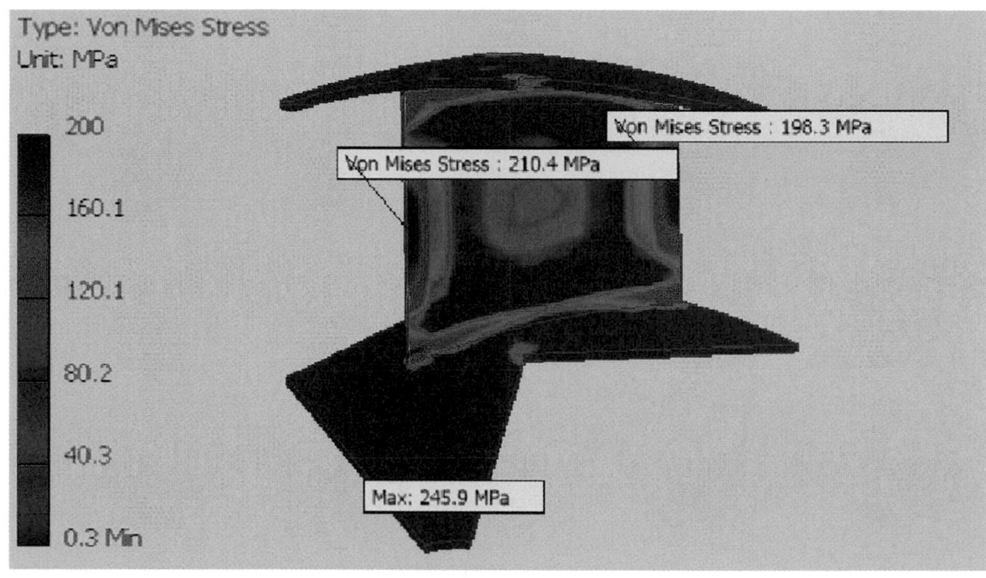

Zoom into the area of interest before selecting the area of interest using **Probe**.

IMPORTANT—The exact stress value of **Probe** is dependant on the location clicked; hence, the value may slightly differ.

Below is a stress plot of a complete model, illustrating similar stresses in all of the blades of the fan.

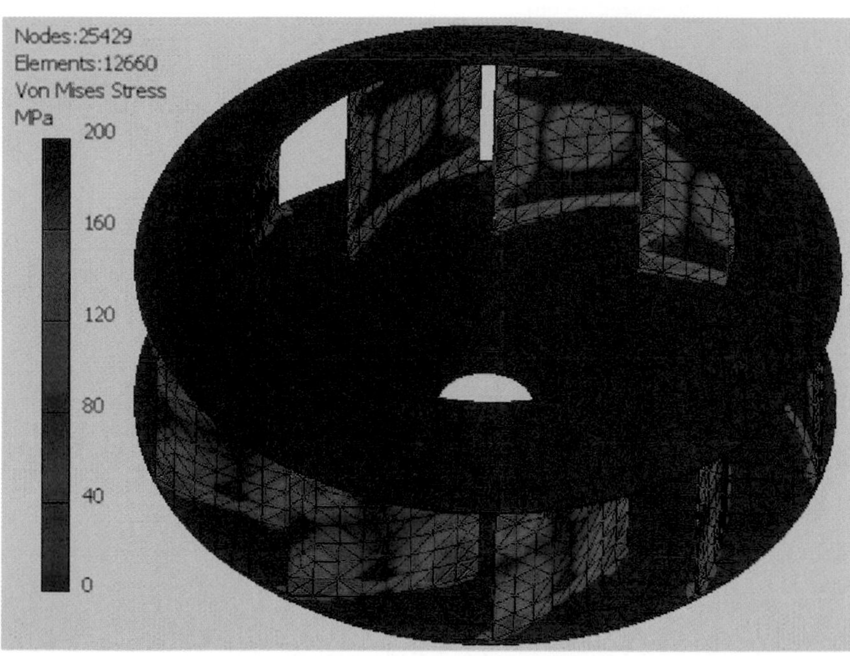

Now, we will increase the mesh to see whether the stress results change in the blades.

15. Select **Mesh Settings** > Change **Average Element Size** to 0.05 > Click **OK**

16. Right click **Mesh** > Select **Update Mesh** > Select **Mesh View**

Reducing the average element size can have a significant impact on the size of the mesh.

Reducing the average element size from 0.1 to 0.05 has increased the number of elements by 285% and it will thus take longer to run the simulation.

17. Rerun the simulation > Deselect **Mesh View**

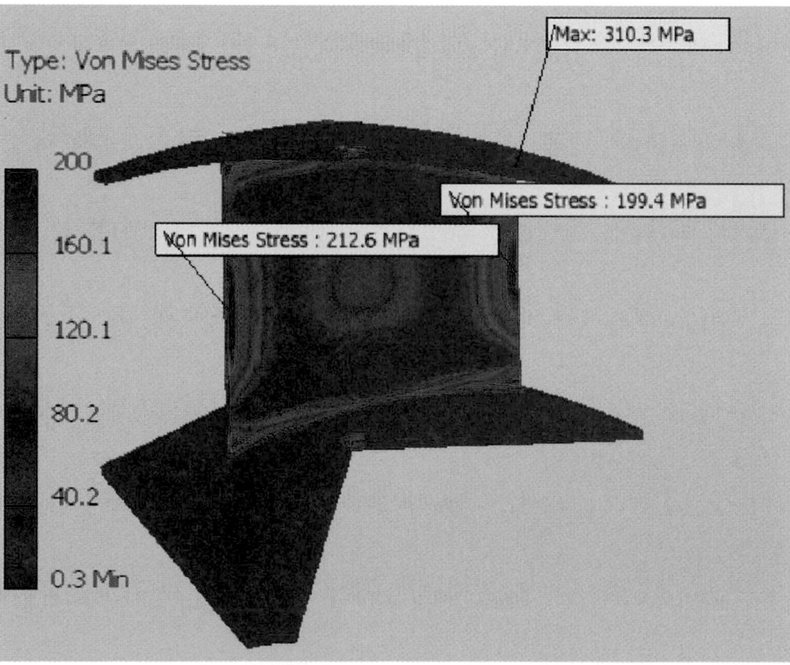

Although the maximum stress has moved to the back of the blade and increased by 26.5%, the stress in the middle of the blade has only changed by 1% to 212.6 MPa. Use your **Probe** value for comparison.

To confirm whether this stress in the middle of the blade has converged, we will rerun one more analysis with a smaller element size.

18. Select **Mesh Settings** > Change **Average Element Size** to 0.025 > Click **OK**

19. Right click **Mesh** > Select **Update Mesh** > Select **Mesh View**

Reducing the average element size from 0.1 to 0.025 has increased the number of elements by 1432%.

A full model with a similar mesh size of 0.025 will create 164 680 elements, as illustrated below.

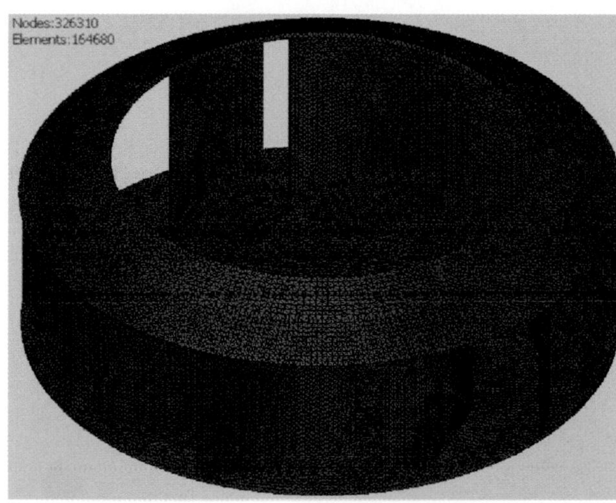

20. Run Simulation > Deselect **Mesh View**

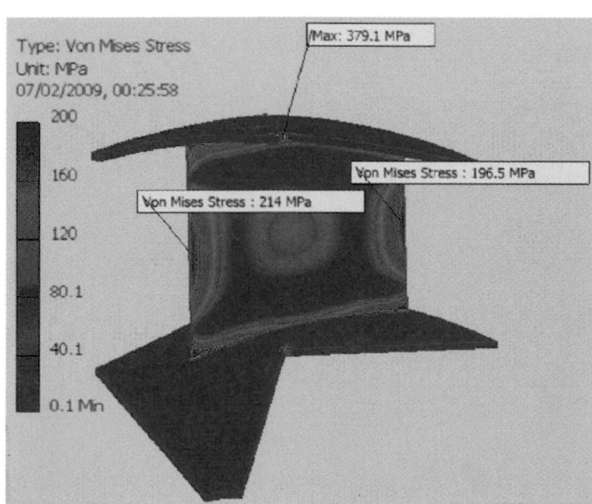

Ignore the maximum stress as it is occurring on the top plate and blade interface due to geometrical discontinuities leading to stress singularities.

Probe positions can be altered by right clicking and selecting Edit Position as shown below

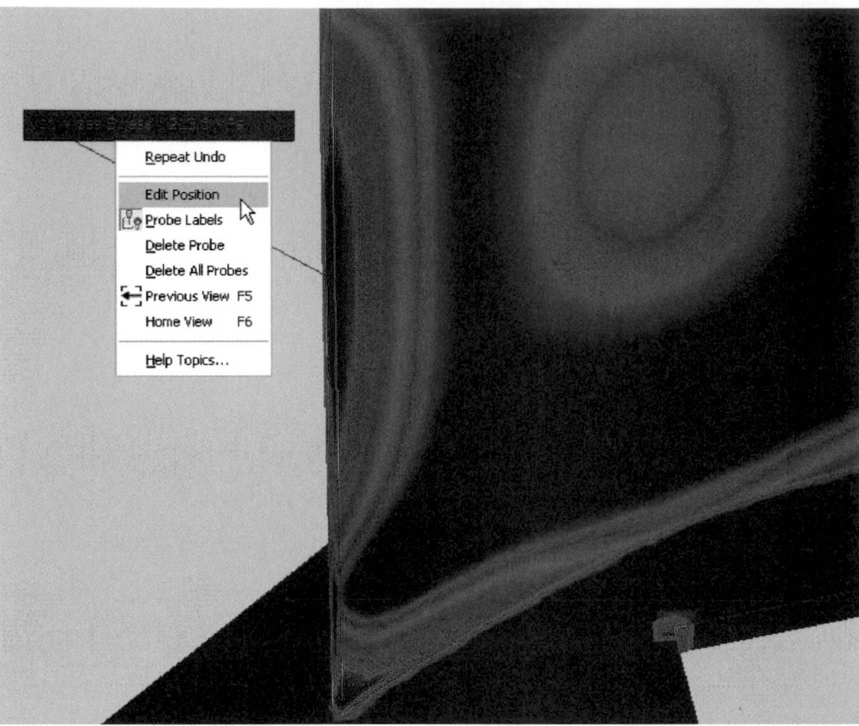

By changing the position, you can display the results in different areas of the model.

Alternatively, you can select multiple areas of the model with the **Probe** option.

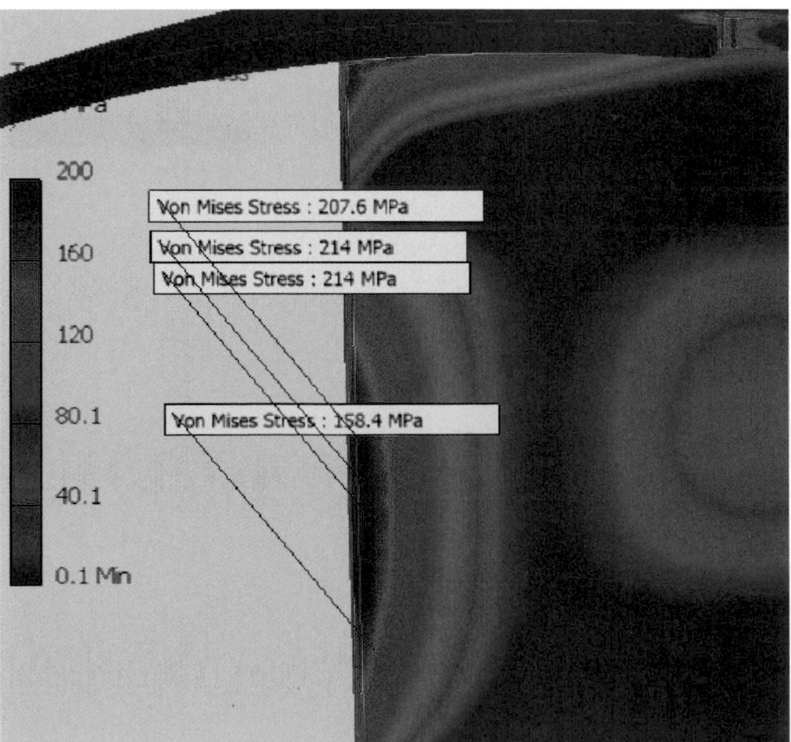

The maximum value in the middle of the blade does not exceed 218 MPa. As we are only interested in this region, we can confidently say that the results have converged in the area of interest.

21. Double click **Displacement** from the **Stress Analysis** browser

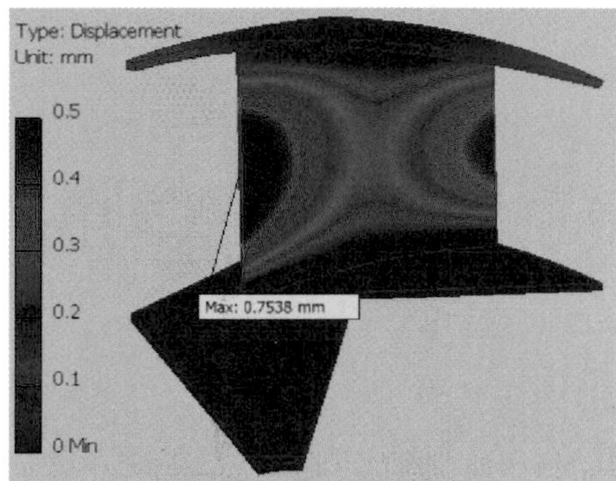

The maximum displacement plots for mesh settings of 0.1 and 0.05 are also shown below.

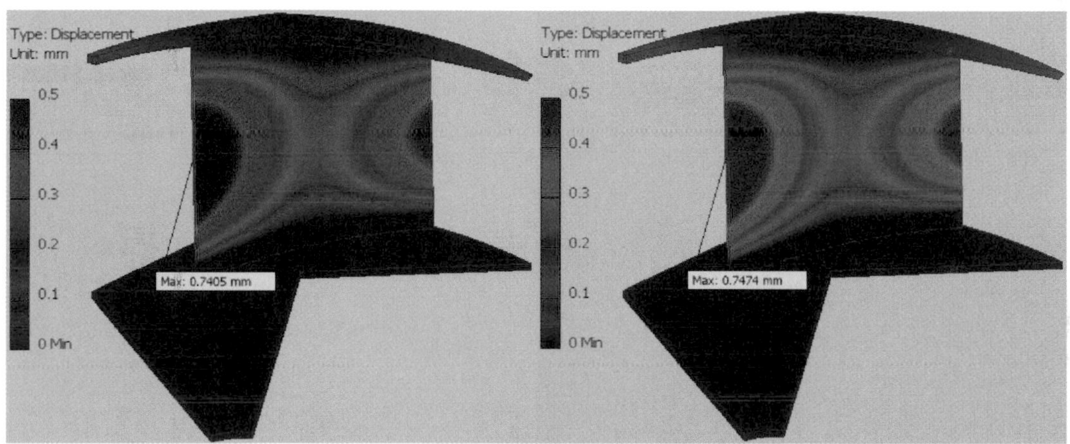

The maximum displacement occurs in the middle of the blade and changes from 0.7405 to 0.7538, a change of 1.8%, such that the displacement values can also be treated as having converged. The values may differ slightly.

22. Double click **Safety Factor**

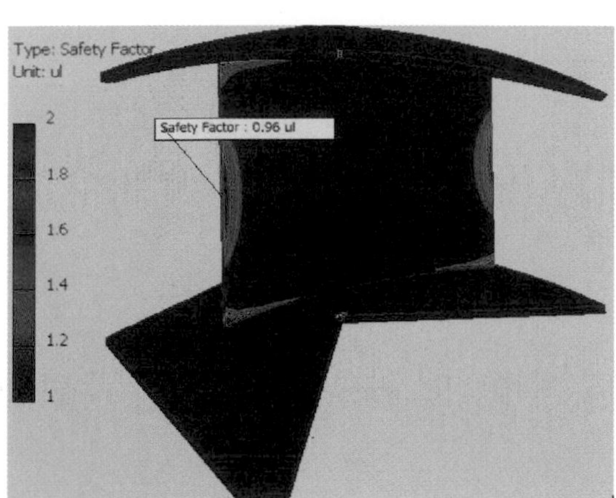

Use the color bar to adjust the range.

Based on the stress in the middle of the blade (214 MPa), we have a safety factor below 1, which suggests that the design has failed as the design limit was 1.5. In the next section, we will perform an optimization study to meet the design limits.

Optimization

In this section, we will alter the blade thickness from 2 to 5 mm using the parametric study and manually alter the material from mild steel to high strength steel.

23. Right click One-Blade > Select **Copy Simulation** > Click **OK**

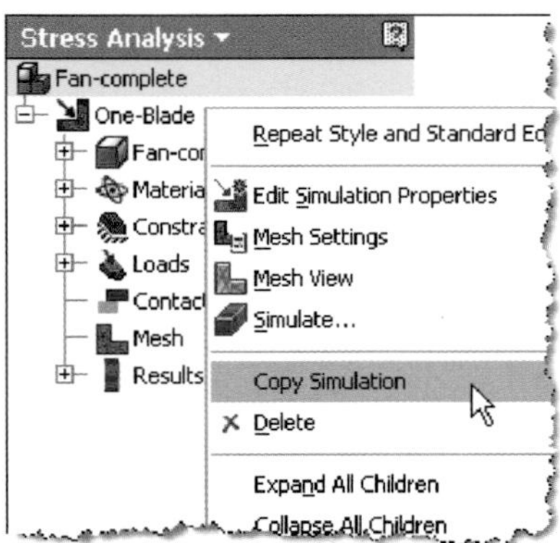

24. Right click copied Simulation:1 > Select **Edit Simulation Properties**

25. Specify 'Blade-Optimization' for **Name** > Select **Parametric Dimension** for **Design Objective** > Click **OK**

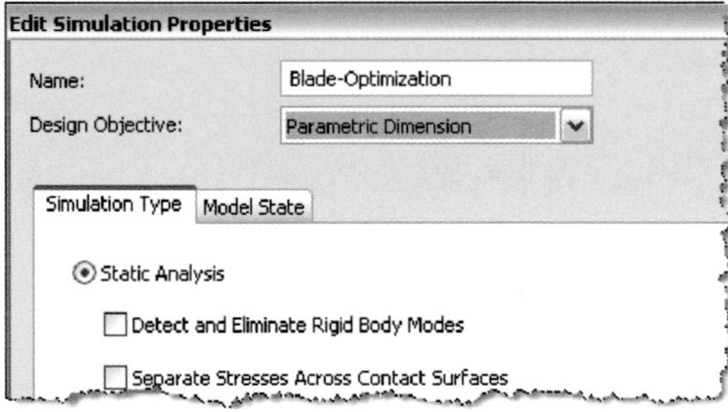

This will now allow us to carry out a parametric study.

26. Right click Fan-complete.ipt in the browser > Select **Show Parameters**

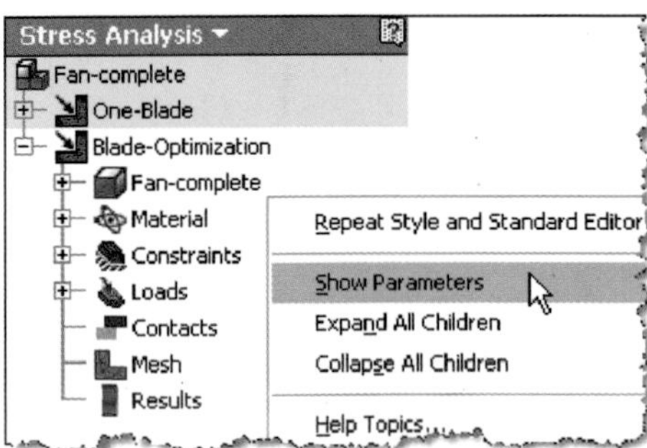

27. Select the **Bladethickness** user parameter > Click **OK**

Reference Parameters				
	d13	mm	266.5 mm	266.5
User Parameters				
☑	Bladethickness	mm	2 mm	2.0

28. Select **Parametric Table**

29. Right click in the **Design Constraints** row > Select **Add Design Constraint**

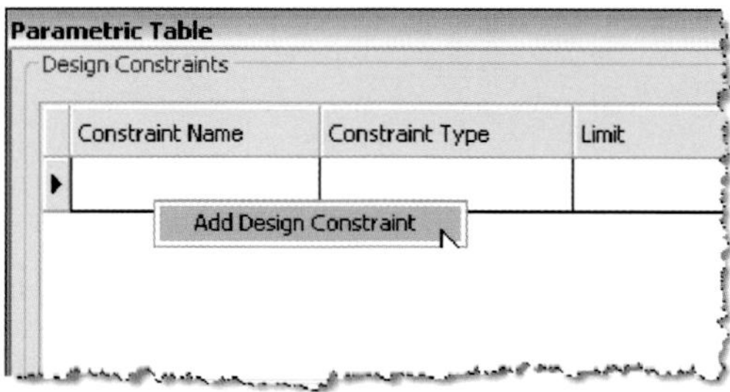
Parametric Table
Design Constraints

Constraint Name	Constraint Type	Limit

Add Design Constraint

30. Select **Max Von Mises** from the list

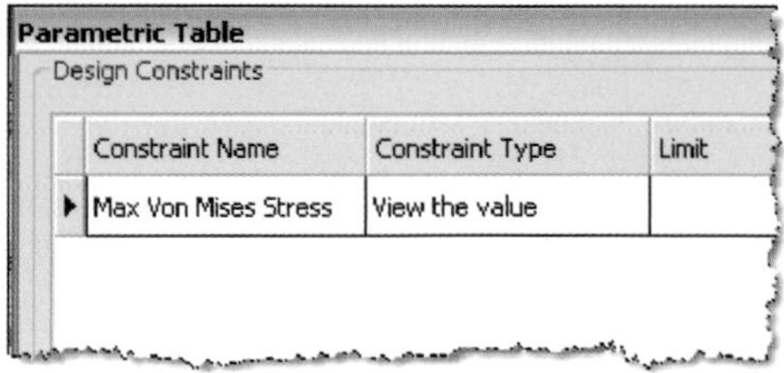
Parametric Table
Design Constraints

Constraint Name	Constraint Type	Limit
Max Von Mises Stress	View the value	

31. Repeat Step 29 to add displacement design constraints

32. Change the **Constraint Type** for Max Von Mises Stress to **Upper limit** > Specify **Limit** as 200

33. Change the **Constraint Type** for Max Displacement to **Upper limit** > Specify **Limit** as 0.5

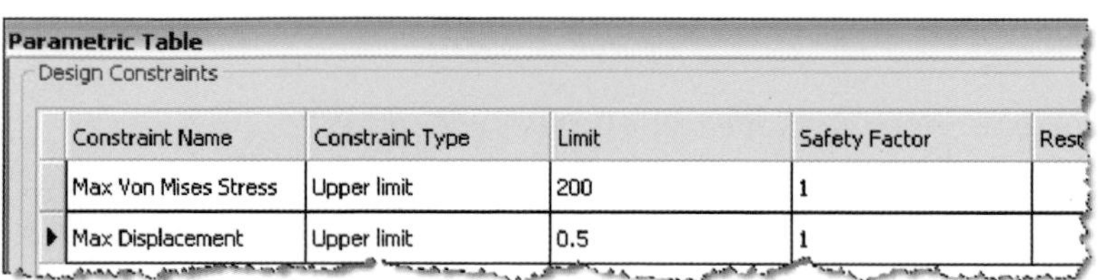
Parametric Table
Design Constraints

Constraint Name	Constraint Type	Limit	Safety Factor	Res
Max Von Mises Stress	Upper limit	200	1	
Max Displacement	Upper limit	0.5	1	

34. Specify 2 - 5:4 in the Bladethickness **Values** field

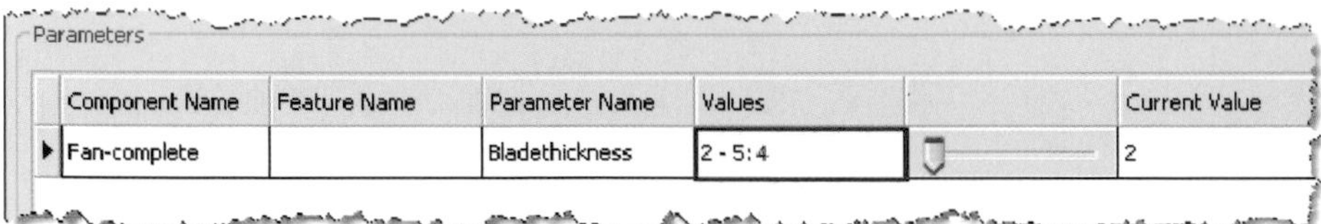

Parameters

Component Name	Feature Name	Parameter Name	Values		Current Value
Fan-complete		Bladethickness	2 - 5:4		2

This will generate values of 2, 3, 4, 5 such that three additional parameters will be created.

35. Right click anywhere in the parameter rows and select **Generate Range Configurations**

36. Move the slider to see the blade changing its thickness > Click **Close**

37. Select **Mesh Settings** > Specify **Average Element Size** > Click **OK**

38. Select **Mesh View**

39. Select **Simulation** > Run the simulation

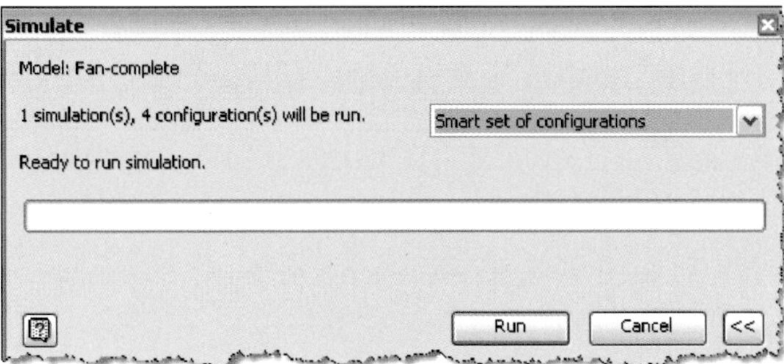

40. Select **Actual** for **Displacement Scale** > Deselect **Mesh View**

41. Select **Parametric Table**

The red icon indicates unacceptable parameters based on the constraint limits.

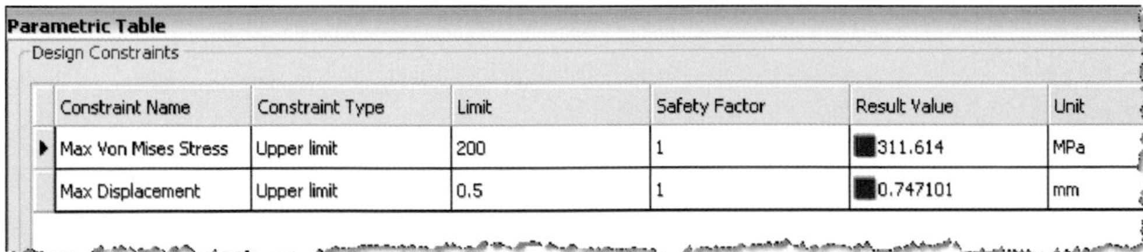

Parametric Table

Design Constraints

Constraint Name	Constraint Type	Limit	Safety Factor	Result Value	Unit
Max Von Mises Stress	Upper limit	200	1	311.614	MPa
Max Displacement	Upper limit	0.5	1	0.747101	mm

The Max Von Mises Stress value is also misleading as this value represents stress singularities in the model. To synchronize the stress limit with the model, change the color bar range to between 0 and 200.

42. Select **Color Bar** > Specify 200 for the **Maximum** value > Click **OK**

Now compare the color plots as you move the slider between 2 and 5. From the color plots, blade thickness values 4 and 5 do not show any red color in the blades, indicating low stress, with thickness 5 showing the least stress.

43. Move the slider to read a value of 5 > Select **Close**

We will now use the probe to determine the exact value of stress in the middle of the blade.

44. Select **Probe** and select the blade at the highest stress point

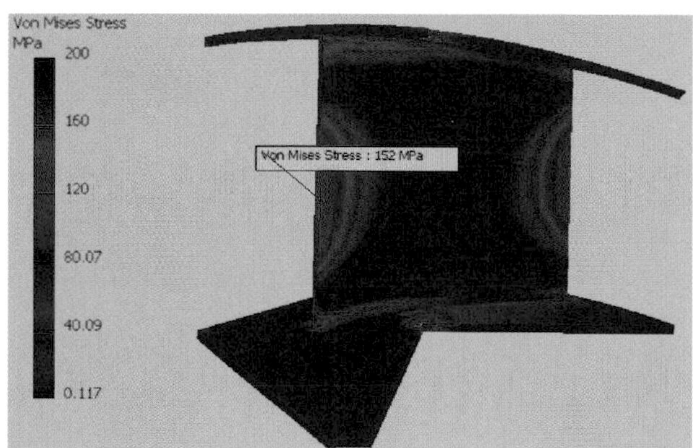

You may need to select several locations to get an indication of the highest stress point.

45. Double click **Safety Factor**

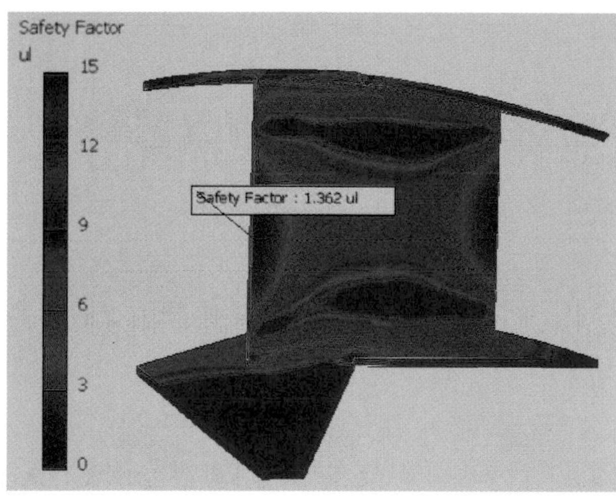

The safety factor is still below the design limit of 1.5, so we will now assign a new material.

46. Select **Assign Material**

47. Select **Steel, High Strength Low Alloy** from the **Override Material** list > Click **OK**

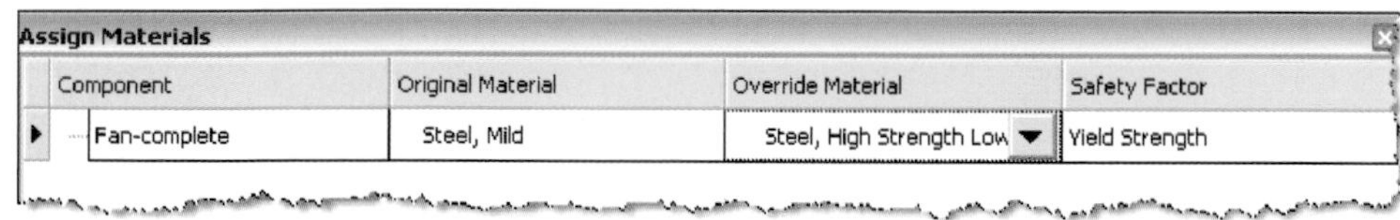

Assign Materials

Component	Original Material	Override Material	Safety Factor
Fan-complete	Steel, Mild	Steel, High Strength Low	Yield Strength

48. Select **Parametric Table** > Move the slider to read a **Current Value** of 5

49. Right click the slider > Select **Simulate this configuration** > Select **Run**

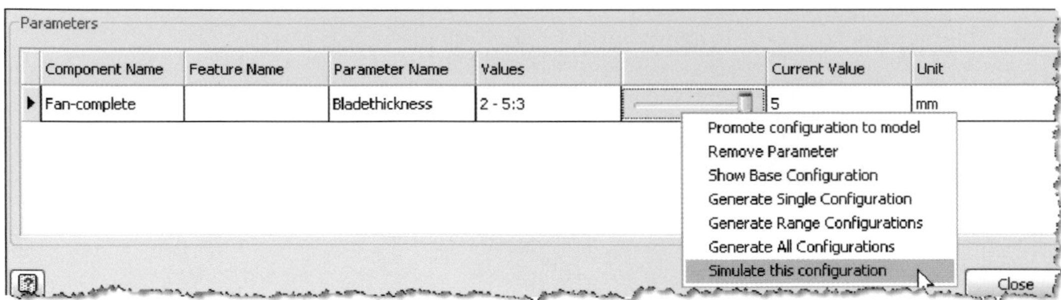

50. Select **Actual** for **Displacement Display** > Double click **Safety Factor**

51. Select **Probe** > Select the area of minimum safety factor

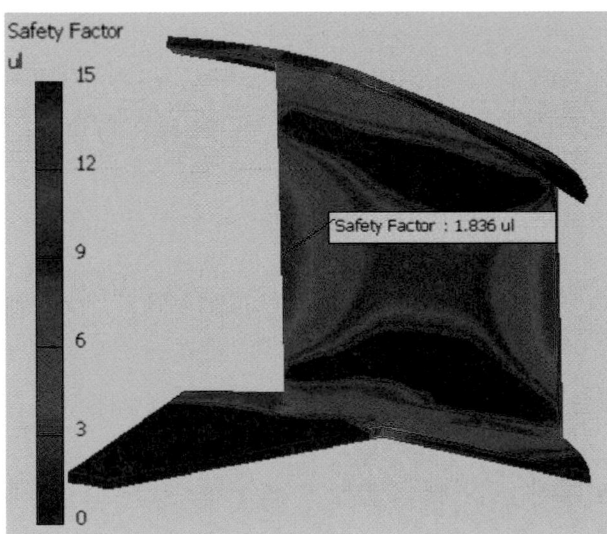

Now, by changing the material, we have reached our goal of having a safety factor above 1.5 and a maximum displacement below 0.5 mm. Ignore maximum stress as it is occurring on the top plate; in reality, this does not exist. Refer to the stress display of the complete fan shown earlier.

52. Close the file

CHAPTER 13

DP11 – Weldment Analysis

Structural Design of Moving Bridge

(Design Problem courtesy of British Waterways Ltd)

KEY FEATURES INTRODUCED IN THIS DESIGN PROBLEM

	Key features
1	Convert assembly to single part using **Shrinkwrap Substitute**
2	Same scale color legend display
3	Planar X, Y, Z stress plots

INTRODUCTION

This design problem is a follow-on from Design Problem 2, in which a British Waterways team were involved in designing a new jack mechanism to open a canal bridge. The jack force of 28729 N determined in Design Problem 2 will be used to validate the structural integrity of the new structure, which is to be incorporated into the existing structure beneath the bridge.

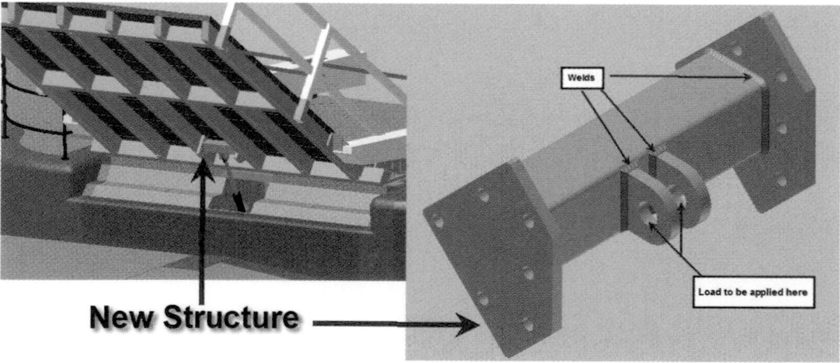

As this new structure is designed with welds incorporated as a weldment assembly, the **Export Loads to FEA** feature within Dynamic Simulation cannot be used automatically. With this in mind, we have two options.

Up and Running with Autodesk Inventor Simulation 2011. ISBN: 978-0-12-382102-7
Copyright © 2010 Elsevier Inc. All rights of reproduction, in any form reserved.

1. Part analysis – Convert the weldment to a single component using
   a. **Shrinkwrap** – This will convert the assembly into a separate single component.
   b. **Shrinkwrap Substitute** – This will create a shrinkwrap of the assembly and creates a new substitute level of detail.

To perform stress analysis, the assembly has to be shrink wrapped to solid components and not surfaces.

2. Assembly analysis

We will use Options 1b and 2 and compare the results.

The main requirements of this design problem are to determine:

- The maximum compressive stress in the structure whilst the bridge is being opened.
- The maximum deflection in the structure.
- The factor of safety of the new design – for example, the fatigue life could be predicted.

In addition to the above requirements, the design criteria to be used for this design problem are:

- Material to be used is mild steel.
- Weld material to be used is mild steel.
- Factor of safety required is 2.5.

WORKFLOW OF DESIGN PROBLEM 11

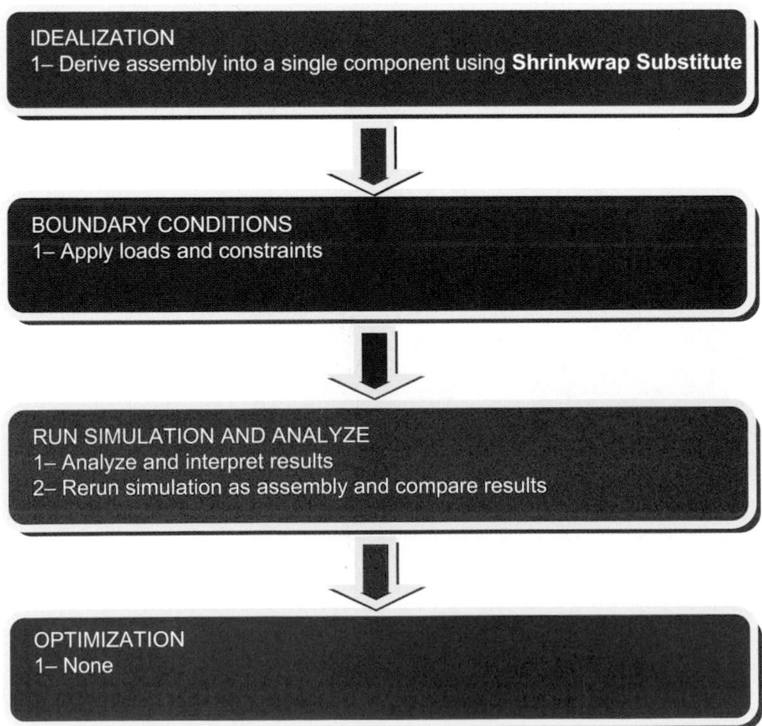

Idealization

In order to perform a comparative study between Options 1b and 2, we initially need to create a single part using **Shrinkwrap Substitute**.

1. Open *Cylinder reaction beam.iam*

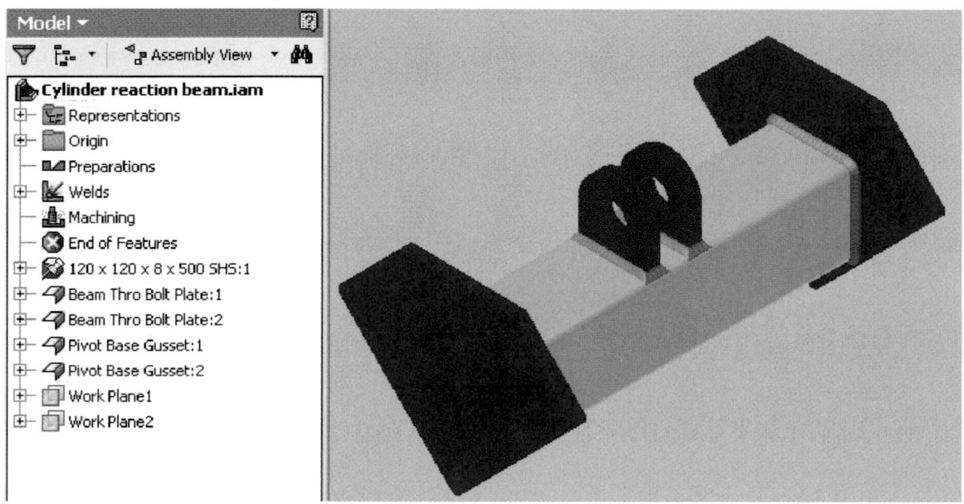

2. Select **Shrinkwrap Substitute** > **OK**

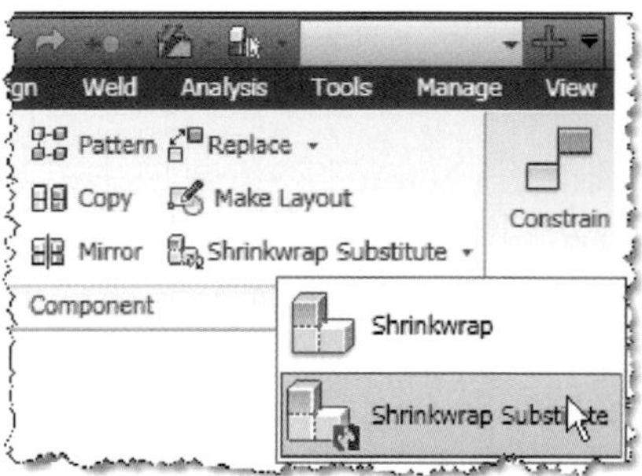

3. Select the **Single Body** option > Select **None** for **Hole patching**

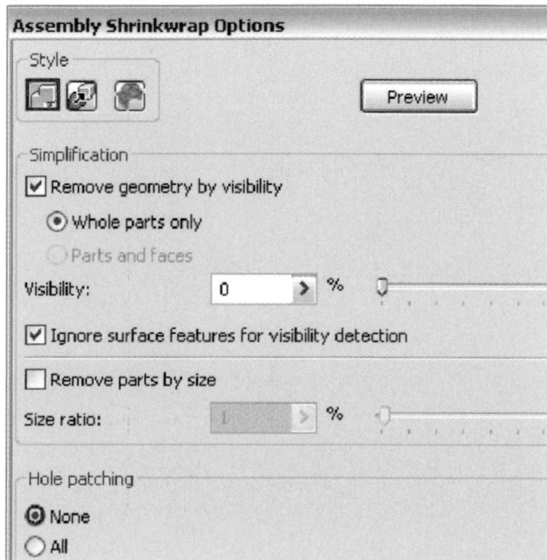

4. Click **OK**

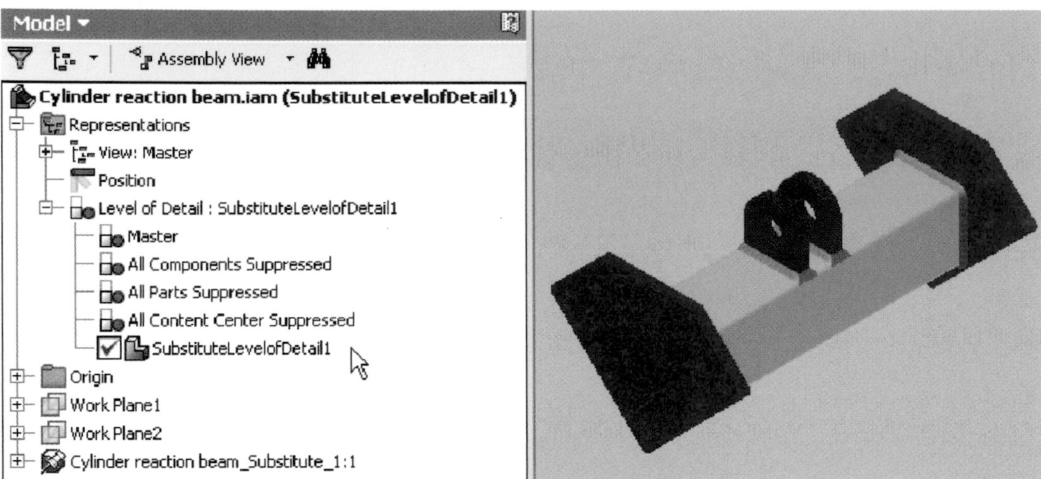

Now, we can begin the second stage of the analysis by applying boundary conditions.

Boundary conditions

5. Select **Environments** tab > **Stress Analysis**

6. Select **Create Simulation** > Specify 'Shrinkwrap-Analysis' for **Name** > Select **Substitute Level of Detail1** within the **Level of Detail** > Click **OK**

7. Right click **Welds** > Select **Exclude From Simulation**

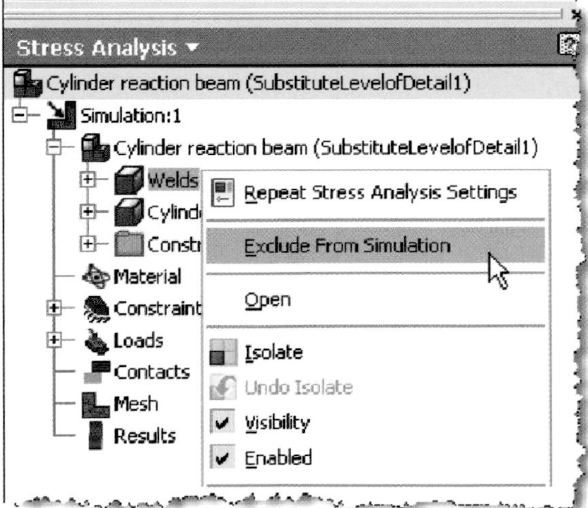

 **Shrinkwrap** has copied the welds and hence we do not need them again.

8. Select **Fixed Constraints** > Select the faces on both sides as shown > Specify 'Bolted-Plates' for **Name** > Click **OK**

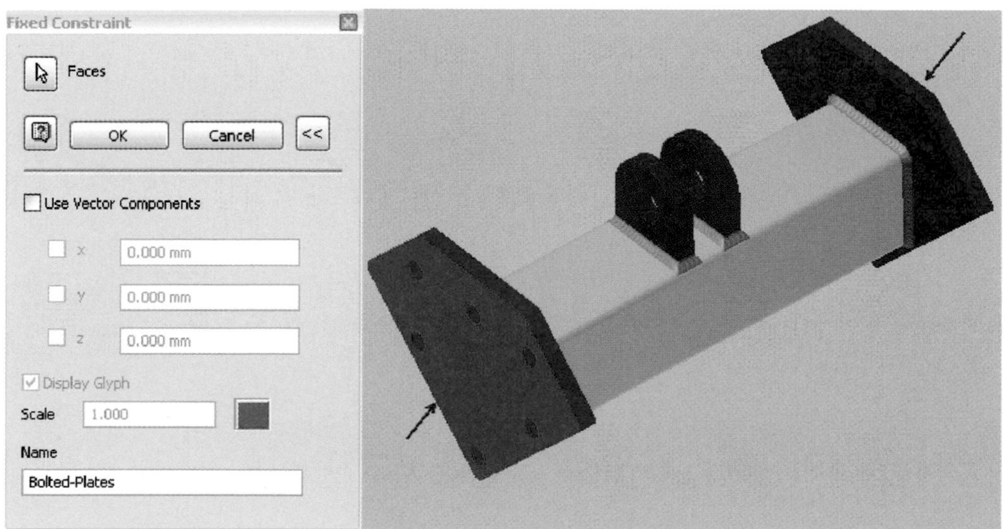

9. Select **Edit Bearing Loads** > Select the internal circular face of both lugs to specify the location > For **Direction**, select the top face of the channel > Specify 28729 N for **Magnitude**

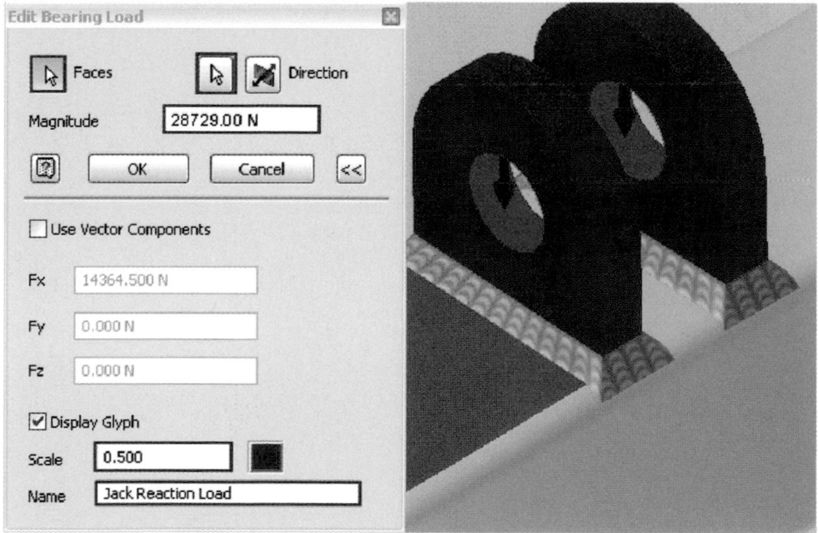

The force will be split for the bearing loads when both faces are selected together using the same bearing load command.

10. Specify 0.5 to reduce the force display size > Specify 'Jack Reaction Load' for **Name** > Click **OK**

11. Select **Assign Materials** > Select **Steel, Mild** from the **Override Material** list > Click **OK**

The newly created part has no valid material and hence we need to define the material.

12. Select **Simulate** > Run the analysis

13. Select the left view as shown > Select **Undeformed** for **Displacement Display** > Select **Show Max value** in **Display**

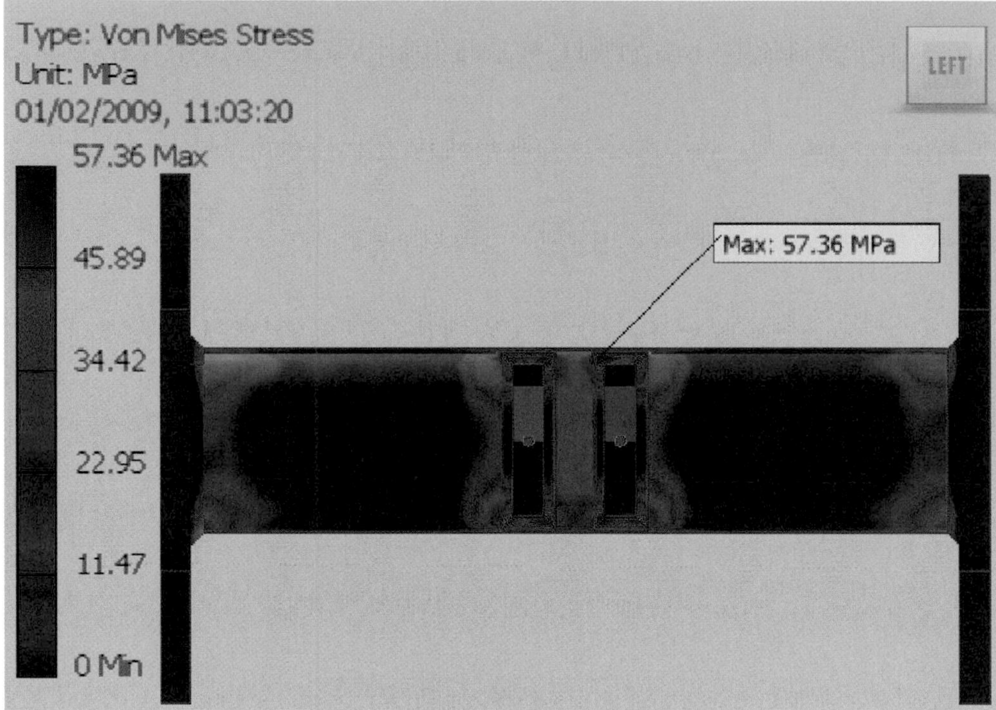

The Max Value is around welds. Now, we will perform the analysis as an assembly and then compare the results.

14. Right click Shrinkwrap-Analysis > Select **Copy Simulation**

15. Right click Shrinkwrap-Analysis:1 > Select **Edit Simulation Properties**

16. Specify 'Weldment-Analysis' for **Name** > Select the **Model State** tab > Select **Master** for **Level of Detail** > Click **OK**

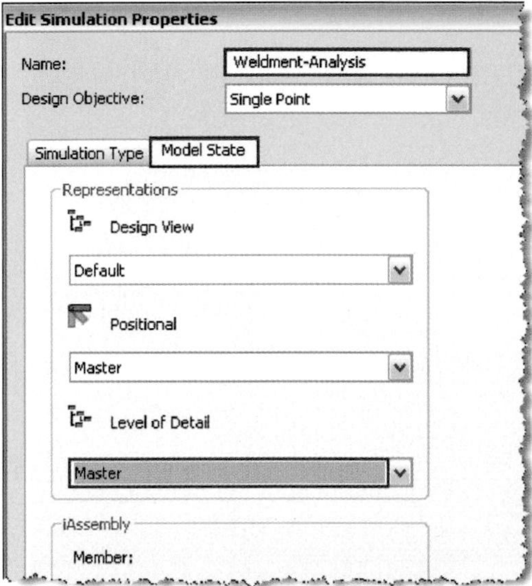

 The welds will be suppressed from the simulation.

17. Right click **Welds** > Reselect **Exclude From Simulation**

 The welds are now included in the simulation.

18. Right click **Contacts** > Select **Update Automatic Contacts**
A total of 38 contacts will be created within the weldment assembly.

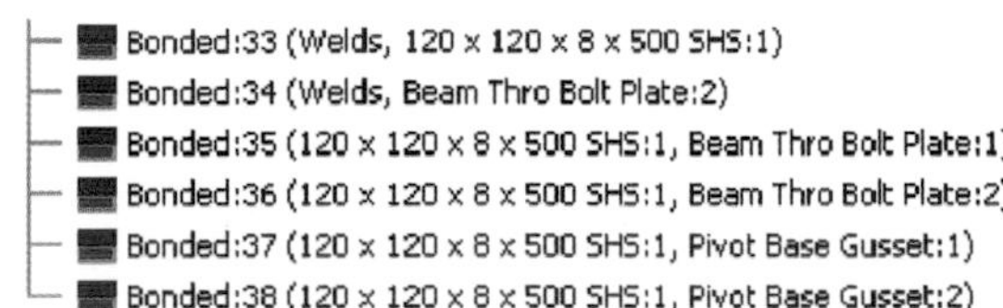

Using fillets will increase the number of contacts produced. For example, the channel below with round edges will result in eight extra contacts.

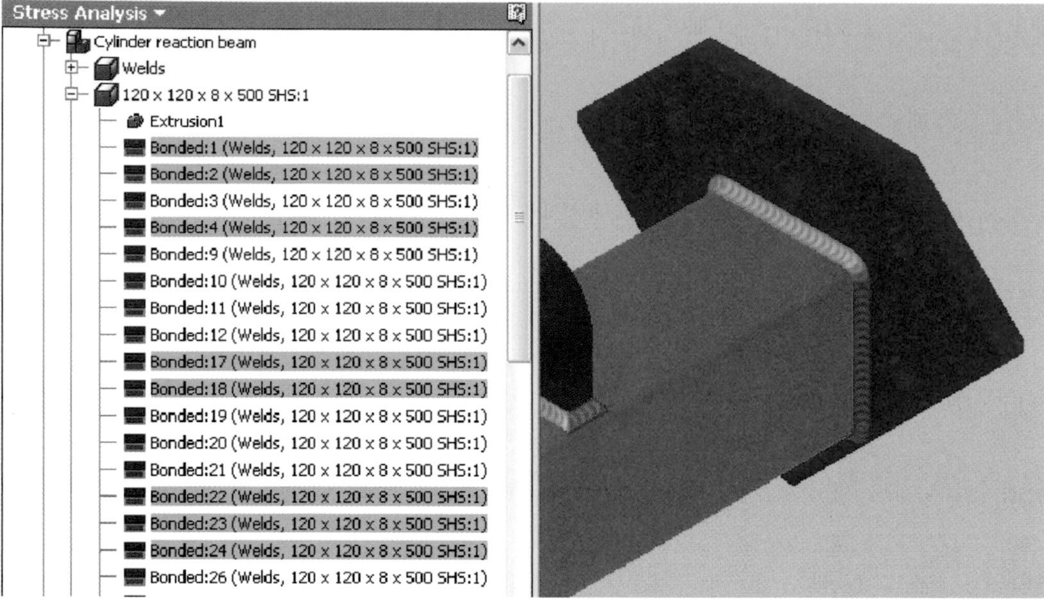

Suppress fillets to reduce the number of contacts produced.

We will continue this design problem without suppressing the fillets, as the assembly is small.

19. Right click the Bolted-Plates constraint > Select **Edit Fixed Constraints** > Reselect the faces to apply the fixed constraint > Click **OK**

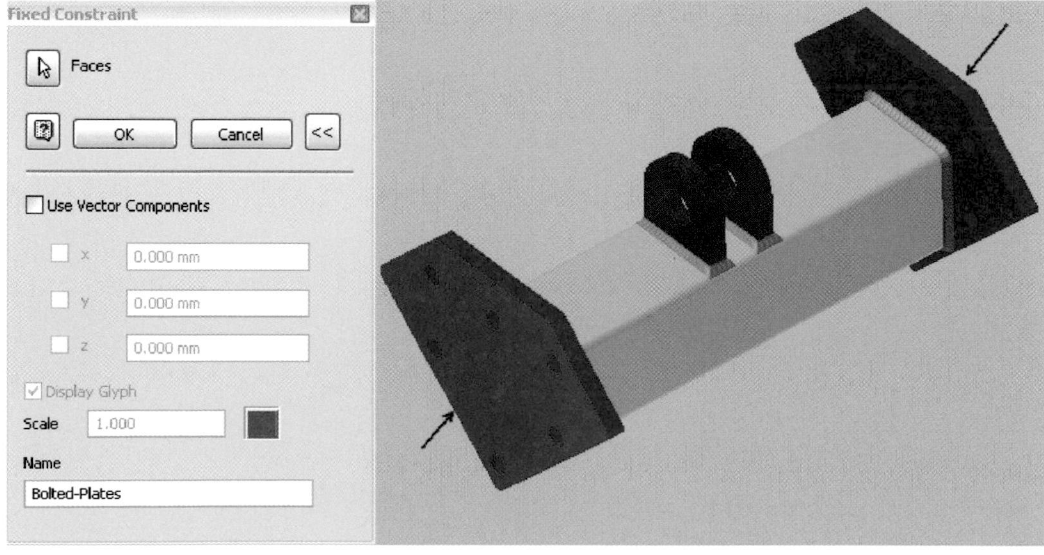

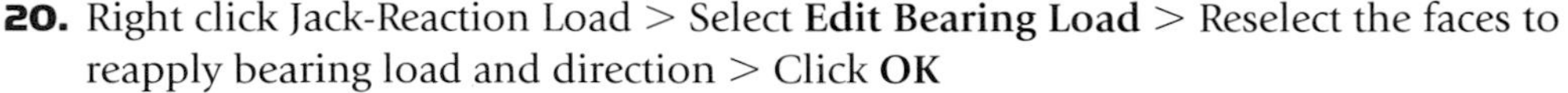

20. Right click Jack-Reaction Load > Select **Edit Bearing Load** > Reselect the faces to reapply bearing load and direction > Click **OK**

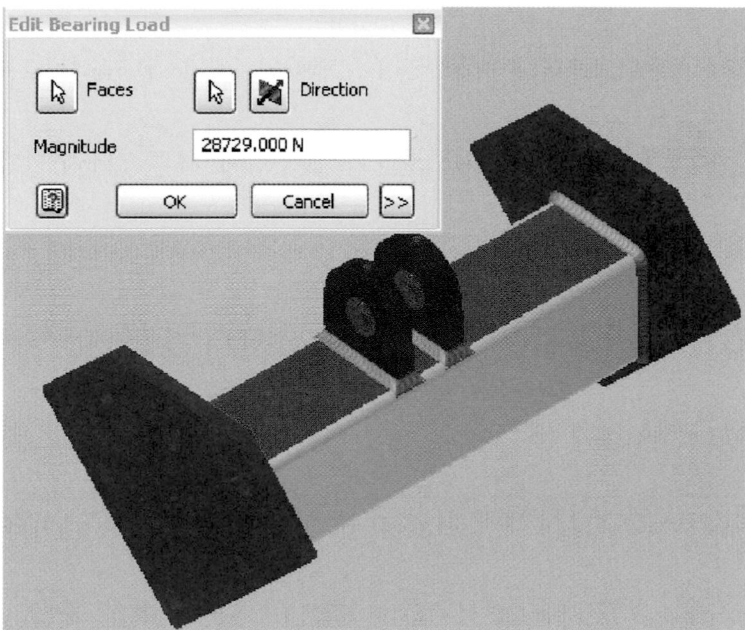

It is necessary to reapply the loads and constraints as the model has changed from a part to an assembly.

21. Select **Mesh View**

Run simulation and analyze

22. Select **Simulate** > Run the simulation

23. Select **Actual** for **Displacement Display** > Select **Show Max value** in **Display**

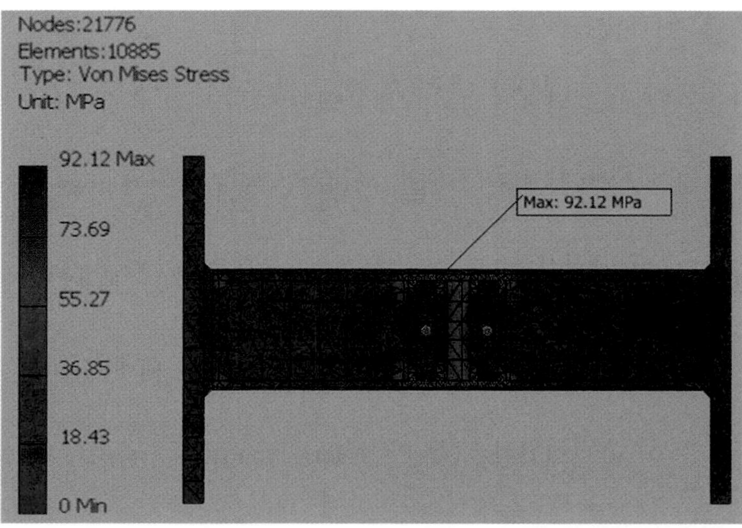

The maximum stress value is around the weld areas and the maximum stress is higher than that found in the single component analysis. Note that the stress is higher than the shrinkwrapped component by 34.76 MPa – an increase by 60%. This is normally due to sliver (or highly distorted) mesh elements between the welds and the components. So, when analyzing weldments, be aware of sliver elements distorting the results.

To gain some confidence in the accuracy of the results, the mesh needs to be refined either by using a smaller global average element size or refining the mesh around the weld areas only using local mesh control. The latter method will be used.

24. Deselect **Show Max value** > Deselect **Mesh View**

25. Select **Local Mesh Control** > Specify 5 mm > Select all eight weld faces > Select four vertical faces of the lugs > Click **OK**

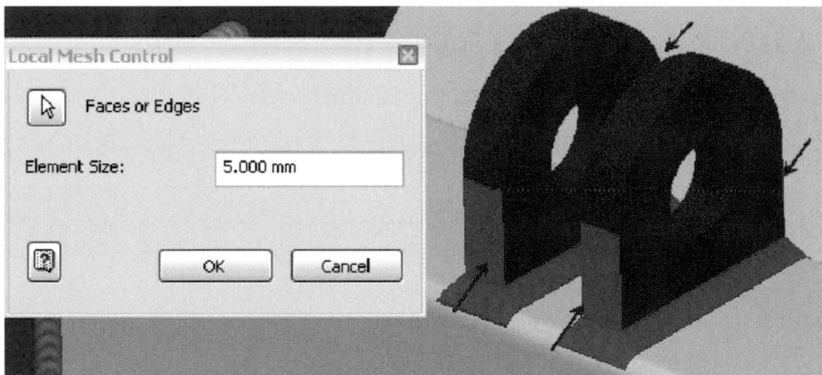

26. Right click **Mesh** > Select **Update Mesh**

27. Select **Mesh View**

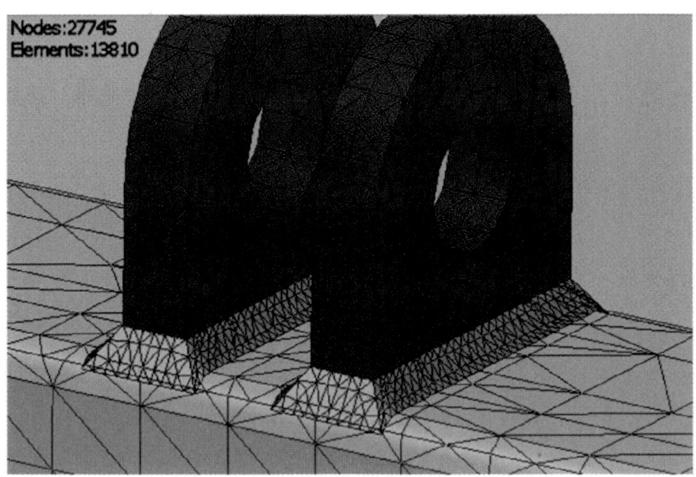

28. Select **Simulate** > Rerun the simulation

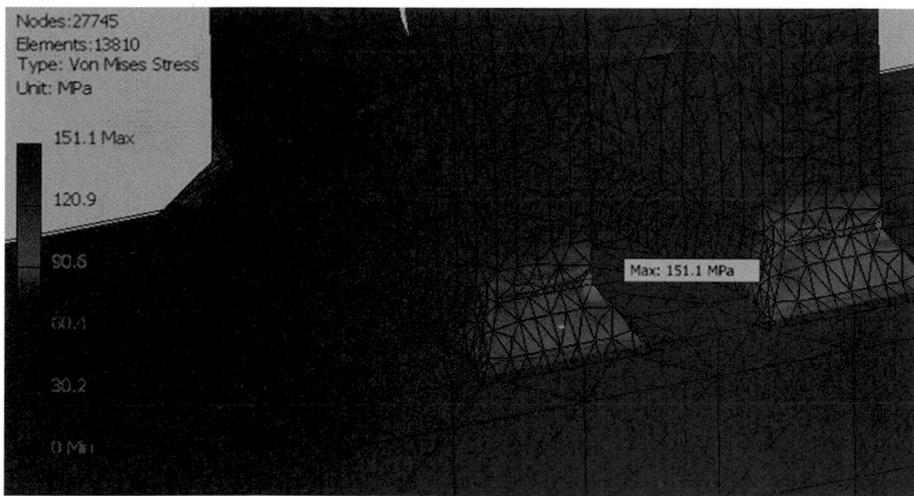

The maximum stress has now increased by 64% to 151.7 MPa. This significant increase is due to stress singularities caused by sharp discontinuity in the geometrical shape around the welds. Using automatic convergence will also not result in convergence of results.

At this stage we can try to further idealize the model by removing the sharp edges around the welds by introducing fillets.

29. Select **Finish Stress Analysis** > Double click **Master Level Of Detail** > Double click Pivot Base Gusset:1

30. Introduce Fillet1 by moving End of Part to the end > Select **Return**

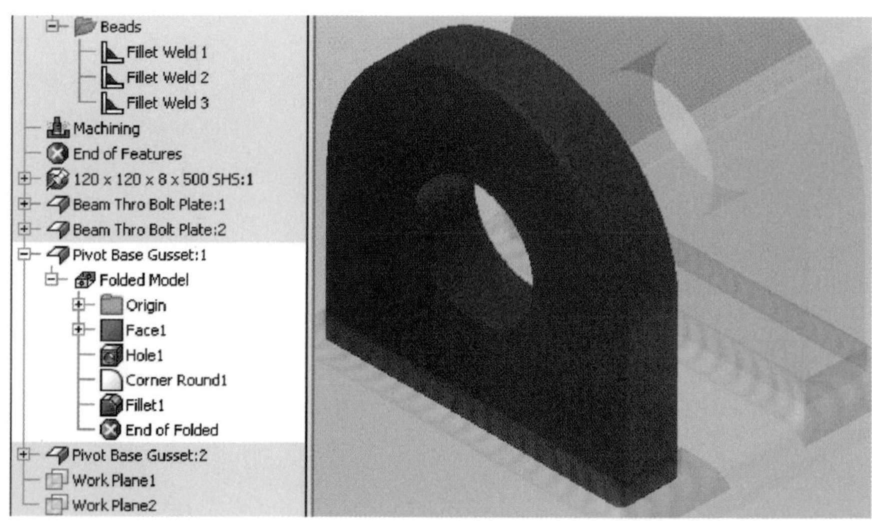

31. Double click Fillet Weld 3 > Select component 2 > Select both lugs > Click **OK**

32. Select **Environments** tab > **Stress Analysis**

33. Right click Local Mesh:1 > Select **Delete** to remove the mesh control

34. Right click **Contacts** > Select **Update Contacts**

35. Right click **Mesh** > Select **Update Mesh**

36. Select **Simulate** > Run the simulation

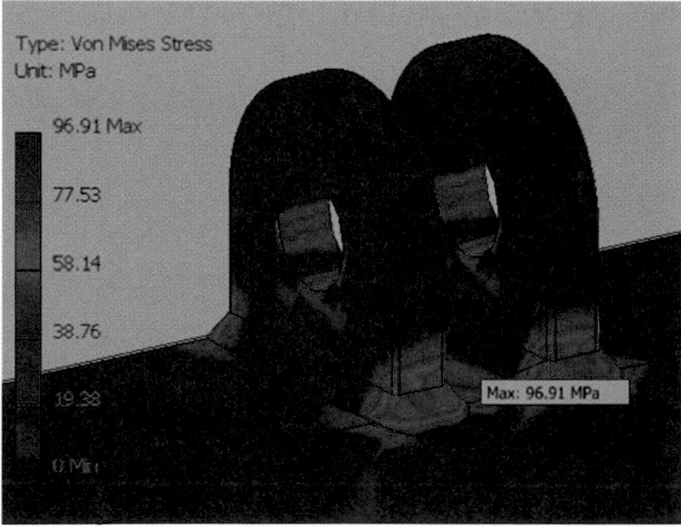

Although the stress has reduced slightly to 96.91 MPa, we still cannot gain convergence in the results due to stress singularities. In these scenarios, by manipulating the color display, we can visualize the areas of stress singularities.

The Max. stress may slightly differ.

37. Select **Color Bar** > Unselect **Maximum** > Specify 55MPa > Click **OK**

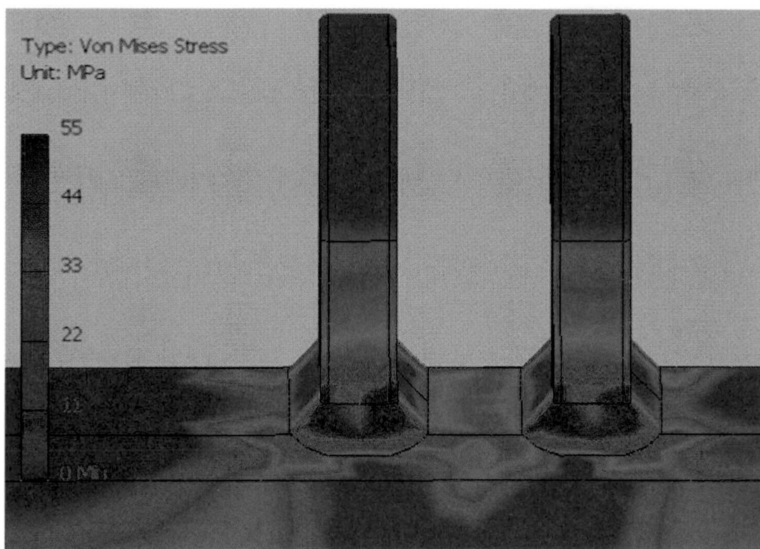

To obtain further confidence in the results, we can analyze the X, Y, and Z planar stresses.

38. Double click Stress XX

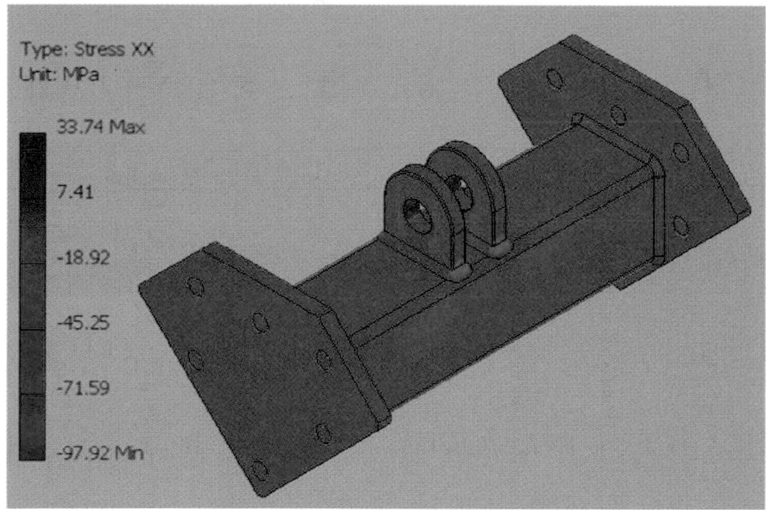

By displaying the Stress XX plot, we have noticed that the maximum stress value of around 98 is a compressive stress and the maximum tensile stress of around 33 MPa is located in the area where the load is applied. It is important to note here that the compressive strength of most materials is higher than the tensile strength, which means that the tensile stress becomes the area of concern rather than the compressive stress.

39. Double click Stress YY

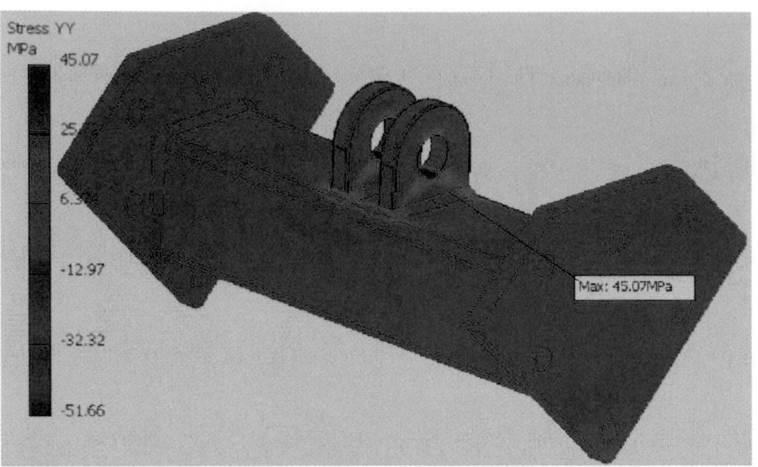

40. Right click 120 × 120 × 8×500 SHS:1 > Select **Isolate** to view the results of the channel only

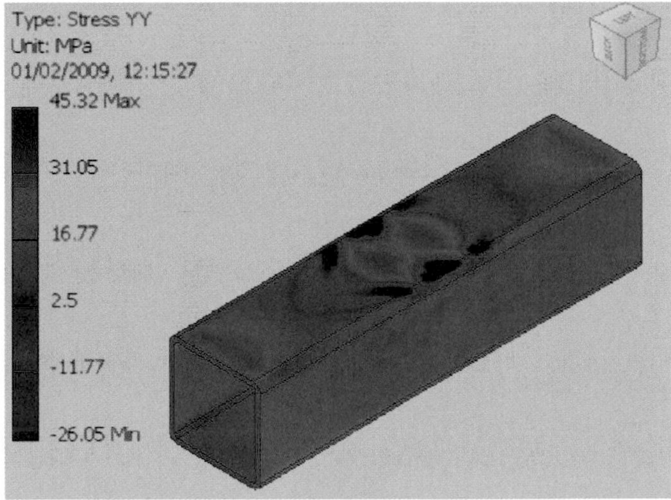

41. Select **Same Scale**

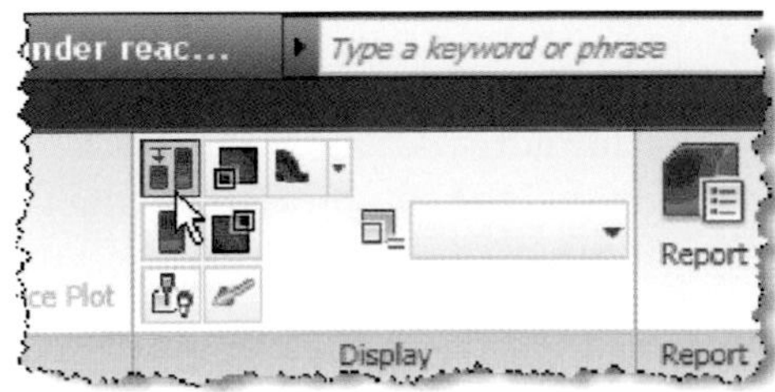

Same Scale does not alter the range of the color bar even when you are looking at individual components in isolation. So, it helps to compare results visually.

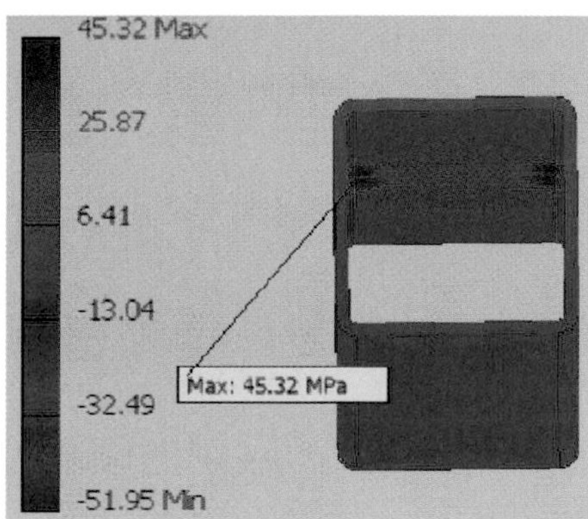

The above image illustrates that the maximum tensile stress of 45 MPa is located inside the channel, directly beneath the lugs.

42. Double click Stress ZZ

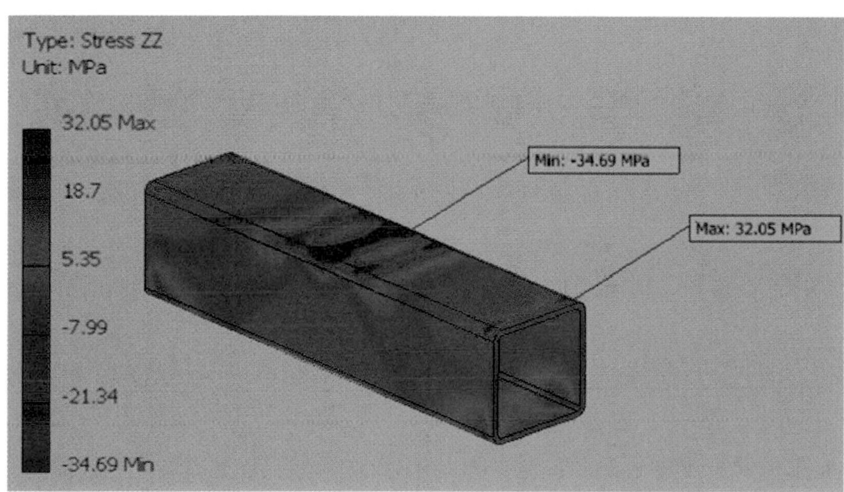

43. Reselect **Undo Isolate** > Double click **Safety Factor**

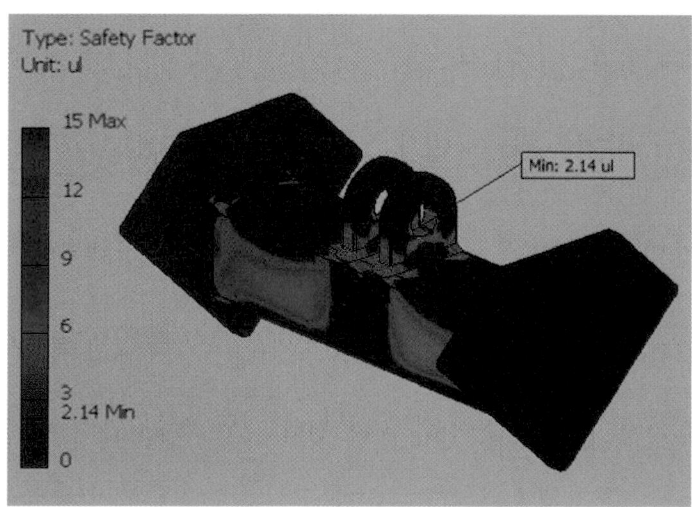

The minimum factor of safety is calculated based on the maximum stress of 96.91 MPa.

Hence,

$$\text{Factor of safety} = \frac{205}{96.91} = 2.1$$

Based on the single shrinkwrapped component

$$\text{Factor of safety} = \frac{205}{57.36} = 3.57$$

Take care in manipulating the results.

By adjusting the color bar, we can display the minimum factor of safety, as illustrated below.

44. Close the file

DP12 – Assembly Analysis with built-in welds

Structural Validation of Trailer Chassis

(Design Problem courtesy of Wright Resolutions Ltd)

KEY FEATURES INTRODUCED IN THIS DESIGN PROBLEM

	Key features
1	Automatic bonded contacts – welded fabricated structure analysis
2	Multiple loads
3	Planar X,Y,Z stress plots
4	Interpretation of results with stress singularities present

INTRODUCTION

Wright Resolutions Ltd is a design consultancy specializing in agricultural cultivation and crop establishment machinery. Current clients include a number of well-known UK and European agricultural machinery manufacturers.

As part of a project to design a new concept for a trailer chassis, it was necessary to determine the loadings on and strength/deflections of a conventionally manufactured trailer chassis, as can be seen in the following picture. Such chassis are generally manufactured from hollow section steel together with flame-cut steel plates and flat bar parts.

Initially, the trailer was modeled in Dynamic Simulation to determine the loads at all critical areas including the drawbar, axle spring mountings, tipping cylinder, and rear hinges to the body. Maximum load situations during tipping were then taken and applied via FEA to determine the parameters listed on the following page. One such situation, simplified, is used for this design problem.

Up and Running with Autodesk Inventor Simulation 2011. ISBN: 978-0-12-382102-7
Copyright © 2010 Elsevier Inc. All rights of reproduction, in any form reserved.

The requirements of this design problem are to determine:

- The maximum compressive and tensile stresses in the chassis.
- The maximum deflection of the chassis under load.
- The factor of safety.
- The key stress zones for potential reinforcement when designing an alternative chassis.

In addition to the above requirements, the design criteria to be used for this design problem are as follows.

- Material to be used is EN 50D/S355J2G3 steel.
- Factor of safety required is 1.5.

WORKFLOW OF DESIGN PROBLEM 12

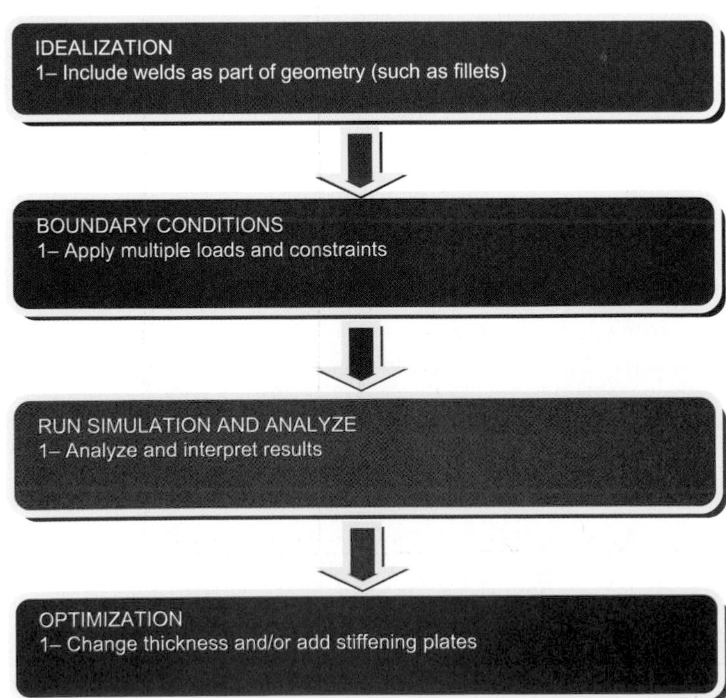

PART 1 – CHASSIS DESIGN WITH WELDS AND RHS CHANNEL RADII

Idealization

To simplify the analysis of the fabricated chassis, the welds have been modeled as fillets within the components, which will greatly help to reduce the number of contacts produced.

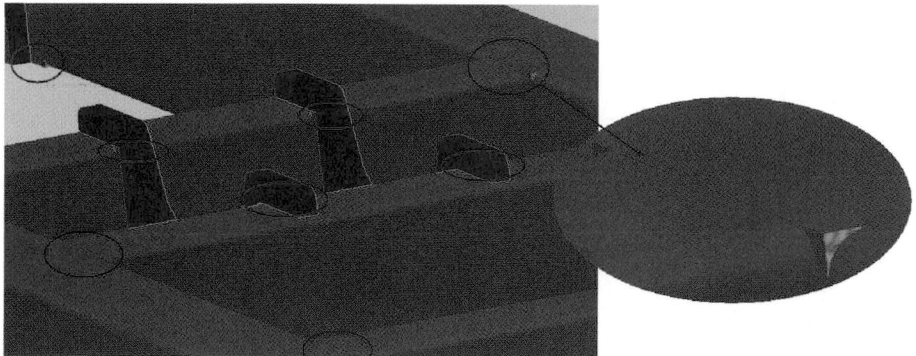

As the strength and characteristics of RHS are dependent on the corner radii, it is important to include these for more meaningful results. If welds are modeled separately to the RHS, the joints created in FEA are often complex and can be based on very thin slivers (highly distorted mesh elements) at the limits of the corner radii. Stress singularities produced can be very high. In practice, provided that the welds are correct and homogenous to the sections to which they are applied, such slivers are not present. Extruding the weld as part of the original section can represent nearer to a realistic situation. *It is important to simulate welds in a manner that represents reality, as closely as possible, for the results to be meaningful.* The use of filler materials to bridge over the joints, or partial V butt welds, for example, would alter the strength and integrity of the structure in practice and lead to different results from those simulated.

1. Open *Chassis.iam*

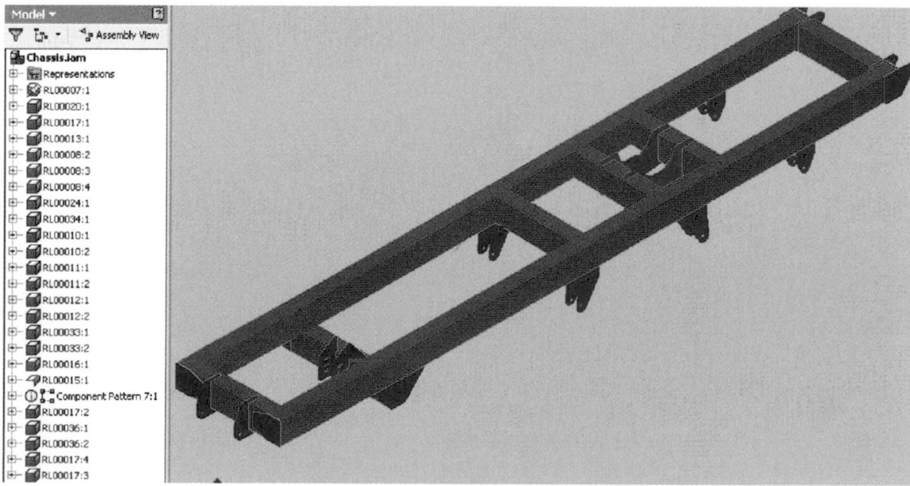

2. Select **Environments** tab > **Stress Analysis**
3. Select **Create Simulation** > Specify 'Chassis-Analysis' for **Simulation Name** > Click **OK**

Boundary conditions

4. Select **Automatic** contacts to detect adjacent faces between components and welds

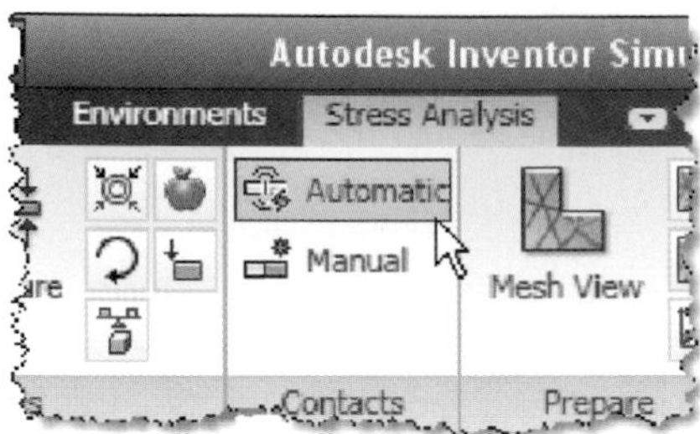

A total of 111 contacts will be created within the weldment assembly.

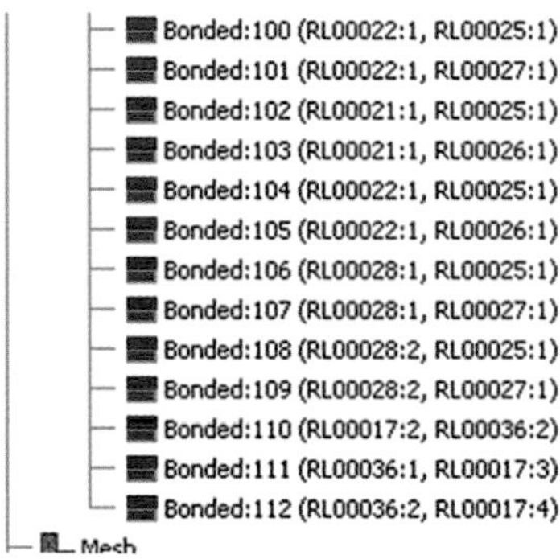

Many more contacts would have been created if the welds had been modeled separately as a weldment assembly.

The chassis is attached to the tractor via a drawbar arm, which is secured to the chassis via locking pins.

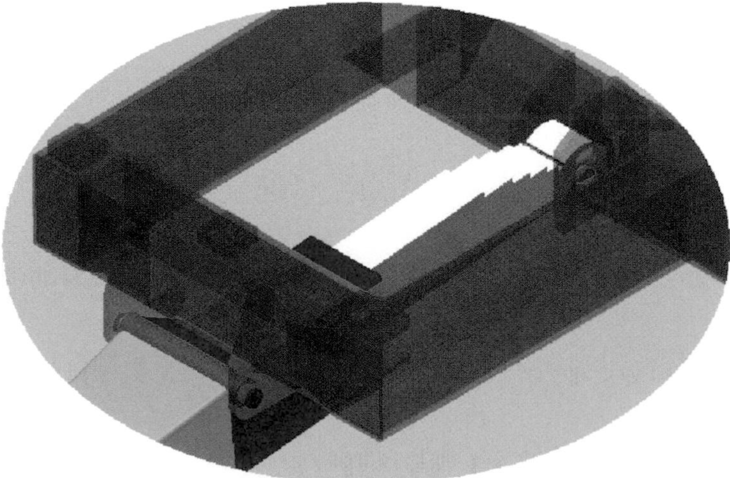

Therefore, we will apply pin constraints to secure the chassis.

The Constraints and Loads dialog boxes in the following pages do not show the apply button, which is being used to apply boundary conditions in the following steps.

5. Select **Pin Constraint** > Select the faces of both holes as shown > Click **Apply**

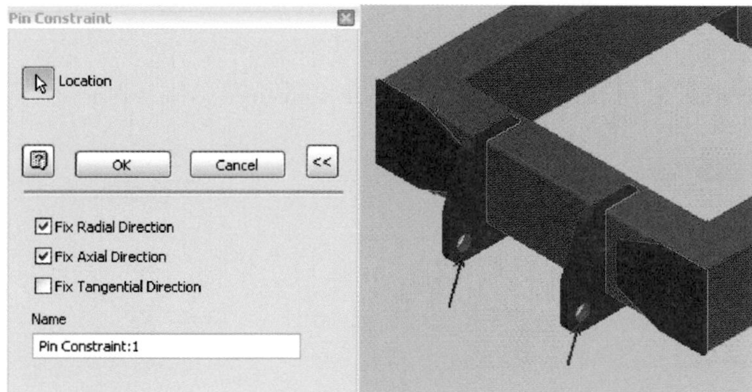

6. Select **Pin Constraints** again > Select the back faces of the two middle slots as shown > Click **OK**

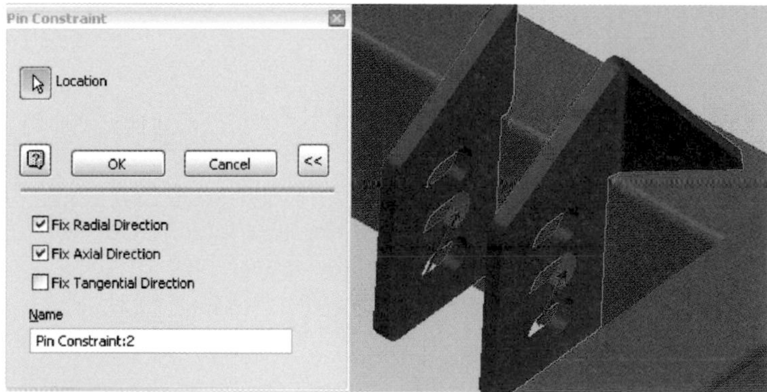

With the aid of Dynamic Simulation, the trailer is used to simulate tipping to determine the maximum reaction forces on the chassis.

As the loads cannot be exported automatically, because this function only supports single parts, we will specify these bearing loads manually.

First, we will apply the forces generated by the load and weight of the chassis.

7. Select **Gravity** > Select **Vector Components** > Specify –9810 in the Z-direction

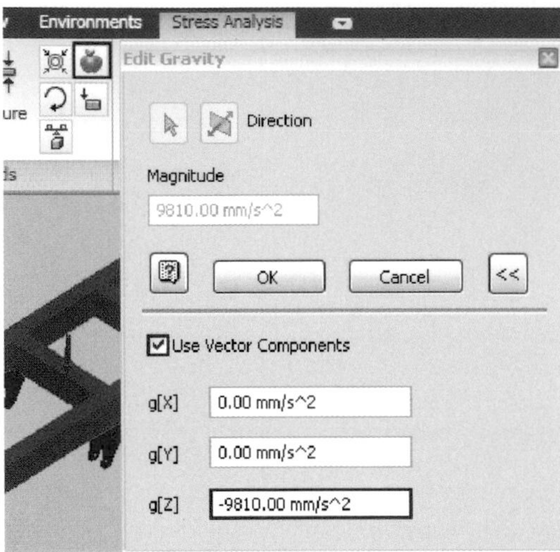

As there are many bearing loads to be applied, we will select multiple bearing load faces, lying on the same axis, to help speed up the creation of loads. As the bearing loads are split equally by the number of faces selected, we will simply multiply the actual loads by the number of faces selected. Alternatively, you can create a load on each individual face.

8. Select **Edit Bearing Load** > Select two internal circular faces to specify **Location** > Specify the top face of the plate as the **Direction** of the force > Specify 1.4e5 * 2 for **Magnitude**

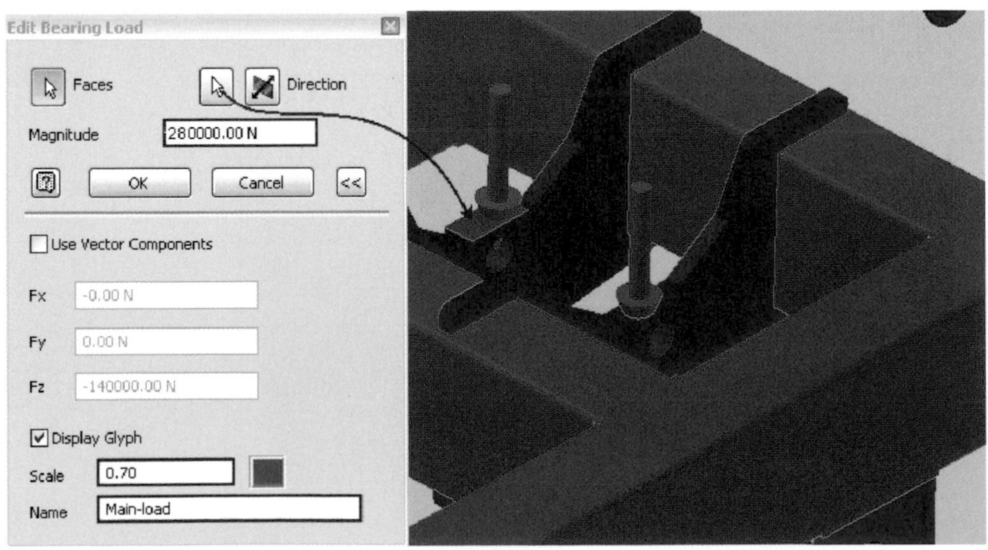

9. Specify 0.7 to reduce the size of the force display > Specify 'Jack Main-load' for **Name** > Click **Apply**

10. Select the two internal circular faces of the bushings > Specify the top face of the plate to specify the **Direction** of the force, as shown > Specify 2.5e4 * 2 for **Magnitude**

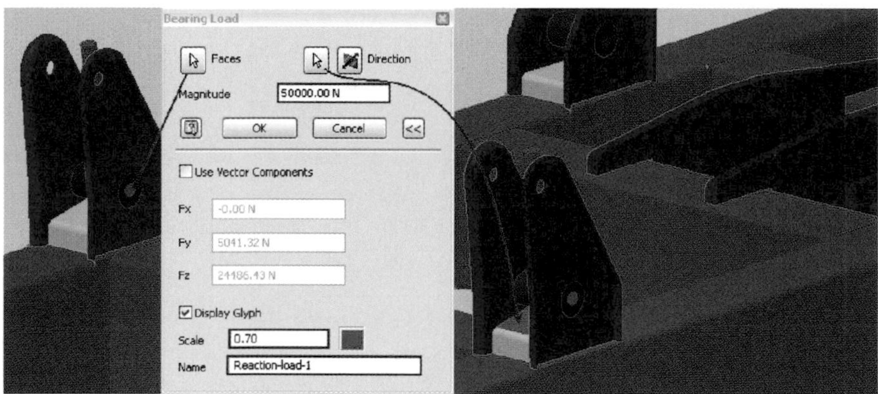

11. Specify 0.7 to reduce the size of the force display > Specify 'Reaction-load-1' for **Name** > Click **Apply**

12. Select the two internal circular faces of the bushings > Specify the top face of the plate to specify the **Direction** of force, as shown > Specify 4.1e4 * 2 for **Magnitude**

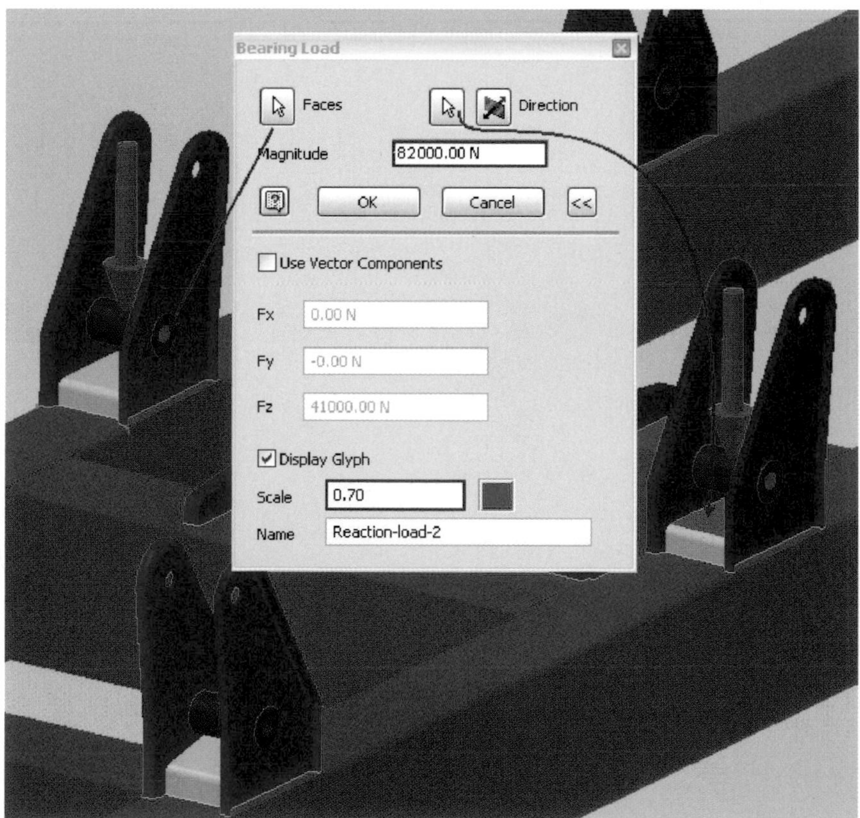

13. Specify 0.7 to reduce the size of the force display > Specify 'Reaction-load-2' for **Name** > Click **Apply**

14. Select the two internal circular faces of the bushings > Specify the top face of the plate to specify the **Direction** of force, as shown > Specify 2.9e4 * 2 for **Magnitude**

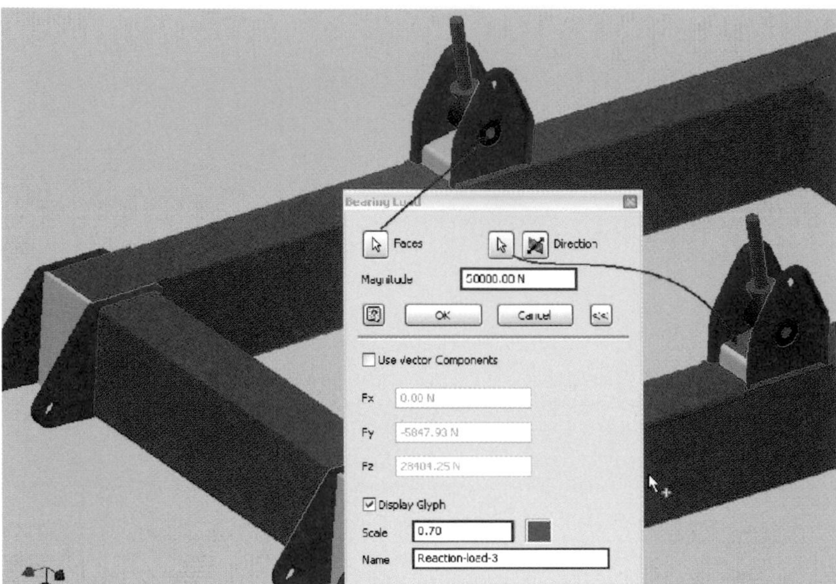

15. Specify 0.7 to reduce the size of the force display > Specify 'Reaction-load-3' for **Name** > Click **Apply**

16. Select the four internal circular faces of the bushings > Specify the top face of the chassis to specify the **Direction** of force, as shown > Specify 2.1e4 * 4 for **Magnitude**

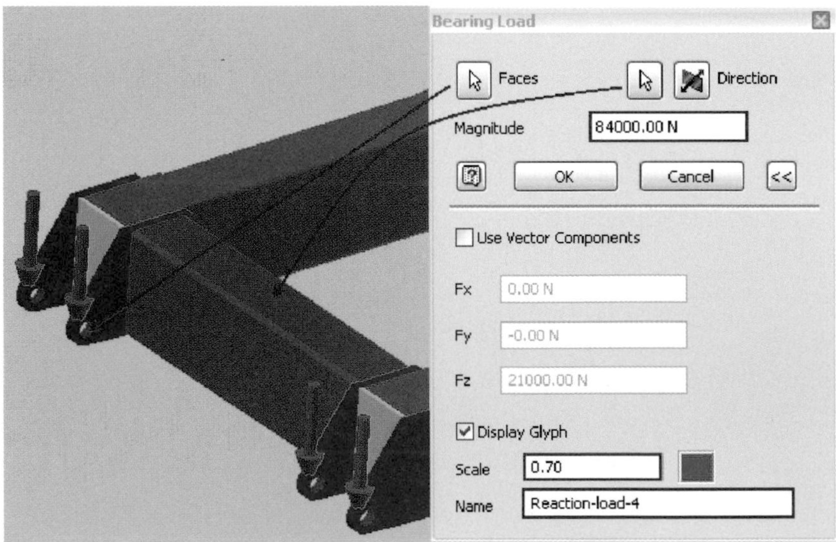

17. Specify 0.7 to reduce the size of the force display > Specify the 'Reaction-load-4' for **Name** > Click **OK**

18. Select **Mesh Settings** > Specify **Create Curved Mesh Elements** > Deselect **Use part based measure for Assembly** mesh > Click **OK**
19. Select **Mesh View**

23 077 elements are generated with the default mesh size. Leaving **Use part based measure** for **Assembly mesh** selected would have resulted in around 75 000+ elements.

The number of elements created may differ.

Run simulation and analyze

20. Select **Simulate** > Run the analysis
21. Deselect **Mesh View** > Select **Undeformed** for **Displacement Display** > Select **Show Max value** in **Display** > Deselect **Boundary Conditions**

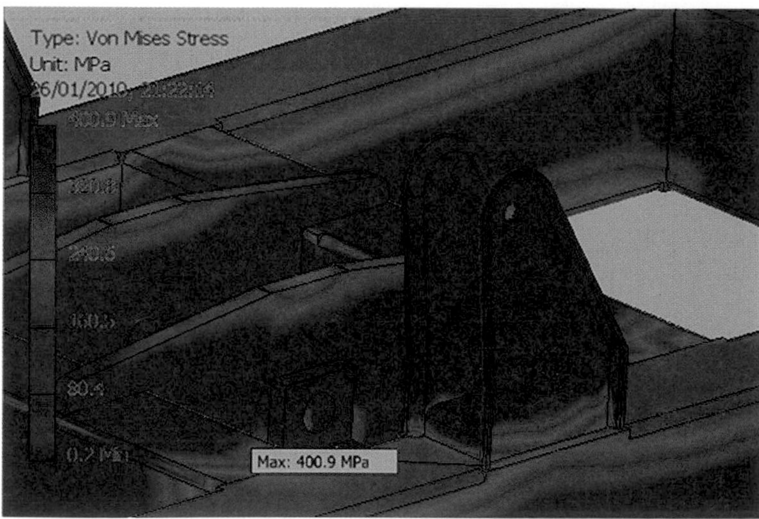

The maximum stress value may differ.

The maximum stress value is located around the weld areas, and mounts, and is largely due to stress singularities as a result of discontinuity in the geometrical shape. Refining the mesh around these areas will not necessarily reduce stresses and in most cases will further increase the stresses.

As the result stands, the safety factor relating to maximum stress indicates failure at a value around 0.86.

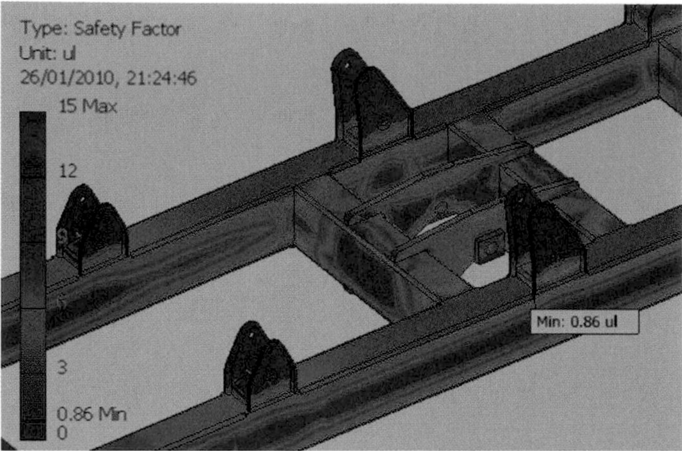

If the stress singularities are a very low percentage of the total joint area, local yielding can occur initially until the load is transmitted at lower stress by the full joint. As a rule of thumb, if such singularities result from static loading and are concentrated in small localized areas, they can be ignored for the purposes of calculating the overall safety factor, for example. *Experience of the effect of such high stress points is needed to ensure that the correct interpretation is made of FEA results.* For example, dynamically loaded situations can have stress reversals; where these occur at welded joints, there is a high chance of fatigue failure occurring. In these situations, we can make use of the color bar to better understand the results, as suggested in the following steps:

22. Reselect **Von Mises Stress**

23. Select **Color Bar** > Unselect **Maximum** > Specify 355 MPa > Click **OK**

355 MPa is the yield limit of the material used for the chassis.

As the concentration of the red color display is extremely low, and in this case stress reversals are unlikely, we can assume the safety factor of the design is above 1 for this case.

To determine what the safety factor actually is, however, and to illustrate zones of high stress requiring design changes, we can further manipulate the color bar. In the first instance, we will change the maximum value of the color bar again to 260 MPa and then to 245 MPa.

Use **Contour Shading** rather than **Smooth Shading** to help isolate stress singularity stresses.

Change the minimum value of the color bar in addition to the maximum value to help identify the areas of high stress.

Change the number of legends to help identify the maximum value to be used for calculating the safety factor, despite having stress singularities present in the model.

24. Select **Contour Shading**

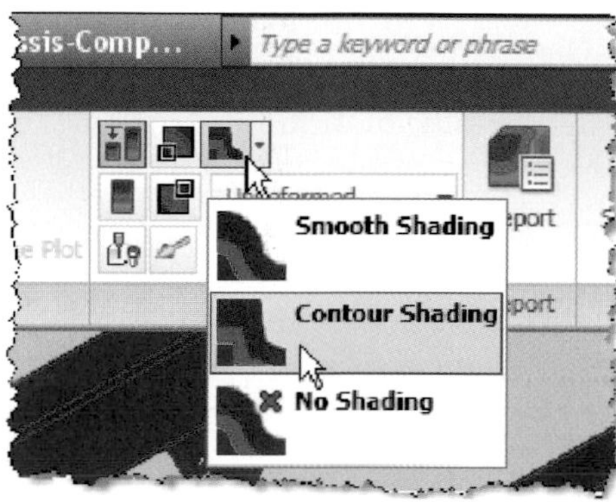

25. Reselect **Color Bar** > Unselect **Maximum** > Specify 260 > Unselect **Minimum** > Specify 220 > Click **OK**

By changing the color bar range from 260 to 220 MPa, we can see localized areas of red color display, indicating where the maximum stress occurs.

26. Reselect **Color Bar** > Specify 245 for **Maximum Value** > Click **OK** > Look at Back View

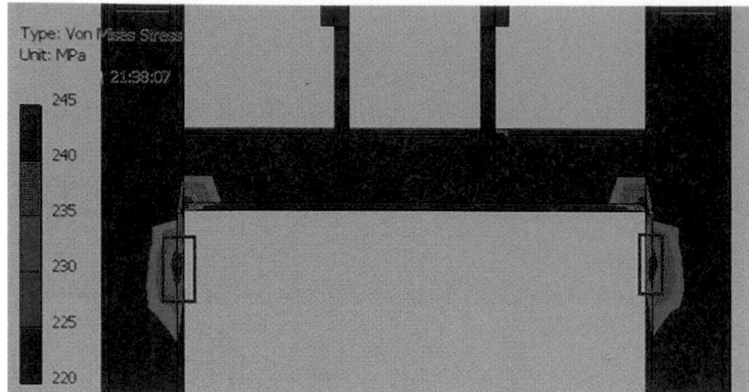

As soon as we change the color bar maximum value to 245, we can see that the stresses occurring above 240 MPa are now appearing away from the welds and a picture of areas requiring redesign to optimize the chassis is becoming clear.

By altering the color bar maximum value, we can pinpoint the value of maximum stress (244) in the area of interest, as shown below.

> **IMPORTANT**—You may need to alter the color bar maximum value so that the red display just starts to appear away from the welds. (The legend value underneath max is the value we are manipulating, by altering max value, and the one to be used to calculate the safety factor).

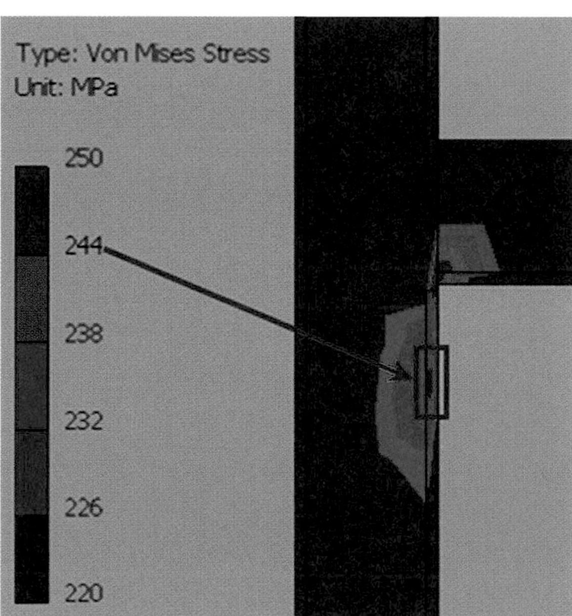

Now, by using the value of 244 MPa, we can manipulate the color bar maximum value and number of legends to achieve the Von Mises plot below. Use your own value to calculate the safety factor, as mentioned above, as it may slightly differ.

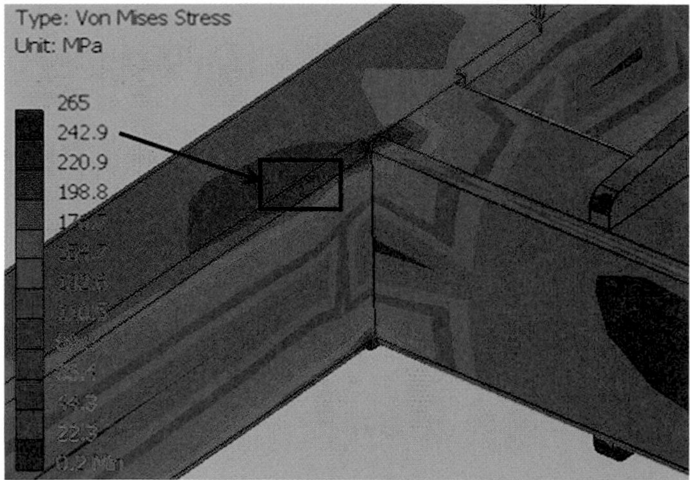

The above Von Mises plot confirms that the stress value around 243 starts to appear away from the weld areas due to bending.

So, for the purposes of calculating the safety factor, we will use the value of 244 MPa (or use your calculated value):

$$\text{Factor of safety} = \frac{355}{244} = 1.45$$

As the value is close to the design limit of 1.5, we will look at the planar stresses primarily occurring on the long channels, due to bending.

27. Double click Stress YY

28. Select **Color Bar** > Specify 292 for **Maximum Value** > Specify –292 for **Minimum Value** > Increase the number of colors to 12 > Click **OK**

29. Select **Back View**

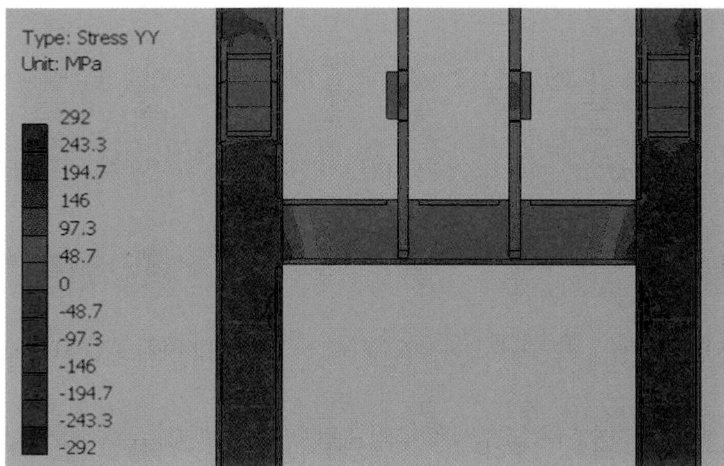

This displays tensile stresses along the long RHS member of the chassis. As indicated earlier, the highest stresses are located along the length of the chassis.

The stress display above 243 MPa appears to be significant relative to the width of the cross member, and also occurs near a weld.

30. Reselect **Color Bar** > Specify 305 for **Maximum Value** > Specify –305 for **Minimum Value** > Click **OK**

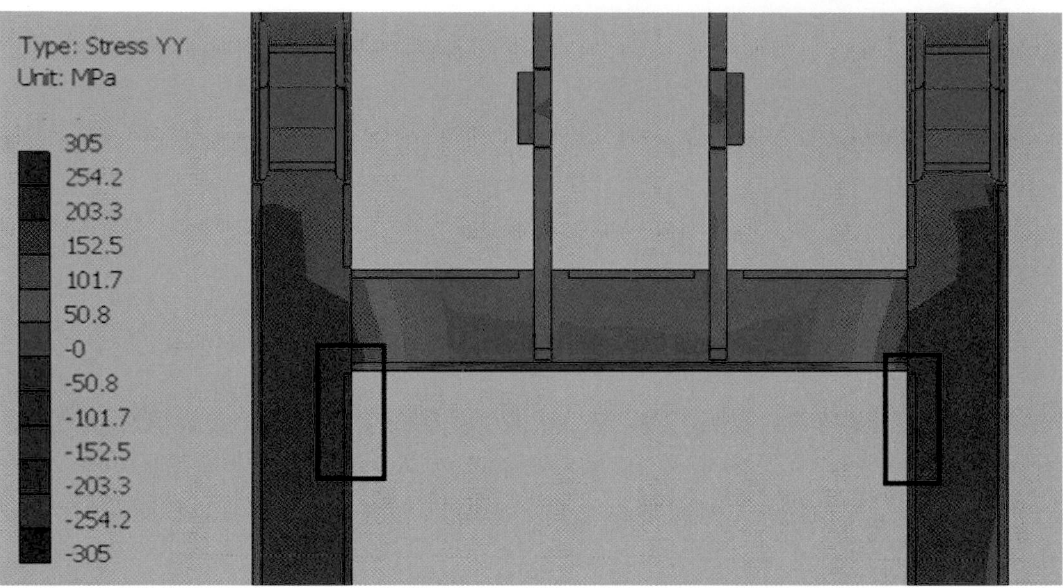

31. Select **Front** view to display compressive stress

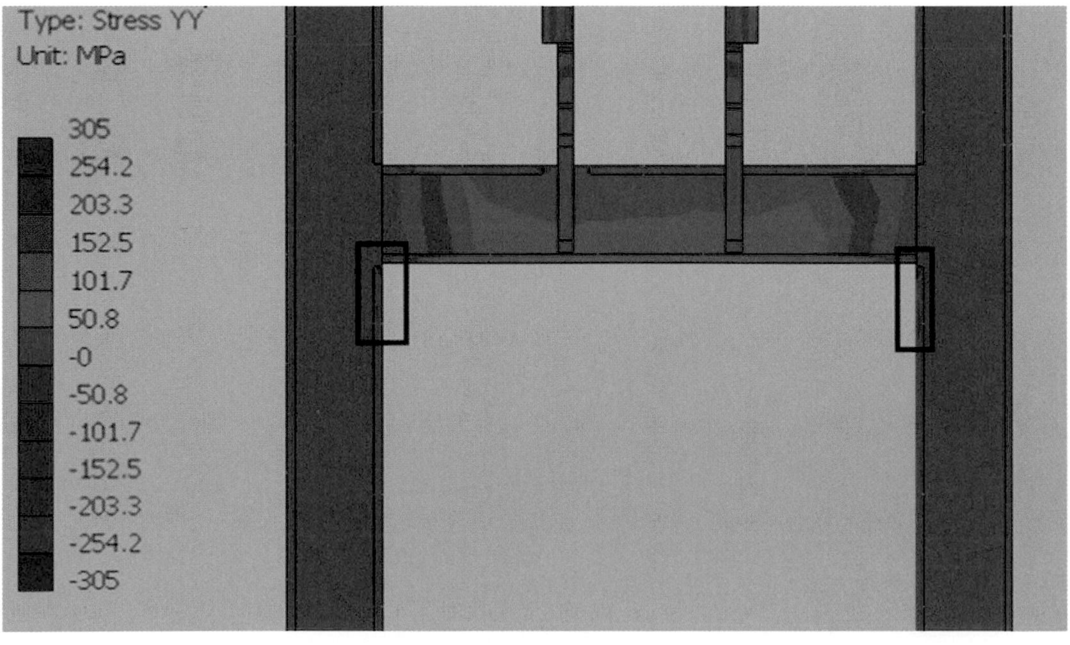

The compressive and tensile stress display shows values above 254.2 MPa, which are relatively small in comparison with the width of the cross member. As loadings in this case are gradually applied and relatively infrequent, we will use this value to calculate the safety factor.

$$\text{Factor of safety} = \frac{355}{254.2} = 1.39$$

This value is below the design limit.

32. Double click **Displacement**

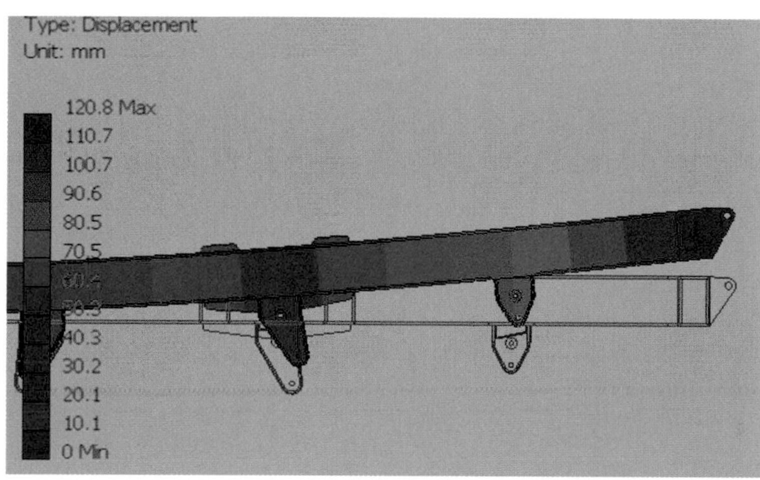

The maximum displacement of the chassis is 120.8 mm; this value is relatively high compared to the overall length of the chassis (approximately 7000 mm). This confirms the stress and factor of safety values determined above.

Basically, this analysis suggests that this design needs to be further stiffened to meet the design goal. In the next example, we will further idealize the chassis by removing the radii from the RHS channel. This will simplify the model further but at the same time will further strengthen the chassis and hence will not represent reality; however, it will give us an indication of whether the chassis is more rigid.

PART 2 – CHASSIS DESIGN WITHOUT WELDS AND RHS MEMBER RADII

At this stage, we will further idealize the model by removing the radii and inbuilt welds. This will help to remove stress singularities at the inbuilt welds. However, it is important to note that this process will stiffen up the chassis, so care must be taken in interpreting results. The boundary conditions will remain unaltered and therefore there will be no need to apply them again.

Idealization

In Autodesk Inventor 2011, part features within an assembly can be suppressed within the Stress Analysis environment. However in the following steps the features will be suppressed from within the Assembly environment.

34. Finish the Stress Analysis

35. Double click the RL00007:1 component > Move End of Part below the Shell1 feature > Suppress Fillet1 > Select **Return**

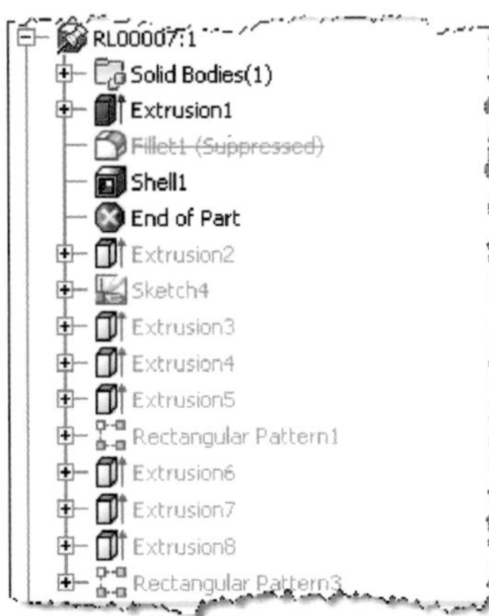

36. Double click the RL00013:1 component > Move End of Part below Shell1 feature > Suppress Fillet1 > Select **Return**

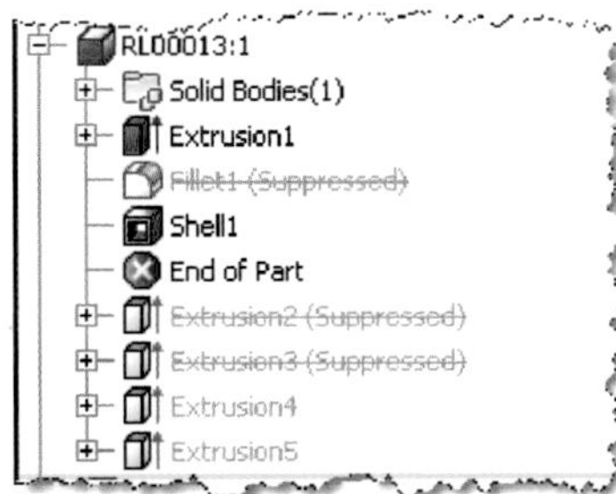

37. Double click the RL00008:2 component > Move End of Part below Shell1 feature > Suppress Fillet1 > Select **Return** > Accept the warning

38. Double click the RL00024:1 component > Suppress Fillet1 > Select **Return**

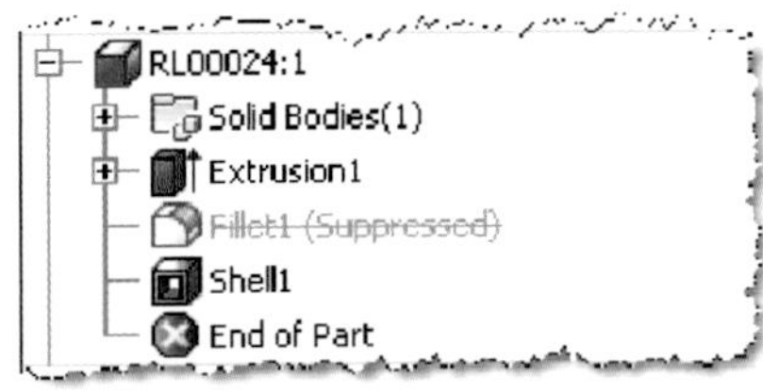

39. Double click the RL00034:1 component > Move End of Part below Shell1 feature > Suppress Fillet1 > Select **Return**

40. Select **Environments** tab > **Stress Analysis**

41. Right click **Contacts** > Select **Update Mesh Automatic Contacts**

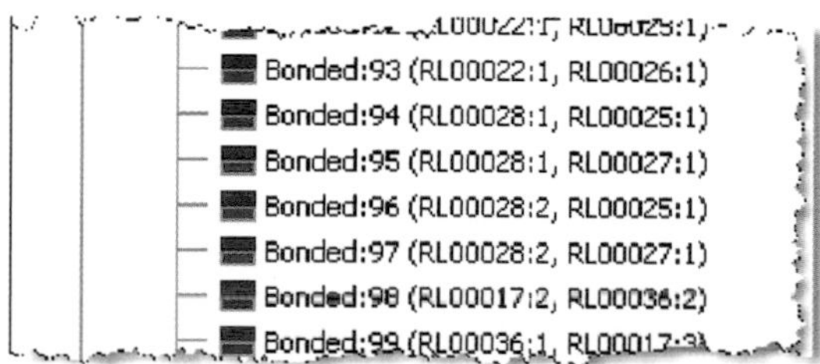

42. Right click **Mesh** > Select **Update Mesh** > Select **Mesh View**

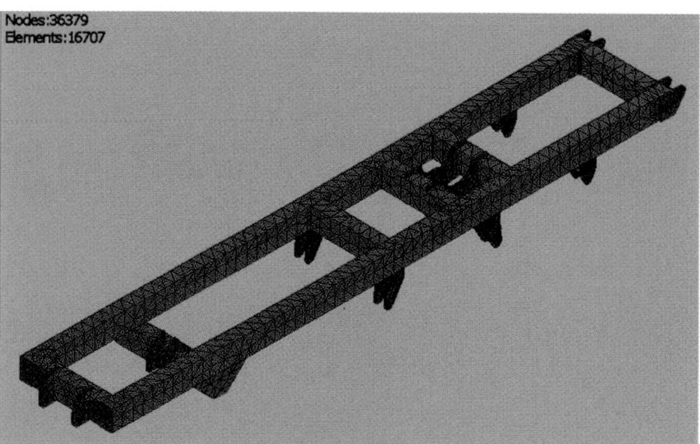

16 707 elements are generated with the default mesh size, a 40% reduction in mesh size with fillets and built-in in welds.

The number of mesh elements created may slightly differ.

Rerun analysis and analyze

43. Select **Simulate** > Run the analysis

44. Select **Undeformed** for **Displacement Display** > Select **Show Max value** in **Display**

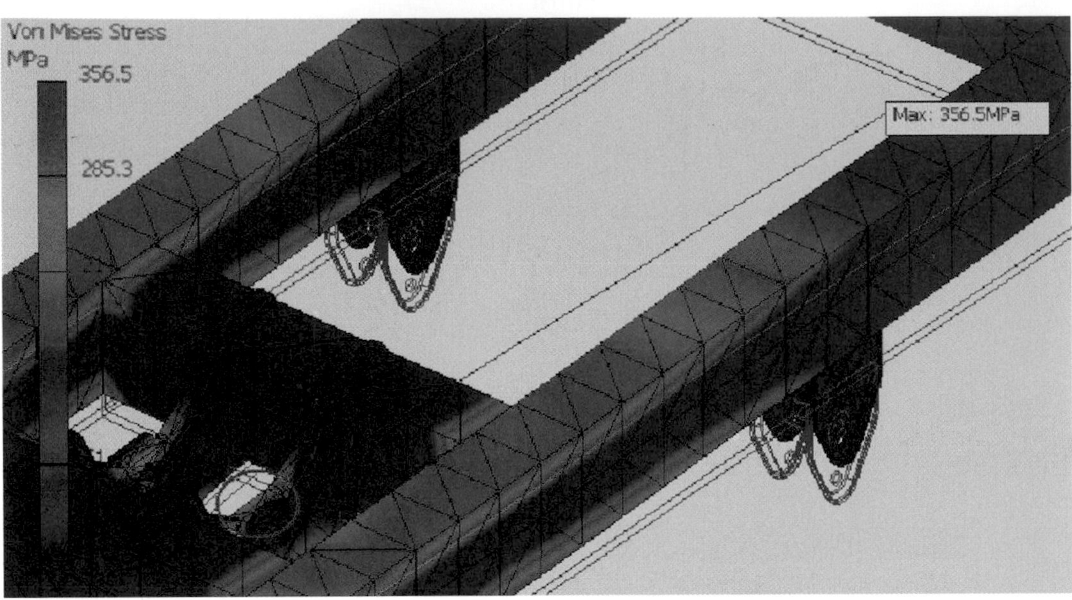

Maximum stress now occurs in the middle of the chassis, around the welds, and mounts, as before.

45. Double click Stress YY > Select **Color Bar** > Reselect **Maximum value** > Reselect **Minimum value**

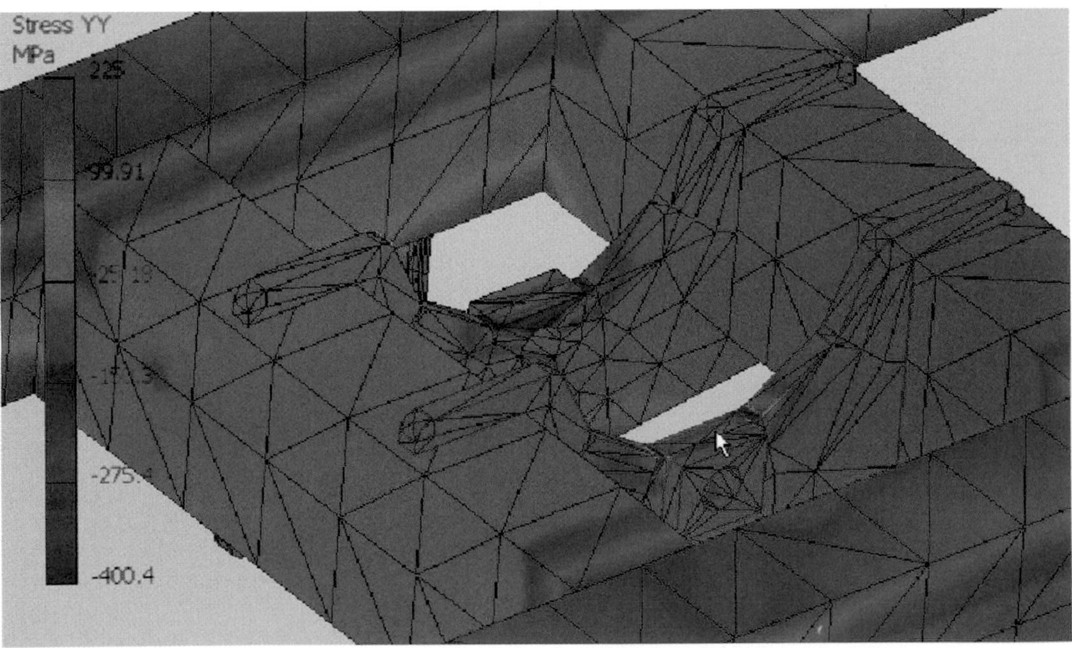

Maximum stress in the middle is compressive. As most components and welds in cases such as this usually fail due to tensile stress (and/or reversals) and not compressive stress, we will look at the tensile stress on the other side.

46. Rotate the component to see the back of the chassis (you may need to alter the maximum and minimum values using the Color Bar to obtain a better understanding of the results).

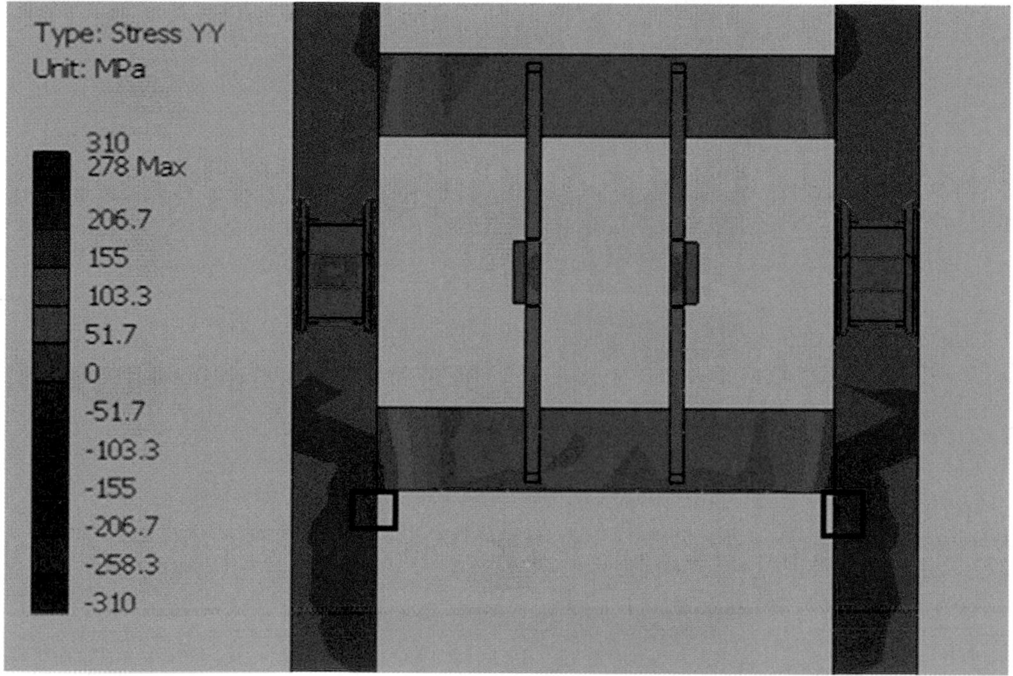

Apart from the stress singularities around the mounts, the maximum stress occurs on the main member, away from the cross member, unlike before.

There are no stress singularities between the cross member and the main member, as before.

> **IMPORTANT**—To calculate factor of safety, use the color legend value below maximum value (or minimum value for compressive stresses)

$$\text{Tensile Factor of Safety} = \frac{355}{278} = 1.28$$

47. Double click **Displacement;** the displacement has reduced to 117 mm

The value has reduced slightly due to the increased stiffness of the model caused by eliminating the radii of the RHS members, as expected.

Now, we will optimize the design in the next section to meet the design goals.

Optimization

Based on the above analyzes of the chassis, the results indicate that the design does not meet the design criteria. To meet the design criteria, we have two options.

- Increase the thickness of the RHS members; however, this approach is not cost-effective when compared to Option 2 or if the component has already been manufactured.

- Place a plate (suggested thickness ≥ 10mm) between the mounts and the RHS members for a distance that can be determined from the FEA results above (see picture below).

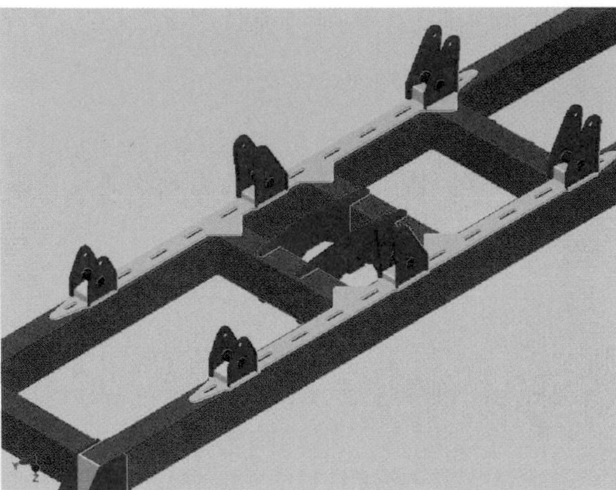

If the chassis design is not built, a combination of Options 1 and 2 can be used to manufacture a more rigid chassis.

The following illustrates another possible design for the new generation of trailer chassis:

In practice, additional loading scenarios are analyzed; for example, when the chassis has a torsional load applied down its length. In this case, stress reversals are often present that need to be taken into account when determining the final design of reinforcements to be made.

48. Close the file

DP13 – Assembly Optimization

Structural Optimization of a Lifting Mechanism

(Design Problem courtesy of Unipart Rail Ltd)

KEY FEATURES INTRODUCED IN THIS DESIGN PROBLEM

	Key features
1	Sliding/no separation contact
2	Manual contacts
3	Planar X, Y, Z stress plots
4	Parametric optimization

INTRODUCTION

Unipart Rail is part of the Unipart Group, one of Europe's leading independent logistic companies, employing more than 9000 people worldwide with annual turnover of more than £1.1 billion.

Unipart Rail combines extensive engineering, logistic, and manufacturing experience with industry-leading supply chain and lean expertise.

Autodesk Inventor is used within the design and development division of Unipart Rail to produce, validate, and document complete digital prototypes. In addition, the simulation suite is used extensively, making it possible to optimize, validate, and predict how designs will work under real-world conditions, before the product or part is even built.

Up and Running with Autodesk Inventor Simulation 2011. ISBN: 978-0-12-382102-7
Copyright © 2010 Elsevier Inc. All rights of reproduction, in any form reserved.

Within Unipart Rail's bogie overhaul facility at Doncaster, UK, there is a requirement for the design of innovative jigs and fixtures to facilitate lean processes. One such requirement identified is the need for a lifting device to handle the secondary suspension units fitted to the rail vehicle bogies as illustrated below. There are two secondary suspension units per bogie, each weighing 78 kg.

Bogie Secondary Suspension System

In this design problem, we need to determine the structural integrity of the lifting mechanism in addition to the following:

- The maximum working stress in the key components.
- The maximum deflection.
- How to reduce the overall weight.

New Lifting Mechanism

- Material to be used is mild steel.
- Factor of safety to be at least 4.

WORKFLOW OF DESIGN PROBLEM 13

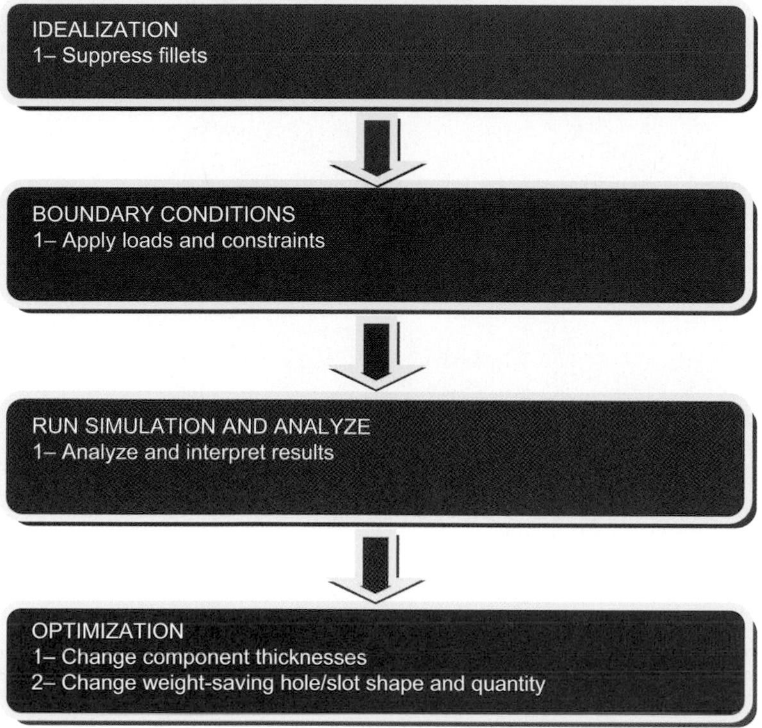

Idealization

In this stage of the FEA workflow, the components and assemblies need to be simplified in terms of having nonstructural features, including holes and fillets, that need to be suppressed. In this design problem, most of the noncritical fillet features have already been suppressed.

Further, nonstructural components will not be suppressed but instead will be excluded from the simulation within the Stress Analysis environment. This will additionally help to simplify the analysis with a reduced number of components and contacts.

1. Open *Airbag lifting jig.iam*

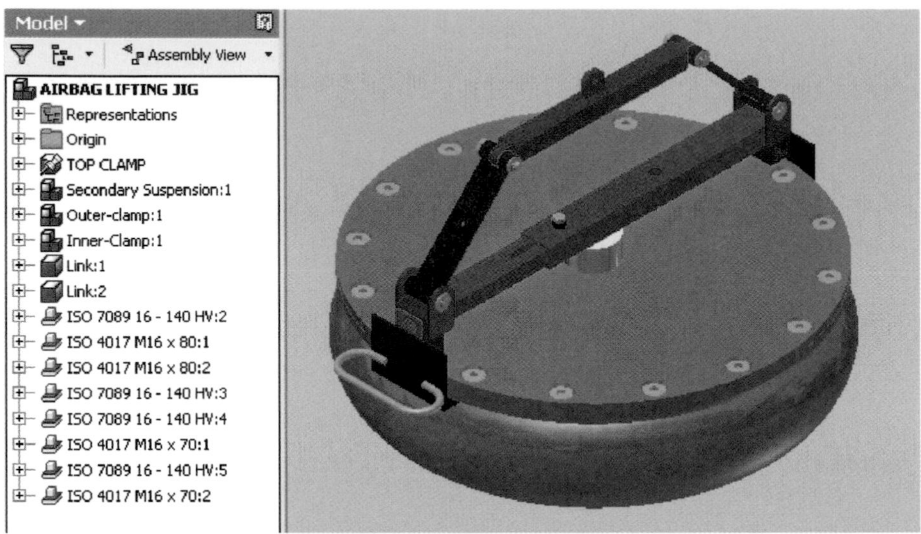

2. Select **Environments** tab > **Stress Analysis**
3. Select **Create Simulation** > Specify 'Optimisation Study' for **Name** > Select **Parametric Dimension** for **Design Objective** > Specify 0.3 mm for **Tolerance** > Click **OK**

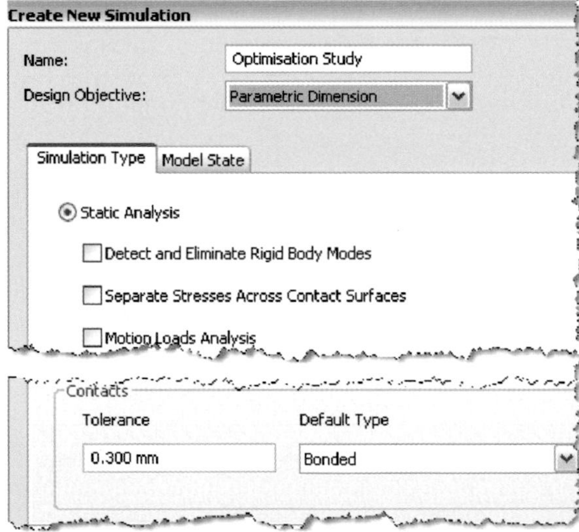

Excluding non-structural components can further simplify the model, the including reducing the number of contacts created.

Changing **Tolerance** to 0.3 mm will create contacts between adjacent components that have gaps of 0.3 mm or less.

4. Select the following components > Right click > Select **Exclude From Simulation**
   - Secondary Suspension
   - Boss
   - ISO 4018 M12 × 30
   - Handle

- ISO 7089 16 - 140 HV:2
- ISO 7089 16 - 140 HV:3
- ISO 7089 16 - 140 HV:4
- ISO 7089 16 - 140 HV:5

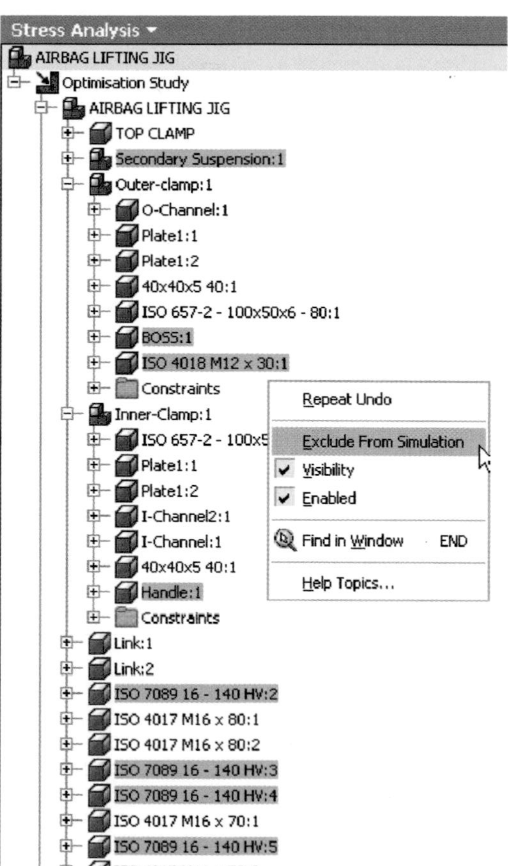

 The excluded components will now become transparent.

At this stage, we can make the components invisible by altering the **Stress Analysis Settings**.

5. Select **Stress Analysis Settings** > Select **Invisible** for **Excluded Components** > Click **OK**

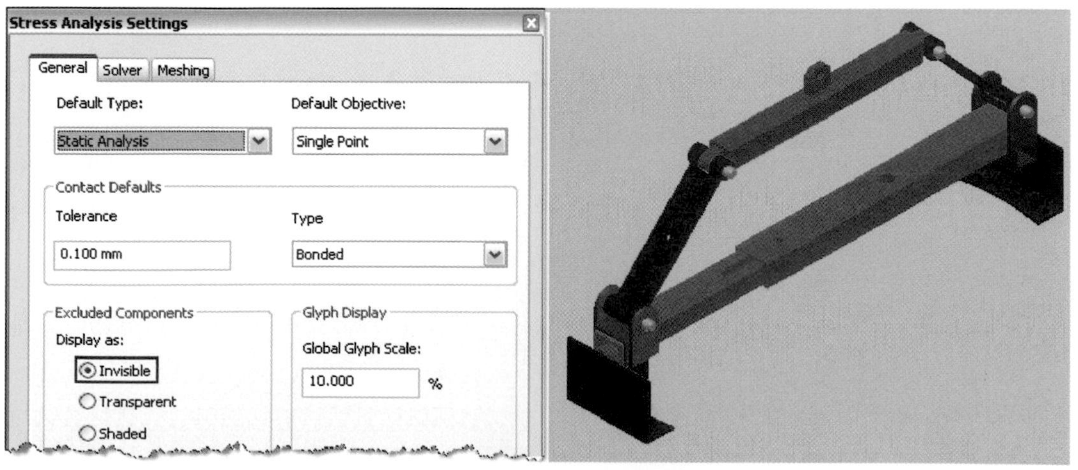

357

The next stage is to apply boundary conditions including loads, restraints, and materials.

Boundary conditions

As the primary goal of this design problem is to determine the structural integrity of the new design, to lift the secondary suspension, we will apply its mass as a force on the new design, as it is excluded from the simulation. The following will be used to convert mass to force:

Force = Mass × Acceleration

where

Mass of unit is 78 kg

Acceleration (Gravity) = 10 m/s^2 (actual value is 9.81 m/s^2)

Therefore,

Force = 78 × 10 = 780 N

If the suspension unit were not excluded from the simulation, there would have been no need to apply force as the weight of the unit would have been transferred via contacts.

6. Select **Force** > Select the faces as shown > Specify 780 for **Magnitude** > Click **OK**

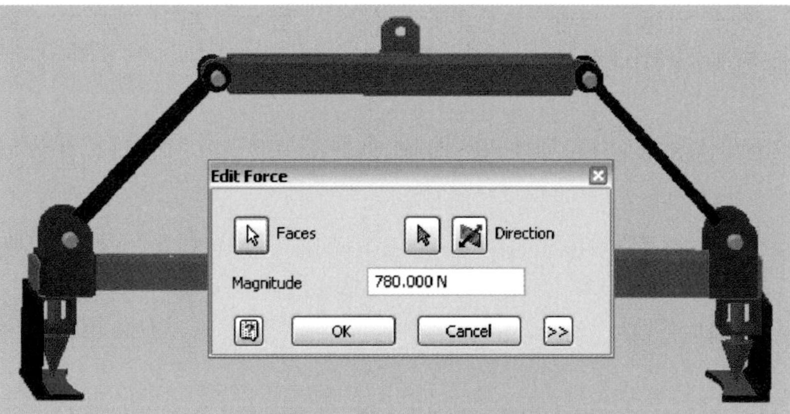

Selecting both separate faces, using the same **Force** tool, will split the force equally to 390 N on each face. Otherwise, the force would need to be halved if two forces were to be applied separately.

In reality, the weight of the suspension unit will be equally distributed through the new lifting mechanism design.

7. Select **Fixed Constraint** > Select the internal hole of the top clamp, as shown > Click **OK**

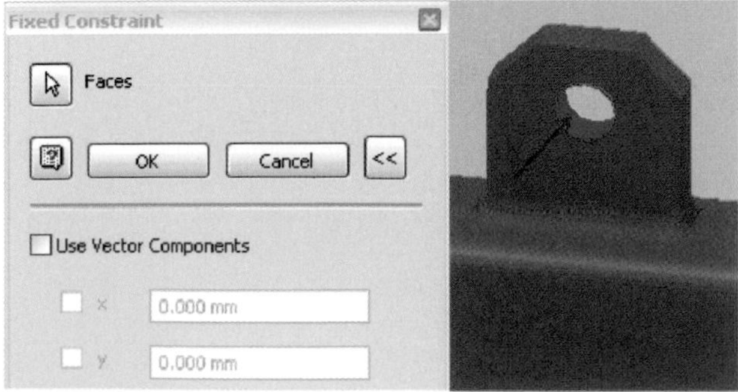

8. Select **Automatic Contacts**

This will create 28 contacts in total.

To easily identify contacts created between components, it is best to expand the components and assemblies within the browser as shown below.

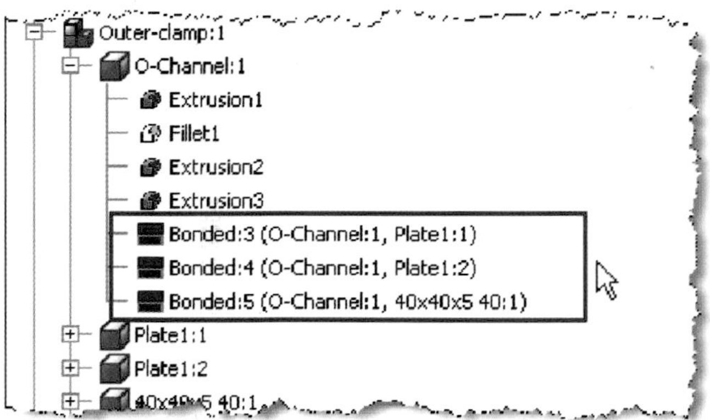

From analyzing the contacts created we can see that four contacts need to be suppressed and all the contacts created between the bolts need to be changed to sliding and no separation contacts, as in reality the bolts can slide and rotate within the holes.

9. Expand Outer-clamp:1 > Expand the 40 × 40 × 5 40:1 component > Select the Bonded:6 and :8 contacts > Right click > Select **Suppress**

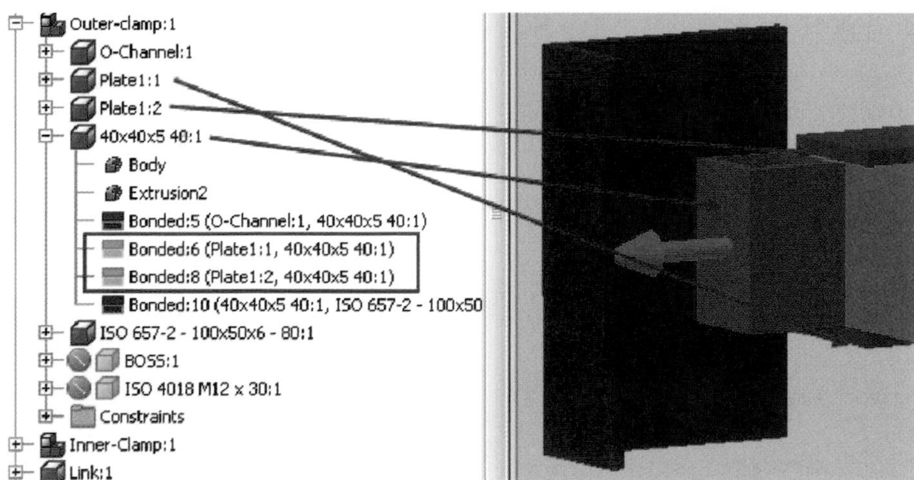

As the adjacent faces of both components are within the 0.3 mm gap, a contact has been automatically created.

Repeat Step 9 for the clamp on the other side.

10. Expand Inner-clamp:1 > Expand the 40 × 40 × 5 40:1 component > Select the Bonded:13 and :16 contacts > Right click > Select **Suppress**

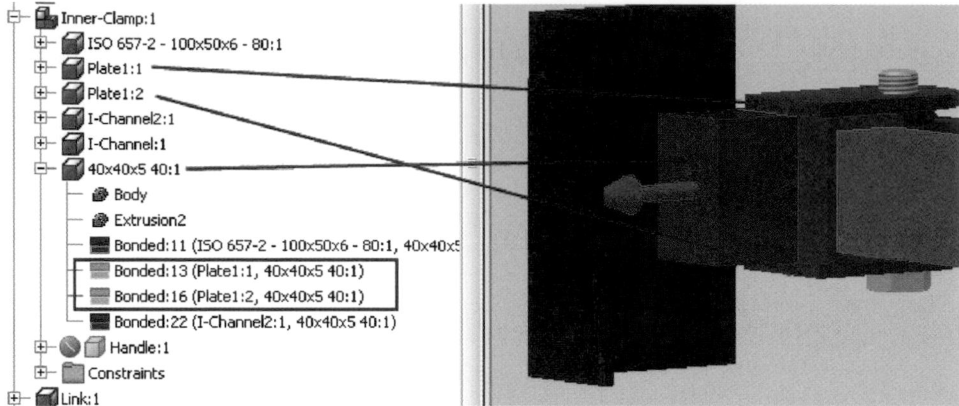

In the following steps, all bonded contacts associated with the four bolts will be changed to sliding/no separation contacts.

11. Expand the ISO 4017 M16 × 80:1 and :2 components > Select the contacts Bonded:14, :17, :26, :7, :9, :23 > Right click > Select **Edit contact**

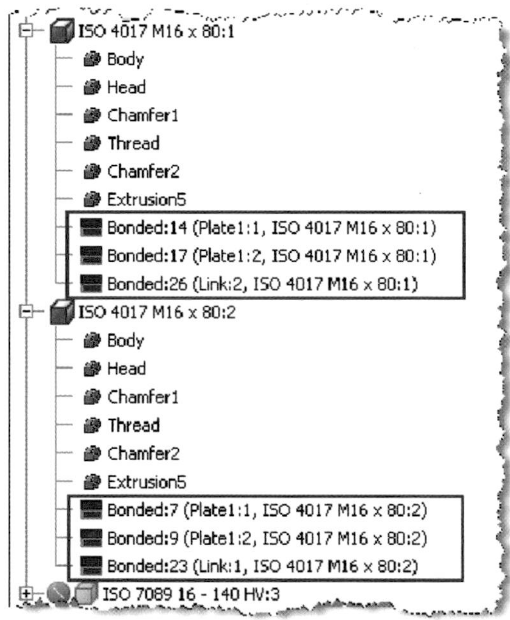

12. Select **Sliding/No Separation** for **Contact type** in the **Edit contacts** dialog box > Click **OK**

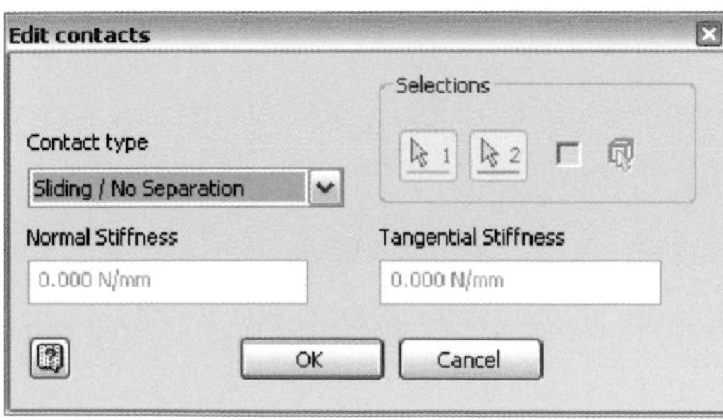

13. Expand the ISO 4017 M16 × 70:1 and :2 components > Select the contacts Bonded:1, :24, :25, :2, :27, :28 > Right click > Select **Edit Contact**

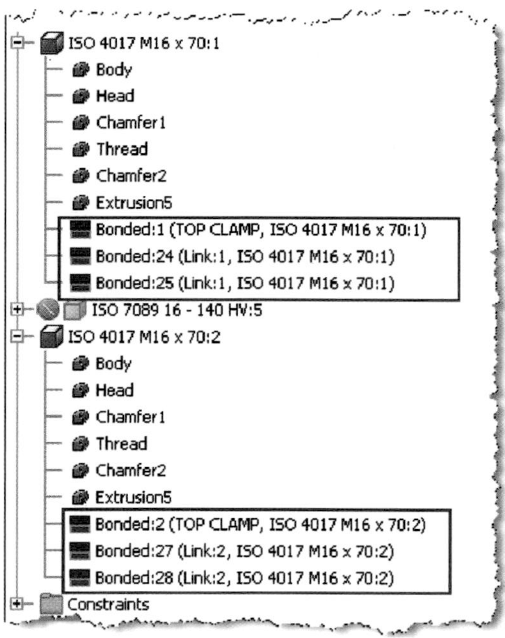

14. Select **Sliding/No Separation** for **Contact type** in the **Edit contacts** dialog box > Click **OK**

In total, 12 sliding/no separation contacts will be created.

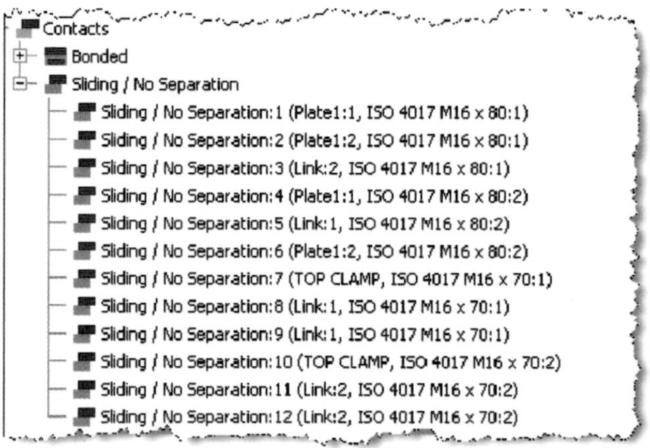

We now need to create bonded contacts between the I-Channel:1 and O-Channel:1 components, as no contacts have been created between these components. This is because the gap between them is 1 mm, which is higher than the default 0.3 mm contact setting.

At this stage, we can edit the simulation properties and change the contact tolerance to 1 mm, as shown on the next page, which will create additional contacts between I-Channel:1 and O-Channel:1.

Alternatively, we can create the contacts manually and, for the following steps, the contacts will be created manually. It may help to change the O-Channel to a transparent color to help with visualizing creating manual contacts.

15. Select **Manual Contact** > Select faces of the I-Channel and O-Channel as shown > Click **Apply**

You may need to click **Select Other** to select the internal faces of the O-Channel.

16. Now select the top faces of the I-Channel and O-Channel, as shown > Click **Apply**

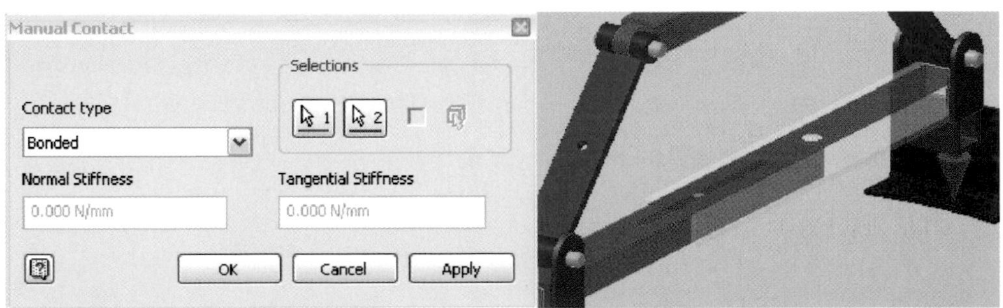

17. Now select the other side faces of the I-Channel and O-Channel, as shown > Click **Apply**

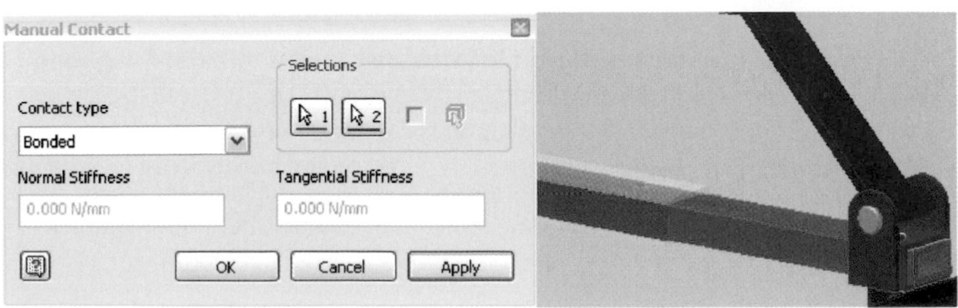

18. Finally, select the bottom faces of the I-Channel and O-Channel, as shown > Click **OK**

The following four manual contacts are created and in total there will be 16 active bonded contacts.

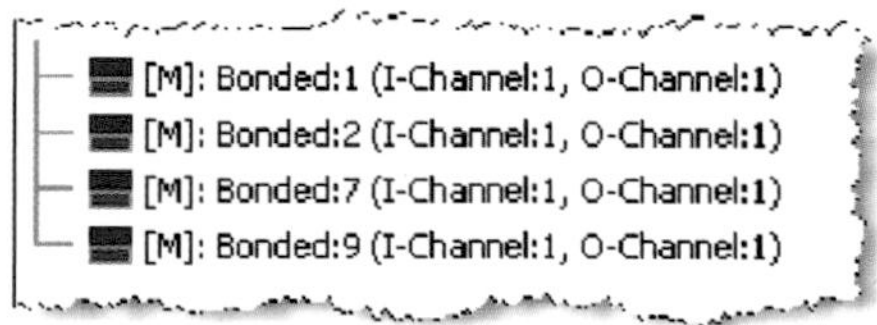

19. Change the O-Channel back to its original color – as defined by the material

20. Select **Mesh View**

This will generate mesh and thus enable us to view the mesh so we can further refine it if required. In this instance, the default mesh seems reasonable for the initial simulation run.

Run simulation and analyze

21. Select **Simulate** > Run the simulation

Ignore the warnings as they relate to sliding/no separation contacts, meaning that the components can slide away from one another. Hence, the software adds a soft spring to stop them sliding away.

The Von Mises Stress is a sum of all the planar stresses.

The maximum Von Mises Stress calculated is approximately 21.07 MPa. To obtain a better understanding of the stresses in the top clamp and the plates, where the suspension unit is held, the planar stress results will be displayed.

22. Now double click Stress YY to display compressive and tensile stresses on the top clamp > Select **Color Bar** > Change the color bar maximum and minimum values to 8 and −8 > Click **OK**

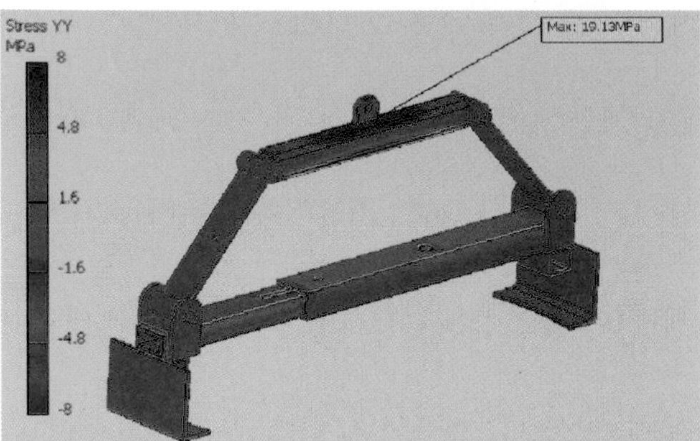

The peak stress is around 19 MPa which is concentrated around the lug. The tensile stress on the beam is in the region of 10–12 MPa and the maximum compressive stress is approximately 9.5 MPa.

Use the probe to find the exact value of stress on the beam.

23. Now double click Stress ZZ to display compressive and tensile stress on the clamp plates where the suspension unit is held

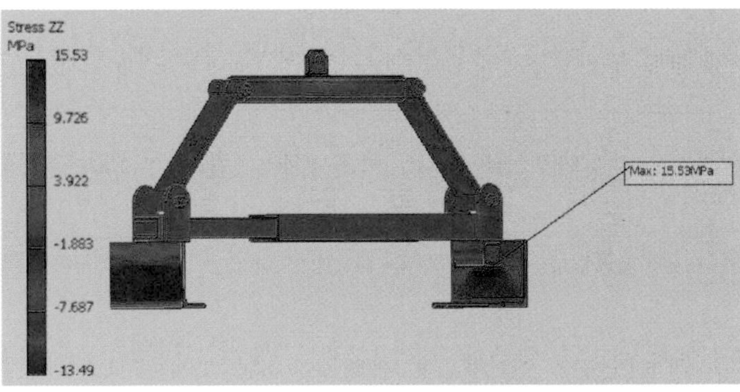

The maximum tensile stress is 15.53 MPa and the maximum compressive stress is 13.49 MPa.

24. Now double click Stress XX to view the third and final planar stress

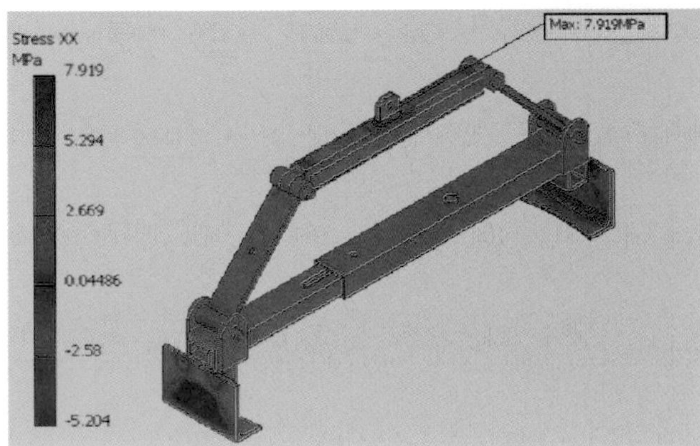

The stresses in the X plane are small when compared to the Y and Z planar stresses. As the maximum stress is at the top face of the top clamp, we will refine the mesh using a local mesh control and then compare the maximum stresses again.

25. Select **Local Mesh Control** > Select the top face and fillet faces around the lug > Specify 5 mm for **Element Size** > Click **OK**

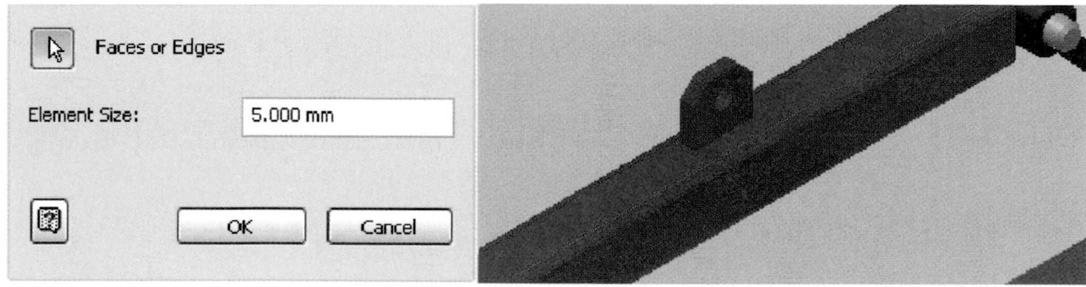

26. Right click **Mesh** > Select **Update Mesh** > Select **Mesh View** > **Simulate** > Run the analysis

27. Double click Stress YY

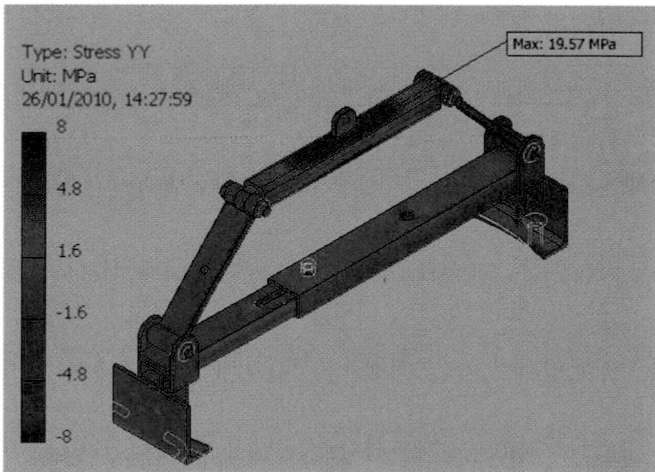

In the top clamp, the stress is significantly higher around the lug in comparison to the stresses along the channel. This localized peak stress is mainly due to the geometrical discontinuity between the lug and the top face of the clamp, also referred to as a stress raiser. Further mesh refinement will potentially increase the stress due to stress singularities and geometrical discontinuities. In this case, as the top clamp geometry is simple and pin jointed, at either end we can treat the top clamp as simply a supported beam and thus use the following equation to calculate bending stress in the beam:

$$\sigma = M = \frac{y}{I}$$

Load is applied centrally on the beam; therefore, to find the maximum bending moment, we can use:

$M = P \times L/4$
$= 780\text{ N} \times 390\text{ mm}/4$
$= 76050\text{ Nmm}$

y = Distance to neutral axis
= Section height/2
= 40 mm/2 = 20 mm

I = 2nd moment of area (for box section)
= outer 2nd moment of area – inner 2nd moment of area
$= (BD^3/12) - (bd^3/12)$
$= (40 \times 40^3/12) - (30 \times 30^3/12)$
$= 2560000 - 675000$
$= 145833\text{ mm}^4$

Therefore, the maximum tensile or compressive stress due to bending is:

$\sigma = 76050 \times 20/145833$
$= 10.43\text{ N/mm}^2 = 10.43 \times 10^6\text{ N/m}^2 =$ **10.43 MPa**

This stress calculation is based on the assumption that the maximum stress is occurring evenly in the center of the beam. In reality, when we examine the model, we can see that the lifting lug acts as a stiffener on the beam section. Although the maximum moment occurs in the center, the maximum stress will be redistributed to either side of the lug on the upper surface of the beam.

We can see below that the section to the side of the lug marked Z–Z exhibits a maximum tensile stress on the upper face in the region of 10 MPa. This is where we would expect to see the redistributed maximum tensile stress within the beam.

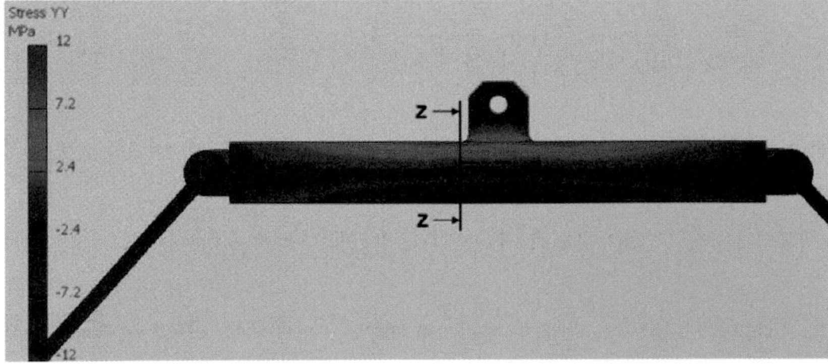

Further, as the top portion of the beam is stiffened by the lug, the neutral axis (zero stress) moves upward, causing a greater internal moment on the underside of the beam. We would expect this to act as a stress raiser on the bottom face of the beam.

On the lower face, the maximum compressive stress is still in the middle but does increase in comparison with the upper tensile stress, as expected.

These stress redistributions show a good correlation between the FEA and the simple approximated hand calculation, giving us a high degree of confidence in the overall integrity of the FEA solution.

28. Double click **Displacement** to display the maximum displacement

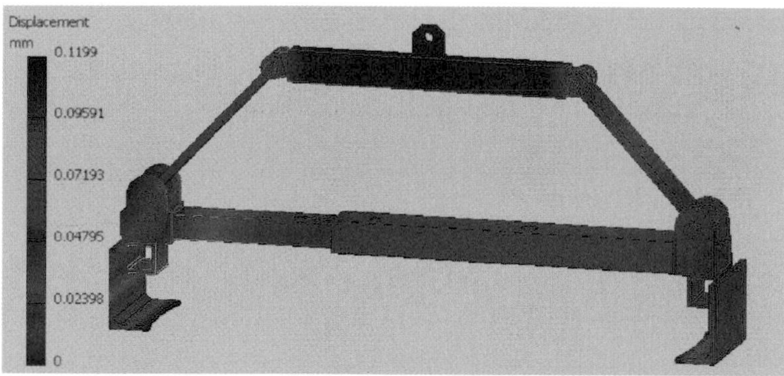

29. Double click **Safety Factor**

The value may differ – due to the different global mesh distribution over the assembly, as it is controlled by the software.

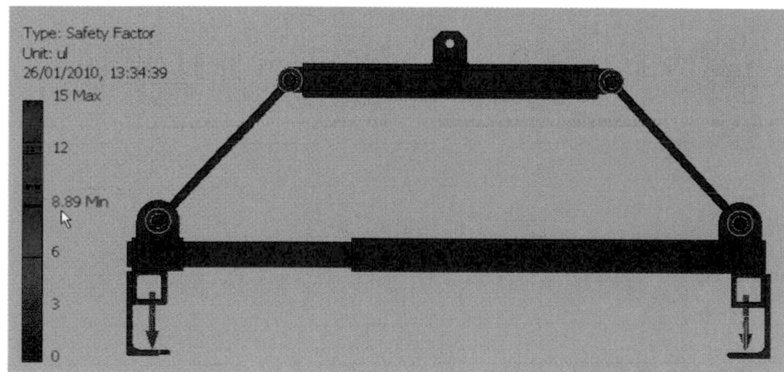

Use the color bar to enhance the clarity of the display. The minimum safety factor is around the lug at the top clamp, the as illustrated below (the position of maximum stress). Change the minimum value of the color bar to obtain a display of the safety factor results.

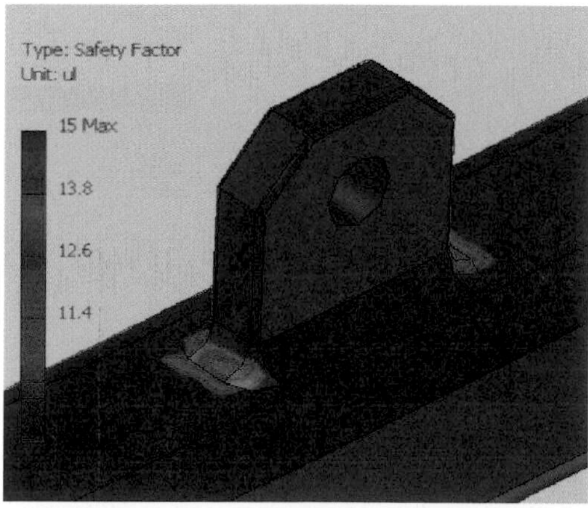

This suggests that the design can be further optimized, as the calculated safety factor of 9 is more than twice the design limit of 4.

Optimization

Here, we will use the parameters from the Component and Assembly environments and alter them to determine the best configuration that satisfies the set design constraints.

30. Select **Parametric Table**

31. Right click in the **Design Constraints** row > Select **Add Design Constraint**

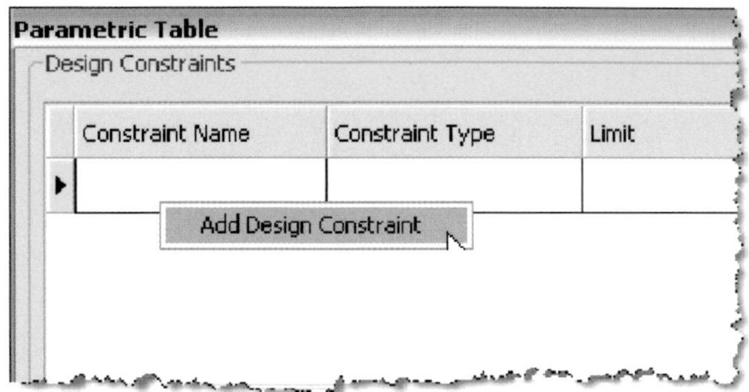

32. Select **Mass** from the list

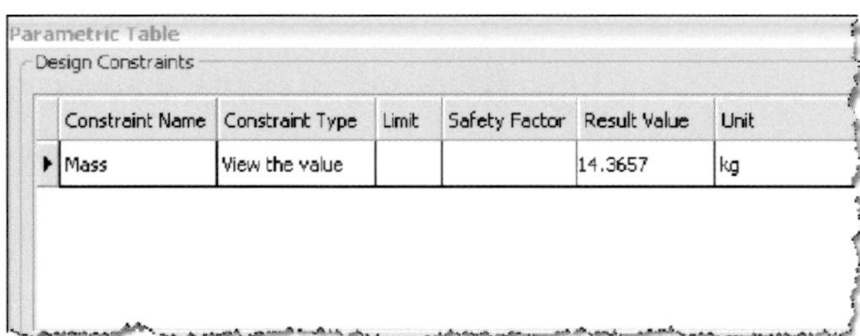

This will help to determine the best design/parametric configuration for minimum weight.

33. Repeat Steps 31–32 to add **Max Displacement** and **Max Safety Factor** design constraints

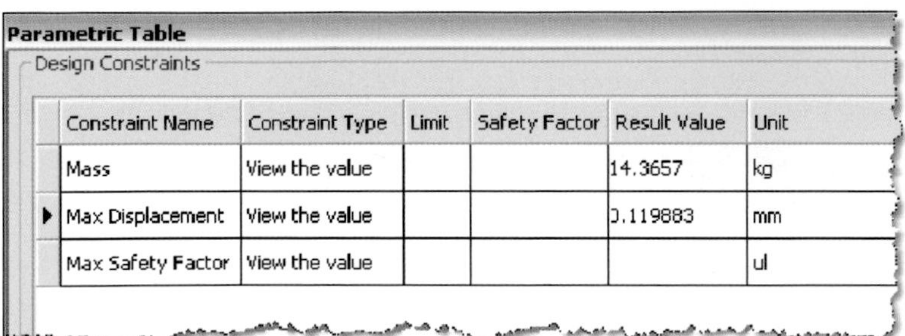

34. Change the **Constraint Type** for Max Displacement to **Upper limit** > Specify **Limit** to be 0.2

This is the maximum allowable deflection of the assembly and a value higher than this will make the design unsuitable. At the moment, the displacement is within the limit at 0.12 mm.

35. Change the **Constraint Type** for Min Safety Factor to **Lower limit** > Specify **Limit** to be 4

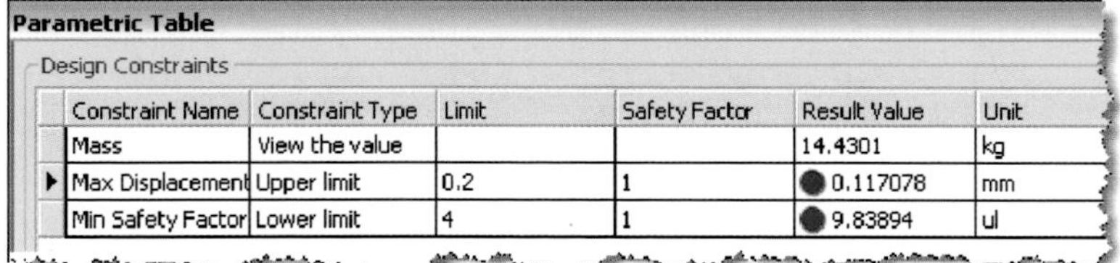

Parametric Table

Design Constraints

Constraint Name	Constraint Type	Limit	Safety Factor	Result Value	Unit
Mass	View the value			14.4301	kg
Max Displacement	Upper limit	0.2	1	0.117078	mm
Min Safety Factor	Lower limit	4	1	9.83894	ul

This is the minimum allowable safety factor of the assembly and a value lower than this will make the design unsuitable.

36. Right click Top Clamp in the browser > Select **Show Parameters**

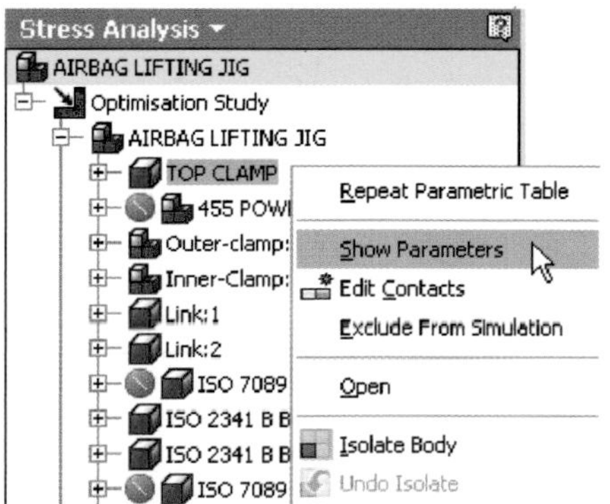

This will display all the parameters associated with this component, which can be selected to be used in the optimization analysis.

37. Select **Thickness** from **User Parameters** > Click **OK**

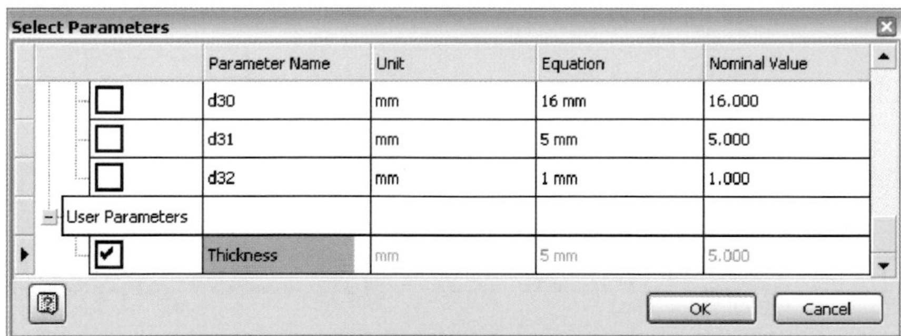

Select Parameters

	Parameter Name	Unit	Equation	Nominal Value
☐	d30	mm	16 mm	16.000
☐	d31	mm	5 mm	5.000
☐	d32	mm	1 mm	1.000
User Parameters				
☑	Thickness	mm	5 mm	5.000

OK Cancel

This will allow us to see the effect of changing the thickness of the top clamp.

IMPORTANT – Only select the linkthickness parameter in the next step if you do not have a powerful machine.

38. Right click Link:1 > Select **Show Parameters** > Select the following user parameters > Click **OK**

Select Parameters

	Parameter Name	Unit	Equation	Nominal Value
Reference Parameters				
	d24	mm	50.000 mm	50.000
User Parameters				
☑	Linkthickness	mm	10.00000000 mm	10.000
☑	Slotthickness	mm	10.00000000 mm	10.000
☑	Slotwidth	mm	12.00000000 mm	12.000
☑	Slotnumbers	ul	1.00000000 ul	1.000

The linkthickness parameter will allow us to see the effect of changing the thickness of the link arms.

The slotthickness and slotwidth parameters together allow us to control the shape of the cutouts in the link arms. The shape of the cutout can be either a slot or circle.

The slotnumbers parameter will allow us to control the weight-saving cutouts in the link arms.

39. Select **Parametric Table**

Parametric Table

Design Constraints

Constraint Name	Constraint Type	Limit	Safety Factor	Result Value	Unit
Mass	View the value			14.4426	kg
Max Displacement	Upper limit	0.2	1	● 0.116862	mm
Min Safety Factor	Lower limit	4	1	● 8.8921	ul

Parameters

Component Na	Feature Name	Parameter Nar	Values		Current Value	Unit
Top-Clamp		Thickness	5		5	mm
Link		Linkthickness	10		10	mm
Link		Slotthickness	12		12	mm
Link		Slotwidth	12		12	mm
Link		Slotnumbers	1		1	ul

All the selected parameters now appear in addition to the design constraints, both of which can be used for the optimization study.

The weight of the lifting mechanism can be reduced by using any of the above combination parameters including the thickness of link arms, the number of weight-saving holes, and the size.

40. Specify 3 – 5:3 in the Top-Clamp **Values** field

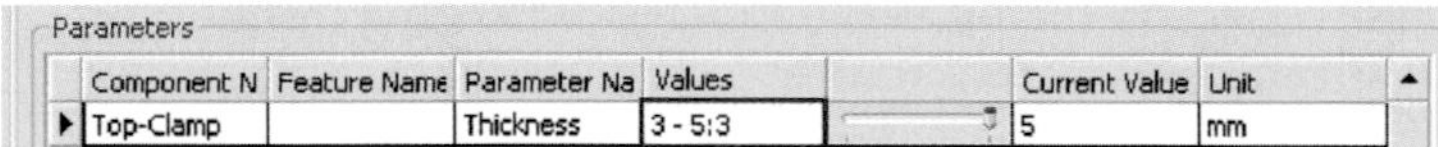

Parameters

	Component N	Feature Name	Parameter Na	Values		Current Value	Unit
▶	Top-Clamp		Thickness	3 - 5:3		5	mm

This will generate values of 3, 4, and 5 such that two more parameters will be created.

41. Specify 3,6,9,10 in the Linkthickness **Values** field

Parameters

	Component Na	Feature Name	Parameter Nar	Values		Current Value	Unit
	Top-Clamp		Thickness	3 - 5:3		5	mm
▶	Link		Linkthickness	3,6,9,10		10	mm

This will only generate the specified parameters.

42. Specify the following values to complete the parametric table

Parameters

	Component Na	Feature Name	Parameter Nar	Values		Current Value	Unit
	Top-Clamp		Thickness	3 - 5:3		5	mm
	Link		Linkthickness	3,6,9,10		10	mm
	Link		Slotthickness	10 - 12:3		12	mm
	Link		Slotwidth	12,20,30		12	mm
▶	Link		Slotnumbers	1 - 9:5		1	ul

1–9:5 will produce four more parameters, equally spaced between 1 and 9. Here, the additional parameters created in addition to 1 are 3, 5, 7, and 9.

The values created above do not encompass all the selected parameters. You can add them later once you have analyzed these chosen parameters. Choosing too many parameter configurations can result in it taking a long time to generate and produce results.

It's better to check parameter configurations with the assembly environment before generating them within the simulation environment.

43. Right click anywhere in the parameter rows and select **Generate all Configurations**

This can take a long time if all parameters are selected.

It is more effective to select **Generate range Configurations** by selecting each row individually.

44. Now move the slider to see the effect of the parameter changes on the lifting mechanism

45. Select **Simulate** > Select **Exhaustive set of configurations** > Select **Run**

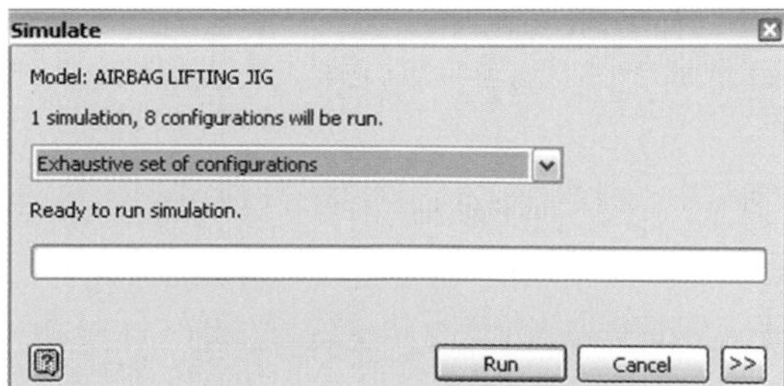

Selecting the **Exhaustive set of configurations** option can result in it taking a long time to analyze, if all parameters are selected.

Select **Smart set configurations** and, for configurations that have been calculated, you can individually create the configurations by moving the slider, to select individual values for each parameter, and then select **Current configuration only**.

46. Now move the slider to see the effect of the parameters on the design constraints

47. Finally, select **Minimize** in the **Constraint Type** for Mass and see the parameter configuration selected

Here, you can add more parameters, change design constraint limits, etc., to further optimize the design.

Based on the parameters that are chosen, which are entirely dependent on the designer, any of the following design configurations can be achieved.

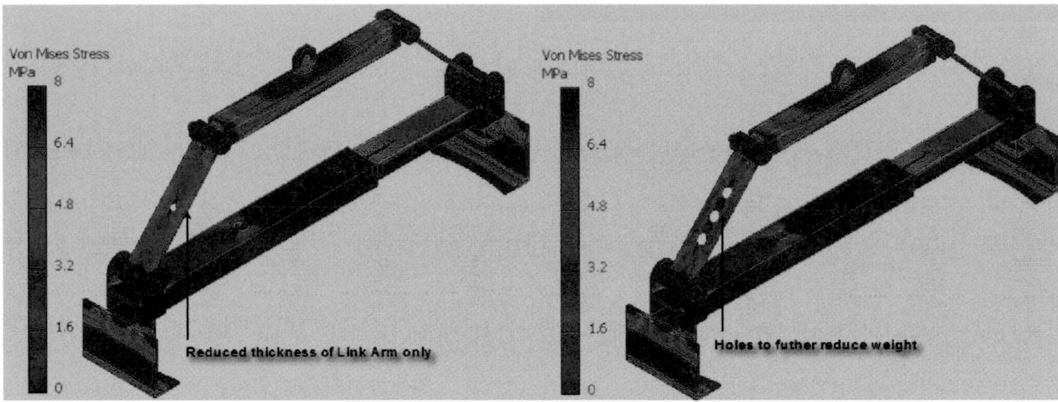

The number of design configurations is unlimited and the above configurations can be further enhanced by adding more holes and reducing the thickness of the other channels of the lifting mechanism.

Another possible configuration is illustrated below.

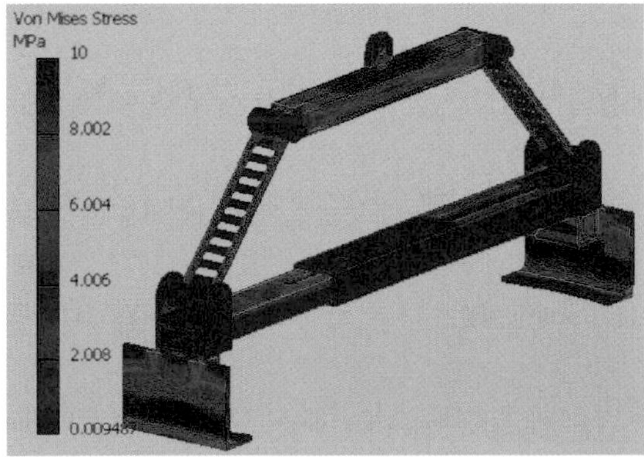

However, it is important to note that the selection of the design is dependent on various criteria including manufacturability, company/designer preferences, best practices, etc.

48. Close the file

CHAPTER 16

DP14 – Modal Analysis

Modal Analysis of TV Camera Arm Attached to Helicopter

(Design Problem courtesy of Aerospace Design Facilities Ltd)

KEY FEATURES INTRODUCED IN THIS DESIGN PROBLEM

	Key features
1	Modal shapes and natural frequencies
2	Modal optimization

INTRODUCTION

Aerospace Design Facilities Ltd, based in the UK, is a European Aviation Safety Agency (EASA) FIC approved design organization for both helicopters and fixed wing aircraft. Aerospace Design Facilities Ltd also supplies bespoke design and manufacturing services to the film industry. One typical example of their work is designing camera mounts to be placed on helicopters, as shown below.

Up and Running with Autodesk Inventor Simulation 2011. ISBN: 978-0-12-382102-7
Copyright © 2010 Elsevier Inc. All rights of reproduction, in any form reserved.

A major consideration during the design of a camera mount on helicopters is the need to ensure that the vibrations produced by the dynamics of the airframe (rotating components) are not amplified. Generally, the camera systems have been designed to be either hard mounted (i.e. they can cope with the amplitude and frequencies) or attached via an isolation mount. The basic design of the structure to support the camera must be evaluated for its natural frequency response with the camera system installed, and without if the structure is to be flown without a camera system attached.

The camera arm design is attached to a helicopter having a three-bladed main rotor with a nominal speed of 393 rpm. Therefore, the dominant frequencies will be 393 rpm/60 s = 6.55 Hz and, as there are three rotor blades, there will be three passes of the blade in one revolution, producing a further dominant frequency of 6.55 Hz $\times$ 3 = 19.65 Hz.

The design of the camera mount needs to cope with the structural aspects of aerodynamic crash loads; however, the first part of the analysis should be an evaluation of natural frequencies.

Although there are many other considerations involved in this design, this chapter will only investigate two of the natural frequencies that need to be avoided during the design of this camera mount:

Frequency 1 (R_1) = 6.55 Hz
Frequency 2 (R_3) = 19.65 Hz

The major design restrictions for the camera mount are as follows:

- The positions of the fixing points for the camera mount and the camera are fixed.
- The maximum weight of the camera mount should not exceed 50 kg.
- The camera offset position relative to the neutral axis of the camera arm is restricted by clearance to both the ground and the helicopter. In an ideal situation, the position of the center of gravity of the camera needs to be in line with the neutral axis of the mount arm to minimize vibration.

WORKFLOW OF DESIGN PROBLEM 14

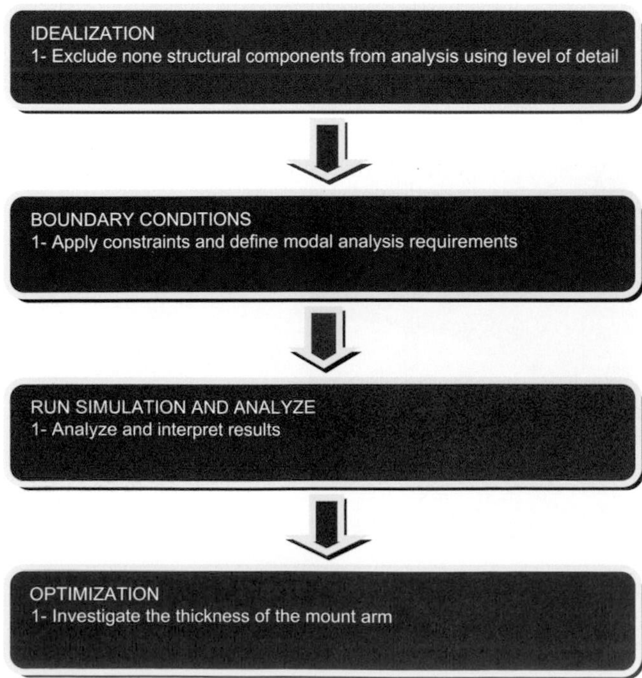

Idealization

The main weight of camera mount assembly comprises the camera arm (30.5 kg) and mounts. The mass of the fasteners is small and to simplify the analysis further all fasteners and associated components are excluded from the simulation. Furthermore, the model of the camera and connecting plate has been simplified by suppressing radii, holes, etc. All this idealization of the camera mount is saved as a level of detail, avoiding the need to suppress non-key components and features in the Stress Analysis environment. Two levels of detail have been created: one without the camera and the second with the camera. This will allow us to analyze and compare the modal shape, including natural frequencies, of both models. Initially we will have a look at the natural frequencies of the arm.

1. Open *Mount Assembly.iam*

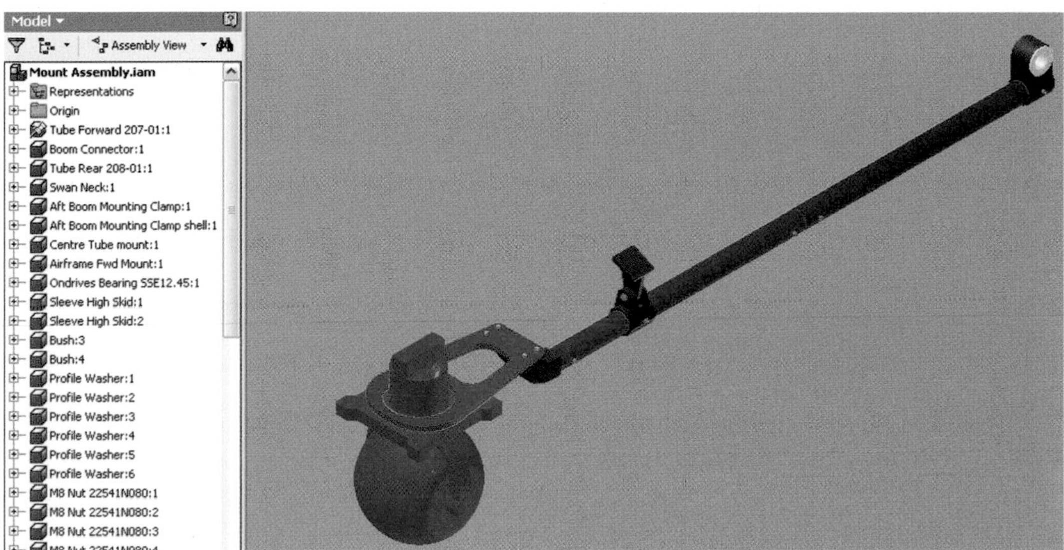

Boundary conditions

2. Select **Environments** tab > **Stress Analysis**
3. Select **Create Simulation** > Specify 'Arm' for **Name** > Select **Modal Analysis** > Specfic 4 for **Number of Modes** > Select **Enhanced Accuracy**

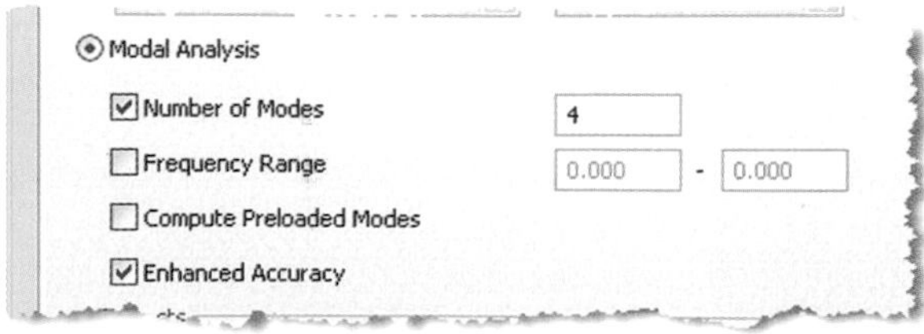

4. Select the **Model State** tab > Select **Arm** for **Level of Detail** > Click **OK**

5. Select **Fixed Constraint** > Select the face as shown > Click **OK**

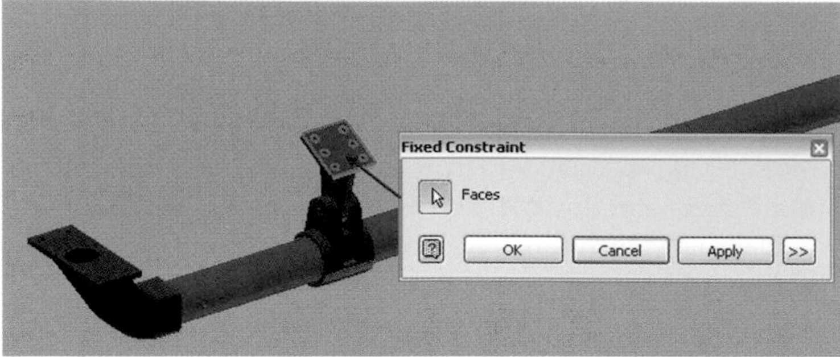

6. Select **Pin Constraint** > Select the face as shown > Click **OK**

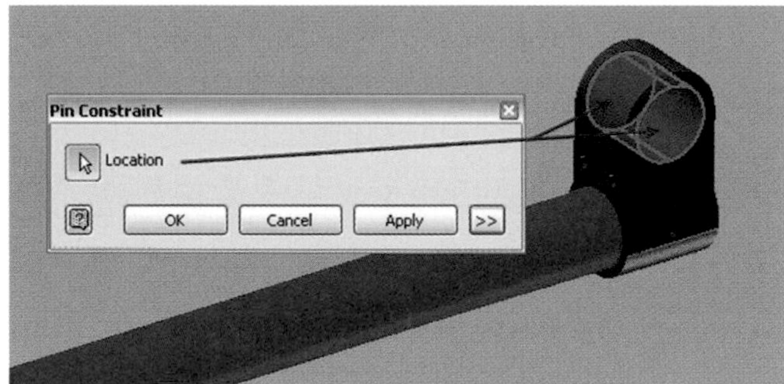

7. Select **Automatic Contacts** to create contacts between components

Five contacts will be created in total.

Some contacts will need to be created manually between components that have larger clearances due to excluding components from the simulation (washers, connectors, etc.)

8. Select **Manual Contact** > Select the faces as shown > Click **Apply**

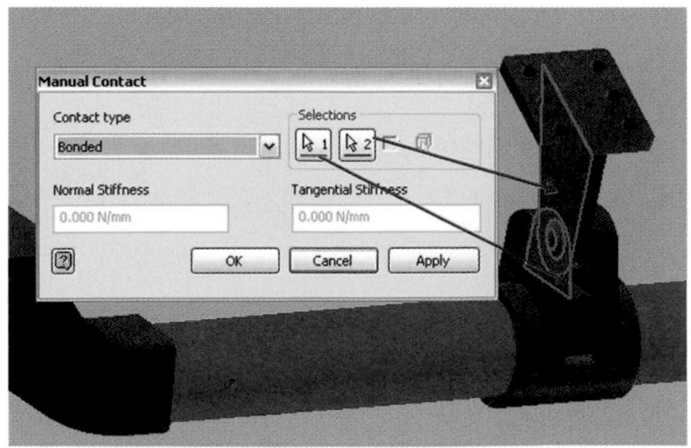

9. Repeat Step 8 for the faces on the other side of the clamp > Click **Apply**

10. Select the two faces to connect the tubes together > Click **OK**

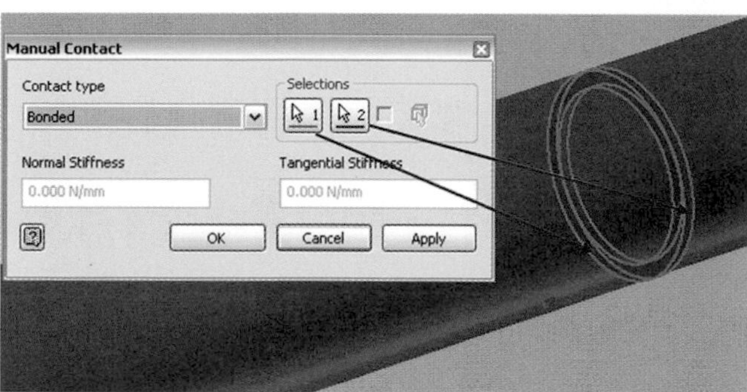

The following contacts should be created:

11. Select **Mesh Settings** > Specify 0.02 for **Average Element Size** > Check **Create Curved Mesh Elements** > Uncheck **Use part based measure for Assembly mesh** > Click **OK**

12. Select **Mesh View**

A total of 64 564 nodes and 33 316 elements will be created. There may be slight variation.

Use a relatively small mesh size to produce a more accurate modal analysis.

Run simulation and analyze

13. Select **Simulate** > Run the analysis > Unselect **Mesh View**

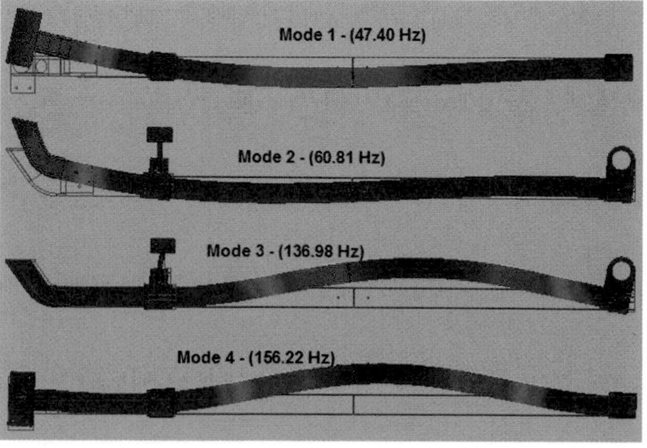

These are the first four modes, defined by the modal shape and natural frequencies, of the arm without the camera. It is important to note that all calculated frequencies are much higher than the R_1 and R_3 frequencies, which means that the arm without the camera will produce minimal vibration under normal helicopter operating conditions.

Now we will see the effect of adding the camera to the arm and see the first four modes again.

14. Right click Arm:1 > Select **Copy Simulation**

15. Right click Arm:1 > Select **Edit Simulation Properties** > Specify 'Arm-with-Camera' for **Name** > Select **Model State** > Select **Arm-with-Camera** for **Level of Detail** > Click **OK**

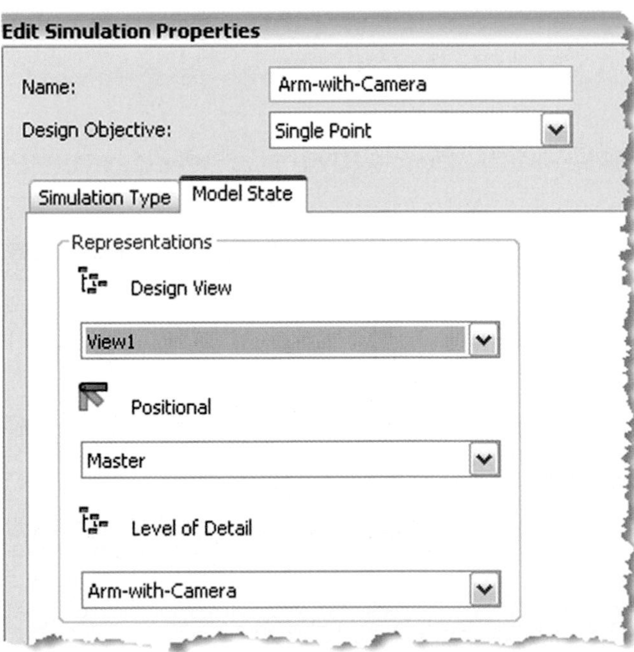

16. Right click **Contacts** > Select **Update Automatic Contacts**

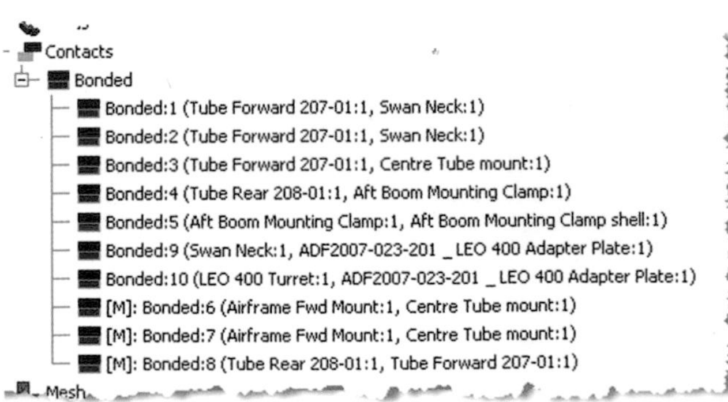

Two more contacts will be created between the camera assembly and arm; 10 in total.

17. Select **Mesh View**

Ignore any mesh warnings, that may occur, as they relate to distorted mesh elements in the area of less importance in the model. Alternatively, you can refine the mesh.

18. Select Simulate > Run the analysis > Unselect **Mesh View**

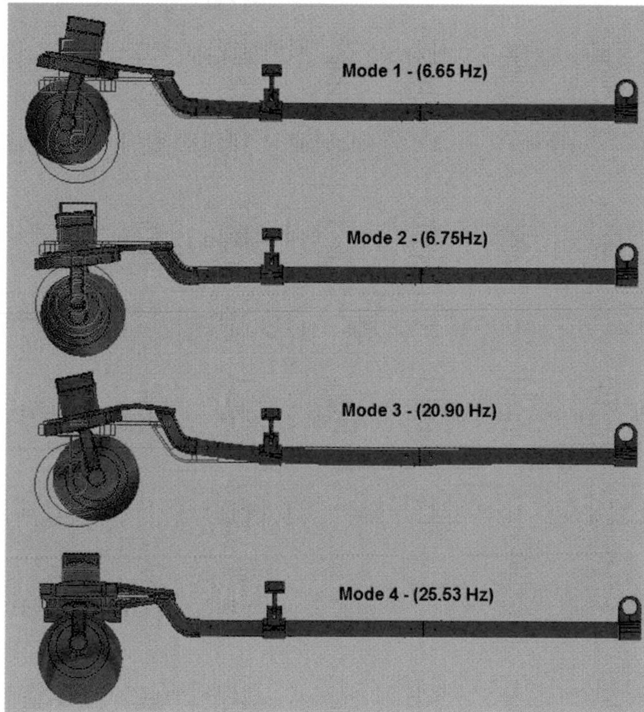

We can now clearly see that the fundamental frequency (Mode 1) has significantly reduced from 47.40 to 6.65 Hz, and this applies to all three other modes as well. Also, it is important to note that the Mode 1 and Mode 2 frequencies are within 1% range of the R_1 frequency and that the Mode 3 frequency is within 1% range of the R_3 frequency. This means that we need to increase the frequencies of the first three modes.

This can be achieved in a number of ways, including specifying different material properties, increasing the tube thicknesses, moving the mount positions, and altering the geometry of the swan neck part to move the position of the center of the gravity of camera in line with the neutral axis of the tubes. Some of these options are limited by the camera ground clearance, interference with the helicopter airframe, and the overall weight of the assembly.

Here, we will pursue the option of increasing the thickness of the arm to alter the natural frequencies of the first three modes.

Young's modulus, density, and Poisson's ratio are the only material properties taken into account when performing a modal analysis.

Optimization

To change tube thicknesses we can either go back into each component or change the sketches, or alternatively we can use parametric optimization within stress analysis to see the effect of tube thicknesses. The latter option is more efficient and, depending on the results, can allow the ability to replace and update existing model parameter values with new stress analysis parameters.

Create user parameters when using the parametric option.

19. Right click Arm-with-Camera > Select **Copy Simulation**
20. Right click Arm-with-Camera:1 > Select **Edit Simulation Properties**
21. Change **Design Objective** to **Parametric Dimension** > Click **OK**
22. Right click the Tube Forward 2071-01:1 part > Select **Show Parameters** > Check the **Tube_thickness** user parameter > Click **OK**
23. Right click the Tube Forward 208-01:1 part > Select **Show Parameters** > Check the **Tube_thickness** user parameter > Click **OK**
24. Right click the Swan Neck:1 part > Select **Show Parameters** > Check the **Tube_thickness** user parameter > Click **OK**
25. Select **Parametric Table** > Add **Mass Design Constraint**

This will help us keep a control on the mass limit of 50 kg.

26. Specify the following values for Tube_thicknesses

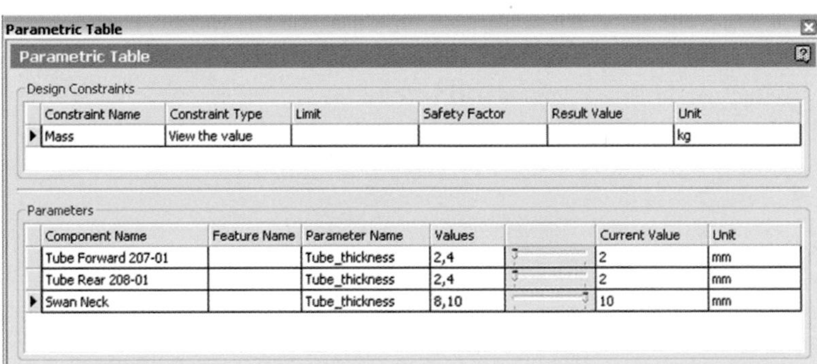

Parametric Table

Design Constraints

Constraint Name	Constraint Type	Limit	Safety Factor	Result Value	Unit
Mass	View the value				kg

Parameters

Component Name	Feature Name	Parameter Name	Values		Current Value	Unit
Tube Forward 207-01		Tube_thickness	2,4		2	mm
Tube Rear 208-01		Tube_thickness	2,4		2	mm
Swan Neck		Tube_thickness	8,10		10	mm

We are going to increase the thickness values of the main tubes of the camera mount assembly from 2 to 4. The swan neck component is attached to the inside of the tubes and, to avoid interference, we will need to reduce its thickness by the same amount (2 mm).

27. Right click any of the **Parameters** rows > Select **Generate All Configurations**
28. Set the **Current Values** for each parameter as shown

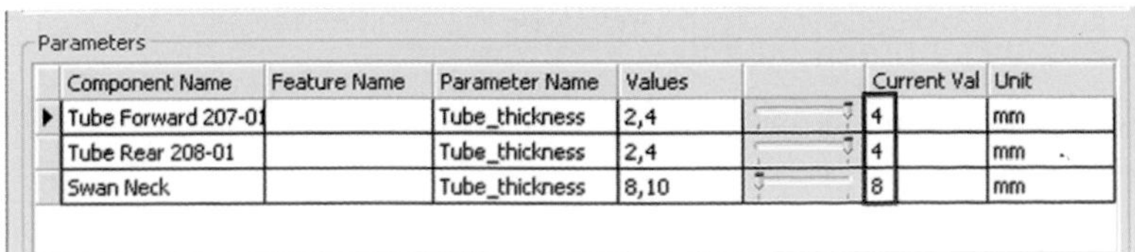

Parameters

Component Name	Feature Name	Parameter Name	Values		Current Val	Unit
Tube Forward 207-01		Tube_thickness	2,4		4	mm
Tube Rear 208-01		Tube_thickness	2,4		4	mm
Swan Neck		Tube_thickness	8,10		8	mm

29. Select **Simulate** > Select **Current configuration only**

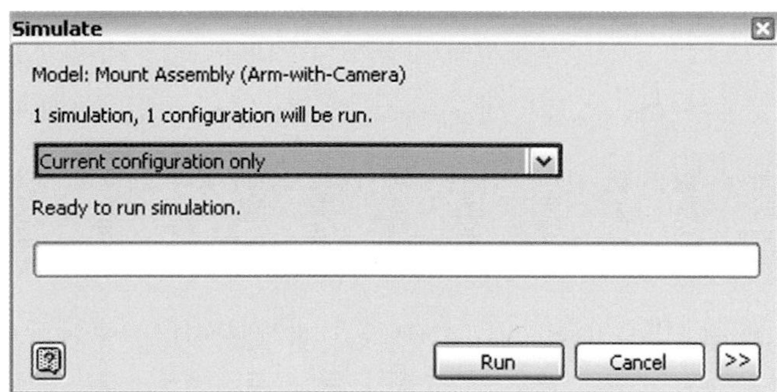

This will analyze model based on the current values set by the user, with the aid of the slider.

30. Select **Run**

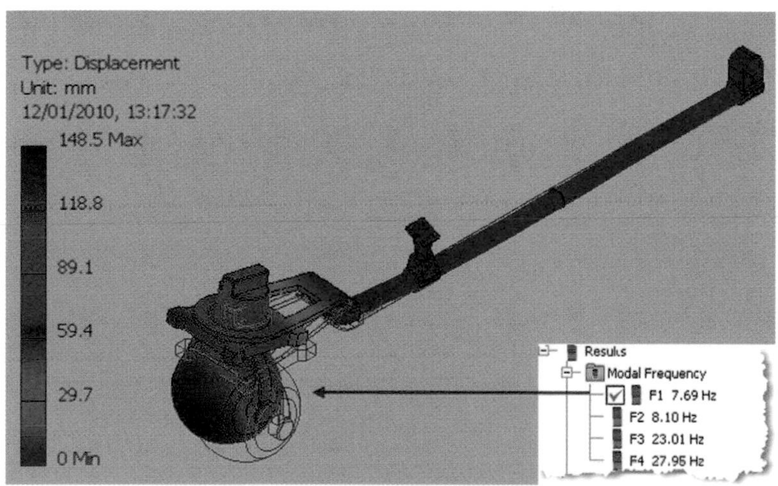

31. Select **Parametric Table**

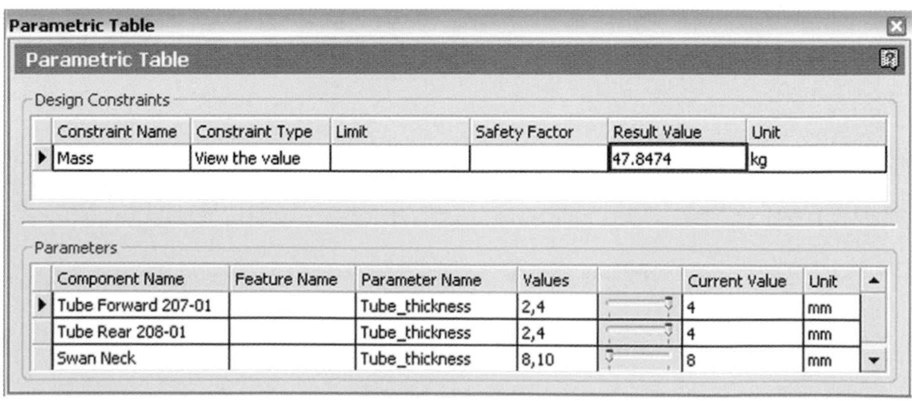

We can clearly see that the first three modes have changed by doubling the wall thickness of the tube and the mass of 47.84 kg, from within the parametric table, is below the 50 kg limit.

32. Close the file

CHAPTER 17

The Frame Analysis Environment

FRAME ANALYSIS OVERVIEW

Frame analysis is normally associated with analyzing large structures mainly comprising uniform cross-section channels/frames. Typical examples include bridges structural platforms, towers, etc. Some of the examples are illustrated below.

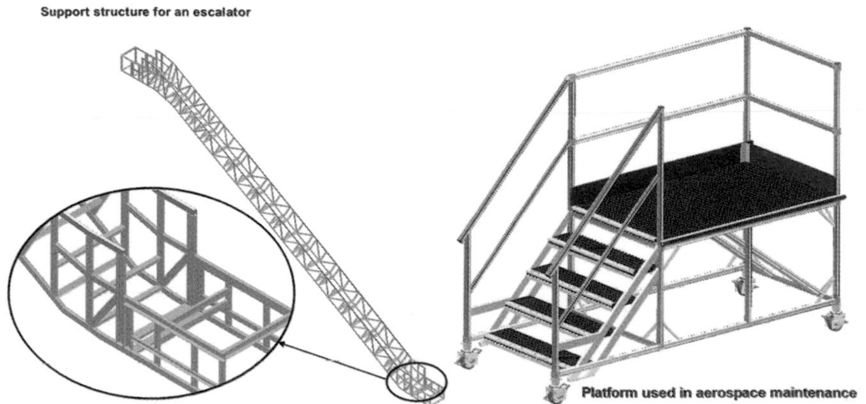

Frame analysis, within Autodesk Inventor Simulation, allows us to define criteria for static and modal analysis, including prestressing. In addition, frame analysis uses beam elements instead of the 3D tetrahedron elements used within the Stress Analysis environment. A simple beam element comprises two nodes, one at each end, and has three translational and three rotational degrees of freedom (DOF); six in total.

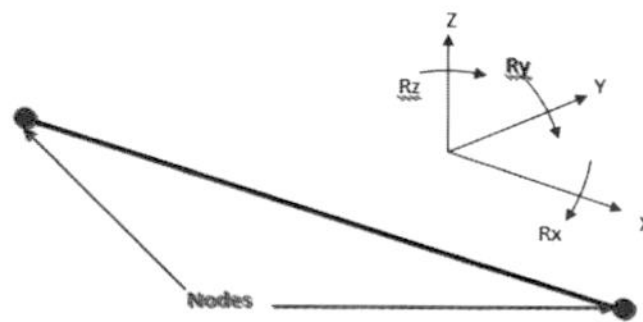

Some reasons for using beam elements rather than tetrahedron elements are reduced model sizes, reduced number of elements used, and faster analysis times. This can be demonstrated by the following example in which a beam is fixed at one end and a 1000 N load is applied at the other end.

Up and Running with Autodesk Inventor Simulation 2011. ISBN: 978-0-12-382102-7
Copyright © 2010 Elsevier Inc. All rights of reproduction, in any form reserved.

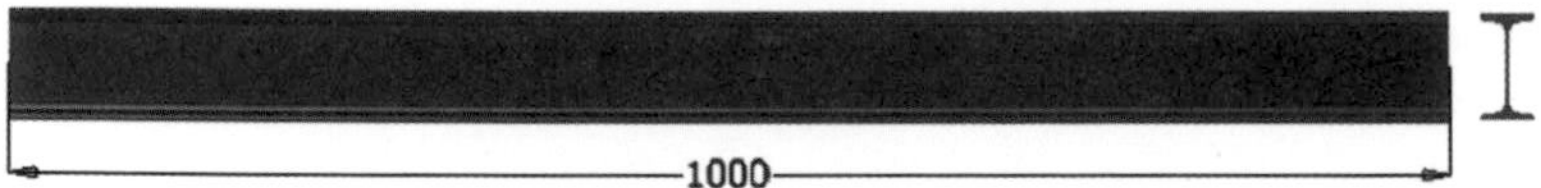

The results of both stress and frame analysis are summarized below (the results are dependent on factors such as computer speed).

	Displacement (mm)	Stress (max) (Mpa)	Mesh time* (s)	No. of elements	Analysis time (s)
Stress analysis	1.97	53.32	8	5780	6
Frame analysis	1.949	51.46	1	10	1
Theoretical results	1.949	51.46	–	–	–

*A default mesh of 0.1 **Average Element Size** was used within the Stress Analysis environment.

These results illustrate that a simple structure, such as an I-beam, can have a significant impact on model sizes and analysis times. For this simple reason it is normal practice to analyze a thin structure, with a uniform cross-section, with beam elements. Another advantage of using beam elements is that there are no stress singularities/stress concentrations to overcome. These stress concentrations probably cause the slight difference in the stress results when compared with the theoretical result (less than 5%).

FRAME ANALYSIS WORKFLOW

The process of creating an analysis (both stress and modal) involves four core steps:

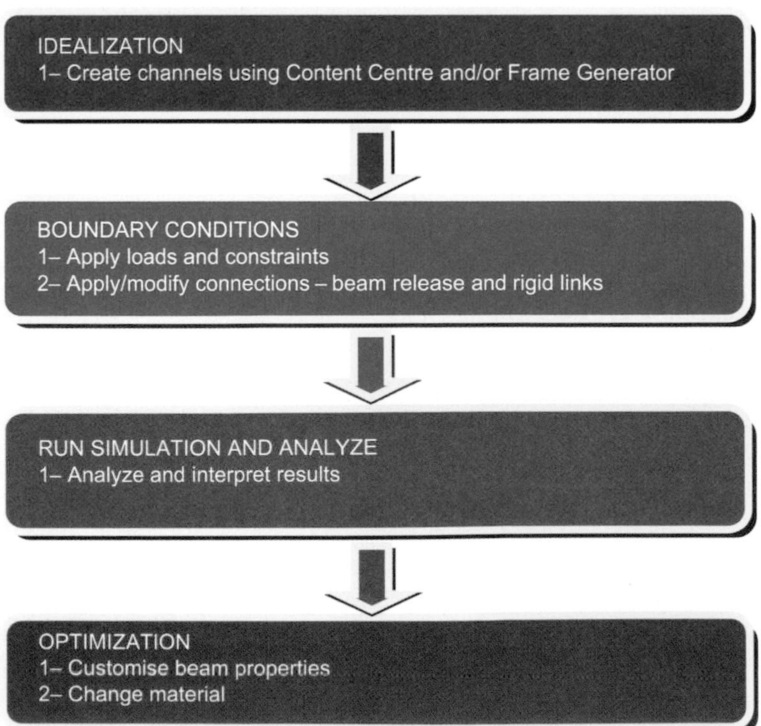

FRAME ANALYSIS USER INTERFACE

Frame Analysis can only be accessed from the Assembly environment via either the **Environments** or **Design** tab.

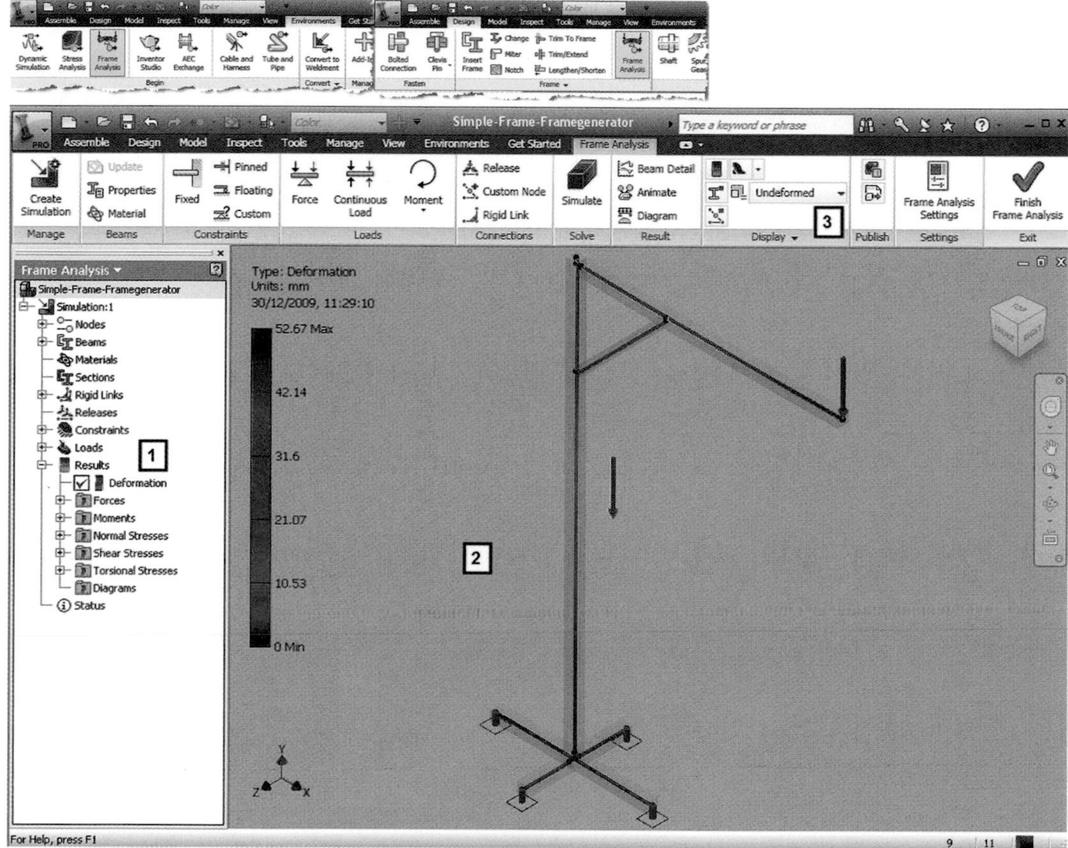

1. **Frame Analysis** browser
2. **Frame Analysis** graphic window
3. **Frame Analysis** panel

Frame analysis browser

Displays the simulations with the part or assembly and simulation parameters in a hierarchical view with nested levels of features and attribute information. You can:

- Copy whole simulations or simulation objects between simulations.
- Right click on a node for context menu options.
- Expand the folders, select the nodes, and see the selection cross-highlight in the graphic region.

Frame analysis graphic window

Displays the model geometry and simulation results. Updates to show the current status of the simulation, including applying boundary conditions and loads with the help of view manipulation tools.

Frame analysis panel

Frame Analysis tab	Workflow stage	Description
Create Simulation Manage	Step 1	**Create Simulation** – Here you decide whether you need to create a stress, modal, or parametric analysis
Update Properties Material Beams		**Beams** – Create and apply materials for the components if not already defined in the Part environment
Fixed Pinned Floating Custom Constraints	Step 2	**Constraints** – Represent how a part is fixed or attached to other parts in reality, and thus how their motion is restricted
Force Continuous Load Moment Loads		**Loads** – Represent the external forces that are exerted on a component. During normal use, the component is expected to withstand these loads and continue to perform as intended
Release Custom Node Rigid Link Connections		**Connections** – Create contacts between components automatically or manually. There are seven types of contact, including bonded
Simulate Solve		**Solve** – Run the simulation to analyze results as a consequence of defining materials, constraints, and loads
Beam Detail Animate Diagram Result	Step 3	**Results** – View the stress and deformation results to provide an informed decision regarding whether the component will function under the defined loads and constraints
Color Bar Smooth Shading Beam Labels Undeformed Node Labels Display		**Display** – Modify color plots including displaying maximum and minimum values
Report Export Publish		**Publish** – Generate an html report of the results to share
Frame Analysis Settings Settings		**Settings** – Predefine initial settings including contact tolerance and mesh settings

MANAGE TAB

This is the first step in creating a stress analysis study.

Create simulation

Here you can define whether you want to carry out a single static analysis, a modal analysis, or a parametric study – including the option of selecting different levels of detail.

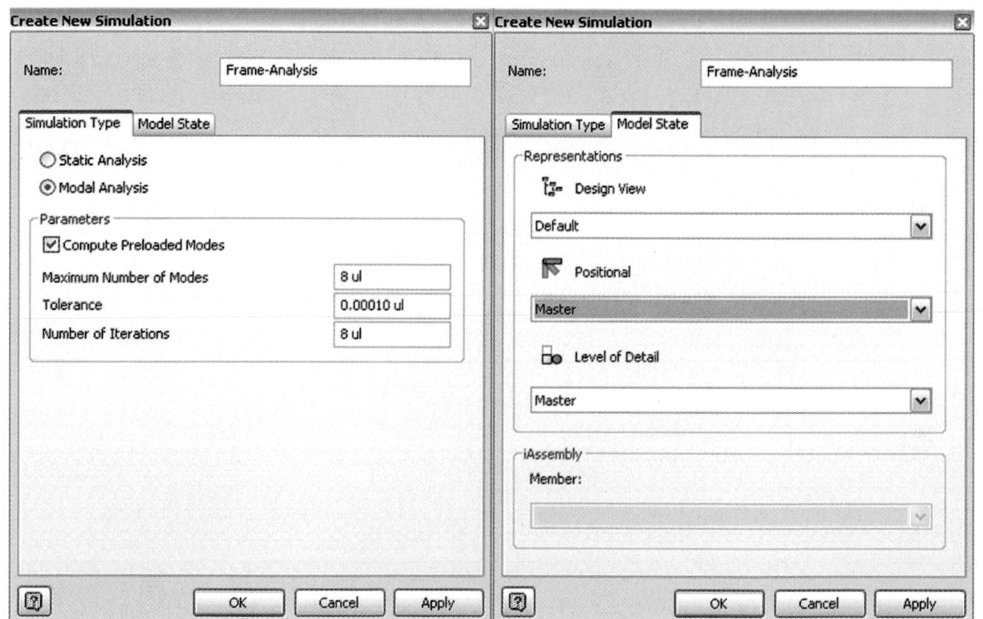

Simulation Type – Here you define whether a stress or modal analysis is to be carried out.

Model State – For an assembly you can choose any design view and level of detail on which to perform the analysis.

Parameters (modal analysis only) – When performing modal analysis, there are three settings that can be defined.

Compute Preloaded Modes – This is checked by default and calculates the modes of the model when in the loaded condition. Modes under the preloaded condition tend to have higher natural frequencies than models that are not preloaded.

Maximum Number of Modes – This calculates the number of modes, including modal shape and natural frequencies, required; the default is set to 8.

Tolerance – Here you can define the accuracy of the results required. The iterative solver goes through repetitive analysis until the difference in the results is within this tolerance setting.

Number of Iterations – Here you define the number of repetitive analyses required. The default value is set to 10.

BEAMS TAB

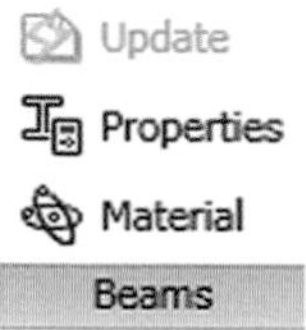

Update

This becomes active when the model has been changed, for example in the assembly environment, and thus requires an update.

Properties

Normally most components will have their materials assigned within the Part environment, thus removing the need to assign a material, as they will come across directly from the Part environment.

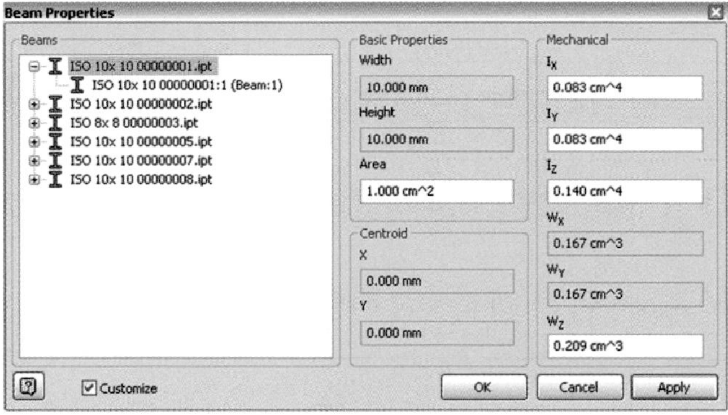

Basic Properties – The width and height are the overall dimensions of the cross-section of the beam; for example, a diameter of 10 mm will have a width and height of 10 mm. The area is the true cross-section area of the beam.

Centroid – The position of the centroid with reference to the beam coordinate system.

Mechanical – These values provide details of the second moment of area (I_x, I_y), polar second moment of area (I_z), and section modulus (W_x, W_y, W_z).

Below are two examples of simple shapes with their associated cross-sectional mechanical properties:

	Second moment of area	Section modulus
Y, X, X, D	$\frac{\pi D^4}{64}$	$\frac{\pi D^3}{32}$
b, $\frac{d}{2}$, $\frac{d}{2}$, X, Y, X, d	$\frac{bd^3}{12}$	$\frac{bd^2}{6}$

Material

The materials of the beams are normally preassigned when you create frames using Frame Generator and Content Center. These are the material properties that are read by the Frame Analysis environment:

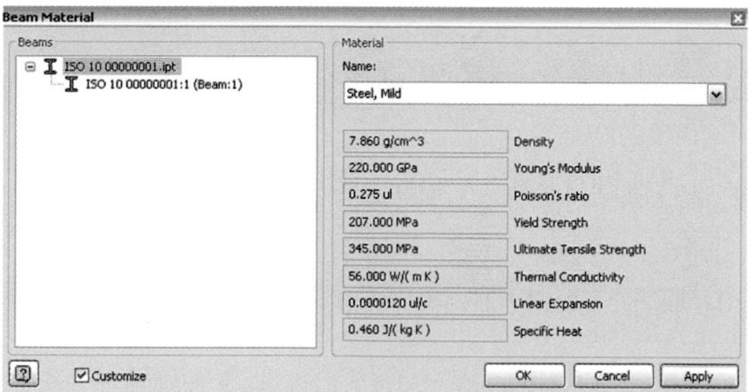

If the material is inadequately defined then the simulation will not run and an error message will be displayed in the Status folder in the browser. The material can be changed within Frame Analysis using the **Customize** button.

The **Customize** button will not let you alter the material properties, including density, Young's modulus, etc.

CONSTRAINTS TAB

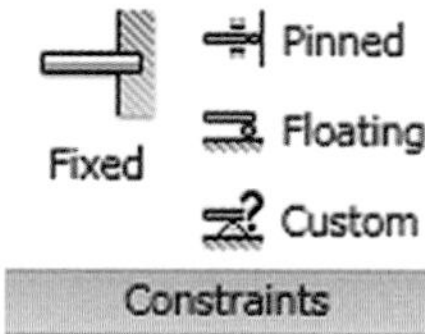

Constraints can be created by either using the heads up display (HUD) or constraints dialog box, as illustrated below.

When a beam is selected, to place any constraint, an internal node is created that connects the constraint to the beam.

Fixed constraint

Fixed Constraint removes all translational and rotational DOF of the selected node or beam. Both nodes and beams can be selected to specify the origin. The **Offset** parameter will only be activated if a beam is selected to define **Origin**. The **Offset** can be specified either in absolute or relative values. A value of 0.5, when **Relative** is selected, will place the constraint in the middle of the beam. Fixed constraints simulate bolted and welded connections, and joint as illustrated on the next page.

Bolted Connection

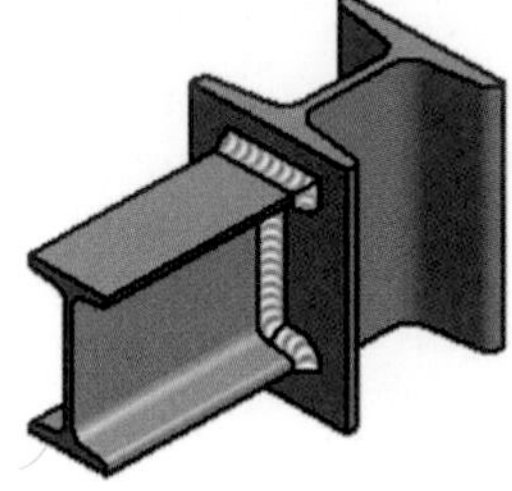
Welded Connection

Pinned constraint

Pinned Constraint only removes all translational DOF of the selected node or beam. Both nodes and beams can be selected to specify the **Origin.** The **Offset** parameter will only be activated if a beam is selected to define **Origin**. The offset can be specified in either absolute or relative values. An absolute value of 100 mm will place the constraint in the middle of a 200 mm beam. Pin constraints simulate hinge and pin-hole connections, and joints as illustrated below:

Pin Connection

Hinge Connection

A single pin-type constraint will behave more like a spherical joint/constraint, as it has three degrees of rotation. However, in most applications there will be more than one constraint and in such circumstances the constraint will behave more like a pin constraint, as its rotational DOF will be restricted.

Floating pinned constraint

Floating Pinned Constraint restricts rotation and translation in one plane only for a selected node or beam. Both nodes and beams can be selected to specify the **Origin**. Additionally, the direction can be specified by selecting work axes, work planes, or beams. The **Offset** parameter will only be activated if a beam is selected to define the **Origin**. The **Offset** can either be specified in absolute or relative values. An absolute value of 100 mm will place the constraint at one-third of the length along a 300 mm beam. The angle of the plane – where the constraint has one degree of displacement – can also be specified in addition to the angle of the constraint – with reference to the default *Z* axis. Floating pinned constraints simulate rollers, wheels and smooth surface-type joints, as illustrated below.

The table below is a summary of the supports and connections the DOF of which is fixed depending on the type of standard support/constraint used.

		Translational DOF			Rotational DOF		
		X	*Y*	*Z*	R_x	R_y	R_z
1	Fixed support	√	√	√	√	√	√
2	Pinned connection	√	√	√			
3	Floating support*		√		√		√

Finally, below is a summary of the types of results available, including reactions and moments, depending on the type of standard constraint used.

		Reacting forces			Reacting moments		
		R_x	R_y	R_z	M_x	M_y	M_z
1	Fixed support	√	√	√	√	√	√
2	Pinned connection	√	√	√			
3	Floating support*		√				

*Fixed and loaded in a plane normal to the Y axis.

Custom constraint

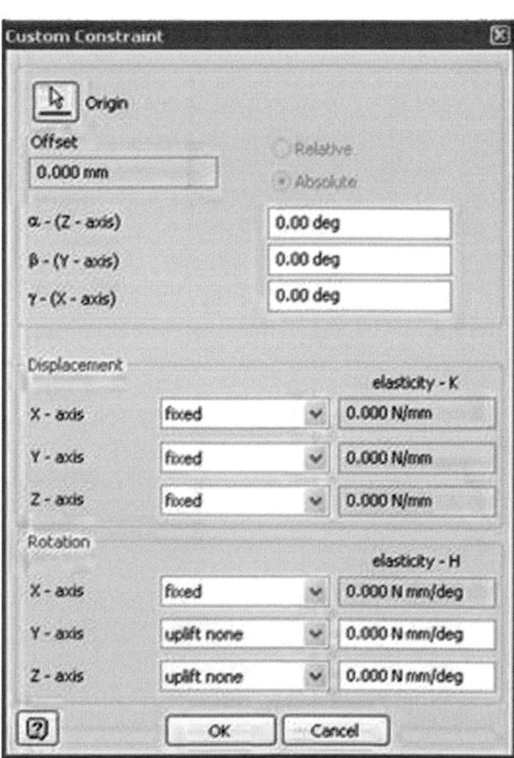

Custom constraint is the only constraint that allows control of its six DOF. For example, a custom constraint could be used to define a pin constraint by fixing its two degrees of rotation and three degrees of translation. **Custom constraint** also allows the specification of stiffness (elasticity), enabling simulation of connections between moving components.

Below are the details of all the options:

α – **(Zaxis)** – Specify the angle of constraint rotation about the Zaxis

β – **(Yaxis)** – Specify the angle of constraint rotation about the Yaxis

γ – **(Xaxis)** – Specify the angle of constraint rotation about the Xaxis

Fixed – Means that the joint behaves like a fixed constraint.

Uplift none – Means that the DOF (rotation and displacement) is free and not restricted.

Uplift+ – Means that the DOF (rotation and displacement) is free only in the positive direction with respect to the beam coordinate system (local).

Uplift− – Means that the DOF (rotation and displacement) is free only in the negative direction with respect to the beam coordinate system (local).

Also refer to Example 2 and further details on the custom node, which is available in the Inventor Simulation 2011 **Help** menu.

LOADS TAB

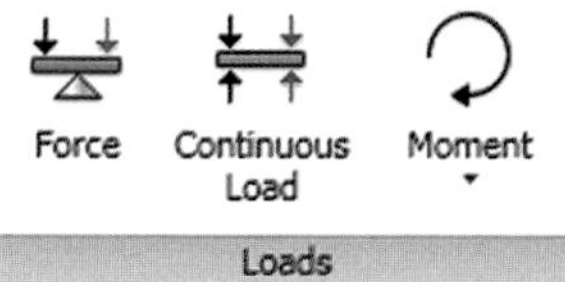

Force

To fully define a force, an origin, magnitude, and direction are required. **Direction** can be specified by selecting either the beam or axis. Alternatively, the direction can be specified by specifying **Angle of Plane** and **Angle in Plane**, where **Angle of Plane** rotates the *XY* plane of where the load is defined and **Angle in Plane** defines the angle of force from the Zaxes. The **Offset** is only available if the beam is both selected and can be defined in absolute and relative values; for example, a relative value of 0.5 will place the force in the middle of the beam.

Below is an example of force being defined using **Vector** values.

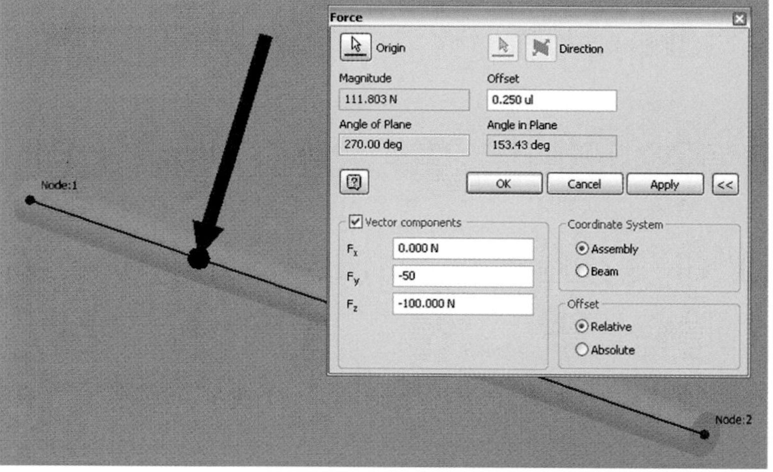

Continuous load

To fully define a continuous load, an origin, magnitude, and direction are required. **Direction** can be specified by selecting either the beam or axis. Alternatively, the direction can also be specified by specifying **Angle of Plane** and **Angle in Plane**, where **Angle of Plane** rotates the *XY* plane of where the load is defined and **Angle in Plane** defines the angle of force from the *Z* axes.

Below is an example of a continuous load being defined using the HUD:

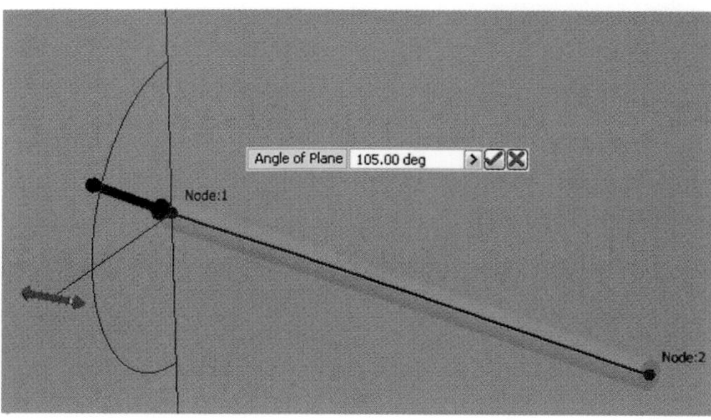

Split the frame into multiple frames if a continuous load needs to be applied on a portion of the beam.

Moment

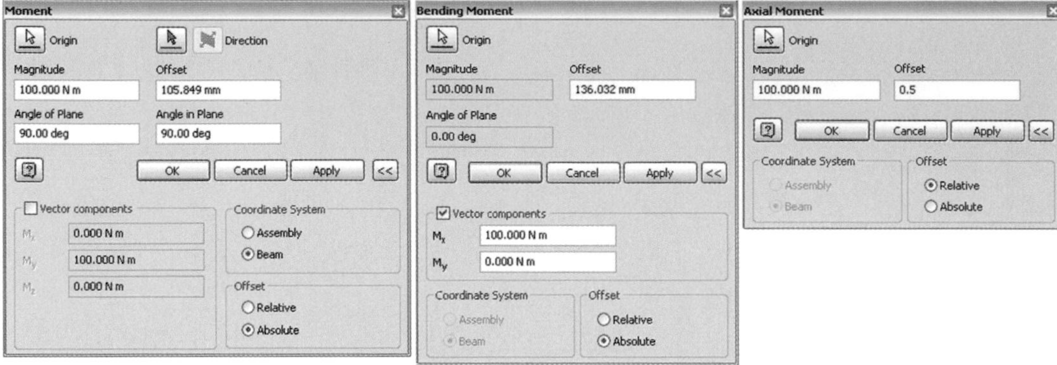

Moment (general)

To fully define a general moment, an origin, magnitude, and direction are required. **Direction** can be specified by selecting either the beam or axis, including specifying moment in the beam and assembly coordinate system. Alternatively, the direction can be specified by specifying **Angle of Plane** and **Angle in Plane**, where **Angle of Plane** rotates the *XY* plane of where the moment is acting and **Angle in Plane** defines the angle of moment from the Z axes. The offset is only available if the beam is selected and can be defined in absolute and relative values; for example, a relative value of 0.5 will place the force in the middle of the beam.

Bending and axial moment can also be defined by using general moment. Below is an example of a bending moment being created using HUD.

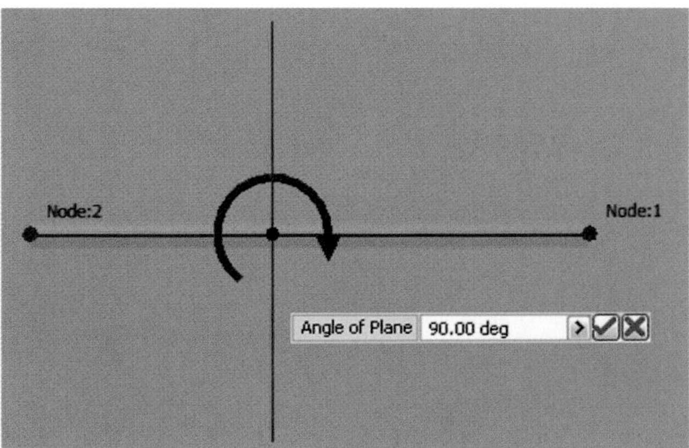

Bending moment

Creates a bending moment on the selected beam and is applied in the plane parallel to the beam axis. It works in the beam coordinate system only and requires fewer input values than the general **Moment** dialog box.

Axial moment

Creates an axial moment on the specified beam and is applied in the plane perpendicular to the beam axis. It works in the beam coordinate system only and, again, requires fewer input values than the general **Moment** dialog box.

EXAMPLE 1 – CANTILEVER MODEL RESULTS COMPARED WITH HAND CALCULATIONS

The following example is of a cantilever loaded at one end and fixed at the other.

The beam, made out of mild steel (E = 220 GPa), is 200 mm long and 10 mm in diameter. With a 100 N load applied, the maximum deflection and bending stress are 2.469 mm and 203.7 MPa, respectively.

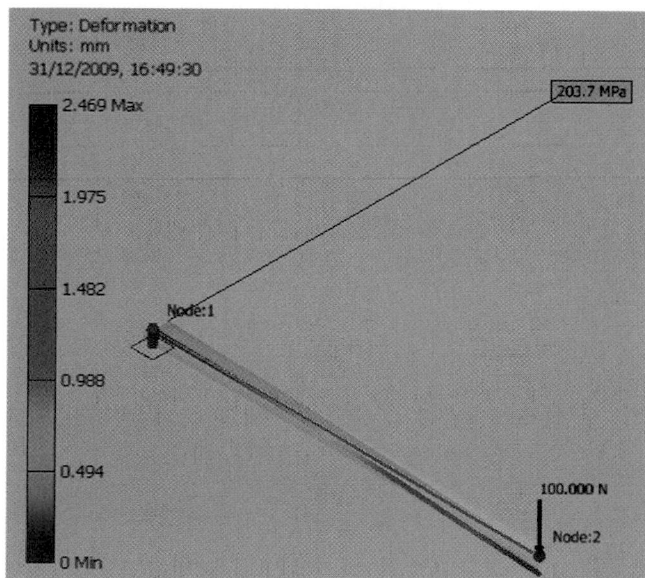

As a comparison, the maximum deflection and bending stress based on theoretical results, using the following formulae, are **2.469 mm and 203.7 Mpa**, respectively.

where:

W = load
L = length
E = Young's modulus
I = second moment of area.

I for circle is

$$\pi D^4/64 = \pi \times 0.04^4/64 = 4.908 \times 10^{-10}$$

Based on the classical bending stress formula

$$\frac{M}{I} = \frac{\sigma}{y}$$

Maximum deflection $= WL^2/3EI = 100 \times 0.2^2/3 \times 220 \times 10^9 \times 4.908 \times 10^{-10} = 0.002469\text{m}$

where:

M = Maximum bending moment = $F \times L = 100 \times 0.2 = 20$ N m
σ = Maximum stress
y = distance from neutral axis

Maximum Stress = $20 \times 0.005/4.908 \times 10^{-10} = 203.7 \times 10^6$ N/m^2

Try it out yourself: Open *Singlebeam.iam*

EXAMPLE 2 – SIMPLY SUPPORTED BEAM CREATED WITH CUSTOM CONSTRAINT

The model used here is the same as in Example 1, except that it is simply supported by pinned constraints at either end, with a load applied in the middle.

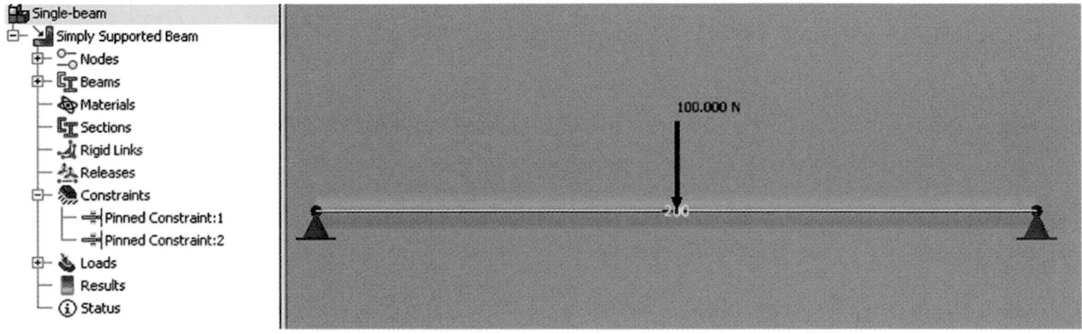

With a 100 N load applied, the maximum deflection and bending stress are 0.1543 mm and 50.93 Mpa, respectively.

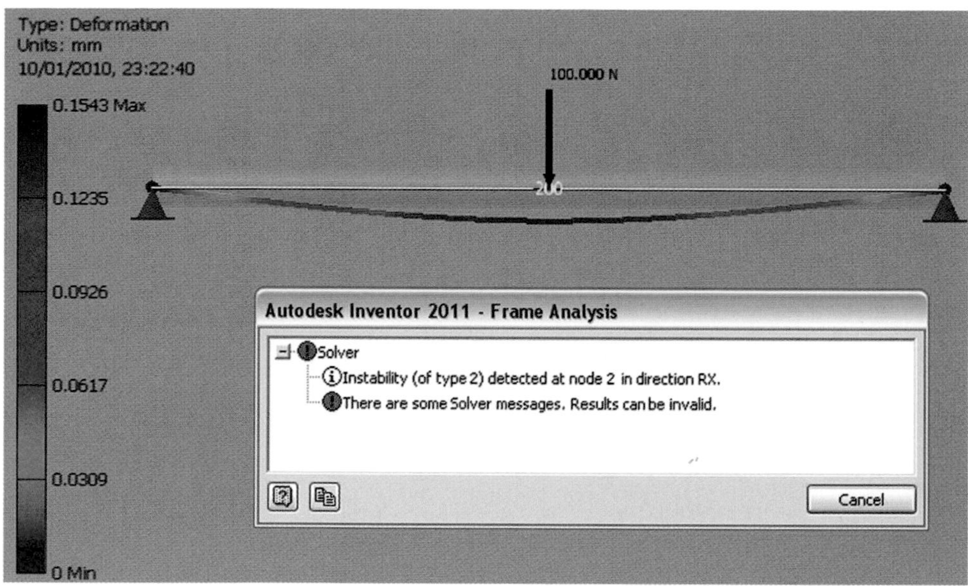

Although the results are correct based on the theoretical results, a warning appears detailing that the beam is free to rotate about its axis. This is due to the fact that applying a second pinned constraint on the other end of the beam has restricted the *Y* and *Z* (global) rotational DOF of both constraints. However, none of the constraints restrict motion about the axis of the beam. To avoid the warning, we can replace one of the pinned constraints with a custom constraint, with all displacements and rotation about the axis of the beam fixed.

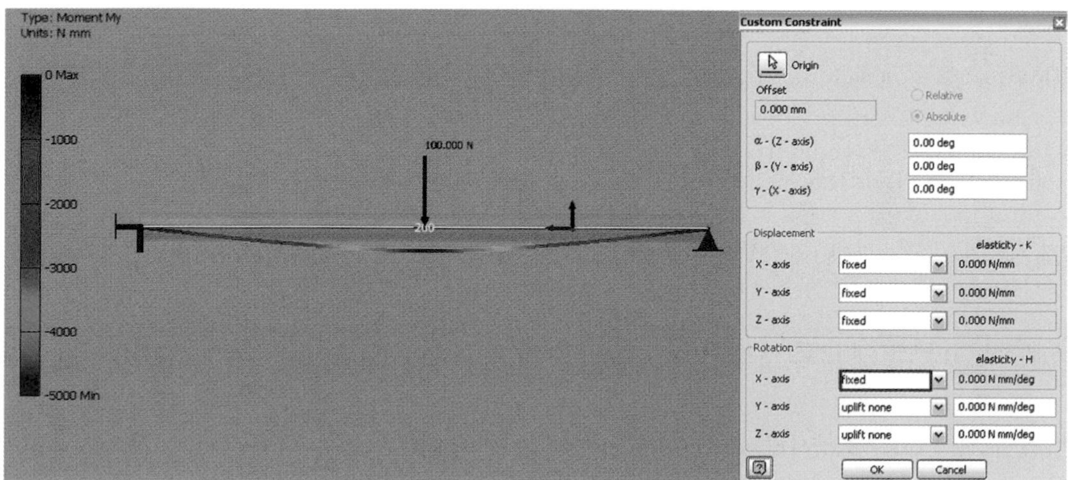

Try it out yourself: Open *Singlebeam.iam*

CONNECTIONS TAB

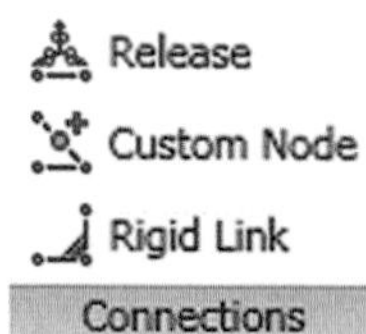

Beam release

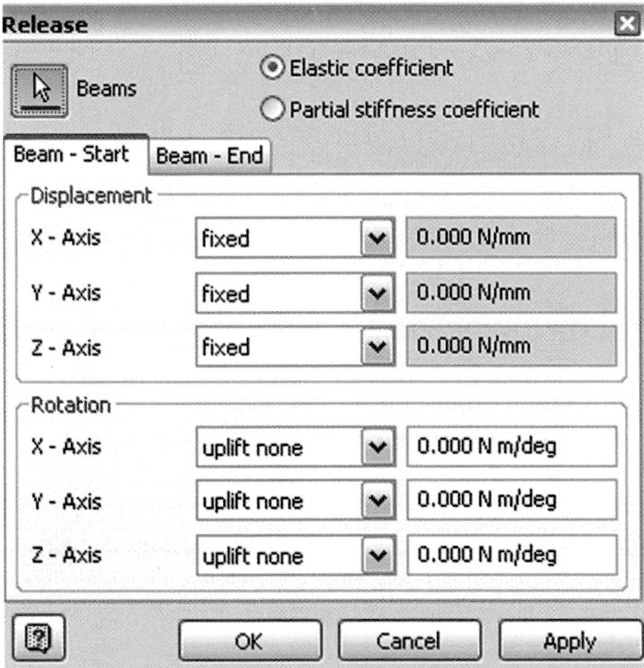

Beam release allows the release of rotational and translational DOF in beam connections that have been created automatically; by default, beam end connections are rigid. For example, by releasing the rotational DOF, the fixed connection will convert to a pinned connection at the beam ends, thus removing moments in the beam. Refer to the beam release example. In addition to releasing DOF, elastic coefficients can also be specified to create stiffness at the beam ends, enabling some flexibility at the connections. When **Partial stiffness coefficient** is selected, a value of 1.0 will mean no release and a value of 0 will mean maximum release.

When there are two or more adjoining beams, one of them should be fixed without a release at that beam end, as the constraint already contains information about the boundary conditions of the adjoining beam.

While editing, the beam coordinate system is displayed near the start end of the beam.

Fixed – Means that the joint behaves like a fixed constraint.

Uplift none – Means that the DOF (rotation and displacement) is free and not restricted.

Uplift+ – Means that the DOF (rotation and displacement) is free only in the positive direction with respect to the beam coordinate system (local).

Uplift– – Means that the DOF (rotation and displacement) is free only in the negative direction with respect to the beam coordinate system (local).

EXAMPLE 3 – RELEASING MOMENTS IN A STRUCTURE USING BEAM RELEASE

In this example, the following frame is loaded as shown.

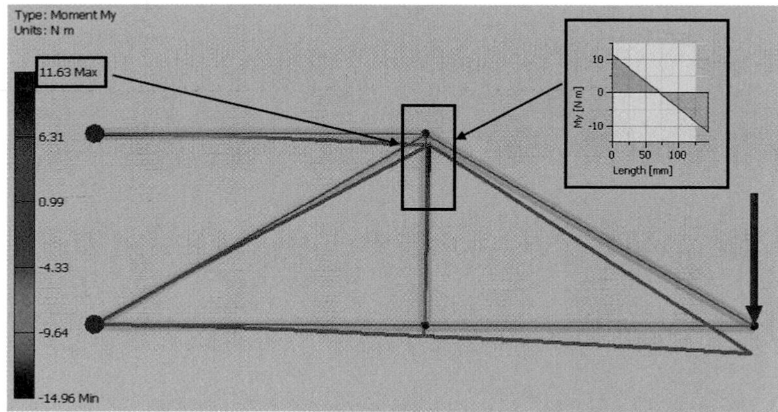

The maximum bending in the vertical member is 11.63 N m and, again, this is due to the frame member not having any rotational DOF. In this example there are no rigid links, as in the previous example, in which the rotational DOF were released. This is due to the fact that this truss frame has members with no end treatments; hence, no gaps resulted between the members. Nevertheless, we can achieve the same effect by using the beam release command.

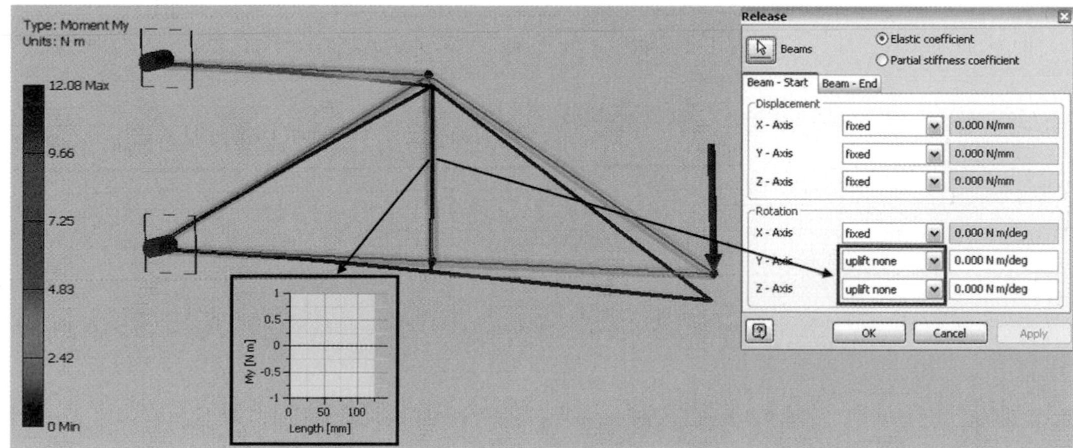

As we can see, releasing rotation in the Y and Z axes for both ends of the beam results in zero moment in the beam.

Try it out yourself: Open *Truss.iam*

Custom node

Custom nodes can be created anywhere along a beam and can be used to place forces and create rigid links. They are not graphically different from automatically converted nodes in the browser. However, we can assign different colors in the **Frame Analysis** settings.

Rigid link

Rigid links are used to join disconnected beams together. A rigid link comprises a parent node and a child node. All displacements and rotations of the parent node are passed on to the child node; for example, compatibility between both nodes is maintained by the parent node. Displacements and rotations defined for a rigid link can be changed; for example, the **Rigid Link** setting below will maintain translation only between the nodes.

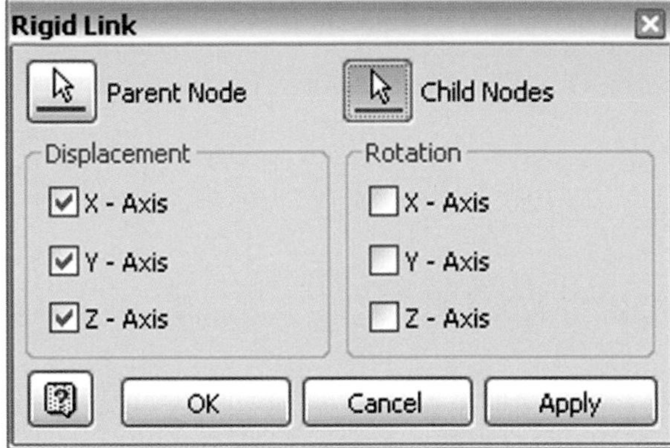

Rigid links only act between nodes.

Rigid links are only defined in assembly coordinate systems.

Rigid links require two nodes: one parent and one child.

Below is an example of how two simple beams are connected together using rigid links.

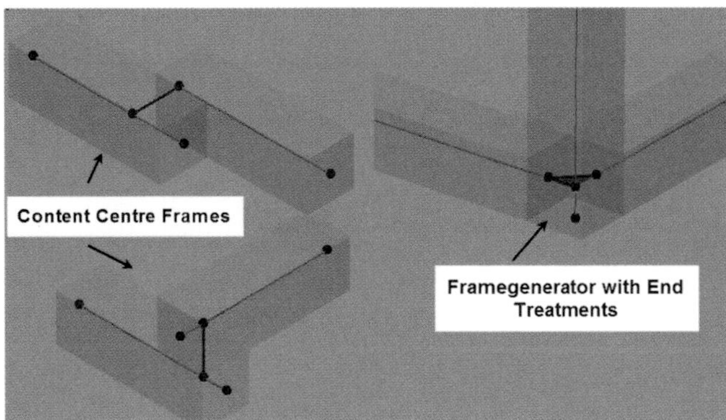

Use Frame Generator to creator frames if the frames are to have end treatments, as this will have an impact on the creation of rigid links between mitered/butt-connected beams.

Below is an example of how four beams with end treatments, using Frame Generator, are connected with rigid links.

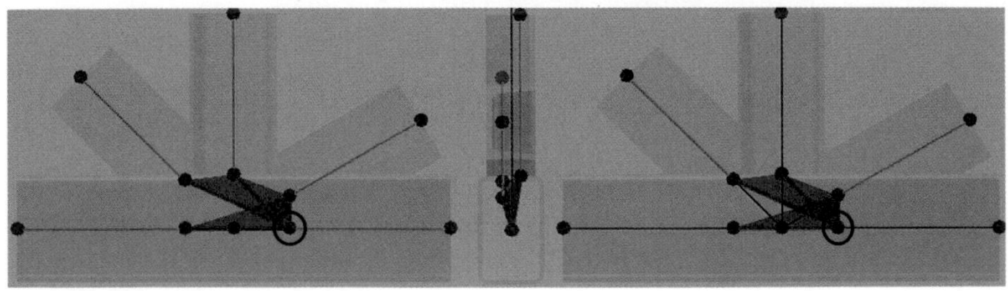

A parent node is one in which all rigid links have a common connection point.

The blue lines are the axes used to create offset frames using Frame Generator and the green lines are the beam elements created in Frame Analysis based on the true neutral axis of the beam. Therefore, it is important to note that the beam elements are not necessarily created on the same axis on which the frames were originally created, with the exception of, for example, circular and square members. The above image shows a typical frame structure connection used in the Kone Escalator Support Structure as shown in Chapter 9.

Further information on how rigid links are connected between beams is available from the Autodesk Inventor Simulation 2011 **Help** file.

EXAMPLE 4 – SIMPLE FRAME IN WHICH BEAMS ARE CONNECTED USING RIGID LINKS

The following support frame example is loaded at one end and fixed at the other end. Further, the example is created using Frame Generator with no end treatments. The deformation results show a maximum deflection of 2.035 mm and a maximum normal stress of 118.8 MPa.

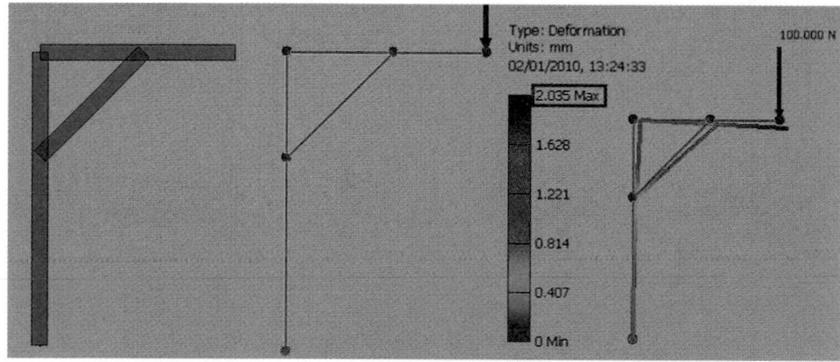

Try it out yourself: Open *Simple-Frame.iam*

Now, the same example with frame generator end treatments is simulated again.

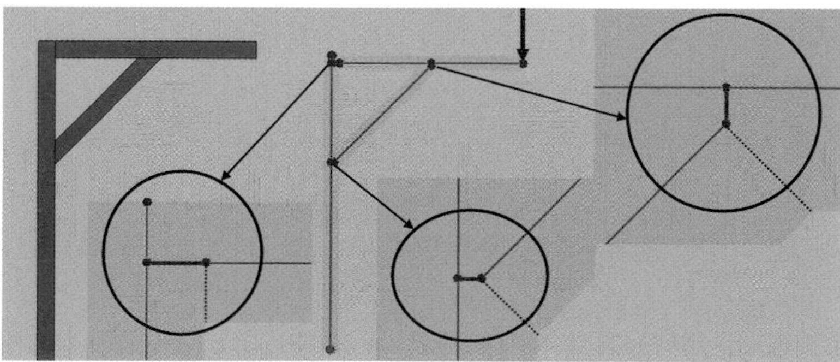

The immediate difference, even before we run the analysis, is the creation of the red rigid links. These links are automatically created and join the disconnected beams, due to trimming. The beams are extracted based on the neutral axis position and their lengths are defined by the intersection of the longest face of the solid beam with a plane perpendicular to the neutral axis (see above). As a result of this, the rigid links are created between the beam elements using the shortest possible distance. However, the difference in the results is negligible as the maximum deformation and stress are 2.039 mm and 118.8 Mpa, respectively.

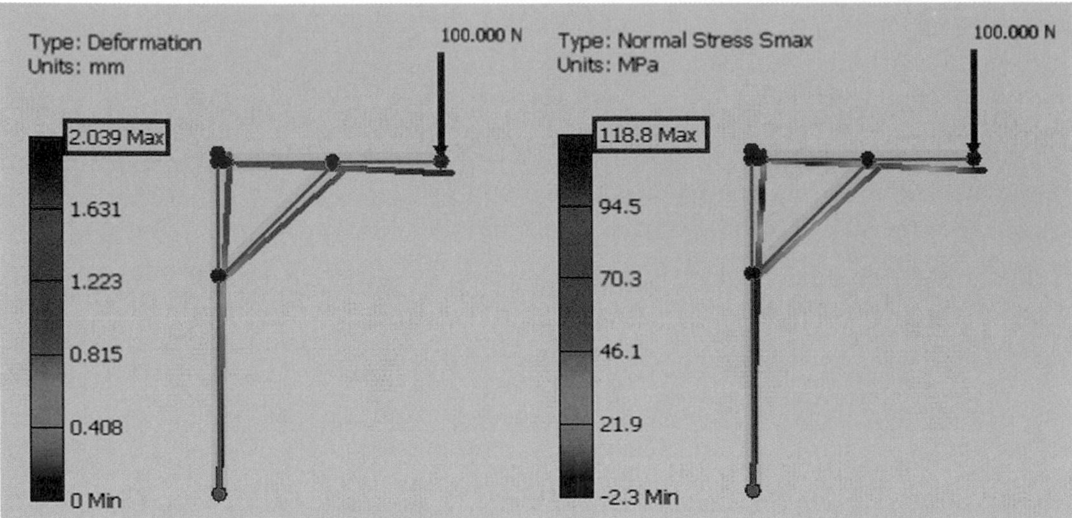

The reason for this similar result is that the cross-section of the beams is very small compared to the length of the beam. In most structural applications this will always be the case, hence the reason for using beam elements rather than solid elements.

If the example is further analyzed, the beam connecting the horizontal and vertical beams is shown to have moments as shown below.

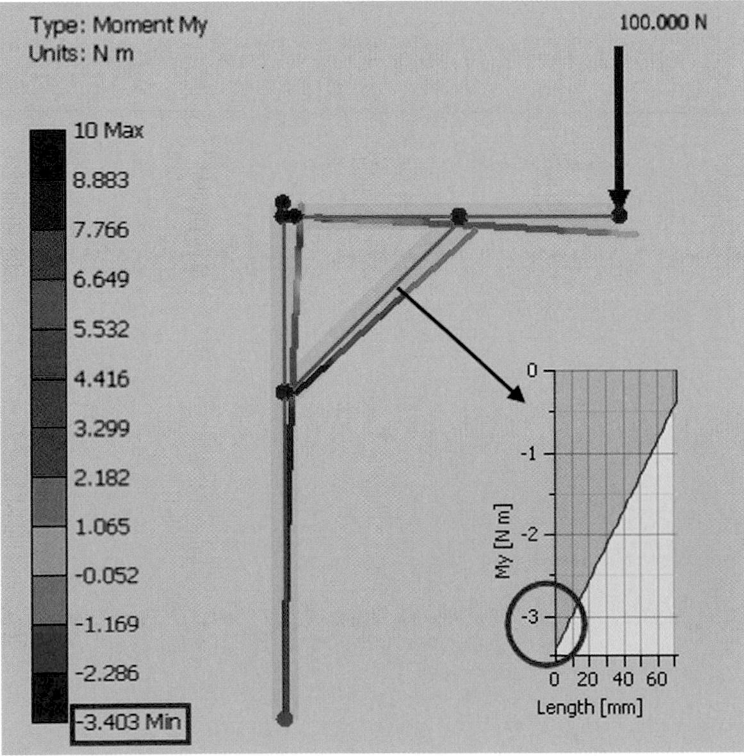

The reason for these moments is that the beam is completely fixed and thus does not allow for beam rotations at the end, as would a pinned connection. To isolate moments in the beam, the rotational DOF can be released.

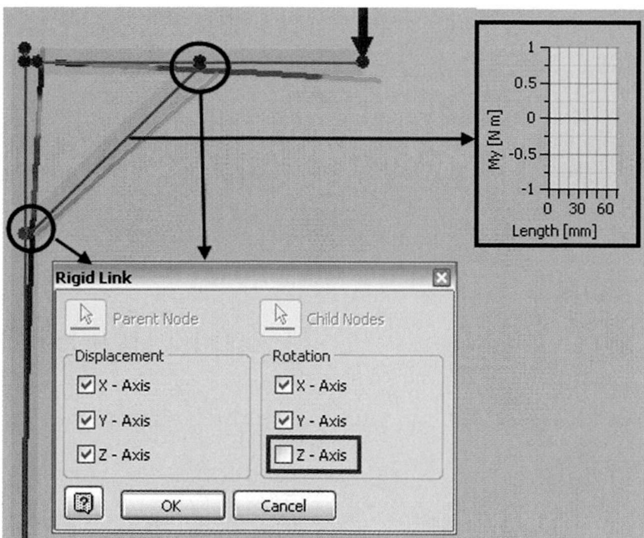

Rotation about the *Z* axis (global coordinate system) is released and thus removes moments completely in the beam. The stress (axial) now induced in the beam is completely a result of axial loading and not bending.

Try it out yourself: Open *Simple–Framegenerator.iam*

RESULT TAB

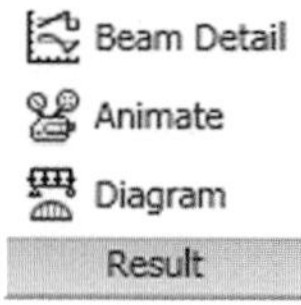

Beam detail

Beam Detail allows quick analysis of the details of the results for selected beams including maximum bending and forces.

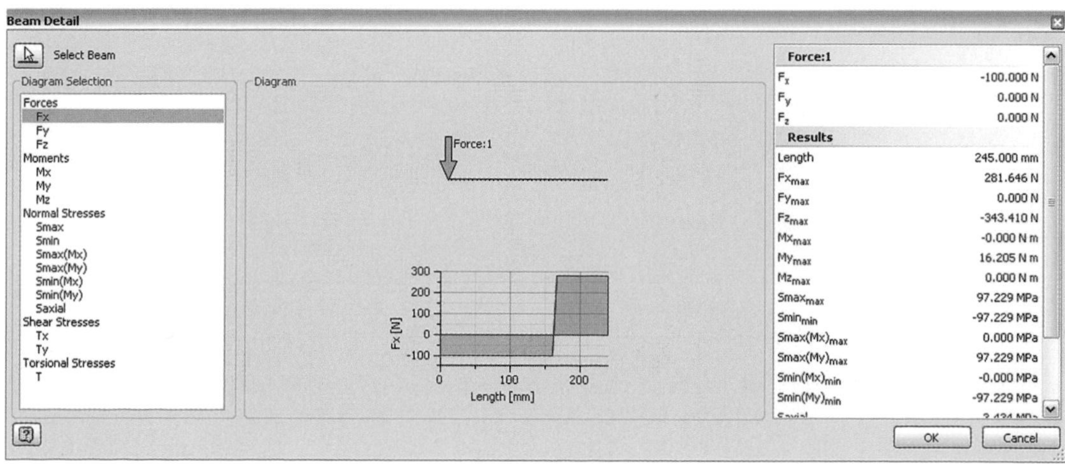

Animate

Creates a video file of the animation.

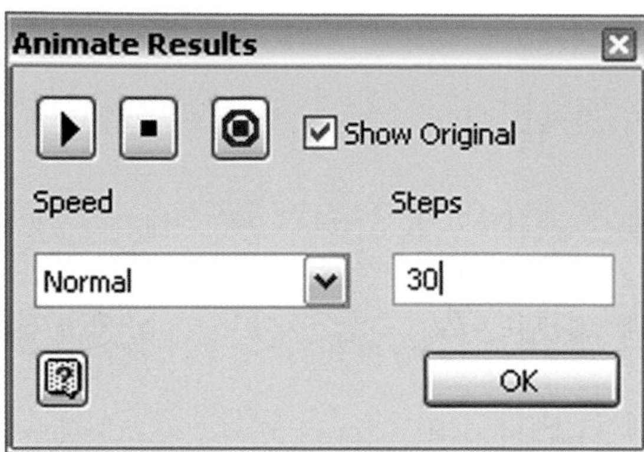

If **Show Original** is selected then the original model shape is visible during animation playback. On the other hand, if the option is not selected, the original wireframe is presented as an overlay on the deformed model. This option is checked by default.

For a smoother display, increase the number of steps. You can specify any value between 1 and 100.

Diagram

Plots specific results, as diagrams, on beam models; result types include maximum forces and moments.

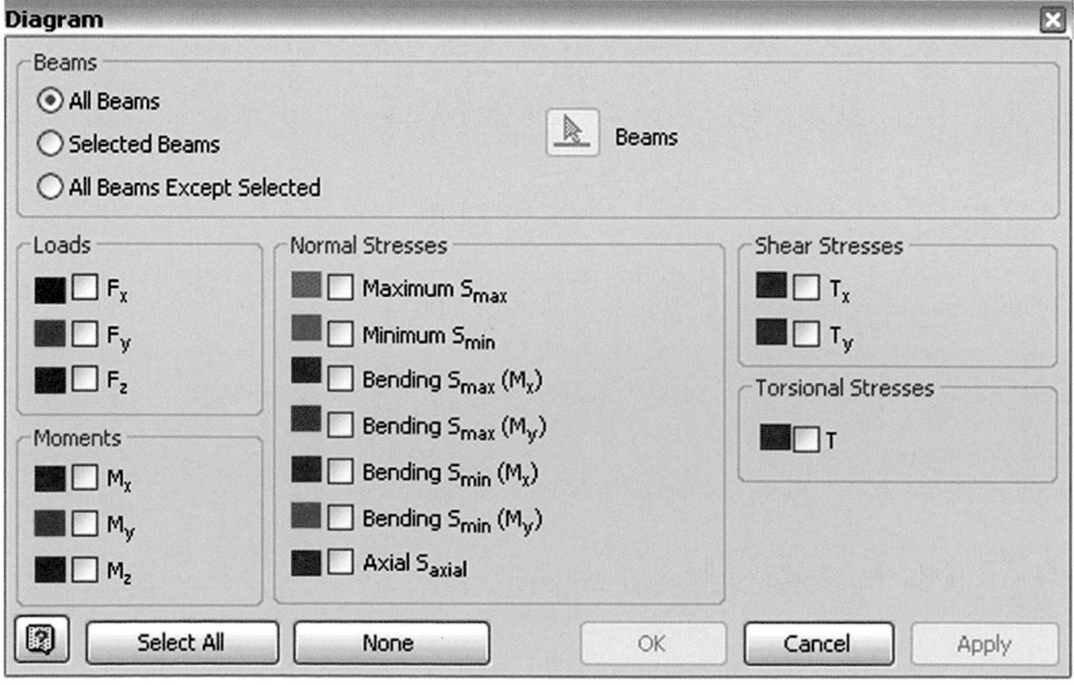

Multiple diagrams can be applied at the same time for the selected beam. The diagrams are plotted according to the beam local coordinate system and on the undeformed beams. The graphs that would normally be displayed along the beam axis are displayed along the Y or Z axis of the beam coordinate system (in plane *XY* or *XZ*). The following shows a diagram of the structural beam model.

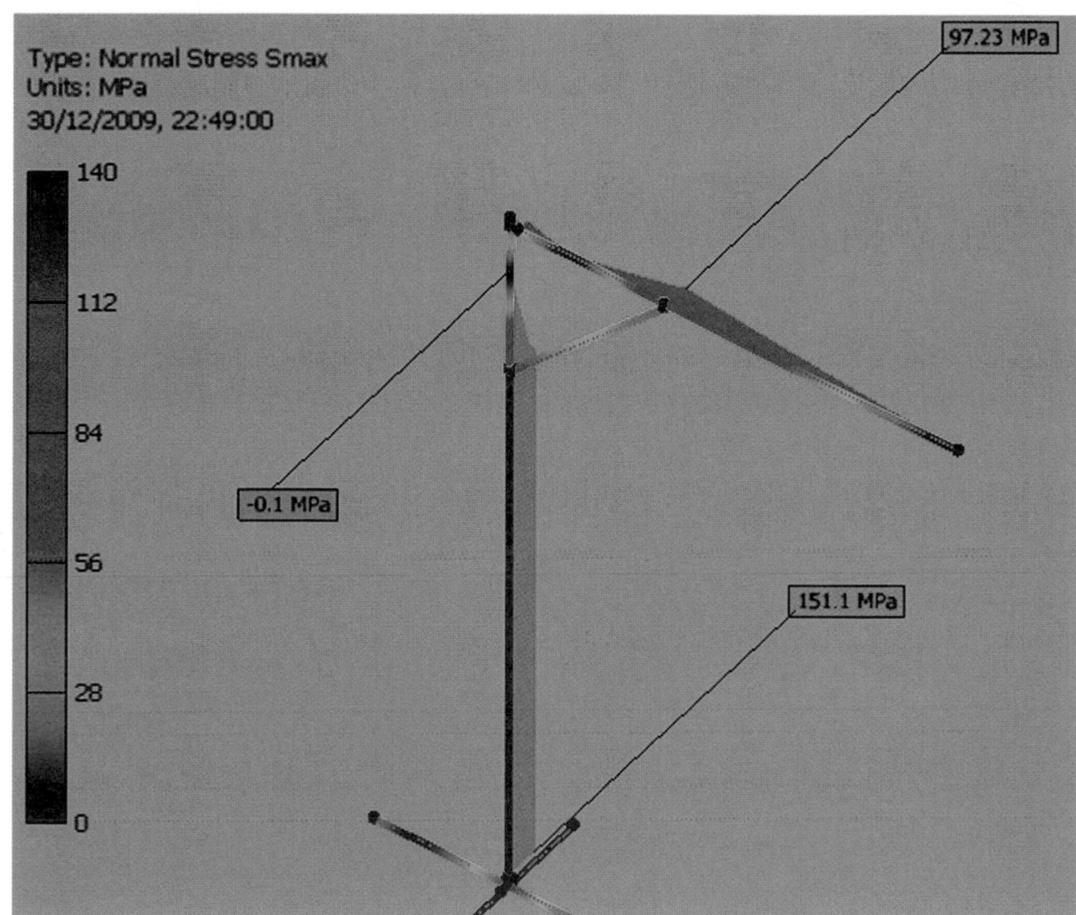

The complete list of result displays available is shown below.

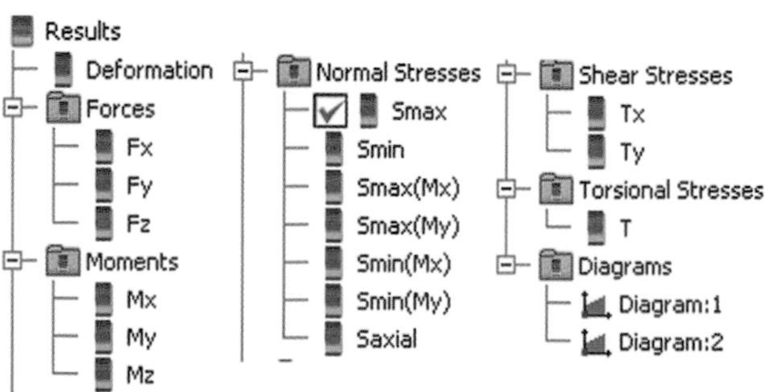

DISPLAY TAB

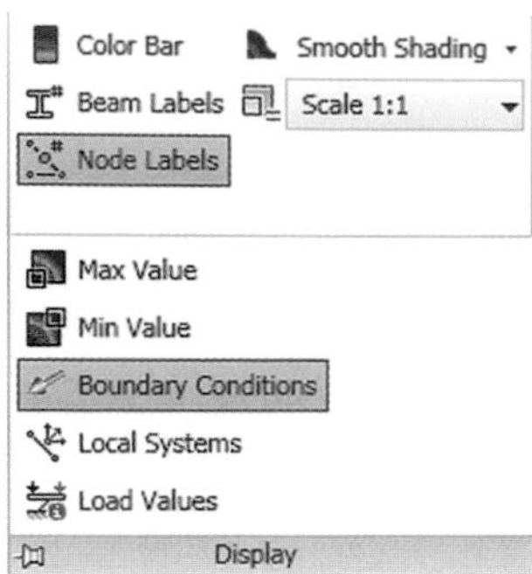

Color bar

The **Color Bar** is probably the most important tool within the **Display** panel and, when effectively used, can help with understanding the results with ease. It can be displayed in various locations in the graphic window using the **Position** setting. The maximum and minimum threshold values can be altered by unchecking the **Maximum** and **Minimum** values.

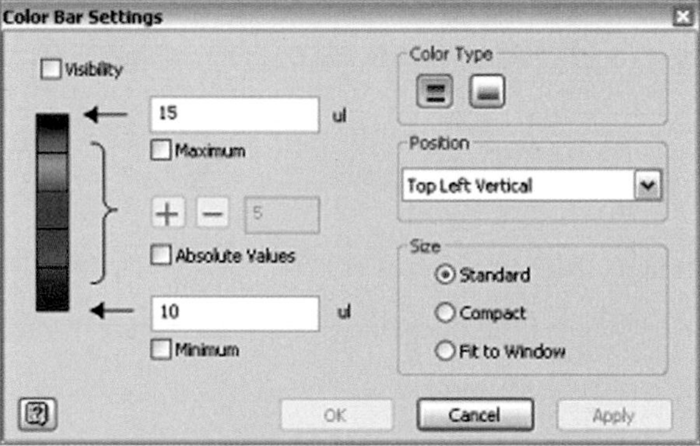

Absolute Values, when checked, displays all result values in absolute values and the color bar reflects those values.

When **Absolute Values** is checked, the negative values for the maximum and minimum thresholds are invalid.

The numbers of the color legend can only be changed when **Contour Shading** is selected. **Smooth Shading** by default will use **Maximum**.

Beam and node labels

These help with visually identifying specific nodes and beams by displaying labels, as illustrated below.

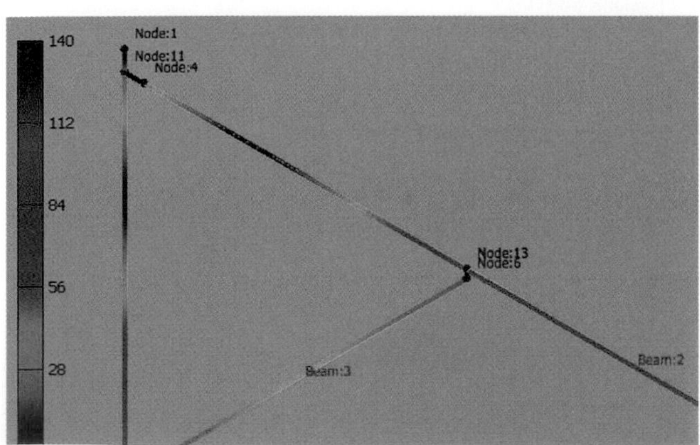

Display results

Here you can decide whether you want the results to be displayed with smooth, contoured, or no shading.

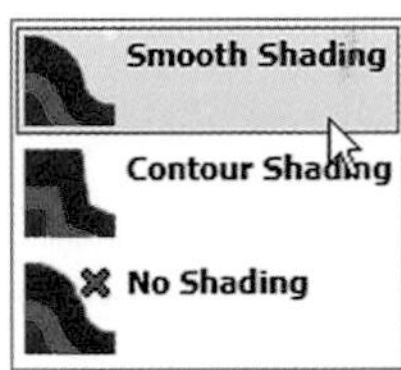

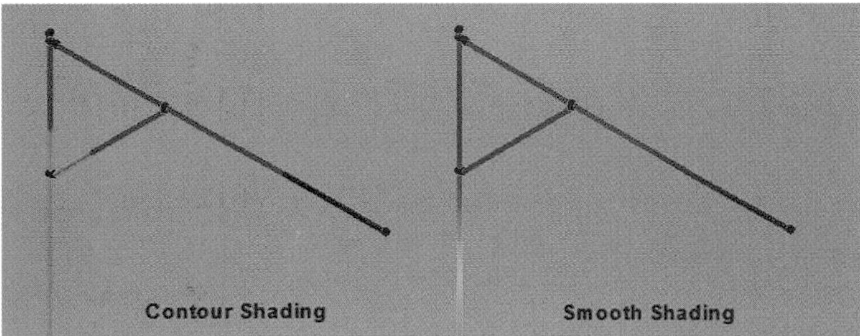

Adjust displacement display

Here you can adjust the scale of the results to obtain a better indication of whether the boundary conditions applied are correct.

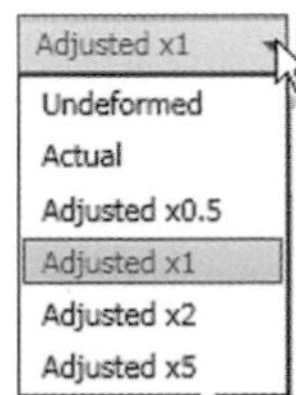

Adjust the scale so that the deformation is visible before selecting **Animate results,** as animations without visible deformation are less visual.

Max and min values

Displays the maximum and minimum values, for the selected results type, aiding in locating their positions on the model, as illustrated below.

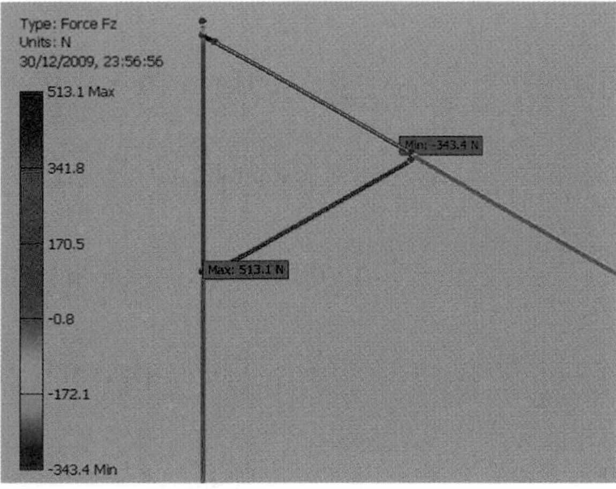

Boundary conditions

Displays all the loads and constraints applied on the model.

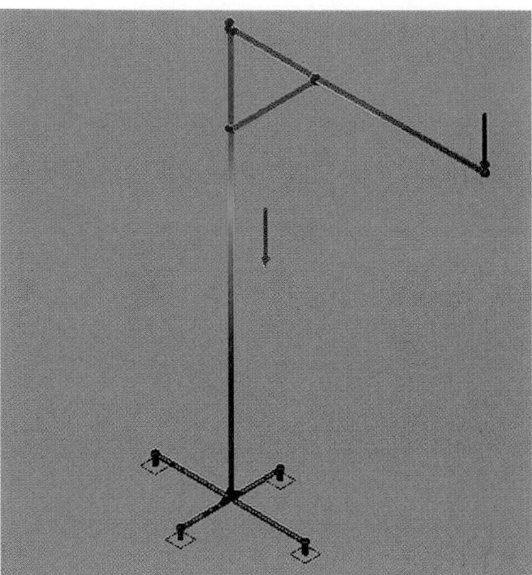

Local systems

Displays the local coordinate systems for all beams.

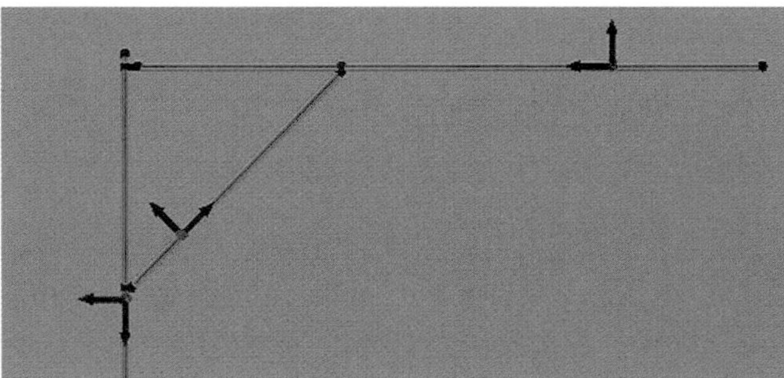

Load values

Displays the load values associated with all the loads applied on the model.

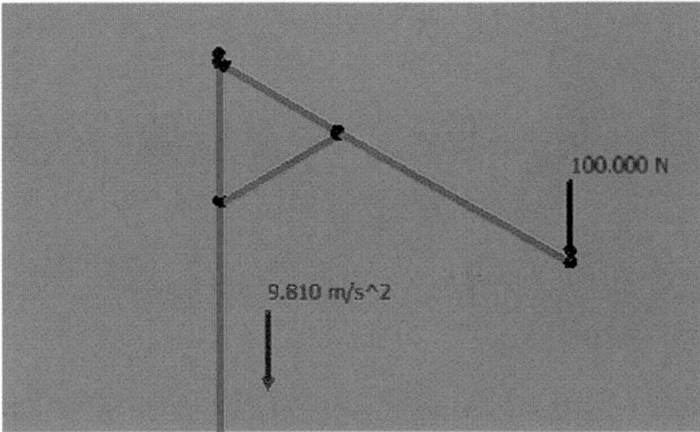

Boundary Conditions, needs to be active in order to be able to see load values.

PUBLISH TAB

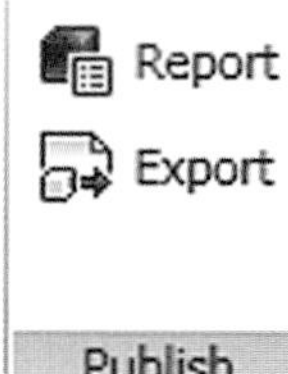

Report

Autodesk Inventor 2011 – in addition to standard html format – now lets you create reports in mhtml (single web page) and rich text formats (Word documents), making it very easy to customize the reports to specific requirements.

Microsoft Word is required to generate the RTF file.

In addition to being able to customize settings using the **General, Properties,** and **Simulation** tabs from the **Report Generator** dialog box, there are now more settings within the **Format** tab.

Use Dynamic Content – Select this to include size buttons for image width and buttons that you can click to collapse or expand the associated sections.

Not available for the RTF format.

Create OLE Link – Select to create an OLE link from the model browser to the report. The report icon displays under the **Third Party** folder in the model browser. To edit the report, double click the icon or right click and select **Edit**.

Frame Analysis Report

Autodesk

Analyzed File:	Single-beam.iam
Version:	2011 Beta2 (Build 150194000, 194) Debug 0
Creation Date:	21/01/2010, 16:29
Simulation Author:	Wasim
Summary:	

⊟ Project Info (iProperties)

⊟ Summary

Author	Wasim

⊟ Project

Part Number	Single-beam
Designer	Wasim
Cost	£0.00
Date Created	20/12/2009

⊟ Status

Design Status	WorkInProgress

⊟ Physical

Mass	0.123 kg

⊟ Static Result Summary

Name	Minimum	Maximum
Deformation		
Deformation	0.000 mm	2.469 mm
Forces		
Fx	-100.000 N	-100.000 N
Fy	-0.000 N	-0.000 N
Fz	0.000 N	0.000 N
Mx	-0.000 N m	0.000 N m
My	0.000 N m	20.000 N m
Mz	0.000 N m	0.000 N m
Normal Stresses		
Smax	0.000 MPa	203.718 MPa
Smin	-203.718 MPa	0.000 MPa
Smax(Mx)	-0.000 MPa	0.000 MPa
Smin(Mx)	-0.000 MPa	0.000 MPa
Smax(My)	0.000 MPa	203.718 MPa
Smin(My)	-203.718 MPa	0.000 MPa
Saxial	0.000 MPa	0.000 MPa
Shear Stresses		

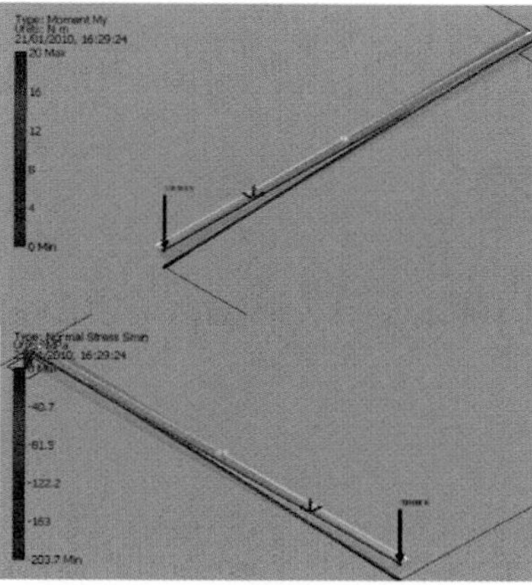

Export

Autodesk Inventor also allows you to export frame analysis data to Autodesk Robot Structural Analysis 2011 data in RTD file format.

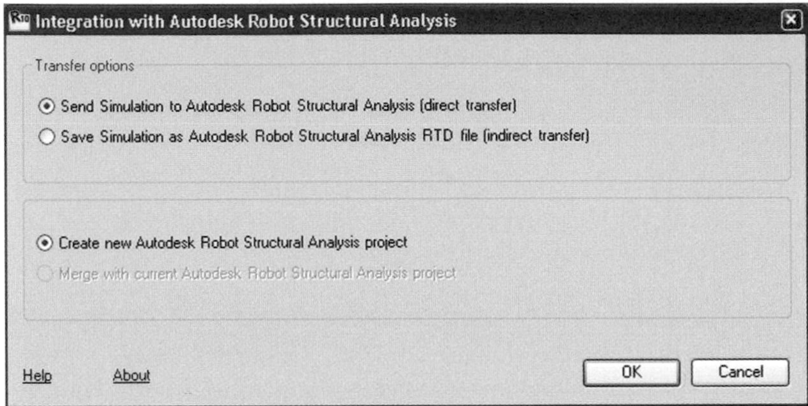

Only available if an active Robot Structural Analysis license is installed.

The RTD file contains all defined loads, constraints, rigid links, releases, beam materials, and beam sections.

The following export options are available:

- Save the simulation as an RTD file.
- **Send Simulation to Autodesk Robot Structural Analysis** – Directly creates a calculation model in Autodesk Robot Structural Analysis based on the current frame analysis data in Autodesk Inventor.
- **Create New Autodesk Robot Structural Analysis project.**
- **Merge with current Autodesk Robot Structural Analysis project.**

Further details on **Export** are available in the Autodesk Inventor Simulation **Help** menu.

FRAME ANALYSIS SETTINGS TAB

Allows the predefinition of settings for current and preceding analysis.

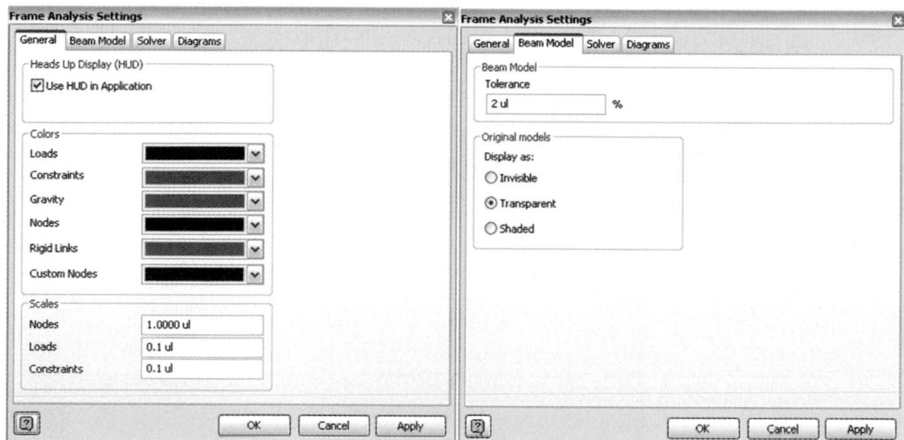

Use HUD in Application – Checked by default. Uncheck it if you want to use the dialog boxes for editing and creating boundary conditions, etc.

While creating boundary conditions, with HUD active, you can right click and select **More Options.** This will allow you to edit using dialog boxes.

Colors – Here you can define specific colors for loads, constraints, nodes, etc.

Scales – Here you can alter the visual scale of the nodes, loads, and constraints.

Beam Model – This is the tolerance that dictates whether a rigid link will be created between the beams that are not connected. The default value is 2%. When the distance between two beams is smaller than the sizes of sections multiplied by this tolerance, a rigid link connection is created via the shortest distance between the nearest nodes. You can specify any value between −100% and 500%, with a negative value meaning that the beams need to be intersecting for a rigid link to be created.

Original models – Here you can define how the original models appear in frame analysis. By default, the original models are set to be transparent.

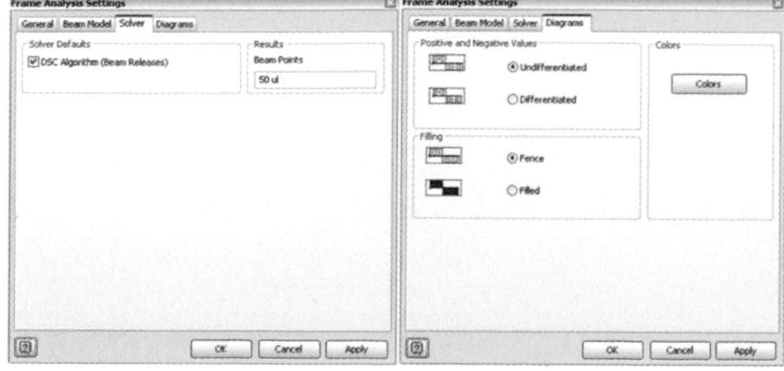

Solver Defaults – DSC Algorithm (Beam Releases) should be checked if the structure contains beam releases. The algorithm carries out the following operations:

1. A new node is generated in the structure (during the structure model generation).
2. The input element with the release is modified in such a way that the new node takes the place of the old one in the element (the old node remains in other structure elements).
3. Between the old and the new node, the program creates the so-called DSC element; see the image below.

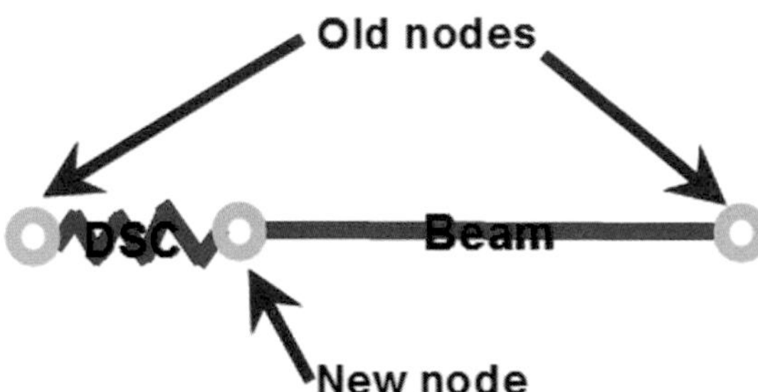

Results – Here you can specify number of beam points that are calculated by solving; this basically splits the beam into smaller linear beam elements. Any value between 5 and 1000 can be specified, with 50 being the default. This value can become important when analyzing a curved beam as the software will need to split the beam into lots of smaller linear elements to provide accurate results, as the software does not support curved beam elements.

Positive and Negative Values – Here you can specify how positive and negative values are visually displayed.

- **Undifferentiated** – No differentiation between positive and negative values
- **Differentiated** – Shows positive and negative values

Filling – Here you can specify how the diagrams will be filled.

- **Fence** – Uses a fence–style shading to display the diagram
- **Filled** – Displays the diagram as completely shaded/filled

Colors – Select to change the colors of the loads and stresses displayed in the graphic diagrams.

CHAPTER 18

DP15 – Frame Analysis Using Content Center Structures

Structural Design of Aerospace Maintenance Platform

(Design Problem courtesy of Planet Platforms Ltd)

KEY FEATURES INTRODUCED IN THIS DESIGN PROBLEM

	Key features
1	Content center frames converted to beam elements
2	Tolerance settings – used to create rigid links automatically
3	Pinned constraints and continuous loading
4	Change material and beam properties

INTRODUCTION

Planet Platforms, established in 1977, is a leading manufacturer and distributor of workplace access solutions. Ranging from podium steps to award-winning intelligent platforms, their access solutions have been keeping people safe for the past 30 years. Through a process of communication, site surveys, CAD-rendered visuals, and proven manufacturing, Planet Platforms delivers the end result – a platform that is safe, reliable, and perfectly suited for the application.

Some of their prestigious clients include Rolls Royce, Thomas Cook, the Orient Express, the Royal Albert Hall, ICI Chemicals, Shell Offshore, and The National Trust.

Up and Running with Autodesk Inventor Simulation 2011. ISBN: 978-0-12-382102-7
Copyright © 2010 Elsevier Inc. All rights of reproduction, in any form reserved.

Above is a picture of a platform, that is used by Rolls Royce personnel to carry out essential maintenance work. The platform is constructed from mild steel and is to be designed such that the platform, including steps, can withstand the load of two people including their maintenance equipment and other essential components. To this effect, a total load of 250 kg will be used to determine:

1. The maximum bending stress in the platform under normal operating conditions.
2. The maximum deflection in the platform.
3. The factor of safety (FOS).

With

- Maximum deflection not above 5 mm.
- Minimum FOS of 4.
- Material construction to be limited to aluminum or mild steel.
- Frame tube construction to be used from Standard ISO Content Center.

WORKFLOW OF DESIGN PROBLEM 15

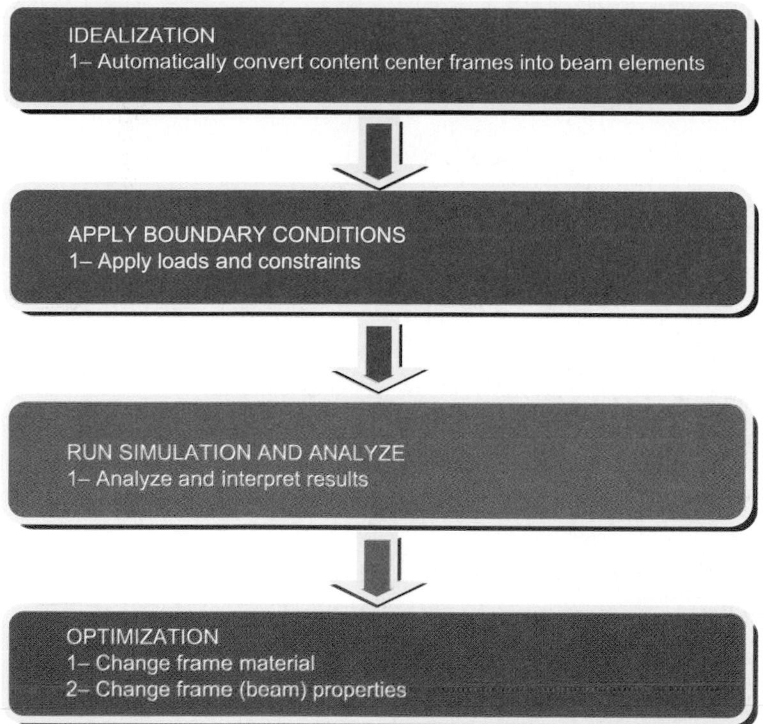

Idealization

Frame analysis within Inventor 2011 automatically converts frames/channels created from both Content Center and Frame Generator. All other components will not be converted and therefore are included in the analysis. Note: if neither Content Center nor Frame Generator is used to create any content then no frames or channels will be idealized into beam elements.

1. Open *Cylinder reaction beam.iam*

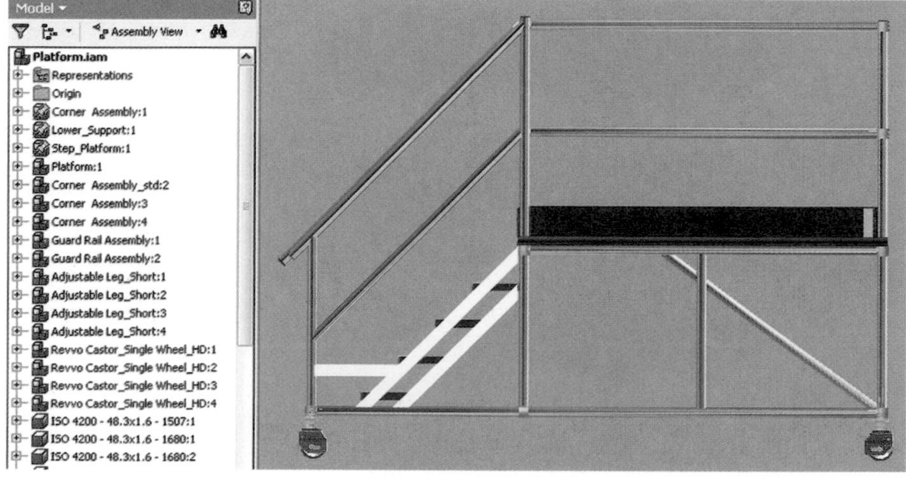

Now we can begin the second stage of the analysis by applying boundary conditions.

Boundary conditions

2. Select **Environments** tab > **Frame Analysis**

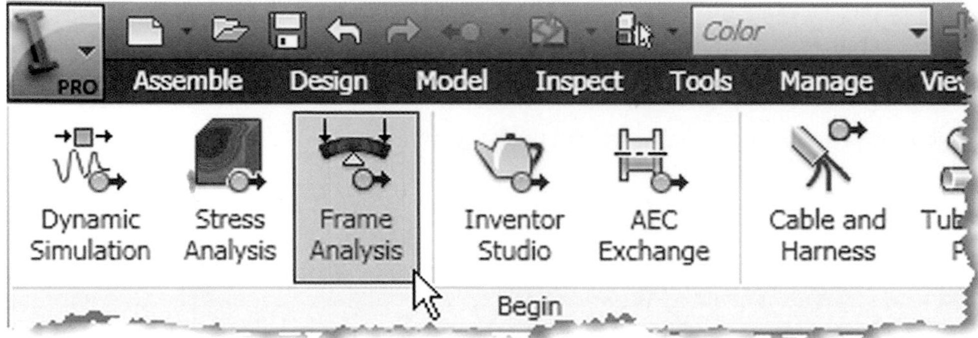

3. Select **Create Simulation** > Select the **Model State** tab > Select **Main-Structure** for **Level of Detail** > Click **OK**

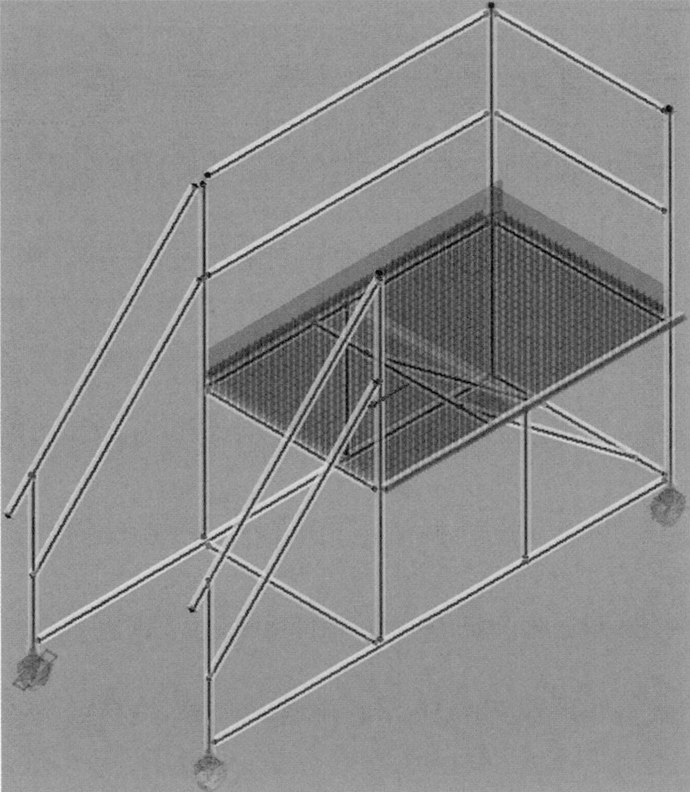

This will idealize all the content center frames into beam elements and exclude all the remaining parts/assemblies from the analysis (shown transparently).

The steps and supporting frames are not included in this beam analysis, as the assembly was not part of the level of detail.

It is also important to note that, as the frames were connected via tee-connectors – which are not included in the analysis – there will be gaps between the ends of the beams, as shown below.

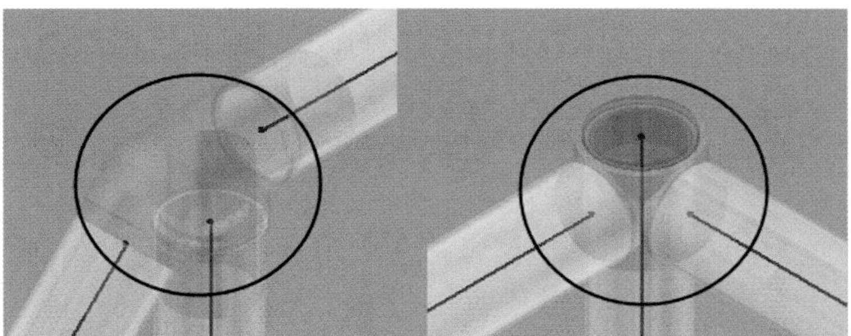

To connect these beams via rigid links, the tolerance needs to be increased.

4. Select **Frame Analysis Settings** > Select the **Beam Model** tab > Change the **tolerance** setting from 2 to 50% > Click **OK**
5. Select **Update**

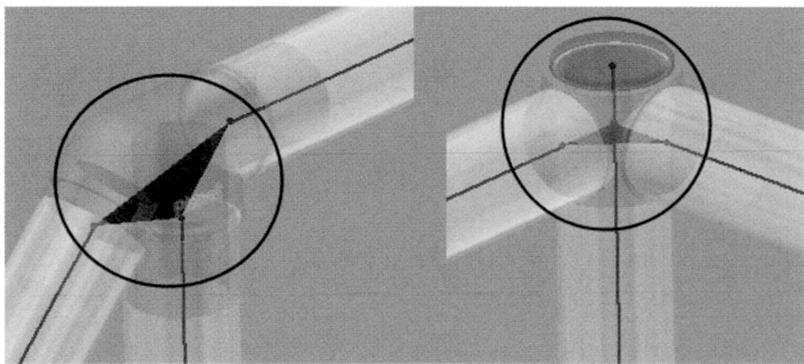

All the beams at the tee connections are now connected via rigid links. This can be verified by simply looking at all the connection points. For large modes this can be tedious and, hence, in these situations it would be quicker to do a modal analysis to check whether all beams are connected by analyzing the deformation results.

Next, we are going to apply the constraints and loads so we can determine the structural integrity of the platform when carrying two people with the necessary maintenance equipment.

6. Select **Pinned Constraint** > Right click > Select **More Options** > Select the node at the bottom near the wheels to define the **Origin** of the constraint > Select **Apply**

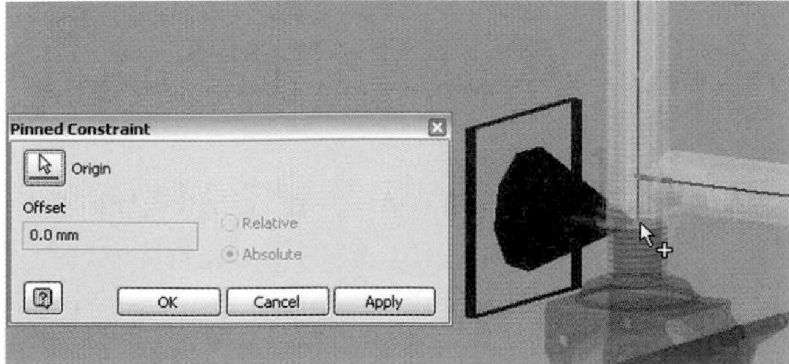

7. Repeat Step 6 to creating pinned constraints at the other three wheels > Click **OK** once the last node has been selected

The platform needs to withstand the load of two maintenance workers, including the weight of the tools, components, etc. This equates to a weight of 125 kg/person. So the total load the platform needs to withstand is 2500, using the value of 10 m/s^2 for gravity

$$\text{Total load} = \text{weight of person} \times \text{number of people} \times \text{gravity}$$

As the platform deck is supported by the frames directly beneath it, this load will be distributed evenly across all these frames. The load we will use is the continuous load in N/mm. To calculate this value, we determine the total length of all the frames directly supporting the platform deck and then divide the total load by the total length of the frames. This will be a good estimate to determine the weight of the two people, in the absence of the platform deck not being analyzed.

$$\text{Continuous load} = \frac{2500}{6.6} = 378\,\text{N/m} = 0.378\,\text{N/mm}$$

8. Select **Continuous Load** > Right click > Select **More Options** > Select the beam as shown > Change the **Angle of Plane** value to 0 > Specify 0.378 N/mm for the load value > Select **Apply**

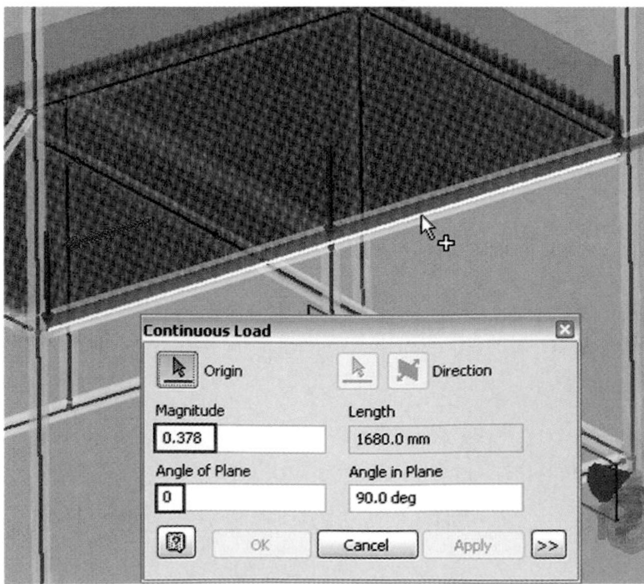

The gravity is acting in the wrong direction. It needs to act in the *X*-direction.

9. Repeat Step 8 for the other four beam elements directly under the platform deck > Click **OK**

10. Right click **Gravity** in the browser > Select **Edit** > Change **Direction** to positive *X*-direction > Click **OK**

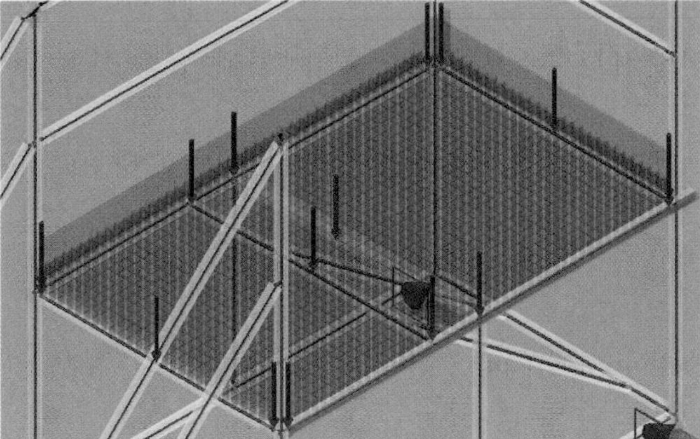

 There should be five continuous loads in the browser.

Run simulation and analyze

11. Select **Simulate**
12. Select Adjust ×1 for **Displacement Display** > Select S_{max} **Normal Stress**

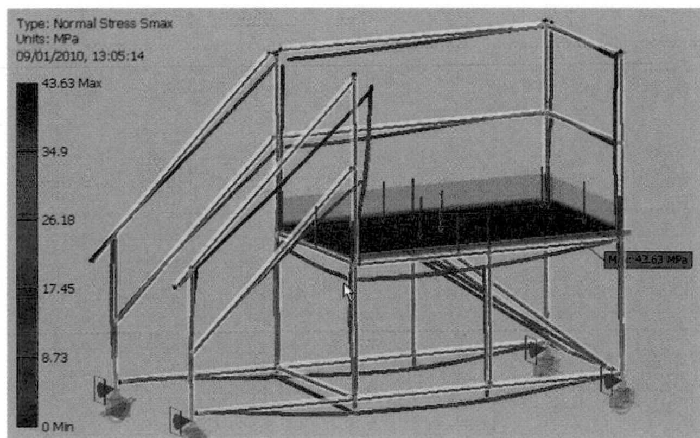

There are two maximum stresses: one at the front, illustrated by the arrow, and one at the back, shown by the Max value, the maximum deflection being 1.464 mm. We will now reanalyze the platform but this time will include additional steps and see whether they have an effect on the overall structural integrity of the platform.

13. Right click Simulation:1 > Select **Copy Simulation**
14. Right click Simulation:2 > Select **Edit Simulation Properties**
15. Select the **Model State** tab > Select **Master** for **Level of Detail** > Click **OK** > Accept Warning

 The warning refers to that of some beam elements which have a zero value for torsional modulus

16. Select **Frame Analysis Settings** > Select the **Beam Model** tab > Select **Invisible** for the original models not included in the analysis > Click **OK**
17. Select **Simulate** to rerun the analysis again

The maximum deflection has reduced by half to 0.7614 mm.

Select **Display Load values** to visually display load values. This helps us to see whether any loads have been applied incorrectly.

18. Select S_{max} **Normal Stress** > Unselect **Display Load values** and **Boundary Conditions** > Select **Color Bar** > Unselect **Maximum Value** > Specify 30 as the new max value > Click **OK**

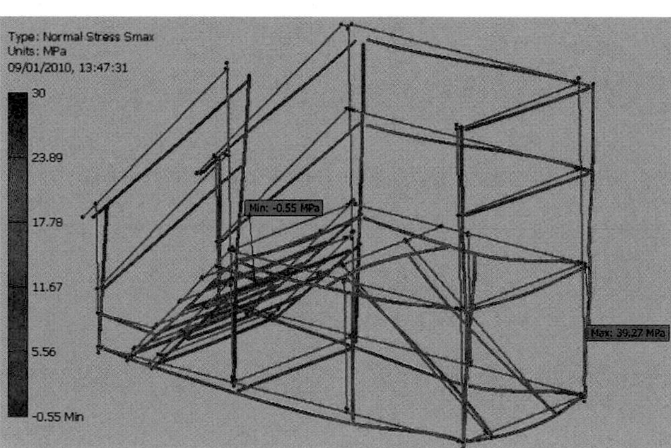

Although the maximum stress value has reduced, it is worth noting that the stress due to bending at the front has considerably reduced by adding the steps structure. We can now confidently say that the steps structure is fundamental to that structural integrity of the platform. You will also note that the axial, shear, and torsional stresses are insignificant compared to the stresses caused by bending. This can be investigated by looking at different stress plots using the axial, torsional, and shear stress results. Based on the maximum stress value, the minimum FOS is

$$\text{FOS} = \frac{\text{Yield stress}}{\text{Operational stress}} = \frac{209}{39} \times 27 = 5.27$$

This meets the minimum design FOS of 4.

Optimization

As the structure is made from mild steel, I would like to investigate whether the platform could be constructed from aluminum. In Frame Analysis, it is very easy to multi-select all frames and override the material specified within Content Center. This is ideal to check the suitability of different materials before changing the material via Content Center – which can be tedious, especially when you have a lot of frames/components.

19. Right click Simulation:2 > Select **Copy Simulation**

20. Select **Material** from the **Beams** section of the **Frame Analysis** panel

21. Highlight all beams > Select **Customize** > Change **Material** to **Aluminum-6061** > Click **OK**

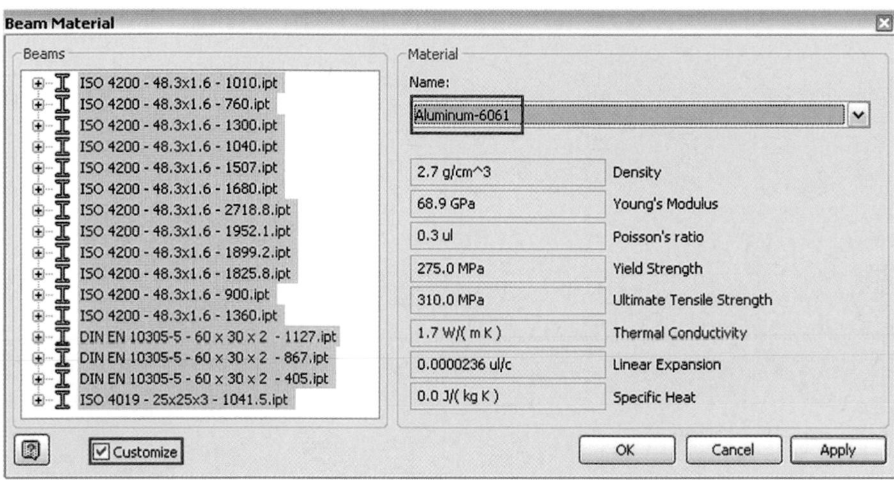

22. Select **Simulate** > Select **Actual** for **Displacement Display**

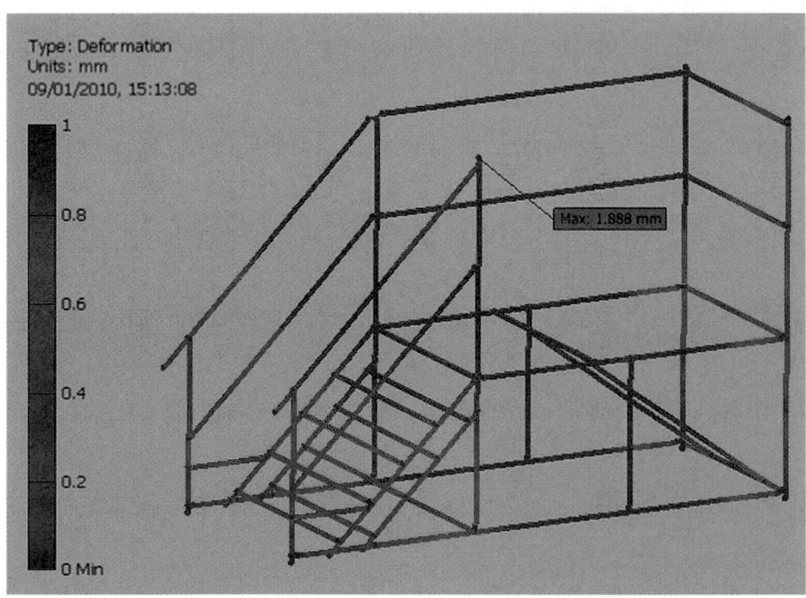

The maximum deflection has more than doubled to 1.888 mm; this is due to the fact that aluminum is more flexible than steel. Although the deflection has increased, it is negligible and insignificant in terms of the overall dimensions of the platform. Another thing to note is that the maximum stress has reduced slightly to 34.48 MPa because aluminum has a higher yield stress value. The FOS now becomes

$$\text{FOS} = \frac{\text{Yield stress}}{\text{Operational stress}} = \frac{275}{34} \times 48 = 7.96$$

As the FOS is almost twice the design FOS, a value of 4, we will investigate using a smaller tube with twice the thickness; tube ISO 4200 33.7 × 3.2. Again, it is a considerable task to change the frames for the purposes of frame analysis. Here, we will consider overriding the key mechanical properties of the original tube with those of the proposed tube, illustrated below.

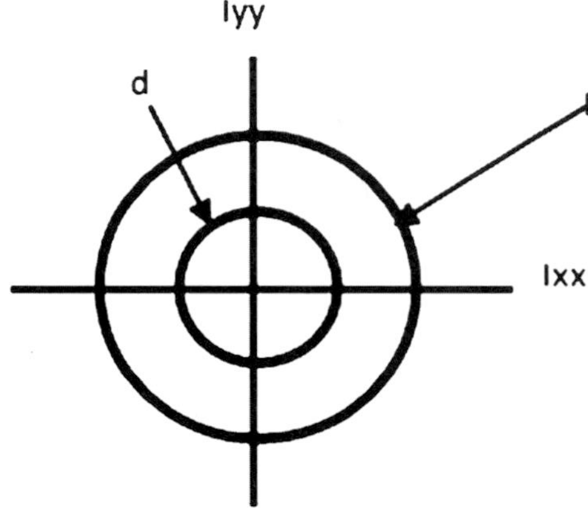

Note that calculations are in cm in the following

$$\text{Area} = \frac{\pi[(D)^2 - d^2]}{4} \text{ and } Ixx = Iyy = \frac{\pi[(D)^4 - d^4]}{64} \text{ and } Ixx + Iyy = Izz$$

$$\text{Area} = \frac{\pi[(33.7)^2 - (27.3)^2]}{4} = 306.619\text{mm}$$

$$Ixx = Iyy \frac{\pi[(D)^4 - d^4]}{64} = \frac{\pi[(33.7)^4 - (27.3)^4]}{64} = 36046.565$$

We will override the default values with the above to represent new tube ISO4200 33.7 × 3.2.

23. Right click Simulation:3 > Select **Copy Simulation**

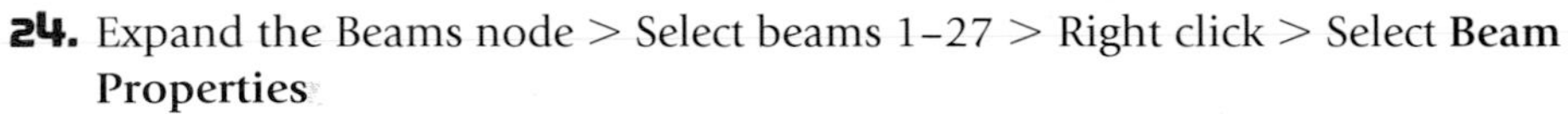

24. Expand the Beams node > Select beams 1–27 > Right click > Select **Beam Properties**

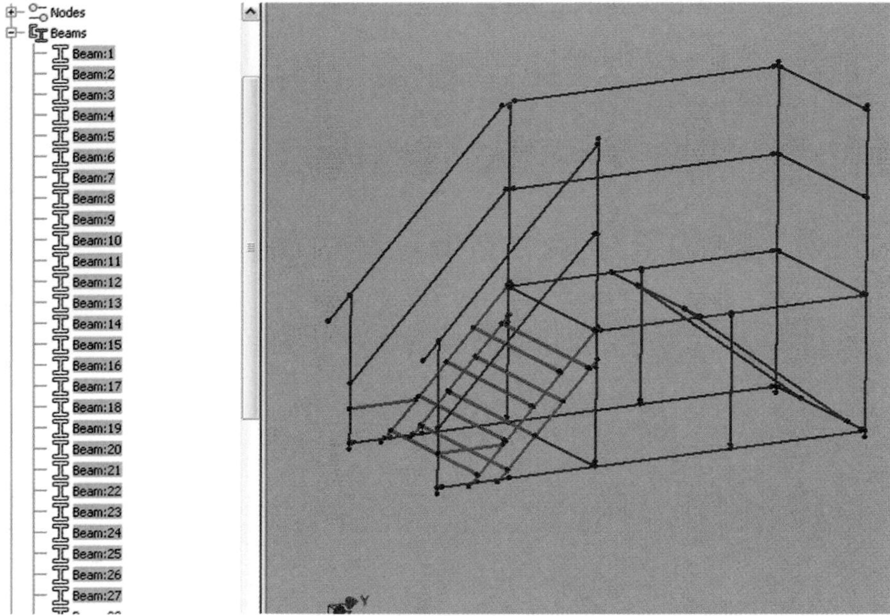

We are not including the step structure as the steps are constructed from box sections and not tubes.

25. Highlight all beams > Select **Customize** > Specify the new calculated values > Click **OK**

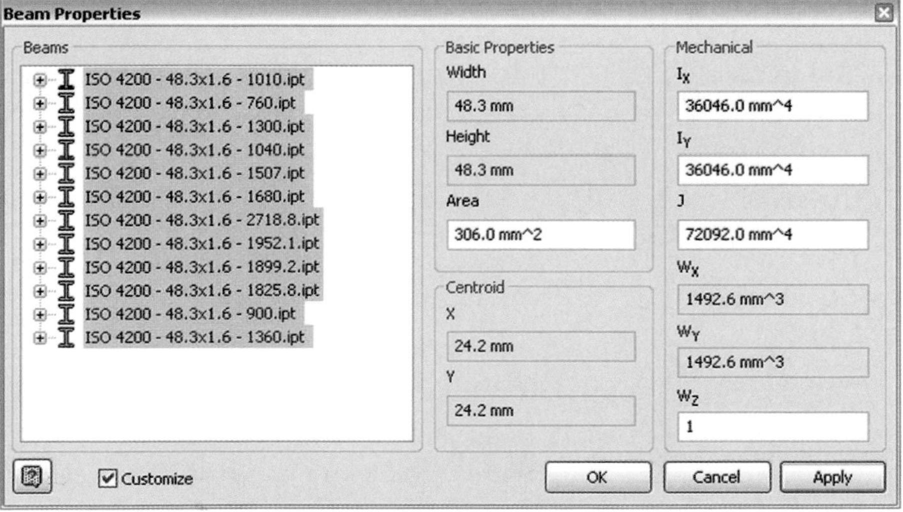

The W_z value has not been calculated or specified and should not affect the results under the current loading.

26. Select **Simulate** > Accept the warning

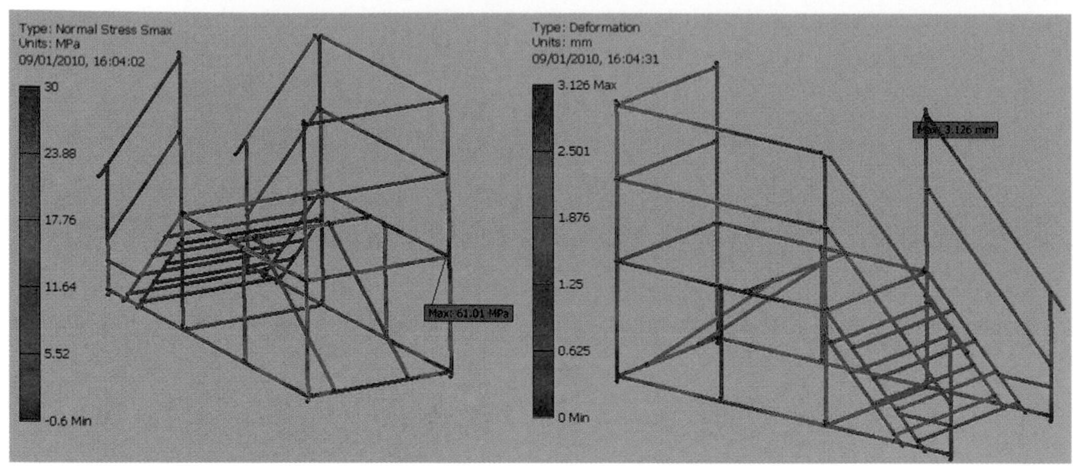

The results show that the deflection has increased to 3.126 mm and the maximum stress to 61 MPa. Taking this new value the FOS now is 4.5

$$\text{FOS} = \frac{\text{Yield stress}}{\text{Operational stress}} = \frac{275}{61} = 4.5$$

As both values are within the specified design criteria, we can safely replace the existing tubes, ISO 4200 43.8 × 1.6, with tubes ISO 4200 33.7 × 3.2.

27. Select **Finish Frame Analysis** > Close the file

CHAPTER 19

DP16 – Frame Analysis Using Frame Generator Structures

Analysis of an Escalator Support Structure

(Design Problem courtesy of KONE plc)

KEY FEATURES INTRODUCED IN THIS DESIGN PROBLEM

	Key features
1	Frames, created using Frame Generator, converted to beam elements
2	Tolerance settings – used to create rigid links automatically
3	Custom constraints and multiple forces
4	Beam diagram, detail, and scales

INTRODUCTION

KONE plc is the world's leading manufacturer of escalators and caters for various markets including retail, infrastructure, leisure, and offices. Retail markets include supermarkets and shopping malls, whereas the infrastructure market serves underground tubes, train stations, and airports. Kone supports its customers every step of the way, from design, manufacturing, and installation to maintenance and modernization.

Up and Running with Autodesk Inventor Simulation 2011. ISBN: 978-0-12-382102-7
Copyright © 2010 Elsevier Inc. All rights of reproduction, in any form reserved.

Above is a picture of a typical passenger escalator, within a shopping mall. The escalator design goes through intensive tests to make sure that it is safe. A typical requirement-in escalator design is to make sure that the support structure, holding the escalator, can hold the weight of passengers and of key components including steps and balustrades, etc. In this example we will determine the strength of the structure in relation to the weight of the structure, passengers, and balustrade:

1. The structure weight is based on the density of the material, which is mild steel.
2. For the passenger load, a value of 200 kg or 2000 N will be used.
3. The weight of the balustrade will be applied as a continuous load of 1.2 N/mm.

Note: In practice a typical escalator can go through up to 30 different load cases to fully validate its design.

The design requirements are:

1. Maximum deflection not to exceed 3 mm.
2. Minimum factor of safety (FOS) to be 4.

WORKFLOW OF DESIGN PROBLEM 16

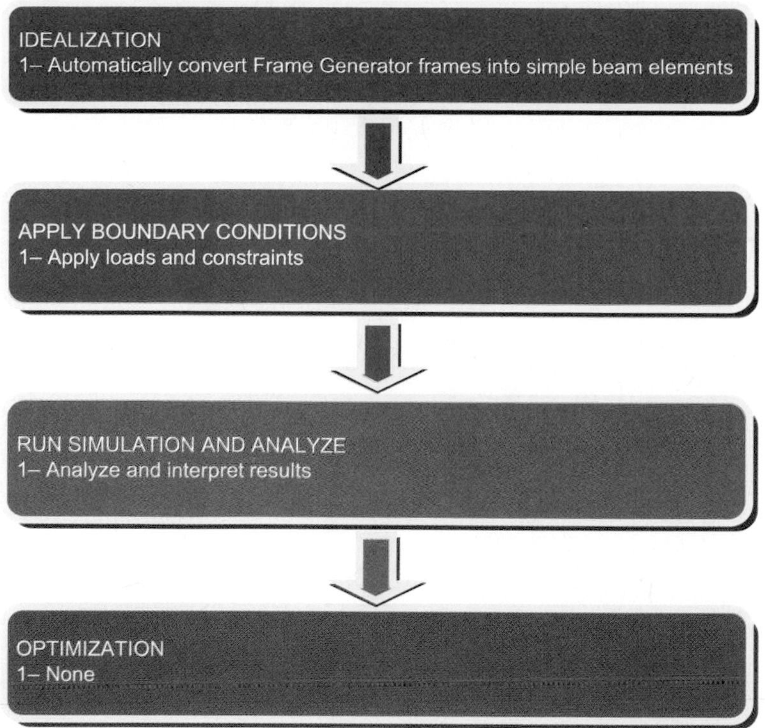

Idealization

As mentioned in Chapter 18, idealization is done automatically by converting frames created by Frame Generator and Content Center into simple line beam elements. It is important to note, however, that frame analysis cannot be used if neither Content Center nor Frame Generator is used to create the frames and channels, as frame analysis will have nothing to analyze.

1. Open *Kone.iam*

Now we can begin the second stage of the analysis by applying boundary conditions.

Boundary conditions

2. Select the **Design** tab > **Frame Analysis**

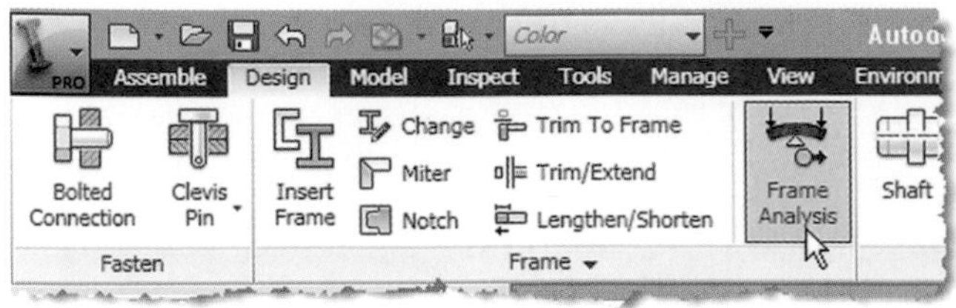

3. Select **Create Simulation** > Click **OK**

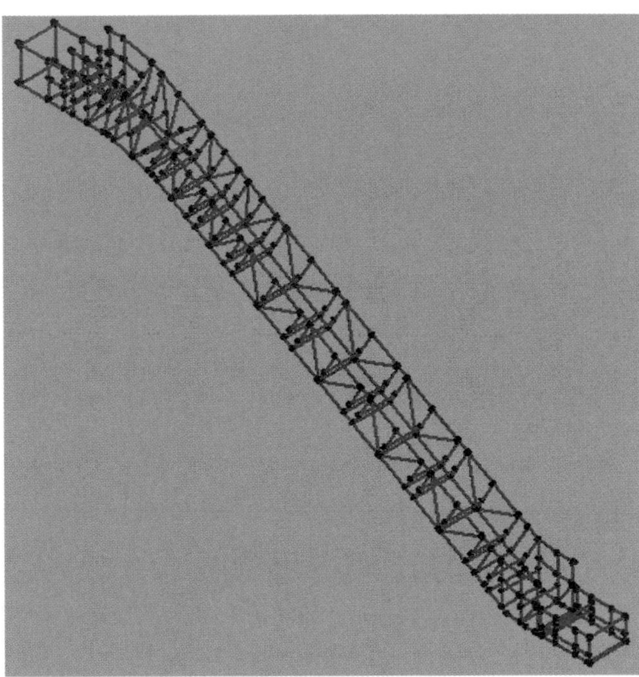

This takes a little while as a total of 230 beams and 240 rigid links are created.

Modal analysis is a very quick way to easily investigate unconnected beams, especially when the model has a large number of beams. This will help in identifying the need to create rigid links manually or automatically by increasing the beam model tolerance. Note: only constraints are required to run a modal analysis.

4. Select **Frame Analysis Settings** > Change **Rigid Links** color to red > Change the scale of **Nodes** to 1 > Change the scale of **Constraints** to 1 > Click **OK**

In the following steps, fixed constraints will be applied to the top end of the escalator structure.

5. Select **Fixed Constraint** > Right click > Select **More Options** > Select the parent node as shown > Select **Apply**

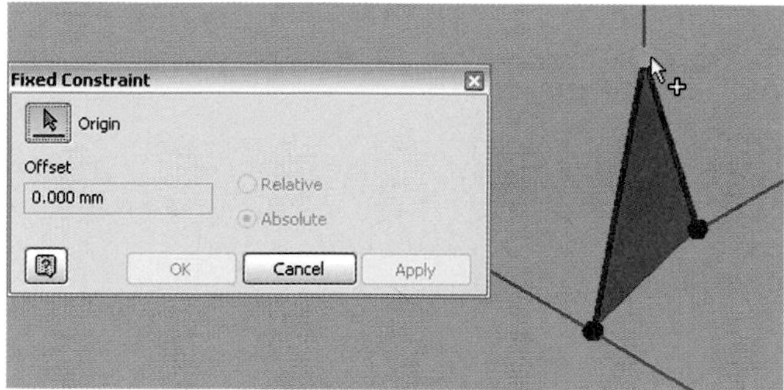

Only the parent node, within the rigid links, is selectable.

6. Repeat Step 5 to apply constraints on the other locations defined below.

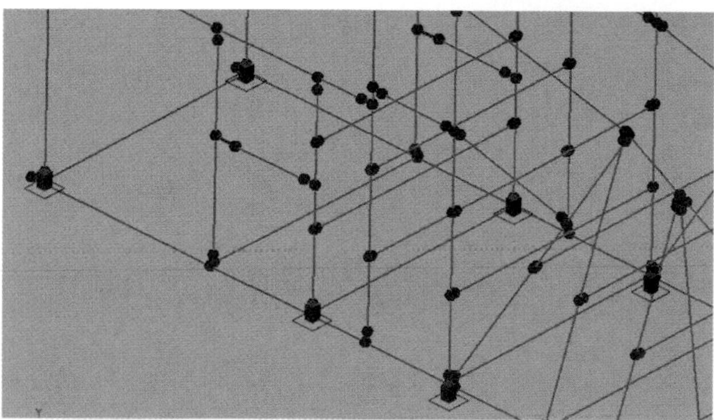

Now we need to apply floating pin constraints at the V frame junctions as indicated below.

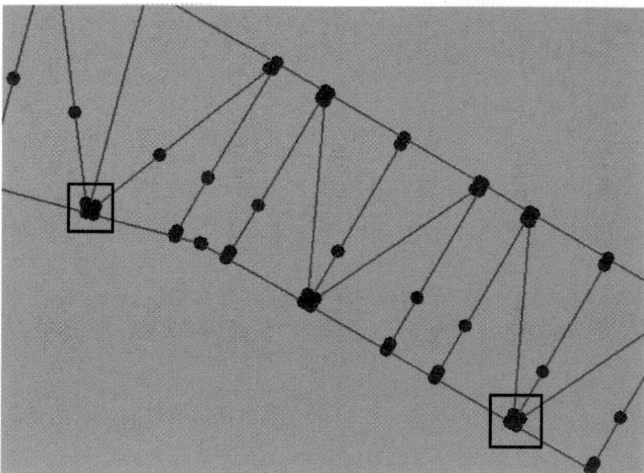

The constraints are to be applied at each alternating V junction on both sides of the structure on the incline of the escalator support structure. However, as we cannot apply rotational constraints, including pinned and floating constraints, on the parent nodes of rigid links, we will apply a custom constraint simulated as a floating constraint. The floating constraint will restrict the motion of the escalator in the vertical direction only.

7. Select **Custom Constraint** > Select **Uplift none** for all displacements except that on the Y axis > Select **Uplift none** for all rotations > Select **Apply**

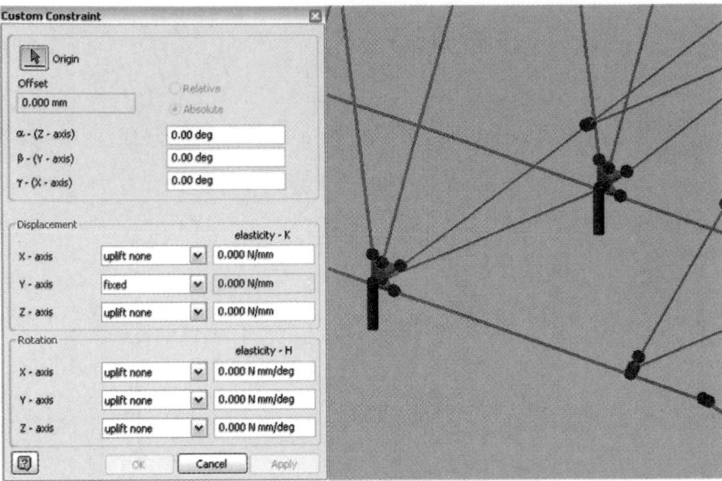

8. Repeat Step 7 to apply custom constraints on the other locations illustrated below.

Do not close the dialog box once all eight custom constraints have been created.

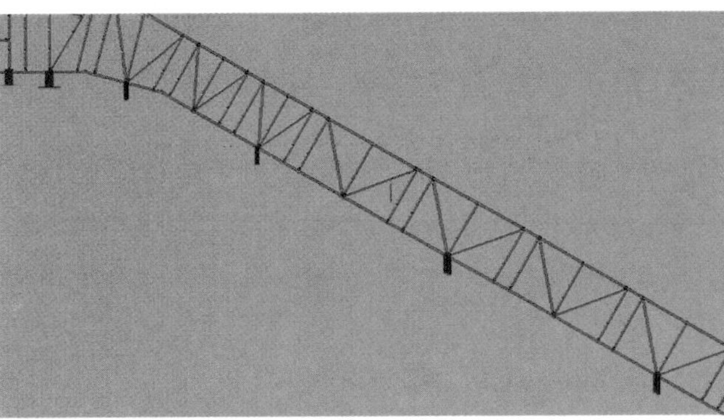

Change scale of constraints to perhaps four or more to visually see if the constraints have been applied at correct locations.

Finally, the custom constraint needs to be applied at the bottom end of the escalator and again the constraints should be such that the escalator is restricted in the vertical direction only.

9. Select the parent nodes of the rigid links at the following locations to create custom constraints > Click **OK** once all six constraints have been created

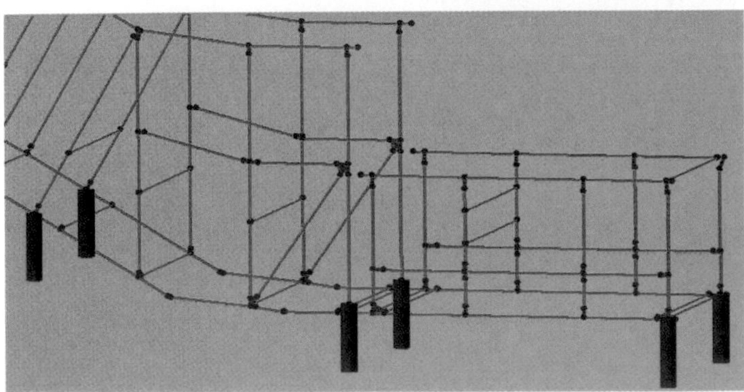

As all constraints have now been applied, the loads need to be defined on the support structure.

The first load to be taken into account is the balustrade weight, as shown above, which will be applied as a continuous load of 1.2 N/mm.

10. Select **Continuous Load** > Right click > Select **More Options** > Specify 1.2 N/mm for **Magnitude** > Select the beam as shown > Select **Apply**

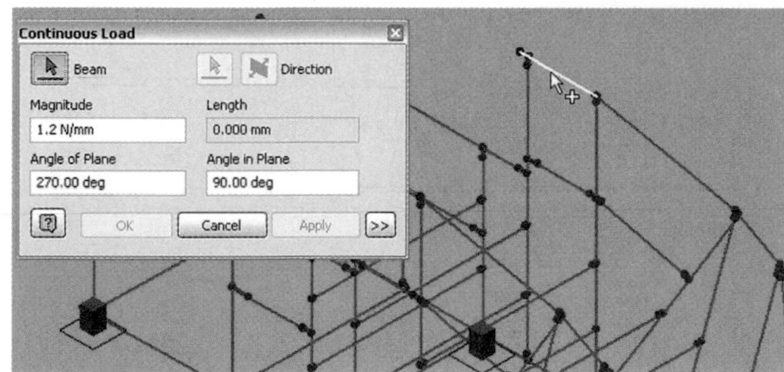

11. Repeat Step 10 until all beams on the top, on both sides, have been selected as shown below > Click **OK**

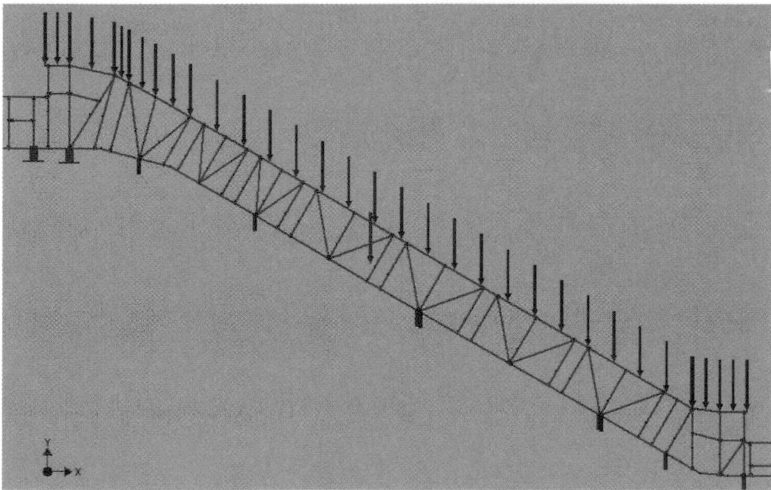

Change scale of loads to perhaps four or more to visually see if the loads have been applied in the correct direction.

Run simulation and analyze

12. Select **Simulate** > Accept the warning

13. Select **Display** > Deselect **Boundary Conditions** > Select **Maximum Value**

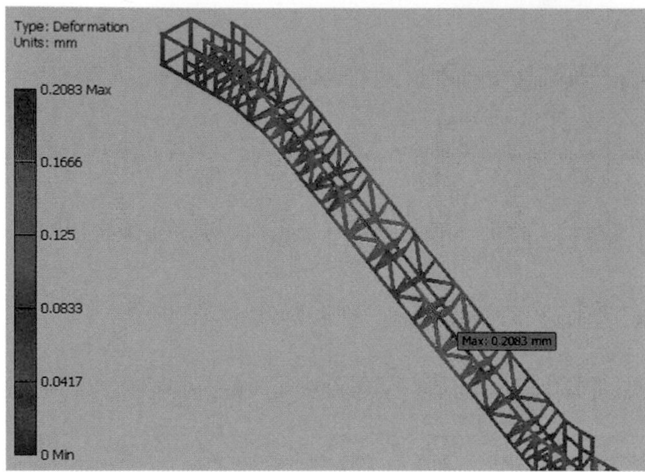

The maximum deflection is only 0.2 mm under the balustrade load and is well below the limit. We will next look at the maximum bending stress in the structure.

14. Select S_{max} **Normal Stress** > Select **Color Bar** > Unselect **Maximum Value** > Specify 8 as the new **Maximum Value** > Select **Absolute Values** > Click **OK**

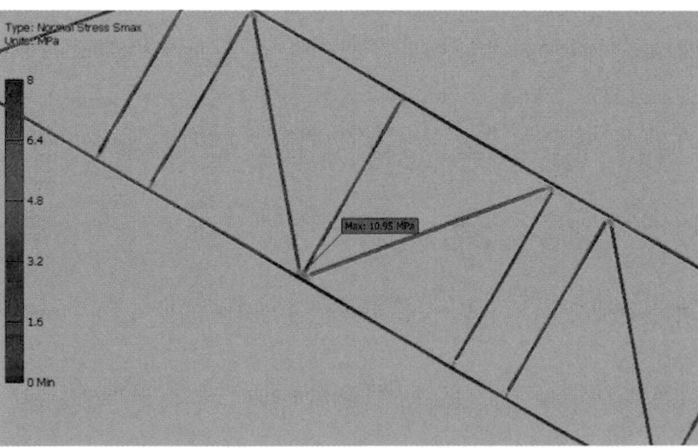

Next, we are going to look at the bending stresses across the lower long beam of the escalator.

15. Select **Diagram** > **Activate Selected Beam** > Select the beam as shown

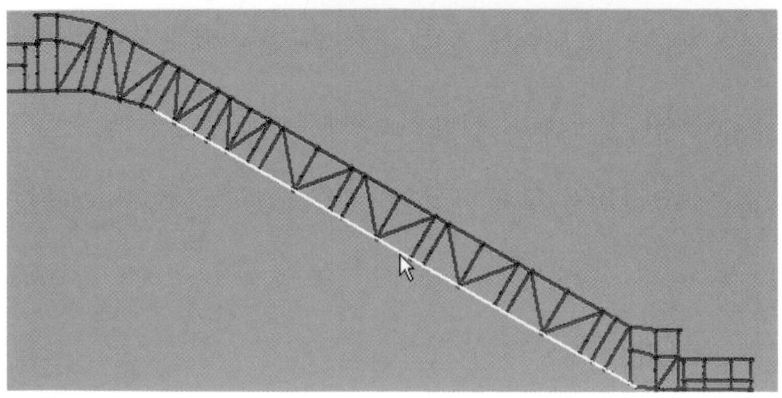

16. Select **Maximum** S_{max} > Click **OK**

17. Select **Frame Analysis Settings** > Select the **Diagrams** tab > Select the **Differentiated** and **Fill** options > Select **Colors** > Select **Red** for S_{max} > Click **OK** twice

18. Right click Diagram:1 > Select **Diagram Scales** > Change **Normal Stresses** to 0.005 00 MPa/mm > Click **OK**

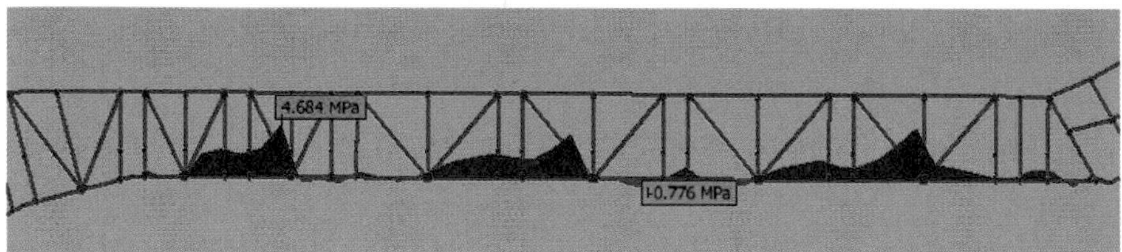

Beam Detail will provide a more comprehensive summary of results, including the reactions of the selected beam.

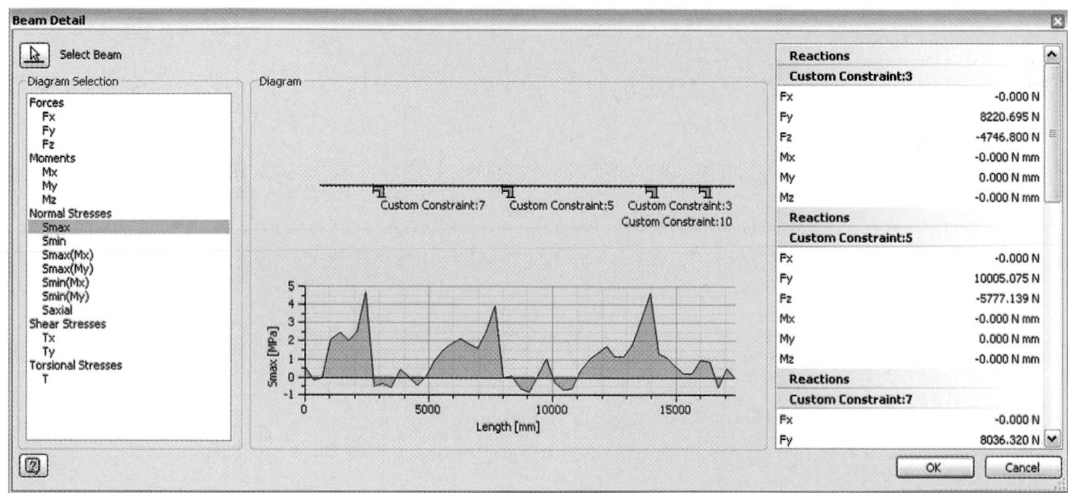

A typical escalator analysis can have up to 30 load cases, including passenger and step load cases. The analysis of all these load cases is beyond the scope of this exercise and therefore only the passenger load case will be considered in addition to the balustrade weight (including the weight of the escalator structure). Each step of the escalator needs to carry two people, including any extra weight that they may be carrying, for example shopping goods etc. Therefore, the weight to be used for the purposes of analysis is 200 kg (or 2000 N) per person.

The load of the passenger is transferred to the structure via escalator tracks, as shown below.

The tracks are positioned 200 mm from the sides and therefore the forces need to be applied on the cross members 200 mm offset from either end. The length of these cross members is ≈1540 mm, which means that the values for offsetting the forces will be 200 and 1340 mm.

19. Right click Simulation:1 > Select **Copy Simulation**

20. Select **Force** > Right click > Select **More Options** > Select 2000 for **Magnitude** > Select the beam as shown > Specify 1340 mm for **Offset** > Select **Apply**

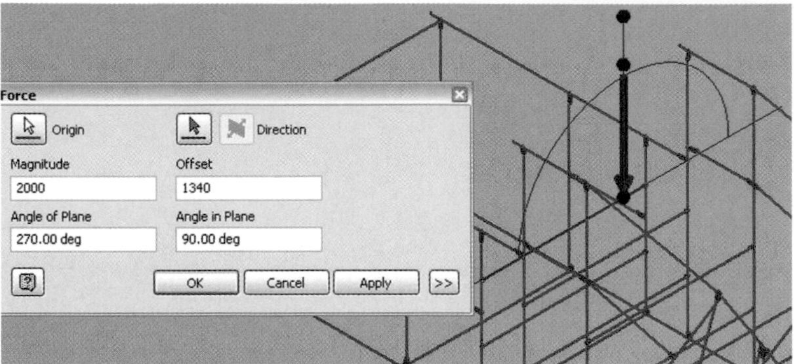

21. Select the same beam again > Select 200 mm for **Offset** > Click **Apply**

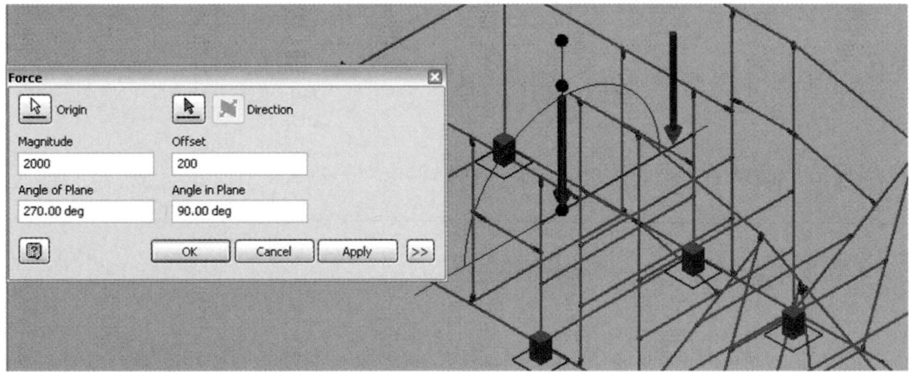

22. Repeat Steps 20 and 21 for all cross member beams on the escalator

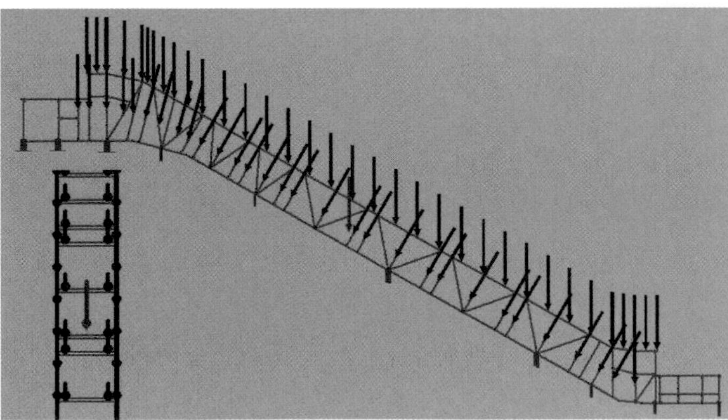

Sixty forces in total are created.

The forces on the incline are 30° to the vertical; for example, the **Angle of Plane** is 240°.

The four forces on the top between the horizontal and the incline, as shown below, have an **Angle of Plane** value of 255°.

23. Select **Simulate**

The maximum value has now increased from 0.2 to 1 mm.

24. Select S_{max} **Normal Stress** > Select **Color Bar** > Unselect **Maximum Value** > Specify 15 as the new **Maximum Value** > Select **Absolute Values** > Click **OK**

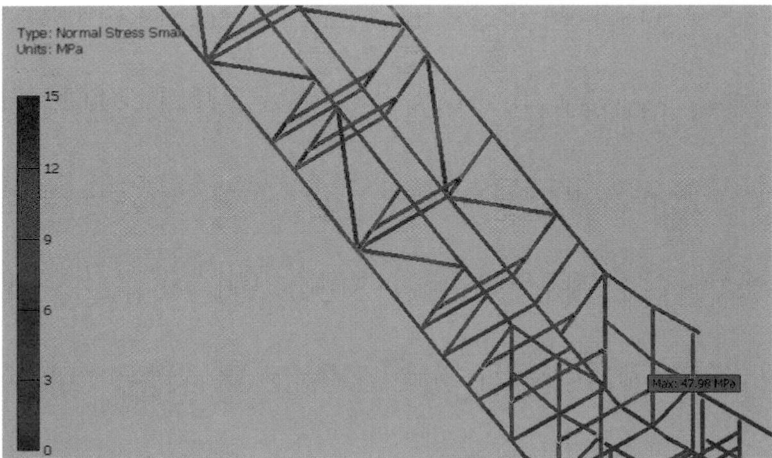

The maximum stress value has increased from 10.95 to 47.98 MPa. Based on this new value the minimum FOS is

$$\text{FOS} = \frac{\text{Yield stress}}{\text{Operational stress}} = \frac{207}{48} = 4.3$$

A typical escalator, as mentioned earlier, will go through several more load case analyses.

25. Select **Finish Frame Analysis** > Close the file

INDEX

D

E

N

O

P

R

S

T